Lecture Notes in Physics

Springer
Berlin
Heidelberg
New York
Hong Kong
London
Milan
Paris
Tokyo

Physics and Astronomy

springeronline.com

The Editorial Policy for Edited Volumes

The series *Lecture Notes in Physics* (LNP), founded in 1969, reports new developments in physics research and teaching - quickly, informally but with a high degree of quality. Manuscripts to be considered for publication are topical volumes consisting of a limited number of contributions, carefully edited and closely related to each other. Each contribution should contain at least partly original and previously unpublished material, be written in a clear, pedagogical style and aimed at a broader readership, especially graduate students and nonspecialist researchers wishing to familiarize themselves with the topic concerned. For this reason, traditional proceedings cannot be considered for this series though volumes to appear in this series are often based on material presented at conferences, workshops and schools.

Acceptance

A project can only be accepted tentatively for publication, by both the editorial board and the publisher, following thorough examination of the material submitted. The book proposal sent to the publisher should consist at least of a preliminary table of contents outlining the structure of the book together with abstracts of all contributions to be included. Final acceptance is issued by the series editor in charge, in consultation with the publisher, only after receiving the complete manuscript. Final acceptance, possibly requiring minor corrections, usually follows the tentative acceptance unless the final manuscript differs significantly from expectations (project outline). In particular, the series editors are entitled to reject individual contributions if they do not meet the high quality standards of this series. The final manuscript must be ready to print, and should include both an informative introduction and a sufficiently detailed subject index.

Contractual Aspects

Publication in LNP is free of charge. There is no formal contract, no royalties are paid, and no bulk orders are required, although special discounts are offered in this case. The volume editors receive jointly 30 free copies for their personal use and are entitled, as are the contributing authors, to purchase Springer books at a reduced rate. The publisher secures the copyright for each volume. As a rule, no reprints of individual contributions can be supplied.

Manuscript Submission

The manuscript in its final and approved version must be submitted in ready to print form. The corresponding electronic source files are also required for the production process, in particular the online version. Technical assistance in compiling the final manuscript can be provided by the publisher's production editor(s), especially with regard to the publisher's own LaTeX macro package which has been specially designed for this series.

LNP Homepage (springerlink.com)

On the LNP homepage you will find:
−The LNP online archive. It contains the full texts (PDF) of all volumes published since 2000. Abstracts, table of contents and prefaces are accessible free of charge to everyone. Information about the availability of printed volumes can be obtained.
−The subscription information. The online archive is free of charge to all subscribers of the printed volumes.
−The editorial contacts, with respect to both scientific and technical matters.
−The author's / editor's instructions.

M. Karttunen I. Vattulainen A. Lukkarinen (Eds.)

Novel Methods
in Soft Matter Simulations

 Springer

Editors

Mikko Karttunen
Helsinki University of Technology
P.O.Box 9203
02015 HUT, Finland

Ari Lukkarinen
CSC - Scientific Computing Ltd.
P.O Box 405
02101 Espoo, Finland

Ilpo Vattulainen
Helsinki University of Technology
P.O.Box 1100
02015 HUT, Finland

M. Karttunen, I. Vattulainen, A. Lukkarinen (eds.), *Novel Methods in Soft Matter Simulations*, Lect. Notes Phys. **640** (Springer-Verlag Berlin Heidelberg 2004), DOI 10.1007/b95265

Cataloging-in-Publication Data applied for

A catalog record for this book is available from the Library of Congress.

Bibliographic information published by Die Deutsche Bibliothek
Die Deutsche Bibliothek lists this publication in the Deutsche Nationalbibliografie;
detailed bibliographic data is available in the Internet at http://dnb.ddb.de

ISSN 0075-8450
ISBN 3-540-20916-6 Springer-Verlag Berlin Heidelberg New York

Springer-Verlag is a part of Springer Science+Business Media

springeronline.com

© Springer-Verlag Berlin Heidelberg 2004

Typesetting: Camera-ready by the authors/editor
Data conversion: PTP-Berlin Protago-TeX-Production GmbH
Cover design: *design & production*, Heidelberg

Printed on acid-free paper
54/3141/du - 5 4 3 2 1 0

Preface

Soft matter and biological systems pose many challenges for theoretical, experimental and computational research. From the computational point of view, these many-body systems have huge variations in relevant time and length scales. One can illustrate this through a simple example: Biological (and most other soft matter) systems operate in the presence of water. Besides displaying complicated behavior on its own, a water molecule is characterized by a very small size of approximately 10^{-10} m, and by vibrations in time scales of about 10^{-15} s. On the other hand, cellular length scales are much larger, typically being around 10^{-5} m, and the numerous cellular processes covering a variety of time scales, spanning at least $10^{-9} - 10^5$ s.

From the above it is clear that modeling and computer simulations of soft matter systems are far from trivial. The standard workhorse, classical molecular dynamics on the atomic level, reaches time scales of about 10^{-7} s at best, but then is limited to a moderate number of particles and system sizes of the order of 10^{-8} m. Even from this rather trivial example it is clear that there is a lot of demand for new simulation techniques since computer simulations have proven invaluable in studies of soft matter systems, including applications to numerous processes such as drug transport, self-organization in nanoscale materials, and rheological properties. The challenge is therefore to develop methods capable of reaching time and length scales much larger than those accessible by brute force computer simulations on the atomic level. The feat is not a simple one, since it requires a major effort over a wide range of activities, including the development of coarse graining techniques, novel simulation methods, and ways to link the different regimes to each other.

The aim of SoftSimu2002 was to bring together leading scientists working on these issues in soft matter research, and to offer young researchers and graduate students an in-depth review of the most recent developments in this field. The lectures were complemented by hands-on exercises to get a better feeling of the practical aspects as well. From the organizers' part, we were very pleased with the relaxed atmosphere and lively discussions during the summer school. We would also like to express our gratitude to all our speakers, who delivered truly excellent sets of lectures and took the time to write up detailed and pedagogical lecture notes. This is, to our knowledge, the first book to cover most of the novel methods developed for soft matter simulations in the 1990's and early 21st century.

SoftSimu2002 received generous support from individuals and organizations alike. We would like to thank the European Science Foundation network SIMU – Challenges in Molecular Simulations, the Finnish National Graduate School for Materials Physics, the Helsinki Institute of Physics, the Helsinki University of Technology (HUT), CSC – Scientific Computing Ltd, and the City of Espoo for making this summer school financially possible. We would also like to express our appreciation to Hewlett–Packard Finland for providing part of the computational resources for the summer school, and to the Finnish Sauna Society for allowing us to use their wonderful facilities.

Many individuals helped and supported us before, after and during the meeting. In particular, we would like to extend our thanks to professors Olli Ikkala (Laboratory of Optics and Molecular Materials), Risto Nieminen (Laboratory of Physics), and Jukka Tulkki and Kimmo Kaski (Laboratory of Computational Engineering) in Helsinki University of Technology for their encouragement and support, as well as to all of our students and postdocs who provided a helping hand whenever needed, and to Mr. Juha Juvonen for assisting with the photos and graphics. Finally, we would like to express our most heartfelt thanks to conference secretary Ms. Eeva Lampinen, without whom this meeting would never have been possible.

Note: Some of the software used in hands-on exercises during the summer school is available at *http://www.softsimu.org/downloads/*.

Helsinki   *Mikko Karttunen*
Finland   *Ilpo Vattulainen*
September 2003   *Ari Lukkarinen*

Contents

List of Contributors

Pep Español
Departamento de Física Fundamental,
UNED, Apartado 60141,
28080 Madrid, Spain
pep@fisfun.uned.es

Raymond Kapral
Chemical Physics Theory Group,
Department of Chemistry
University of Toronto, Toronto, ON M5S
3H6, Canada
rkapral@chem.utoronto.ca

Anatoly Malevanets
Flow Software Technologies,
3070 Jefferson Blvd., Windsor, ON,
Canada N8T 3G9
amalevanets@cisrobotics.com

Christopher P. Lowe
Department of Chemical Engineering,
University of Amsterdam, Nieuwe
Achtergracht 166, 1018 WV Amsterdam,
The Netherlands
lowe@science.uva.nl

Menno W. Dreischor
Department of Chemical Engineering,
University of Amsterdam, Nieuwe
Achtergracht 166, 1018 WV Amsterdam,
The Netherlands
menno@science.uva.nl

Giovanni Ciccotti
INFM and Dipartimento di Fisica,
Universita degli Studi di Roma La
Sapienza, Piazzale Aldo Moro 5,
Roma, Italy
Giovanni.Ciccotti@roma1.infn.it

Galina Kalibaeva
INFM and Dipartimento di Fisica,
Universita degli Studi di Roma La
Sapienza, Piazzale Aldo Moro 5,
Roma, Italy
galia_mex@yahoo.com

Eirik G. Flekkøy
Dept. of Physics, University of Oslo, PB
1048, Blindern, 0316 Oslo, Norway
flekkoy@fys.uio.no

Knut Jørgen Måløy
Dept. of Physics, University of Oslo, PB
1048, Blindern, 0316 Oslo, Norway
k.j.maloy@fys.uio.no

Jens Feder
Dept. of Physics, University of Oslo, PB
1048, Blindern, 0316 Oslo, Norway
feder@fys.uio.no

Sean McNamara
ICA1, University of Stuttgart,
Pfaffenwaldring 27,
70569 Stuttgart, Germany
sean@ica1.uni-stuttgart.de

Geri Wagner
School of Astronomy and Physics,
Raymond and Beverly, Sackler Faculty of
Exact Sciences, Tel Aviv University,
Ramat Aviv, 69978 Tel Aviv, Israel
Geri.Wagner@fys.uio.no

Robert D. Groot
Unilever Research & Development,
Olivier van Noortlaan 120, 3133 AT
Vlaardingen, The Netherlands
Rob.Groot@unilever.com

Alexander P. Lyubartsev
Division of Physical Chemistry,
Arrhenius Laboratory,
Stockholm University, S 106 91,
Stockholm, Sweden
sasha@physc.su.se

Aatto Laaksonen
Division of Physical Chemistry,
Arrhenius Laboratory,
Stockholm University, S 106 91,
Stockholm, Sweden
aatto@physc.su.se

André G. Moreira
Max-Planck-Institut für Kolloid- und
Grenzflächenforschung
14424 Potsdam, Germany,
and Materials Research Laboratory,
UCSB Santa Barbara, CA. 93106, USA
amoreira@mrl.ucsb.edu

Roland R. Netz
Max-Planck-Institut für Kolloid- und
Grenzflächenforschung,
14424 Potsdam, Germany, and Sektion
Physik, LMU,
Theresienstr. 37,
80333 München, Germany
netz@theorie.physik.
uni-muenchen.de

Ignacio Pagonabarraga
Departament de Física Fonamental,
Universitat de Barcelona, C. Martí i
Franqués, 1, 08028-Barcelona, Spain
ignacio@ffn.ub.es

Florian Müller-Plathe
Max-Planck-Institut für
Polymerforschung, Ackermannweg 10,
55128 Mainz, Germany,
and International University Bremen,
P.O. Box 750561, 28725 Bremen,
Germany
f.mueller-plathe@iu-bremen.de

Patrice Bordat
Max-Planck-Institut für
Polymerforschung, Ackermannweg 10,
55128 Mainz, Germany
bordat@mpip-mainz.mpg.de

Séverine Girard
Max-Planck-Institut für
Polymerforschung, Ackermannweg 10,
55128 Mainz, Germany
girard@mpip-mainz.mpg.de

Tapio Ala-Nissila
Laboratory of Physics, Helsinki
University of Technology, P.O. Box 1100,
02015 HUT, Espoo, Finland
Tapio.Ala-Nissila@hut.fi

Sami Majaniemi
Laboratory of Physics, Helsinki
University of Technology, P.O. Box 1100,
02015 HUT, Espoo, Finland
majaniem@pcu.helsinki.fi

Ken Elder
Department of Physics,
Oakland University,
Rochester, MI 48309-4401, U.S.A.
elder@oakland.edu

Introduction

By the term "soft matter", one understands a variety of complex fluids and the solids which can be formed from them, such as liquid crystals, colloidal suspensions, solutions and melts of linear or branched polymers, gels and rubbers, supramolecular structures formed from amphiphilic molecules such as liquid monolayers and bilayers, vesicles, biological membranes and cells, etc. The adjective "soft" emphasizes that rather weak external fields (such as mechanical stresses, electrical fields, laser light, etc.) suffice to cause strong deformations of structure and accompanying changes of dynamical properties and hence yield very desirable opportunities to manipulate the system on scales ranging from the nanoscale to macroscopic scales. For this reason, soft matter systems find increasing applications in various branches of technology – and nature has made use of these favorable aspects of soft matter all the time, after all living matter is soft matter throughout, and living organisms provide superb examples how efficiently self-organized structure formation is possible on a multitude of length scales, and transport of matter and of information is effected in a very efficient way as well.

In view of this importance of soft matter systems for various branches of technology and of biology, considering also the great intellectual challenges that the understanding of soft matter systems requires, research in this area has been truly exploding in the last few decades. However, the theoretical understanding of such systems is particularly difficult, given the fact that there is typically a rather subtle interplay occurring between enthalpic and entropic effects, and one has a long way to go to bridge the gap from the electronic structure of constituent molecular groups and their interactions on the Ångström scale and effective interactions of predominantly entropic origin on mesoscopic scales of 10 or 100 nanometers or more. A typical example, for instance, are mixtures of colloidal particles (stabilized by endgrafted flexible linear polymers, forming a layer of stretched chains called a "polymer brush" on the surface of each particle) with polymers in a common solvent. In addition, ions may be present in the solution and at the surface of the colloidal particles, and hence the long range van der Waals forces can compete with screened but relatively long ranged Coulomb interactions, depletion interactions of entropic origin etc. Similarly, when a protein interacts with a biological membrane in aqueous solution, a multitude of such interactions also needs to be considered. Of course, although such problems in principle fall in the field of statistical thermodynamics, it is clear that analytical theory can always only address restricted partial aspects of such problems. Notwithstanding the fact that such focussed treatments can be very valuable, there is a great need to test them by computer simulation methods, to investigate the validity of very limited models by a simulational approach that avoids too simplifying approximations, and which is suitable for extracting detailed information that can be neither gotten from analytical theory nor from experiment.

In fact for many systems (simple atomic and molecular fluids, pure metals and alloys, ionic and oxidic crystals, semiconductors, magnetic crystals, etc.) the approach of computer simulation has become a third scientific method for investigating both the general laws of nature and specific properties of materials, complementing both theory and experiment[1,2]. One has learned to overcome systematic problems such as finite size effects on phase transitions, one has learned to realize various ensembles of statistical thermodynamics and by suitable choice of boundary conditions one can explore either bulk behavior or interfaces between coexisting phases or free surfaces and small clusters etc. Last but not least, enormous progress has occurred in deriving either suitable effective potentials from quantum chemistry or other methods of electronic structure calculation, or one may avoid the use of simplified effective interatomic potentials altogether by the Car–Parrinello "ab initio Molecular Dynamics" method, which combines the density functional approach to electronic structure and classical Molecular Dynamics in an elegant hybrid approach.

However, all this impressive progress for the most part refers to systems where the relevant structure occurs on the atomic scale of a few Ångströms or nanometers, and correspondingly the associate time scale to reach equilibrium is in the scale below 1 nanosecond. When one attempts to extend these techniques of Monte Carlo and Molecular Dynamics methods to relatively simple soft matter systems, such as melts of synthetic homopolymers[3], one already encounters the problem that a flexible polymer coil in the melt exhibits structure from the length of a covalent bond (≈ 1 Ångstrom) to the gyration radius (≈ 100 Ångstrom), and scales of collective phenomena (unmixing in partially compatible polymer blends, mesophase ordering of block copolymers, etc.) require even much larger length scales for their study.

Even more restrictive is the spread of time scales, which typically ranges from 10^{-13} s (time of bond angle vibrations) to 10^{+3} s (time on which the biphasic structure formed via spinodal decomposition of a polymer blend coarsens). It is clear that the standard methods of computer simulations, i.e. atomistically detailed Monte Carlo or Molecular Dynamics methods, encounter dramatic limitations when one wishes to treat such soft matter systems, and novel methods of simulation are needed to bridge such gaps in time-and length scales[4].

It is the purpose of the present book to assess the state of the art of such novel methods in computer simulations that are better suited for dealing with these problems of bridging length and time scales for soft matter systems. Considering to the large variety of systems and physical problems, of course there is not a unique approach that

[1] *Monte Carlo and Molecular Dynamics of Condensed Matter Systems*, edited by K. Binder and G. Ciccotti, Societa Italiana die Fisica, Bologna (1996)

[2] *A Guide to Monte Carlo Simulation in Statistical Physics*, D. P. Landau and K. Binder, Cambridge University Press, Cambridge (2000)

[3] *Monte Carlo and Molecular Dynamics Simulations in Polymer Science*, edited by K. Binder, Oxford University Press, New York (1995)

[4] *Bridging Time Scales: Molecular Simulations for the Next Decade*, edited by P. Nielaba, M. Mareschal and G. Ciccotti, Lect. Notes Phys. **605**, Springer-Verlag, Berlin Heidelberg (2002)

can handle all these problems. Depending on the system at hand and the physics which one wishes to describe, different approaches may be adequate. To learn which methods now exist, to be able to compare both their merits and their disadvantages, and hence to answer the question "when should I apply which method?" belong to the primary purposes of the present book.

A number of leading experts have cooperated to describe the main facets of this rapidly developing field. Rob Groot describes "Dissipative Particle Dynamics (DPD)": this method is analog to classical Molecular Dynamics where one solves Newton's equations of motion for a many-atom system, but now each "particle" represents a group of many neighboring atoms, hence the forces between these effective "particles" are much softer, therefore one can use much larger time-steps than in a corresponding atomistic simulation, larger length scales are also accessible. C. P. Lowe and M. W. Dreischer discuss this problem of coarse-graining by systematic elimination of irrelevant degrees of freedom from a somewhat different point of view, starting from two classical examples: dynamics of dilute colloidal dispersions (this leads to the classic problem of Brownian motion and the associated description via Langevin and Fokker–Planck equations) and of polymer solutions (where the analogous treatment yields the classic Rouse–Zimm models). A third perspective on the statistical mechanics of coarse-graining then is provided by P. Espanol, comparing different levels of the description and discussing their interrelation. A very interesting theoretical framework in this context is the development of the so-called GENERIC formalism to establish the Fokker–Planck description in the general case. A further powerful approach specifically suited to deal with hydrodynamic flow of soft matter systems is Multi-Particle Collision Dynamics, reviewed by A. Malevanets and R. Kapral.

A different class of problems is discussed in the article by G. Ciccotti and G. Kalibaeva on the Molecular Dynamics simulation of complex systems. Non-Hamiltonian systems are discussed, as well as the proper incorporation of constraints (remember the well-known SHAKE algorithm to simulate molecules with rigid bonds), and mixed quantum-classical systems. At this point we emphasize that this article – as well as the other articles in this book – can only provide us with a kind of "snapshot picture"of what is known now, it is by no means implied that all the problems addressed are already fully solved. As the authors of this chapter put it, "realistic quantum subsystems – real solvents are still a dream"!

Much larger length scales are in the focus of the article by E. G. Flekkøy, S. Mc-Namara, K.-J. Måløy, J. Feder, and G. Wagner, where the hybrid models that bridge particle and continuum scales in hydrodynamic flow simulations are discussed. On the continuum scale, the appropriate description are the well-known Navier–Stokes equations of fluid dynamics, while on the atomistic scale a fluid may be described by an assembly of particles interacting with Lennard–Jones forces. The article describes how such descriptions can be coupled.

A. P. Lyubartsev and A. Laaksonen return to the problem of coarse-graining for static properties, discussing the problem of effective potentials for coarse-grained models: when one wishes to deal with specific systems quantitatively, qualitative potentials as sometimes used by DPD will not do. The authors emphasize the role of "Inverse Monte Carlo"-methods to construct more realistic effective potentials on a coarse-grained level.

A. G. Moreira and R. R. Netz focus on a problem that very often arises for soft matter systems, namely when one deals with counterions in a solution close to charged objects: the "electric double layer". By a combination of Methods (Monte Carlo simulation with advanced techniques to treat with Coulomb ineractions in $d = 2$ dimensions, and analytic methods such as Poisson–Boltzmann or strong coupling theories) they achieve progress in the understanding of these phenomena.

I. Pagonabarraga then addresses an approach that is very powerful for the dynamics of colloidal suspensions, where one has an extreme disparity between the large size of the colloidal particles and the small size of the solvent molecules, which nonetheless are needed to propagate the hydrodynamic interactions: namely the Lattice–Boltzmann method. Also unmixing of binary fluids on mesoscopic length scales is well accessible with this approach.

A problem often encountered already in standard Molecular Dynamics (MD) work is the accurate estimation of various transport coefficients. A nice solution to this problem, also applicable beyond standard MD (e.g. for DPD simulations) is the "Reverse Non-Equilibrium Molecular Dynamics (RNEMD)" method explained by F. Müller-Plathe and P. Bordat.

S. Girard and F. Müller-Plathe describe the state of the art for coarse-graining methods of polymers along the backbone of the polymer chain (replacing the atomistic chain by an equivalent chain of larger effective monomeric units, and constructing effective potentials between these units in a systematic and well-controlled way). Related methods of coarse-graining for polymers have been tried about a decade already, but the present version has the advantage that it should be relatively easy to implement and works rather automatically.

The final chapter, by T. Ala-Nissila, S. Majaniemi and K. Elder, deals with the phase-field modelling of dynamical interface phenomena in fluids. This description then really is already on a more macroscopic level, close in spirit to the finite element methods of engineers, but based on (deterministic or stochastic) equations of the type of time-dependent Ginzburg–Landau equations. When statistical fluctuations are neglected, equations result that find wide-spread application to model dendritic crystal growth and related mesoscopic phenomena of atomic systems, but related equations can also be used for soft matter systems.

Thus this book provides a very broad survey over a variety of methods, their inter-relations, their achievements as well as their unsolved problems. Clearly, soft matter simulations will become a mainstream of research for many years to come. The present book intends to introduce the reader to this fascinating field of research and also motivate him/her to contribute to it!

Mainz   Kurt Binder
June 2003

Applications of Dissipative Particle Dynamics

Robert D. Groot

Unilever Research & Development, Olivier van Noortlaan 120, 3133 AT Vlaardingen, The Netherlands

Abstract. Dissipative Particle Dynamics (DPD) is one of the most promising simulation techniques for studies of mesoscopic properties of soft matter systems. Here, we discuss DPD, its parameterisation in simple systems, as well as in polymeric systems using the Flory–Huggins theory, and generalisations of DPD. Block copolymer mesophase separation, polymers and membranes in surfactant solutions, and biomembrane morphology and rupture will shown as specific examples.

1 Why Mesoscopic Simulation?

Over the last two decades most simulation studies have concentrated on the motion of individual atoms in systems of a few nanometers and a few nanoseconds. Other simulation methods concentrate exclusively on the macroscopic world of planes, trains and automobiles. However, between the nano- and macroscopic scale ranges some forty decades in volume and time. The holy grail of theoretical physics is to bridge this gap. This is due to the fact that in many cases simulation of this intermediate regime is essential for understanding macroscopic phenomena, e.g. molecules ordering spontaneously on mesoscopic length and time scales. This category of problems includes life and biological phenomena such as membrane structuring, perforation and trafficking. As a matter of fact, this list contains all soft condensed matter including surfactants, polymers and (multi)block copolymers that show microphase separation, or form gels or glassy systems, see Fig. 1.

What could we expect if we would be able to extend the time scale over which we can simulate a physical system? If we take the example of lipid bilayers, we find that new phenomena occur every time we increase the time scale at which we look at our system [1]. On the shortest time scale of a few picoseconds the lipids show bond and angle fluctuations of dihedral angles within the same molecule. On larger time scales of a few tens of picoseconds, trans-gauche isomerizations of dihedrals occur [2]. On a time scale of a few nanoseconds the phospholipids rotate around their axis, and on the time-scale of tens of nanoseconds two lipids switch place within a bilayer, giving rise to lateral diffusion. Within this time scale the individual lipids orient, and lipid membranes show protrusions [3]. Finally, on a time-scale of 100 ns peristaltic motions and undulations occur [4].

By virtue of parallelization over several processors or PC clusters, hardware developments have now pushed the limit of molecular simulations to 100 ns [4]. Nevertheless, there is a limit beyond which hardware developments cannot help us. For instance, phenomena such as co-operative motion in phase transitions, insertion of large molecules

R.D. Groot, Applications of Dissipative Particle Dynamics, Lect. Notes Phys. **640**, 5–38 (2004)
http://www.springerlink.com/

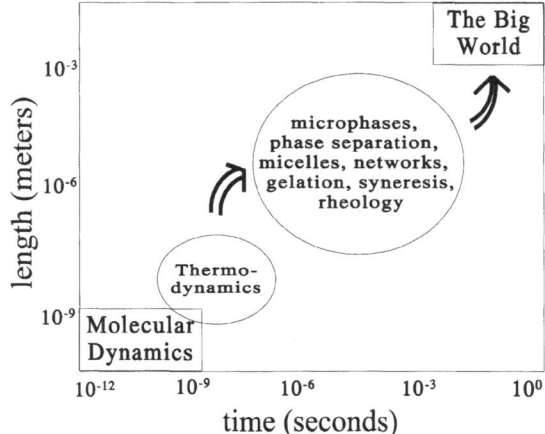

Fig. 1. The mesoscale gap. Time is given in seconds

like proteins into membranes, or membrane fusion occur on much larger time scales and are well outside the range of current simulation power. This requires simulation of the microsecond range, while a new set of phenomena could be studied if we could address the millisecond time scale.

The question thus arises how these phenomena can be modelled. One approach is the dissipative particle dynamics method (DPD). Here, a number of atoms are grouped together into one simulation bead which is used as the new simulation element. The reliability of the result obviously depends on how the underlying atoms translate into the interaction parameters between the DPD beads. Some semi-empirical methods will be discussed here. Then we concentrate on three applications: The mesophase formation of block copolymers, the simulation of polymer-surfactant complexes in bulk solution and the interaction of biological membranes with surfactant.

2   Introduction to DPD

The strategy to simulate molecular motions on length- and time scales that are much larger than what can be achieved with ordinary Molecular Dynamics simulations is based on two main ingredients. First, atoms are lumped together into "united atoms" describing more than one atom. The second ingredient used is that these new particles interact with each other via rather soft forces as the positions of the underlying atoms are smeared out. As we want to describe the correct thermodynamics (and dynamics) on a larger length-scale than an atom, we only need to reproduce the correct compressibility of the liquid and the correct solubilities of the various components into each other [5]. To arrive at this goal, we have the freedom to choose the effective interaction as a rather soft repulsion, provided that we satisfy the criteria discussed above. This means that we can leave out the hard core repulsive interaction between the atoms. Since it is the hard core interaction that forces the use of small time-steps (10^{-15} s), the removal of this core allows a considerable increase of the time-step, typically four orders of magnitude.

2.1 Forces

In DPD a set of interacting particles, whose time evolution is governed by Newton's equation of motion, is considered. Hence, at every time-step the set of positions and velocities, $\{\mathbf{r}_i, \mathbf{v}_i\}$ follows from the positions and velocities at earlier time. The force acting on a particle is given by the sum of a conservative, drag and pair-wise additive random force, i.e. $f_i = \sum_j (F_{ij}^{\mathrm{C}} + F_{ij}^{\mathrm{D}} + F_{ij}^{\mathrm{R}})$ where the sum runs over all neighbouring particles within a certain distance R_{c}. All forces depend on coordinate differences. The conservative force is given by

$$F_{ij}^{\mathrm{C}} = \begin{cases} -a_{ij}\left(1 - |\mathbf{r}_{ij}|/R_{\mathrm{c}}\right)\hat{\mathbf{r}}_{ij} & \text{if } |\mathbf{r}_{ij}| < R_{\mathrm{c}} \\ 0 & \text{if } |\mathbf{r}_{ij}| > R_{\mathrm{c}} \end{cases},$$

where a_{ij} is a maximum repulsion between particle i and particle j, $\mathbf{r}_{ij} = \mathbf{r}_j - \mathbf{r}_i$ and $\hat{\mathbf{r}}_{ij} = \mathbf{r}_{ij}/|\mathbf{r}_{ij}|$ [5,6], see Fig. 2.

Between neighbouring particles on a chain an extra spring force is defined to bind the particles together,

$$F_{ij}^{\mathrm{S}} = 4\mathbf{r}_{ij} \quad \text{if } i \text{ is connected to } j.$$

The drag force F_{ij}^{D} and the random force F_{ij}^{R} act as heat sink and source, respectively, so that their combined effect is a thermostat. The random force is given by

$$F_{ij}^{\mathrm{R}} = \sigma\omega\left(r_{ij}\right)\hat{\mathbf{r}}_{ij}\zeta/\sqrt{\delta t}$$

and the drag force as

$$F_{ij}^{\mathrm{D}} = -\frac{1}{2}\sigma^2\omega\left(r_{ij}\right)^2/k_{\mathrm{B}}T\hat{\mathbf{r}}_{ij}\left(\mathbf{v}_{ij}\cdot\hat{\mathbf{r}}_{ij}\right),$$

where ζ is a random variable with zero mean and unit variance, and $\omega(r) = (1 - r)$ for $r < 1$ and $\omega = 0$ for $r > 1$.

The amplitude of the random force should be taken proportional to $1/\sqrt{\delta t}$. Why is this? Let $\theta(t)$ be the random force exerted on a particle at a particular time step. This force leads to Brownian motion, where the displacement of a particle is proportional to

$$R \sim \sqrt{N_{\mathrm{steps}}}\delta r = \sqrt{\frac{1}{\delta t}}\theta\delta t \sim \sqrt{t} \times \theta\sqrt{\delta t}.$$

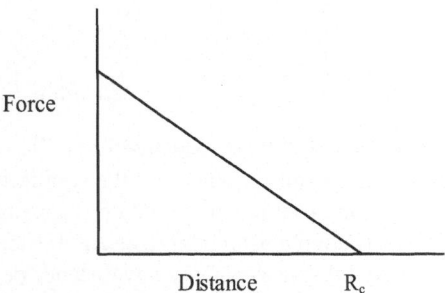

Force

Distance R_{c}

Fig. 2. The conservative force used in DPD

Since the displacement should not depend on the particular time step that we have chosen to simulate the process, $\theta(t)$ should be proportional to $1\sqrt{\delta t}$. This particular thermostat is special in that it conserves (angular) momentum leading to a correct description of hydrodynamics [7]. The reason why this thermostat conserves hydrodynamics is quite profound. All forces acting on particles are exerted on them by other particles nearby. This holds for the conservative forces, as well as for the friction and random forces. Since all particles obey Newton's third law, the sum of all forces in the system vanishes. Moreover, if we take any given volume of liquid, then all forces exerted between particles enclosed by that volume vanish. Consequently, the total acceleration of this volume of liquid equals the sum of all forces that cross the boundary of the volume. This is the very condition that leads to the Navier–Stokes equation. Therefore, whatever interaction force we invent between the particles, as long as it is a local interaction and satisfies Newton's third law we will always have hydrodynamics. If the random force would not be implemented pair-wise, but instead relative to a fixed background, we would break Newton's law. This is the case in Brownian Dynamics. Momentum is no longer conserved, and no hydrodynamic interaction is present in the simulation.

We choose the particle mass, temperature and the interaction range as units of mass, energy and length, hence $m = k_{\mathrm{B}}T = R_{\mathrm{c}} = 1$ and the simulated time is expressed in the natural unit of time

$$\tau = R_{\mathrm{c}}\sqrt{\frac{m}{k_{\mathrm{B}}T}}.$$

The DPD method in general has been shown to produce a correct (N,V,T) ensemble if the fluctuation-dissipation relation is satisfied [5,8]. Why is this important? In general the state of the system can be represented by a vector in 6N-dimensional space, $\{\mathrm{r}^{3\mathrm{N}}, \mathrm{p}^{3\mathrm{N}}\}$. The probability to find the system at any point in phase-space is the density of states $\varrho\{\mathrm{r}^{3\mathrm{N}}, \mathrm{p}^{3\mathrm{N}}\}$. The evolution of the system in phase-space can formally be written via the Liouville equation, which is

$$\frac{\partial \varrho}{\partial t} = \mathcal{L}\varrho = \mathcal{L}_{\mathrm{d}}\varrho + \mathcal{L}_{\mathrm{c}}\varrho. \tag{1}$$

In this equation $\mathcal{L}$ is the Liouville operator, which we can split to operators related to the conservative ($\mathcal{L}_{\mathrm{c}}$) and the dissipative force ($\mathcal{L}_{\mathrm{d}}$). If we turn off all noise and friction in the simulation the latter vanishes, and the evolution is solely governed by $\mathcal{L}_{\mathrm{c}}$. In equilibrium, the density of states does not change, and hence $\mathcal{L}_{\mathrm{c}}\varrho_{\mathrm{eq}}$ must be zero. Here ϱ_{eq} is the Boltzmann distribution:

$$\varrho_{\mathrm{eq}} \propto \exp\left(-\frac{U\left(r^{3\mathrm{N}}\right)}{k_{\mathrm{B}}T} - \sum_i \frac{p_i^2}{2m_i k_{\mathrm{B}}T}\right).$$

If we now check (1), it is clear that the dissipative Liouville operator acting on the Boltzmann distribution must also vanish, otherwise the equilibrium would shift to another distribution when noise and friction are turned on. To maintain the correct Boltzmann distribution, noise and friction must therefore be administered in a particular way. Español and Warren [8] proved that if we choose any distance dependent noise term

$$F_{ij}^{\mathrm{R}} = \sigma\omega(r_{ij})\hat{\mathbf{r}}_{ij}\zeta/\sqrt{\delta t},$$

then the friction term must be taken as

$$F_{ij}^{\mathrm{D}} = -\frac{1}{2}\sigma^2 \omega(r_{ij})^2 / k_{\mathrm{B}} T \hat{\mathbf{r}}_{ij} \left(\mathbf{v}_{ij} \cdot \hat{\mathbf{r}}_{ij}\right).$$

2.2 Simulation Techniques

At every time-step the set of positions and velocities, $\{\mathbf{r}_i, \mathbf{v}_i\}$, is updated from the positions and velocities at earlier time. All update algorithms known from Molecular Dynamics can be used in principle [9], but the presence of the velocity in the forces complicates things. A straightforward method is to use the Euler scheme

$$\begin{aligned}
\mathbf{r}_i\left(t + \delta t\right) &= \mathbf{r}_i(t) + \mathbf{v}_i(t)\delta t, \\
\mathbf{v}_i\left(t + \delta t\right) &= \mathbf{v}_i(t) + \mathbf{F}_i(t)\delta t, \\
\mathbf{F}_i\left(t + \delta t\right) &= f\left(\mathbf{r}_i(t + \delta t), \mathbf{v}_i(t + \delta t)\right).
\end{aligned}$$

However, temperature control is not very accurate in this method. To use a second order update algorithm is not as straightforward as it may seem. A second order algorithm integrates the positions from t to $t + \delta t$ using the velocity and accelerations known at t. To update the velocities, however, we need to know the accelerations at time t and at time $t + \delta t$. In ordinary Molecular Dynamics this is not a problem since the forces at time $t + \delta t$ are known once the new particle positions are calculated. In DPD, however, we need to know the velocity in the next time step in order to calculate the force that we need to update the velocities.

Two solutions to this problem are worth mentioning. The first is a modified version of the velocity-Verlet algorithm [5]:

$$\begin{aligned}
r_i(t + \delta t) &= r_i(t) + \delta t \, v_i(t) + \frac{1}{2}\delta t^2 f_i(t), \\
\tilde{v}_i(t + \lambda \delta t) &= \tilde{v}_i(t) + \lambda \delta t \, f_i(t), \\
f_i(t + \delta t) &= f_i\left(r_i(t + \delta t), \tilde{v}_i(t + \lambda \delta t)\right), \\
v_i(t + \delta t) &= v_i(t) + \frac{1}{2}\delta t \left(f_i(t) + f_i(t + \delta t)\right).
\end{aligned} \qquad (2)$$

The masses of the particles are set to 1, so that the force acting on a particle equals its acceleration. The force is updated once per iteration. The velocity in the next time-step is estimated by a predictor method. This is done in the second step of our algorithm. The velocity is corrected in the last step. If the parameter λ is put at $\lambda = 0.5$ this scheme equals the velocity-Verlet algorithm [10]. It is empirically observed that if we use $\lambda = 0.65$ we find a very accurate temperature control, even at the time-step $\delta t = 0.06\tau$. This is probably due to a cancellation of errors. A more systematic study into the influence of parameter λ was presented by Den Otter and Clarke [11].

The second method, presented by Pagonabarraga et al. [12] can be seen as an extension of this algorithm. In this method the same update scheme as in (2) is used, but the velocity dependent part of the force is iterated until a stable value for the velocity in the new time step is obtained. The scheme is therefore named self-consistent. Because it is self-consistent, the simulation algorithm is also time-reversible. This is found to have an important influence on the temperature control. For most practical applications, however, the predictor method is comparably accurate, but faster.

2.3 Parameterisation

This has two parts, the first is to derive the correct length- and time scales of the simulation, and the second is to obtain the repulsion parameters. DPD can be used either as a flow solver or as a method to simulate molecular dynamics over time scales far beyond what can be reached with Molecular Dynamics. If it is used as a flow solver, the time scale of the simulation is related to hydrodynamic relaxation time of the problem. This must be matched between the simulation and the problem. In practice, this calibration is done by adjusting the viscosity of the fluid. If explicit molecules and their diffusive behaviour are simulated, we need to match, e.g. the diffusion coefficient of water. Here we concentrate on the latter application of DPD. Since water is an important compound we will use it to define the length- and time scales used in 'molecular' DPD [13].

Let a bead correspond to N_m water molecules. The number N_m can be viewed upon as a real-space renormalization factor. Thus, a cube of volume R_c^3 represents ϱN_m water molecules, where ϱ is the number of DPD beads per cubic R_c. From the density of water and its molecular weight, we can calculate the volume per water molecule in liquid water at room temperature as $30 \, \text{Å}^3$. Thus, the physical volume of this cube equals $30 \, \varrho N_m \text{Å}^3$, hence the length scale R_c follows as

$$R_c = 3.107 (\varrho N_m)^{1/3} \ (\text{Å}).$$

To gauge the unit of time, we match the long-time diffusion coefficient of water. Some care must be taken here. The self-diffusion coefficient of a water bead is not the same as the self-diffusion coefficient of water, since the bead represents N_m water molecules. When these move over the vectors $\mathbf{R}_1, \mathbf{R}_2, \ldots \mathbf{R}_{N_m}$, their centre of mass moves over the vector $\mathbf{R}_w = (\mathbf{R}_1 + \mathbf{R}_2 + \ldots + \mathbf{R}_{N_m})/N_m$. Hence the ensemble average of the mean square displacement of the water beads is

$$R_w^2 = \langle \mathbf{R}_w \cdot \mathbf{R}_w \rangle = \frac{(\langle \mathbf{R}_1 \cdot \mathbf{R}_1 \rangle + \langle \mathbf{R}_2 \cdot \mathbf{R}_2 \rangle + \ldots)}{N_m^2} = \frac{R^2}{N_m},$$

where R^2 is the mean square displacement of a water molecule. At the noise and repulsion parameters $\sigma = 3$ and $a = 78$, the diffusion coefficient of water beads in DPD simulation was obtained as

$$D_w = 0.1707(14) R_c^2 / \tau.$$

Equating this to the experimental diffusion coefficient of water [14] $D_{\text{water}} = (2.43 \pm 0.01) \times 10^{-5} \ \text{cm}^2/\text{s}$, leads to the time scale

$$\tau = \frac{N_m D_{\text{sim}} R_c^2}{D_{\text{water}}} = 14.1 \pm 0.1 N_m^{5/3} \quad \text{(ps)}. \tag{3}$$

In this equation it is implicitly assumed that the repulsion parameter between equal beads is fixed to the value $a = 78$, and that the bead density is fixed at $\varrho = 3$.

At this point we can understand why the DPD method is so much faster than straightforward molecular dynamics. There are two combined effects that lead to speed-up. The first contribution comes from the low Schmidt number in the simulation [5]. The Schmidt number is the ratio between viscosity and the self-diffusion coefficient, $Sc = \nu/D$. In an ordinary liquid like water, this ratio is roughly $Sc \approx 1000$, whereas in the DPD method we have $Sc \approx 1$. The origin of this difference can be traced back to the removal of the hard core from the interaction potential. This hard core leads to a caging effect, i.e. an atom undergoes many collisions before it is actually transported. The soft potential used here removes this caging affect, so that the mobility of particles is increased by a factor of 1000. The second factor contributing to the speed-up is the scaling of the physical time with the renormalization factor N_m as in (3). On top of the power 5/3 by which the physical time scale increases, the amount of CPU time will decrease inversely proportional to N_m if we want to simulate a given volume, simply because we have to update the position of fewer objects. Thus, for a given system volume, DPD can be expected to be faster than MD by a factor of roughly $1000\,N_m^{8/3} \approx 2\times 10^4$ for $N_m = 3$ and about 10^5 for $N_m = 6$. This is independent of hardware and disregards the CPU time spent on evaluating the (relatively long ranged) Lennard-Jones potential.

To find the interaction parameters for this model, we need to match the liquid structure function in the limit $k \rightarrow 0$, as this determines the free energy change associated to density fluctuations. This in turn is related to the compressibility and solubilities. Note that the pressure itself drops out in an NVT ensemble, as this is a linear variation of the free energy. It was previously proposed that the following relation should hold [5]:

$$\frac{1}{k_B T}\left(\frac{\partial p}{\partial \varrho}\right)_{\text{simulation}} = \frac{1}{k_B T}\left(\frac{\partial p}{\partial n}\right)_{\text{experiment}},$$

where ϱ is the bead density in the simulation, and n is the density of e.g. water molecules in liquid water. However, this relation only holds if one DPD bead corresponds to one water molecule. In general, the system should satisfy

$$\frac{1}{k_B T}\left(\frac{\partial p}{\partial \varrho}\right)_{\text{simulation}} = \frac{1}{k_B T}\left(\frac{\partial n}{\partial \varrho}\right)\cdot\left(\frac{\partial p}{\partial n}\right)_{\text{experiment}} = \frac{N_m}{k_B T}\left(\frac{\partial p}{\partial n}\right)_{\text{experiment}},$$

where N_m is the number of water molecules per DPD bead. When N_m is chosen as $N_m = 3$, the compressibility of water at room temperature is matched if the repulsion parameter between particles of the same type is determined at $a_{ii} = 78$. Note that it is taken the same for all liquid components, as we actually simulate equal liquid volumes for all components.

The next parameters to determine are the bead-bead repulsions, by matching solubility. In polymer chemistry solubility is usually expressed by specifying the Flory–Huggins χ-parameters. This parameter represents the excess free energy of mixing in the Flory–Huggins model. This is a cell model, where every cell is filled by a fraction ϕ of A molecules and by a fraction $1 - \phi$ of B molecules, i.e. the lattice is completely filled. If A is a polymer occupying N_A cells, and B is solvent that occupying N_B cells, the free energy per cell (disregarding constants and terms linear in ϕ) can be written as

$$\frac{f_\nu}{k_B T} = \frac{\phi \ln \phi}{N_A} + \frac{(1 - \phi)\ln(1 - \phi)}{N_B} + \chi\phi(1 - \phi).$$

Different polymers usually tend to segregate, see Fig. 3. To model this behaviour we impose a larger repulsion between unlike beads than between beads of the same type.

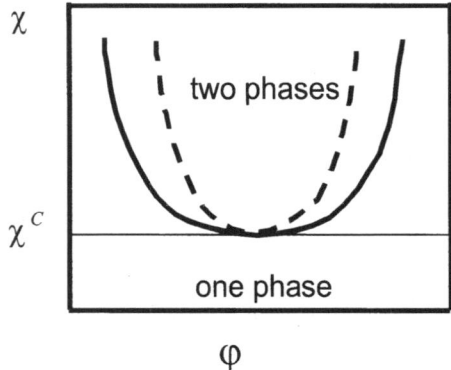

Fig. 3. The demixing curve (*full curve*) and spinodal (*dashed curve*) in Flory–Huggins theory

It has been established that the χ-parameter is linearly related to the *excess* of the AB repulsion over the AA repulsion [5]. When the volume fraction of A in the majority B phase is measured for two liquids, each consisting of molecules of length N, the χ-parameter can be obtained by substituting the simulated volume fraction into the mean-field expression for the binodal:

$$\chi N = \frac{\ln(1 - \phi) - \ln \phi}{1 - 2\phi}. \qquad (4)$$

This expression should be valid far away from the critical point. For $a_{ii} = 78$ this led to the correspondence [13]

$$\chi N = 0.231 \pm 0.001 \Delta a,$$

where $\Delta a = a_{\mathrm{AB}} - a_{\mathrm{AA}}$ is the excess repulsion.

The pertinent χ-parameters can be determined by matching the Flory–Huggins model to relevant experimental solubility data. An alternative to the use of mean-field theory as an intermediate was provided by Wijmans et al. [15]. They simulated the binodal in a mixture of a polymer and a single bead solvent using the Gibbs ensemble Monte Carlo method. This led to the binodal curve:

$$\Delta a \approx 0.516 N^{-0.751} \left| N^{0.435} \ln(1 - \phi) - \ln(\phi) \right|^{1.826} + 2.25 \left(1 + N^{-0.44}\right)^{1.75},$$

where N is the number of beads per polymer. This equation enables us to compare simulations to experiments directly, or alternatively to extract the simulation parameters from experimental data.

2.4   Generalisations and Alternatives

DPD, as described above, is like a minimal version to simulate a molecular liquid. For particular applications, particles can and indeed have been given internal degrees of freedom, such as an internal energy [16,17], angular momentum and orientation [18]. The former generalisation allows constant energy simulations, so that heat flows can be simulated. The latter generalisation describes particles with spin, leading to higher viscosity. Another variation is to use each particle as a centre for a weighted density functional [19]. This gives the freedom to insert any desired free energy functional, and thus alter the equation of state, and simulate free surfaces.

DPD is by no means the only technique by which mesoscale simulations can be performed. One option is to use Lagrangian flow solvers. By adapting this approach a simulation technique for modelling viscoelastic fluid flow has been developed by Yuan, Ball and Edwards [20]. By using a moving Voronoi mesh, the method is able to track the details of fluid behaviour, e.g. deformation and stream lines in viscoelastic liquids. The velocity (and pressure, etc.) is defined on discrete points, which are convected with the flow. The points exchange momentum with their neighbour, and the interactions are chosen by discretising the Navier–Stokes equations.

Smoothed Particle Hydrodynamics (SPH) is a similar scheme without a mesh. It uses an interpolation scheme to calculate spatial derivatives based on weight functions centred around the particles. The particles interact via a pairwise interaction, and pressure is included explicitly. Newton's 3rd law is not obeyed, but the scheme is close to that of DPD [21].

Chris Lowe introduced a variation of DPD [22] in which the interaction potential is the same, but the velocities of the particles are exchanged rapidly via an Andersen Monte Carlo method [23]. New relative velocities are taken from a Maxwell distribution, so that the temperature control is rigorous. When small steps are taken and the velocities are exchanged at every step, this method leads to much higher viscosity than DPD. In fact, any Schmidt number can be chosen. On the other hand, low viscosity is problematic.

Another alternative is the Lattice Boltzmann method, which is used to solve the Navier–Stokes equations on a lattice. The lattice is chosen as a 3D projection of a 4D fcc lattice. This choice minimises lattice artefacts. On this lattice a discrete implementation of the Boltzmann equation is simulated. When a fluid mixture is to be simulated the same lattice may serve as a basis for a Landau expansion of the free energy [24]. Thus, the method contains no explicit molecules, and no noise is needed. Finally self-consistent field theory can also be used to simulate diffusive problems of, e.g. block copolymers on a 3D lattice [25]. Here a lattice is used to calculate the polymer Green functions. From the Green functions follow the local polymer volume fractions. These in turn determine the local chemical potentials of the various segments. The chemical potential gradients are coupled to the polymer mobility via Onsager kinetic coefficients. This leads to a Smoluchowski equation for the density fields which can be solved numerically. Because the polymer statistics is by construction Gaussian, this method is strictly speaking not valid for polymer solutions. Experiments indicate that also block copolymers have markedly non-Gaussian statistics even quite close to their critical point.

All methods mentioned here have positive and negative properties, this also holds for DPD. The unresolved issues in DPD are as follows. First, the Schmidt number problem. The speed by which momentum diffuses is the kinematic viscosity ν, the speed by which particles travel is the diffusion coefficient D. The Schmidt number is $Sc = \nu/D \sim 1000$ in a liquid like water, whereas it is of the order 1 in DPD. This effectively means that the diffusion coefficient is overestimated by a factor of 1000 when the viscous time scale is matched. When viscous flow is to be simulated correctly, an alternative to classical DPD is the Andersen Monte Carlo method by Lowe. For molecular processes that are diffusion controlled, however, fast diffusion is a great help to speed up the simulation. The second problem appears when the method is used for turbulent hydrodynamic problems. The rather soft beads lead to a low sound velocity. This means that at high Reynolds numbers, one may run into unwanted supersonic flow. To repair this flaw, the incompressibility of the liquid has to be built into the method by other means than by soft repulsive particles. Finally, when long polymers and micelles are to be simulated, or breaking oil droplets in a surfactant solution, one may run into a clash of length scales. To resolve a coarse-graining where individual surfactant molecules are simulated (1 nm resolution), and to simulate micron size droplets at the same time (1 μm size) requires a simulation of order 10^{10} particles. This is presently not possible in DPD, but this problem is generic for all mesoscale methods.

Although DPD is a rather new technique it has already been applied to a wide variety of problems including complex two-phase flow, such as the rheology of dense colloidal suspensions [26], the break-up of oil droplets in gravitational and shear fields [27], and spinodal decomposition and domain growth [28–30]. In the next two sections we concentrate on a small number of applications, the phase formation of block copolymers, polymer-surfactant interactions and the simulation of biomembranes.

3 Block Copolymer Mesophase Separation

3.1 Polymers in Melt

Diblock copolymers are polymers consisting of two linear blocks (A and B) of mutually insoluble polymers, chemically connected end-to-end. When a melt of these polymers is quenched (i.e. the temperature is suddenly dropped), the A-blocks and B-blocks tend to phase separate. The connectivity of the polymers prevents macroscopic phase separation, and, consequently, the system can only reduce its free energy by connecting the A-rich and B-rich domains in structures like spheres, rods, sheets, perforated sheets or complicated sponge-like structures. This principle has been known for quite some time, see Bates and Fredrickson for a review [31], but only in recent years our understanding as to which phase is formed under what conditions has increased to a level where we are in the position to predict the phase diagram. The question as to which structure is formed under what condition was first theoretically studied by Leibler [32], who used Gaussian coil statistics to calculate the free energy in a Landau theory. The equilibrium microstructure in this theory depends on the ratio f of the length of the A section relative to the total length of the polymer, and on the mutual solubility of the A and B units, which is usually represented by the Flory–Huggins χ-parameter [31]. We want of a theory, or a simulation method, to be able to resolve the following issues:

1. To predict the phase structure of diblock copolymers as function of f, χ and M_n.
2. To understand the dynamics of formation of a phase after a temperature quench.
3. To describe the transition of a copolymer system from one mesophase structure into another.

Since the driving force for the formation of mesophases comes from the surface tension between phases A and B, this needs to be reproduced correctly. Also the conformation and dynamics of homopolymers in the melt needs to be correct. For homopolymers the theory predicts that the endpoint separation as function of polymer length N in a melt should scale as

$$R_e \sim N^{1/2}.$$

Furthermore, the diffusion coefficient and the relaxation time of the end-to-end vector should scale as [33]

$$D \sim N^{-1} \text{ and } \tau_R \sim N^2.$$

Spenley has checked these scaling relations [34]. He found that

$$R_e \sim (N-1)^{0.498 \pm 0.005}, \; D \sim N^{-1.02 \pm 0.02}, \text{ and } \tau_R \sim N^{1.98 \pm 0.03}.$$

The correspondences are excellent. For the surface tension of an ordinary liquid near its critical point one may expect the scaling law [35]

$$\sigma \sim (1 - T/T_c)^{\mu},$$

where T_c is the critical temperature, and μ is an exponent that takes on the value $\mu = 1.26$ for the Ising model, and $\mu = 3/2$ for the van der Waals liquid. Groot and Warren have simulated the surface tension between two homopolymer melts in the DPD model [5]. They noted that for a polymer-polymer interface, T corresponds to $1/\chi$ and that the critical χ-parameter between two homopolymers is $\chi_c = 2/N$, and thus found the following master equation for the surface tension:

$$\sigma/R_c = 0.58 \, \varrho k T \chi^{0.4} (1 - 2/\chi N)^{3/2}.$$

The power 3/2 is expected, as one often finds mean-field theory to work well for polymers. The prefactor $\chi^{0.4}$ is at variance with mean-field theory, which predicts a factor $\chi^{1/2}$. The polymer length dependence of the surface tension appears to match quantitatively with experimental results, see Fig. 4.

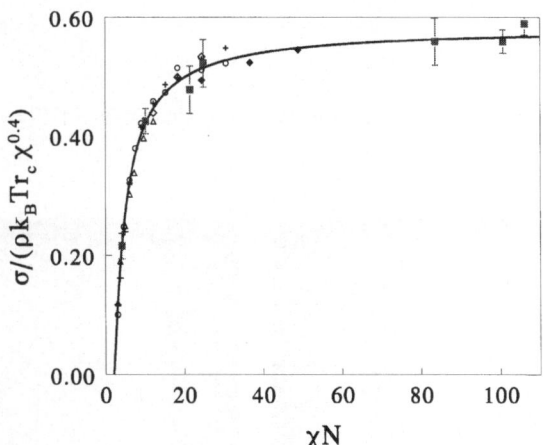

Fig. 4. Simulated polymer-polymer surface tension master curve and experimental data, reproduced from [5,36]

3.2 Expected and Simulated Phase Diagram

To the lowest order, mean-field theory predicts that the block copolymer phase diagram
is determined only by product χN, and by the ratio $f =$ (length of A block) divided
by the total length of polymer, see Fig. 5. Therefore, to lowest order we can rescale
a long polymer down to a small number of segments per chain. DPD simulations of
block copolymers were performed by Groot et al [36,38], who used a polymer length
$N = 10$ and $\chi N \approx 46$. This is well outside the weak segregation limit. Configurations
of A_5B_5 and A_3B_7 polymer systems containing 40 000 particles are shown at time
$\tau = 4\,000\tau \cong 430\tau_R$, where τ_R is the Rouse time of a homopolymer of the same
molecular weight. Due to symmetry of the polymer the A_5B_5 system must be either
lamellar (for large χN) or disordered (for small χN), but when the A:B ratio is changed
away from 1:1 other phases are experimentally found to appear [31]. In the simulation
it is indeed found that the A_5B_5 system converges to a lamellar phase, see Fig. 6. The
A_3B_7 system, in contrast, does not converge to a lamellar phase. The A-domains are
shown as white spots in Fig. 6.

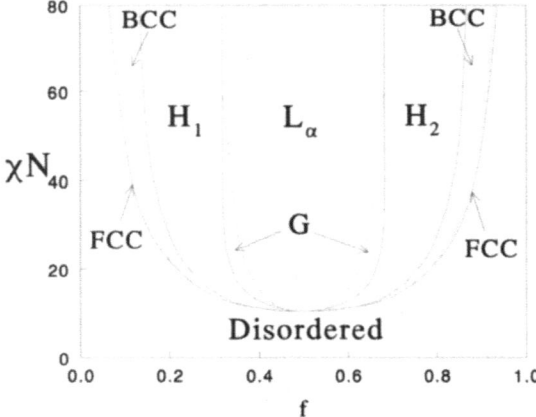

Fig. 5. Expected phase diagram based on work by Matsen and Bates [37] and reproduced from
Groot and Madden [36]

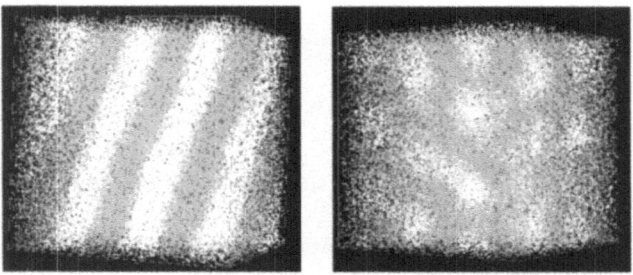

Fig. 6. Conformation of A_5B_5 system (left) and A_3B_7 system (right), after [36]

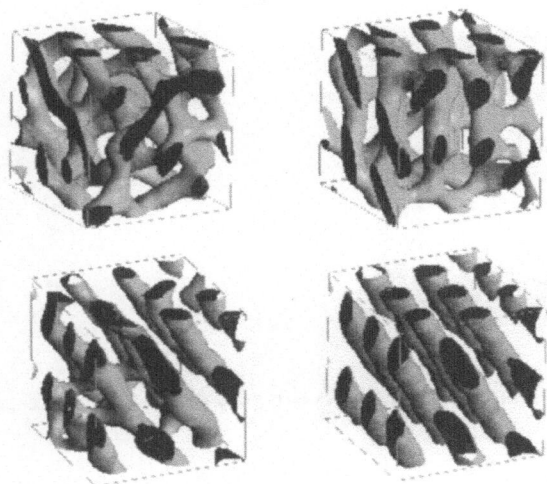

Fig. 7. Evolution of A_3B_7 block copolymer system, after [36]

The time evolution of the A_3B_7 system is shown in Fig. 7. In the top left conformation we see a structure that resembles the gyroid phase, but which is predicted to be unstable. After a further 2 000 time units we find the top right structure. Note that the system of rods has lost symmetry relative to the earlier stage, i.e. the rods tend to align in a cooperative manner. In the next stage, shown in the bottom left picture of Fig. 7, the rods are completely aligned, though some sideward connections are still present. In the final configuration the sideward connections are broken and the system is locked in a state of parallel rods in a perfectly hexagonal arrangement. These results show that the DPD method is capable of changing the topology of a micro-phase structure in an efficient way. Qualitatively, it is found that the A_5B_5 system evolves to the lamellar phase as it should, and that the A_3B_7 system evolves to a hexagonal phase, which is expected to be stable between $0.165 < f < 0.314$ for the present χ-parameter.

In a subsequent study simulations were performed on a range of polymeric systems: A_5B_5 ($f = 0.5$), A_4B_6, ($f = 0.4$), A_3B_7 ($f = 0.3$) and A_2B_8 ($f = 0.2$) and A_1B_9 ($f = 0.1$). The latter remained an isotropic liquid throughout the course of the simulation. Apart from the A_2B_8 system, all of the simulations finally produce a phase structure that is consistent with self-consistent field theory. To further quantify the phase diagram near the H_1-$L\alpha$ phase transition line, simulations of mixed polymers were done. Assuming that for these mixtures the mean value of f is representative of a homopolymer system of the same value of f, A_3B_7 and A_4B_6 polymers were mixed to create systems of *average* value $\langle f \rangle = 0.325$, 0.35 and 0.375. In experiments, Zhao et al. [39] also blended two block-copolymers to obtain a mixture with a preferred (mean) asymmetry, $\langle f \rangle$. These experiments indicate that the mixture behaves as a homopolymer as long as the difference between the two polymers is small. In simulations, all systems within the predicted lamellar phase region did indeed converge to a lamellar phase, and the same holds for the hexagonal phase region. However, between the hexagonal phase and the lamellar phase DPD does not produce a gyroid phase but a perforated lamellar phase

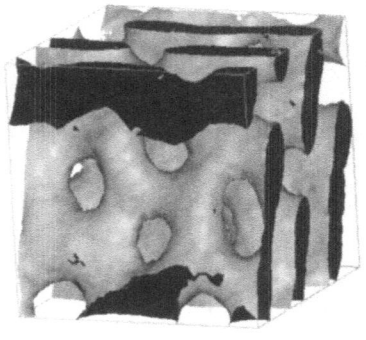
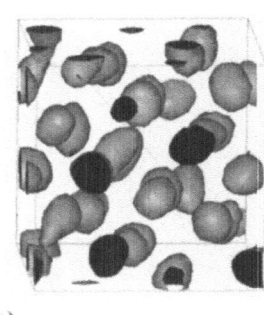

a)   b)

Fig. 8. a) Perforated lamellar system for $\langle f \rangle = 0.35$, after [36]. **b)** Body-centred cubic system at $\langle f \rangle = 0.14$, after [36]

instead, see Fig. 8. This phase has recently been identified in experiments [39,40], but it has not been predicted from self-consistent field theory.

For the A_2B_8 system theory predicts a hexagonal phase. In the simulation this system forms a disordered micellar phase that lasts the total length of the simulation, 32 000 time units. To proceed, polymers of structures A_2B_8 and A_3B_7 were blended. Of these the system with $\langle f \rangle = 0.275$ evolved to the hexagonal phase, and the systems with $\langle f \rangle \leq 0.25$ remained in a liquid-like, entangled tube state during the course of the simulations. Hence at $\chi N = 46$ simulation predicts a phase transition from an entangled tube state to a hexagonal phase near $f_c \approx 0.26 \pm 0.02$. At this point we note that at low f we do not find the expected BCC quasi-crystalline phase, but instead we find a liquid-like ordering in flexible micelles. To understand the differences between theory and simulation, we need to study the influence of the finite chain length.

The simulated polymers are only of length $N = 10$. This increases artificially the importance of fluctuations relative to really long polymers. Thus fluctuations lower the free energy of isolated micelles relative to that of infinitely long rods. For surface tension the finite polymer length is apparently not very important, but for the phase diagram the effects due to finite length can be severe. Weak coupling calculations predict that the order-disorder transition at $f = 0.5$ for small polymers shifts up as [41]

$$(\chi N)_c = 10.5 + 41.0 \bar{N}^{-1/3}.$$

For simulations with small polymer lengths this would imply that the effective χ-parameter (i.e. corresponding to infinite N) is smaller by a factor

$$(\chi N)_{\mathrm{eff}} = \frac{10.5}{10.5 + 41.0 \bar{N}^{-1/3}} \chi N = \frac{\chi N}{1 + 3.9 \bar{N}^{-1/3}}. \tag{5}$$

The decrease of the effective χ-parameter is controlled by fluctuations characterised by a Ginzburg parameter

$$\bar{N} = 6^3 (R_g^3 \varrho_p)^2 = (R_e^3 \varrho_p)^2,$$

where ϱ_p is the polymer concentration and R_g the radius of gyration [31]. It is this $\bar{N}$ which appears at the right hand side of (5). This parameter is determined by the number of other polymers in the volume that a polymer occupies. Substituting the end-point separation that we obtained for homopolymers, and polymer density $\varrho_p = \varrho/N$ we find

$$(\chi N)_{\mathrm{eff}} = \frac{\chi N}{1 + 4.3\varrho^{-2/3}N^{-1/3}} \approx 0.51\chi N$$

for our simulations at $N = 10$. The effective χ-parameters would thus be given by $(\chi N)_{\mathrm{eff}} = 23.4$. From the simulations we find a reasonable match with mean-field theory at $(\chi N)_{\mathrm{eff}} = 20 \pm 2$, though the location of the H-G transition is slightly off.

The consequence of this is that these simulations should be compared with the theoretical phase diagram at $\chi N \approx 20$. If this assertion is correct we should find a BCC phase for the $f = 0.14$ system when we considerably increase χN over the value that we currently used, as this would put us in the middle of the cubic phase. Therefore, a number of runs at various values for f and χ were performed so as to follow the theoretically predicted BCC phase boundary. Intermediate values of f were obtained by blending A_1B_9 with A_2B_8. The structure at $f = 0.14$ and $\chi N = 98$ is shown in Fig. 8b. This system rapidly forms spherical micelles, which afterwards form a quasi-crystalline phase on a much larger time-scale.

If we compare the theory to the simulation results at $\chi N = 20$, we actually find a matching correspondence. Theory predict the transitions from disordered-FCC, FCC-BCC, BCC-hexagonal, hexagonal-gyroid and gyroid-lamellar at $f = 0.210, 0.214, 0.240, 0.340$ and 0.374. The DPD results for the equilibrium structure of block-copolymers are in line with this, and are summarised in the schematic phase diagram shown in Fig. 9. The "effective" Flory–Huggins parameter is obtained by extrapolation to infinitely long chains using finite chain simulations. This diagram is based on 27 systems and should only be seen as a rough indication of which phase we find where. The diagram compares well to the diagram that Larson produced for short lattice chains in a monomer solvent [42]. In accordance with mean-field theory the simulated diagram shows the classical quasi-solid body centred cubic (BCC), hexagonal (H), and lamellar phases (L). However, we also find melted structures like a liquid micellar phase (LM), a liquid rod

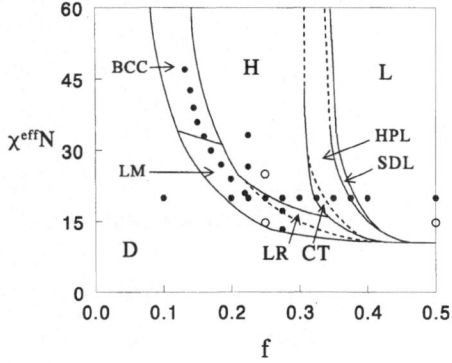

Fig. 9. Rough phase diagram coming forward from DPD simulations, after [38,44]

phase (LR) and a connected tube phase (CT). These melted structures agree with experimental observation [43] and with Monte Carlo simulations of block copolymers. The DPD simulations also predict a hexagonally perforated lamellar phase (HPL) which has been observed in experiments [39,40], and a small region where screw dislocations in a lamellar phase are stabilised (SDL).

3.3 Evolution Pathways

An important advantage of the DPD method is its explicit results for time-dependence. This is very relevant for polymer microphase separation, since for long polymers the typical evolution time can be long, especially when the polymers are branched. For grafted polymers, the time that a side-branch needs to disjoin from one micelle sets a natural scale for the time of topological rearrangements. If a polymer melt is quenched from a high temperature into the ordered phase, the pathway through which the final structure is reached is relevant if the time of interest is months rather than minutes or hours. To introduce the formation process of the mesophases we briefly repeat the qualitative findings from DPD simulations that have been reported elsewhere [36,38,44]. Processes on three different length- and time- scales can be distinguished by the formation of polymer micro-phases:

1. phase separation on the mesoscopic bead level,
2. organisation of polymers into micelles,
3. organisation of micelles into a superstructure with its own particular symmetry.

A schematic diagram summarising the different effects is shown in Fig. 10. The evidence for this scheme comes from observing the time evolution of polymer systems of various compositions at a fixed value of χN, and capturing the qualitative effects

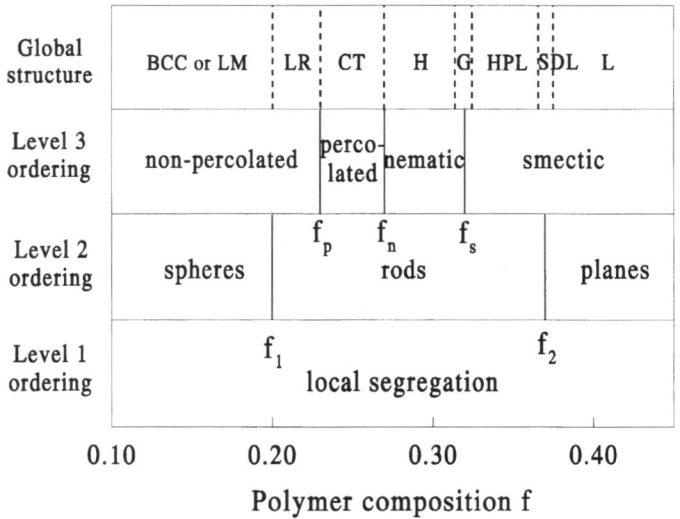

Fig. 10. Schematic diagram of evolutionary processes, after [38,44]

of the evolution in a simple picture. This is a conceptual framework, which helps to rationalise the evolution, rather than an exact description of the location of various transition points. These obviously depend on the precise value of χN. Effects on different length-scales interplay in both the final structure and in the pathway to form it. On level 2, the dimensionality of the micelles (spherical, rod-like or planar) is the dominating factor. At the AB segregation parameters used in these simulations, the transitions between these structures are found at $f_1 \approx 0.20$ and $f_2 \approx 0.37$. On a global level (level 3) the important transition points are the percolation transition, where the rods form an interconnected tube network, the nematic transition and the smectic transition. These are located respectively at $f_p \approx 0.23$, $f_n \approx 0.27$ and $f_s \approx 0.32$. For compositions where $f_p < f < f_2$ a percolating interconnected tube phase is formed as precursor of the final phase. Experimental evidence comes from time-resolved X-ray scattering, see Balsara et al. [45] and references therein. These experiments reveal the presence of two processes, a fast process that is believed to be related to the local segregation of the blocks (ordering levels 1 and 2) and a slow process that leads to long-ranged order (level 3).

3.4 Importance of Hydrodynamics

A clear comparison to establish the role of hydrodynamics can be made when simulations are performed with and without hydrodynamics. Two continuum simulation methods have therefore been compared. Both describe the same Hamiltonian system, but they differ in their evolution algorithm. The first method is the Dissipative Particle Dynamics method, and the second is the Brownian Dynamics method. The only difference between the two is that all hydrodynamic interactions are taken into account in the former method, but not in the latter. The polymer architecture, connectivity, interactions and the liquid compressibility are explicit in both methods. Thus we can make a very pure comparison, to see what happens if only hydrodynamics is turned off while all other physical effects are included. For symmetric polymers the soft sphere model is found to predict the formation of lamellar domains of some eight lamellae across, irrespective of the presence of hydrodynamic interactions. Without external shear experimental samples remain globally disordered, but local order does appear spontaneously. Experimental systems also form domains of some eight lamellae across, hence they order on the length-scale seen in the DPD simulations.

Since different compositions lead to aggregates of different topology, it is not clear beforehand if the influence of hydrodynamic interactions is equally important in the different regions of the phase diagram. For this reason we have performed simulations both for asymmetric polymers ($f = 0.3$), and for symmetric polymers ($f = 0.5$). In the former the system has to go through a percolated state and a nematic transition to find its equilibrium structure and in the latter system domains of local lamellar order have to grow together to form a macroscopically homogeneous phase. We first discuss the results obtained for the A_3B_7 copolymer system. The simulations were performed in a box of $V = 20 \times 20 \times 20$ using periodic boundary conditions. At time $t = 0$, 2400 copolymers of structure A_3B_7 were arranged randomly in the box and the systems were allowed to evolve.

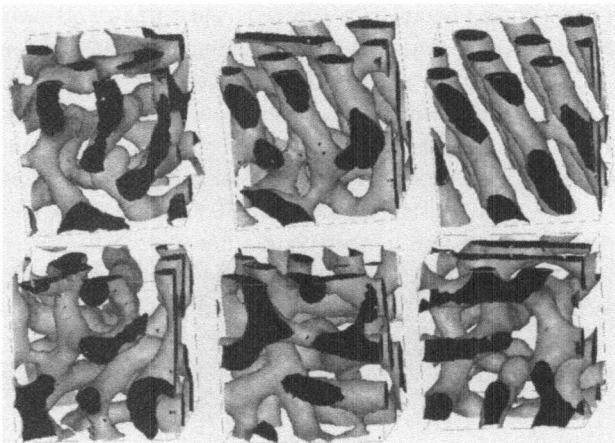

Fig. 11. Evolution of A_3B_7 system with hydrodynamics (DPD, top row) and without hydrodynamic interaction (Brownian Dynamics, bottom row), after [36]

Figure 11 shows three stages in the evolution of the simulated system. The DPD simulation quickly forms micro-phase separated regions that percolate into interconnected tubes. These tubes form a globally disordered fluid phase with tubes changing shape and moving relative to each other. After approximately $7\,500\tau$ a domain of hexagonal order is formed, which grows at the expense of the disordered phase. The subsequently formed hexagonal phase is stable for the rest of the simulation. On the basis of self-consistent field calculations it has recently been put forward [46] that the hexagonal phase is formed from the gyroid by a process where first five-fold connection points are formed, that subsequently break into a three-fold connection and two unconnected tubes. We did not find evidence for this mechanism in our simulations. Instead, we find only three- and four-fold connection points linked by short liquid bridges that sever by a necking mechanism. In the last stages of evolution, where the sample is almost completely hexagonally ordered, we find local defects in the form of liquid bridges between otherwise parallel rods. The dominant mechanism for topological transitions in that stage is the scission of these liquid bridges, see the top-right picture in Fig. 11.

The path taken by the Brownian Dynamics (BD) simulation is very similar to that of the of the DPD simulation in its early stages: the formation of a phase of interconnected tubes. In the BD simulation we also find the tubes to align locally in a hexagonal structure, but this phase is subsequently destroyed again. In many places throughout the simulation box small hexagonal domains arise and disappear. None of these domains manage to grow out to a globally ordered hexagonal phase, even when the simulation is extended to $24\,000\tau$. One may argue that there could be a subtle bug in the BD program, which makes the hexagonal phase unstable [47]. If that would be the case then it is obvious that the hexagonal phase does not form in the BD simulation. To check this loophole, the hexagonal structure, as generated by the DPD simulation, was used as a starting configuration and was evolved in a BD simulation over $50\,000$ time steps ($3\,000\tau$). The hexagonal phase remained stable. In fact the shape fluctuations of the tubes are smaller

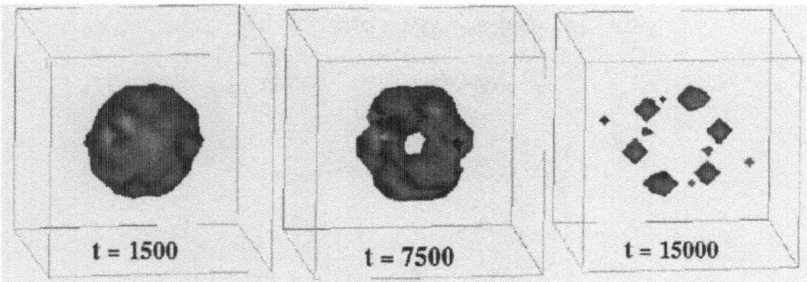

$t = 1500$ $t = 7500$ $t = 15000$

Fig. 12. Evolution of A_3B_7 structure function in DPD, after [36]

than they are in the DPD simulation. So either the hexagonal phase is metastable, but the BD method cannot break it apart, or it is stable and the BD method cannot form it. In either case it is demonstrated that there is a kinetic barrier that the BD method cannot cross. Since the DPD method can cross this barrier, and since the only difference between the two simulation methods is the conservation of momentum leading to a correct description of hydrodynamics in the DPD method, we conclude that hydrodynamic interactions are important in order to cross this barrier.

To define an order parameter for the structure we calculate the structure function:

$$S(\mathbf{k}) = \varrho_A(\mathbf{k})\varrho_A(-\mathbf{k})/N_A,$$

where N_A is the number of A-particles in the simulation. Its time evolution for the DPD system is shown in Fig. 12. What we observe is that the system in Fourier-space first peaks in a spherical shell around the origin (left). This already corresponds to level 2 ordering (see Fig. 10) as the real-space structure (top-left in Fig. 11) is an isotropic network of tubes; level 1 ordering takes place on a much shorter time-scale. When level 3 ordering sets in ($t \approx 7500\tau$) the spherical symmetry is broken, and a ring structure emerges. In real-space this ring corresponds to a hexagonal domain embedded in a network of tubes, see top-middle structure in Fig. 11. This ring subsequently breaks in two halves, that thereupon each break up in three peaks, Fig. 12 middle and right.

The time dependence of the structure function demonstrates that the ordering mechanism goes through various stages, where fewer and fewer modes contribute to the structure. It is this decreasing number of modes contributing to $S(k)$ that is characteristic for the increasing amount of order. Therefore we would like to count the number of k-vectors that contribute to the structure. Since $S(k)$ can be interpreted as a density of states in Fourier space, we define an order parameter by analogy to the entropy of particles distributed in real space as

$$P = \int S(\mathbf{k}) \ln S(\mathbf{k}) \, \mathrm{d}^3 \mathbf{k}.$$

Since this is a non-linear functional of the structure function, it distinguishes between systems having a different number of peaks, but the same overall segregation, i.e. it is a measure of the number of independent modes that contribute to the structure.

In Fig. 13 this order parameter is shown for the DPD simulation (with hydrodynamics) and for the BD simulation (without hydrodynamics). The A_3B_7 simulation results

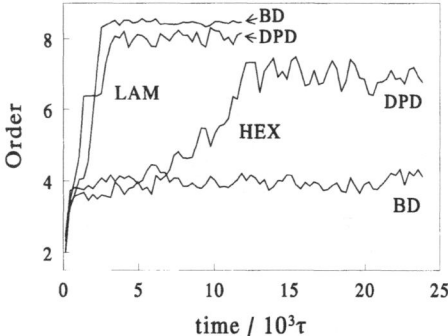

Fig. 13. Evolution of order parameter for A_3B_7 and A_5B_5 systems in DPD and BD, after [36]

are marked HEX. Whereas the DPD simulation shows a continuous increase in order (i.e. self-structuring of the system), the other simulation shows no clear trend. The behaviour of the order parameter demonstrates that hydrodynamic interactions are essential in driving this system to the structure of lowest free energy for this particular point in composition space.

To study the importance of hydrodynamics to the formation of the lamellar phase, a melt of A_5B_5 block copolymers was studied with DPD and BD. As reported previously [36], the DPD simulation swiftly finds its lamellar equilibrium structure. In the light of the previous observations one might expect that the BD simulation does not find the correct equilibrium, because hydrodynamic interactions are absent. However, the BD simulation does converge to the correct equilibrium, following exactly the same dynamics as the DPD system does. Both with and without hydrodynamics the system orders into a single lamellar domain, hence hydrodynamics is not essential for the formation of a lamellar phase. The increase of the order parameter in these simulations is also shown in Fig. 13; the curves are marked LAM. Note that here the time-scale of evolution is much shorter than for asymmetric polymers (marked HEX), where a connected tube structure is formed in the second stage of evolution. The time to form the hexagonal phase is about a factor 8 larger than the time to form the lamellar phase.

For asymmetric copolymers the DPD simulation, which includes hydrodynamics, produces the hexagonal phase predicted by theory and other simulation studies. However, the Brownian Dynamics simulation, which does not include hydrodynamics, does not produce the expected phase but remains trapped in an intermediate structure of interconnected tubes. From these results we conclude that hydrodynamics is important in driving the kinetics of micro-phase separation when an interconnected tube phase is formed as an intermediate structure. This intermediate structure is formed as a precursor for the hexagonal phase and the perforated lamellar phase. Indeed in the formation of the HPL structure [36,44] we found a similar slow evolution as in the formation of the hexagonal phase. The result presented here is a typical example; we have found a very similar pathway and slow evolution in other points within the hexagonal and HPL phases. For symmetric block copolymers that evolve along a pathway that avoids the intermediate connected tube structure, the system evolves quite efficiently if no hydrodynamic interactions are included. Hence hydrodynamic interactions are not critical in this case. The observed mechanism for micro-phase separation is one of the simultaneous formation of domains of lamellar order throughout the box, whereas the nucleation-and-growth mechanism is pertinent to form the hexagonal phase.

Why is this the case? The nature of the symmetry change between isotropic and hexagonal requires the transition to be of the first order: the Landau expansion contains a non-zero cubic coefficient. This is not the case for the isotropic to lamellar transition, which (in the Landau expansion) is second order, but becomes weakly first order when fluctuations are taken into account [32]. Hence there is a natural tendency for a nucleated process in the former transition, whereas this is not the case in the latter. Therefore, the isotropic to lamellar transition must be spinodal. Nucleation-and-growth can be expected to occur when the disordered phase is *meta-stable*, i.e. when a free energy barrier separates the two phases. Now the hexagonal phase arises from a disordered network of tubes. We speculate that this phase is meta-stable because it resembles the gyroid structure (one might refer to it as a melted gyroid phase), and because of the previous symmetry argument. This implies a (strong) first order transition. Hence the hexagonal phase can be expected to grow via a nucleation-and-growth mechanism. The lamellar phase is formed from a structure of disordered lamellae, which is topologically different from the gyroid phase. There is no stable phase that resembles a disordered lamellar system. Therefore this structure is unstable with respect to the lamellar phase (i.e. the isotropic to lamellar transition is second order or weakly first order), and thus the lamellar phase must form via a spinodal growth law.

4   Polymers and Membranes Interacting with Surfactant Solutions

4.1   Polymers and Surfactants in Solution

The DPD model has first been applied to polymers in solution by Kong et al. [48] and the precise scaling relations were checked by Spenley [34]. These results show that even polymer chains as short as $L = 10$ beads follow the correct endpoint distribution and are characterised by the correct scaling exponents. For a well soluble polymer in solution, theory predicts the endpoint separation and relaxation time to scale as

$$R_e \sim N^{0.59} \quad \text{and} \quad \tau_R \sim R_e^3 \sim N^{1.77}.$$

The simulation results by Spenley are [34]

$$R_e \sim (N-1)^{0.58 \pm 0.04} \quad \text{and} \quad \tau_R \sim N^{1.80 \pm 0.04},$$

which is a very good correspondence between theory and simulation.

For the same model the binodal has been simulated by Wijmans et al. [15], using the Gibbs Ensemble Monte Carlo method, see Fig. 14. In these simulations the polymer volume fraction at the critical point scales as

$$\phi_c \approx \frac{1.53}{2.06 + N^{0.38}} \; ; \; \Delta a_c \approx 2.25 \left(1 + N^{-0.44}\right)^{1.75}.$$

This should be compared to the mean-field Flory–Huggins expressions for the critical volume fraction and the critical χ-parameter as function of the polymer length:

$$\phi_c^{\text{FH}} = \frac{1}{1 + \sqrt{N}} \quad \text{and} \quad \chi_c^{\text{FH}} = \frac{1}{2} \left(1 + \frac{1}{\sqrt{N}}\right)^2.$$

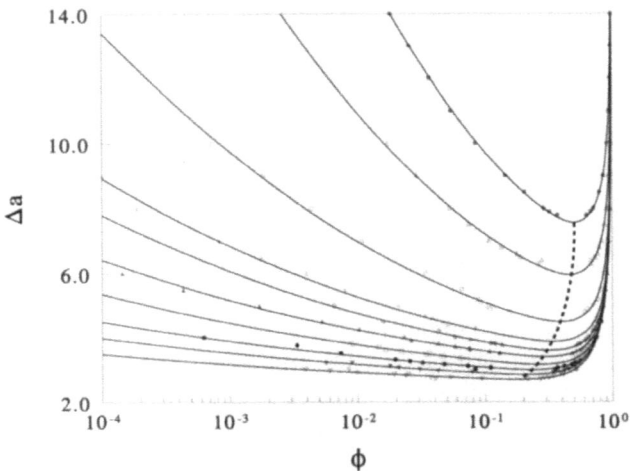

Fig. 14. Binodal curves for soft sphere model, data obtained from Wijmans et al. [15]

Experiments cited by Wijmans et al. indicate a scaling behaviour $\chi_c \sim N^{-0.37}$. Again the correspondence between simulation and experiment is very good, much better than the correspondence between experiment and mean-field theory.

To simulate a surfactant solution with the DPD model Jury et al. [49] used a minimal model. Surfactant was represented by two beads, each representing head (H) and tail (T) parts. When this is dissolved in a solvent (W), we have a model of a symmetric non-ionic surfactant like $C_{12}EO_6$ in solution. The repulsion parameters were fixed at $a_{HH} = a_{TT} = a_{WW} = 25$, $a_{HT} = 30$, $a_{HW} = 0$, and $a_{TW} = 50$. The temperature in the simulation was changed from $k_BT = 0.5$ to $k_BT = 2.5$ and the surfactant concentration from 10 % to 100 %. Very similar phase separation kinetics was observed as in the block copolymer systems described above. They find a micellar phase, a hexagonal phase, a lamellar phase and a disordered structure, in line with the experimental phase diagram of $C_{12}EO_6$. This indicates that the DPD model can indeed be used to study the phase behaviour of complex liquids.

The above results also suggest that DPD is a good candidate to simulate the interaction of polymers with a surfactant solution. The generally accepted picture is that complete micelles adsorb on the polymer [50–52], leading to a necklace of micelle pearls on a polymer backbone [53]. However, small angle neutron scattering (SANS) data on the poly(ethylene oxide) and sodium dodecyl sulfate (SDS) system by Chari et al. [54] suggest that the polymer resembles a swollen cage, rather than a necklace around SDS micelles. Fluorescence measurements on the same system indicate that the aggregation number of SDS is low at the onset of binding, but increases with surfactant concentration where the aggregate forms an elongated rod [55]. For PEO/SDS (PEO = polyethyleneoxide) mixtures it is also found that on increasing SDS concentration the polymer initially reduces in size, but when the surfactant concentration is increased beyond a certain point the polymer swells again [56,57].

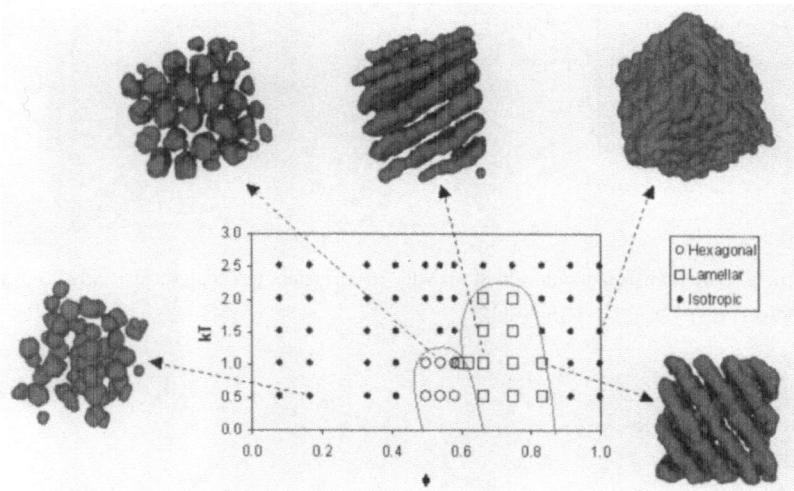

Fig. 15. Phase diagram for minimal surfactant model, after Jury et al. [49]

To predict when in a polymer-surfactant system such molecular bottlebrushes are formed, and when the surfactant adsorbs as micelles, Groot employed the DPD technique [58]. Both the polymer and the surfactant molecules are represented by strings of soft spheres. For this model the chemical potential of surfactant in the presence of polymer can be obtained relatively easy using the Widom insertion method. In the present work a number of examples of polymer-surfactant interactions are described, from which we can deduce when we have micelle binding, and when a continuous binding process is pertinent. To model a two-bead surfactant that readily agglomerates in spherical micelles, the head-head repulsion was increased and the tail-tail repulsion was decreased relative to the water-water repulsion. To model a range of polymer-surfactant interactions, various repulsions between the polymer beads and the surfactant tails and head-groups were studied. When the polymer is attracted towards the surfactant tail, the surfactant can be characterised as hydrophobically interacting, when it is not hydrophobically interacting with the polymer the surfactant can still interact via its head-group.

The simulations comprised of one homopolymer (length $L = 50$) in a box of size $10 \times 10 \times 10$, with various amounts of added surfactant. Pictures of typical polymer conformations with 10 surfactant molecules added (less than one micelle) and 100 surfactant molecules added (more than one micelle) are shown in Fig. 16. The conformations shown are at 100 and 300 ns, respectively. In the $N_s = 10$ system (on the left) all surfactant molecules are already adsorbed on the polymer at 70 ns. What is observed in a movie of the $N_s = 100$ simulation is that sometimes individual micelles are discernible and the polymer coils from one micelle to another. This textbook state is alternated with a state where the polymer-surfactant complex forms a sausage where all surfactant molecules run across the polymer backbone collectively in a wave-like motion. This break-up of micelles is related to the strong attractive interaction between the polymer backbone and surfactant tails.

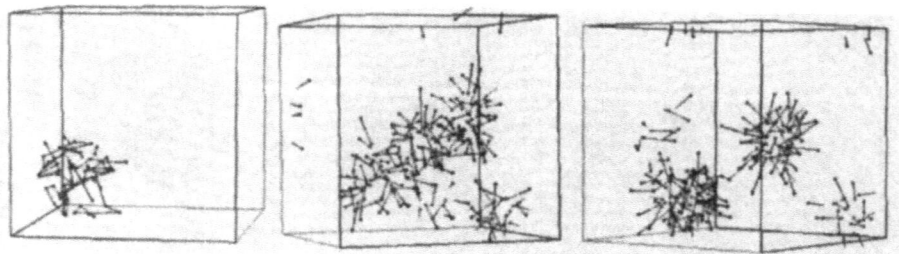

Fig. 16. Polymer-surfactant conformations with 10 surfactant molecules (left) and 100 surfactant molecules (middle and right), after [58]

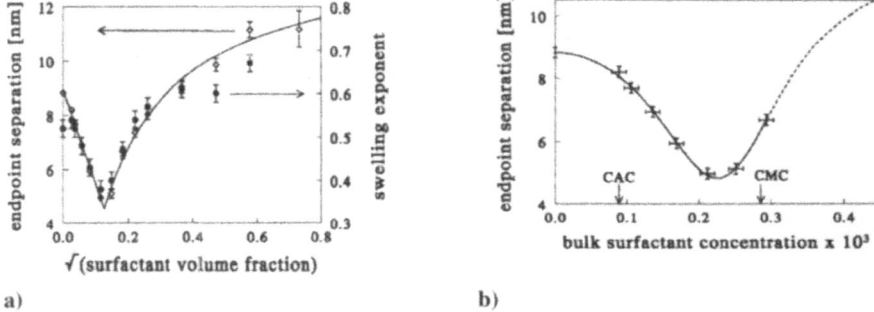

a)   b)

Fig. 17. a) Endpoint separation and swelling exponent passing through minimum, after [58]. **b)** Endpoint separation as function of bulk surfactant concentration, after [58]

Note that in the presence of 10 surfactant molecules the polymer is collapsed, while it is swollen when 100 surfactant molecules are added. All parameter sets studied show the same qualitative behaviour. The polymer endpoint separation is shown in Fig. 17a as a function of the number of surfactant molecules. This figure indicates a dramatic decrease in size of the polymer as the surfactant concentration increases, up to a certain point where precisely one micelle has formed at the polymer. From then on the polymer starts to swell again.

To further analyse the system the endpoint distribution has been fitted to the scaling function [59,60]

$$\ln\left(\Psi(r)\right) = a + \left(\frac{1.026\nu - 0.5}{1 - \nu}\right)\ln(r) - br^{1/(1-\nu)},$$

where a and b are arbitrary fit parameters, and ν is the swelling exponent. Upon addition of surfactant the distribution firstly narrows ($N_s = 20$) but for high surfactant concentration the polymer swells again. In Fig. 17a the swelling exponent that is obtained this way is compared with the endpoint separation. The curves are very similar. This plot indicates that an initially marginally soluble ($\nu = 0.5$) polymer undergoes a coil-globule transition ($\nu < 0.4$) when surfactant is added in a particular ratio. When yet more surfactant is added the polymer swells again, even more than a self-avoiding chain, $\nu = 0.65$. This should be contrasted to experimental observations. Chari et al. [54]

obtained the swelling exponent $\nu = 0.65$ for a saturated PEO/SDS system as we find here. Thompson et al. [57] study the thickness of an adsorbed layer of PEO, and find an initial decrease, followed by a subsequent increase when the SDS concentration is increased, very similar to the present simulation results.

In the previous simulations, the number of surfactant molecules in the system was varied. This number takes rather small values, so that in some cases all surfactant molecules are aggregated into the same cluster. What we want is to describe a single polymer in equilibrium with an infinitely large surfactant solution. To determine this equilibrium concentration, the first thing to establish is which quantity determines this equilibrium. For this purpose we study two boxes, one contains polymer, surfactant and water (I), and the other contains only surfactant and water (II).

The total Gibbs free energy of the system comprising the sub-systems I and II is

$$G = \mu_p^I N_p^I + \mu_s^I N_s^I + \mu_w^I N_w^I + \mu_s^{II} N_s^{II} + \mu_w^{II} N_w^{II},$$

where I and II refer to boxes I and II, and N_p, N_s and N_w are the number of polymer, surfactant and water molecules present in the respective boxes. This thought experiment is an example of the Gibbs ensemble in which the total number of molecules in the two boxes is fixed, but molecules are allowed to move from box I to box II and vice versa. However, as an extra constraint, we impose that the total number of particles in each box is constant. The only allowed moves are swaps of a surfactant molecule from I to II and a simultaneous swap of an equal number of water beads from II to I, and vice versa. The variation of the Gibbs free energy under these swaps is

$$\delta G = \left(\mu_s^I - \mu_s^{II} \right) \delta N_s - L_s \left(\mu_w^I - \mu_w^{II} \right) \delta N_s,$$

where L_s is the number of beads in a surfactant. As in equilibrium $\delta G = 0$, we find

$$\mu_s^I - L_s \mu_w^I = \mu_s^{II} - L_s \mu_w^{II} = \mu$$

from which the proper chemical potential follows as

$$\mu = k_B T \ln \left(\varrho_s \right) - L_s k_B T \ln \left(\varrho_w \right) + \Delta \mu_s - L_s \Delta \mu_w.$$

Here ϱ_s and ϱ_w are the concentrations of surfactant and water molecules respectively. Hence, to calculate the equilibrium concentration of surfactant in a system without polymer we need to measure the excess chemical potentials of both water and surfactant. The bulk surfactant concentration in equilibrium with the polymer-surfactant complex has the same chemical potential. Combining this with the polymer endpoint separation results, we find the curve shown in Fig. 17b.

For hydrophobically interacting surfactants, the chemical potential is a continuously rising function for $0 < N_s < 50$, see Fig. 18a. This behaviour is characteristic for continuous adsorption. In the simulation we find *bottlebrush* conformations at the higher surfactant concentrations. When the head-group repulsion is changed from a net repulsion into a net attraction, the chemical potential curve shifts down, but still has the same initial slope at small surfactant concentration.

However, when the interaction of the polymer with the tail is switched to be repulsive (χ_{pt} is increased from -1.5 to $+2.0$) a dramatic change is observed: the chemical

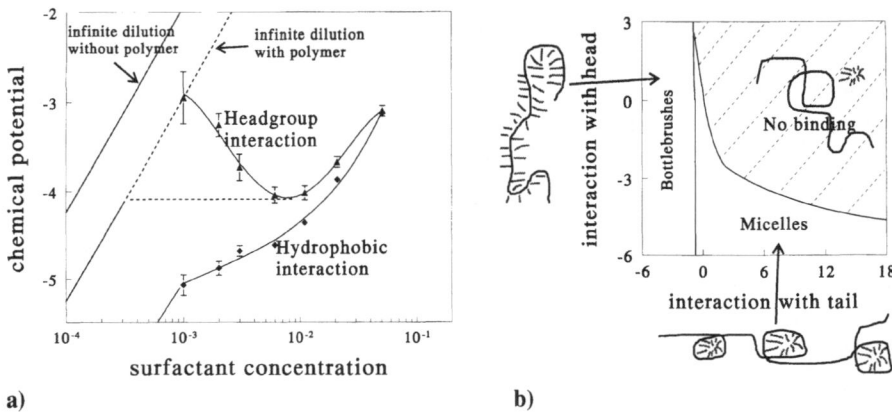

Fig. 18. **a)** Surfactant chemical potential for headgroup/hydrophobic interaction, after [58]. **b)** "Phase diagram" of polymer-surfactant aggregation, after [58]

potential first decreases with surfactant concentration, passes through a minimum, and then increases again. This behaviour signals *micelle adsorption*. When the surfactant concentration in the bulk is slowly increased, the adsorbed amount will follow the infinite dilution limit in the presence of polymer, up to the point where on average 0.3 molecules are adsorbed. Here the chemical potential equals the minimum chemical potential at which adsorbed micelles can exist. At this point the adsorbed amount jumps up to the value corresponding to one adsorbed micelle, see Fig. 18a. Such behaviour is predicted to occur when a strong interaction between surfactant head-group and polymer is pertinent, for instance for the binding of a cationic surfactant to an anionic polymer. This result also implies that when the total surfactant concentration is continuously increased in a system with many polymers present, the bulk surfactant concentration will remain constant at the value corresponding to the critical aggregation concentration (CAC), until a micelle is adsorbed on every polymer in the system. At the CAC, the system contains a mixture of polymers without any surfactant molecules adsorbed, and polymers with a complete micelle adsorbed. These micelles are smaller than the micelles that are formed in the bulk.

To roughly map out when each of the binding modes is pertinent, a number of short runs has been performed throughout the (χ_{pt}, χ_{ph}) parameter space, for one polymer and 30 surfactant molecules. The result is shown in the "phase-diagram" in Fig. 18b. This diagram should be considered as a qualitative picture, capturing the relevant physical phenomena. To predict such a diagram for a particular surfactant requires a careful tuning of the parameters that have been kept fixed here, viz. the head-group/head-group and the head-group/tail interactions. The qualitative agreement between the available experimental data and the presently studied dumbbell surfactant model warrants a further study to map chemically specific surfactant systems onto the dumbbell model.

4.2 Biomembrane Morphology

Another area in which the DPD method has been applied successfully is the interaction of surfactant with biomembranes. Non-ionic surfactants have traditionally been considered as mild. However, alcohol ethoxylates are shown to be capable of inhibiting bacterial growth [61]. There is evidence that non-ionic surfactant can interact with *in vitro* lipid membranes by the formation of channels through the membrane [62]. The occurrence of "hole" formation in bilayers of long chain surfactants has been demonstrated for certain non-ionic surfactants by small angle x-ray scattering studies [63]. Similar structures have been found in experiments on block co-polymers [39,40] and in simulations thereof (see Fig. 7). Finally, addition of cationic surfactants to lipid membranes leads to hole formation [64,65]. It therefore seems reasonable to enquire whether the interaction of alcohol ethoxylates and phospholipids typical of bacterial membranes would naturally lead to such mesh phase which would make the bacterial cell leaky and leading to bacterial stasis and ultimately to cell death. No evidence has been found for a structured perforated phase in deuterium NMR and x-ray diffraction studies on the interaction of a phosphatidylethanolamine extract of *Escherichia coli* with an alcohol ethoxylate formulation. For this reason a simulation study was undertaken, to find evidence of moving or temporary holes. Since hole formation and disappearance is expected to occur beyond the time scale that can be reached with molecular dynamics, the DPD simulation method was employed by Groot and Rabone [13].

In the simulated system, three carbon atoms are taken together into one bead. The Flory–Huggins theory was used to derive the relevant χ-parameters from experimental solubilities. To reproduce the correct solubility of hexane, heptane and octane in water, they found $\chi_{\text{hydrocarbon}-\text{water}} \approx 6.0$, which appears to be relatively independent of temperature. The second χ-parameter to match that is that between polyethyleneoxide and water. The problem here is that PEO and water at room temperature mix in all ratios, hence the solubility does not lead to a parameter value. At elevated temperatures ($T >$ 100° C), water and PEO no longer mix ideally. Hence an alternative route is to describe this demixing at higher temperatures, and to extrapolate the temperature dependence of the χ-parameter back to room temperature. This way Barneveld et al. [66] estimated this χ-parameter as function of temperature and found the value $\chi_{\text{ew}} \approx 0.30 - 0.38$ at room temperature. However, he only took the cloud-point into account where mean-field theory is least reliable. Taking the shape of the whole binodal into account and extrapolating back to room temperature, Groot and Rabone [13] find from the experimental data by Seaki et al. [67]: $\chi_{\text{ew}} \approx 0.30 \pm 0.04$, which is close to the value obtained by Barneveld.

A third important χ-parameter is the interaction between PEO and hydrocarbons. What experimentally is available is neutron scattering data of $C_{12}E_6$ at the air-water interface [68]. Assuming that this is not too dissimilar from the oil-water interface, this data can be compared to DPD simulation data of an oil-surfactant-water system. The experiment shows a significant overlap between the surfactant head group and the surfactant tails. To arrive at the same amount of overlap in simulation as in the experiment, a χ-parameter much smaller than the hydrocarbon-water parameter needs to be used. A good agreement between the width and overlap of the head and tail peaks as seen in experiment and simulation is found at χ-parameter between hydrocarbons and EO beads $\chi_{\text{ce}} = 2.0$.

Finally the χ-parameters describing the head-group of the lipid molecule were defined. Since these groups contain more oxygen than EO does, and also have partial charges, it has been treated as if it were water, with respect to C and EO. The result-

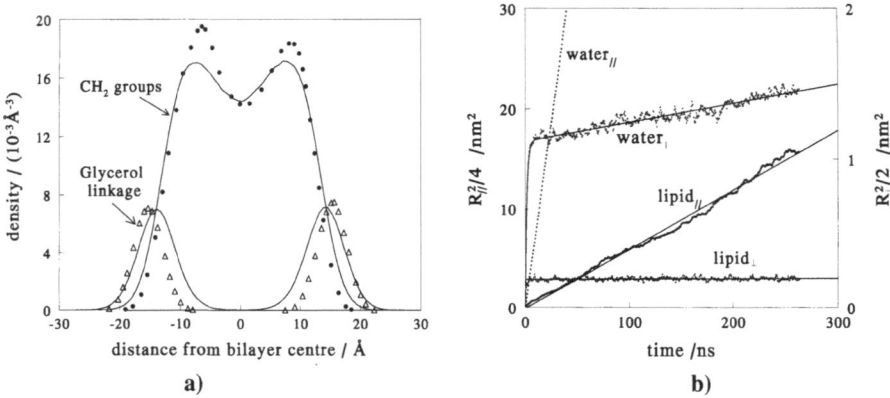

Fig. 19. **a)** DPD bilayer density profile compared to MD density profile, **b)** Displacement of a water and lipid parallel and perpendicular to membrane, after [13]

ing bilayers have been compared with molecular dynamic simulations of 1-palmitoyl-2-oleoyl-sn-glycero-3-phophatidylcholine (POPC) [2] for two sets of parameters, covering the range of uncertainty. It is found that the head-group parameters are not critical for the result. The density profiles obtained from DPD simulations of lipid bilayers containing 200 lipid molecules are compared to previous MD simulation results [2] in Fig. 19a. The two peaks in the CH_2 density profile arise because the CH_3 chain endpoint is localised in the centre of the bilayer, and is not included in the average. In the DPD simulation the average is over all but the last c-bead, i.e. the last three carbon atoms are excluded from the average. Good correspondence with the MD simulation for either set gives confidence in the reliability of the model and parameters.

To validate the time scale a bilayer simulation is used to obtain the lateral diffusion coefficient of the lipid molecules, which can be checked against experimental and MD simulated values. For POPC the simulation result is $D_\parallel = 0.073 \times 10^{-5}$ cm^2/s. Some care must be taken, since the MD simulation was too short for two molecules even to swap places in the layer. Experiments [69,70] indicate lateral diffusion coefficient of DOPC to be in the range $D_\parallel = 0.036 \times 10^{-5}$ cm^2/s and $D_\parallel = 0.02 \times 10^{-5}$ cm^2/s. In the DPD simulation the mean square displacements parallel (and also perpendicular) to the bilayer were averaged. During this run each lipid molecule travelled 10 times the nearest neighbour distance on average. The diffusion coefficient in the DPD simulation is $D_\parallel = 0.06 \times 10^{-5}$ cm^2/s. This lateral diffusion coefficient compares well with the experimental $(0.02 - 0.04 \times 10^{-5}$ cm^2/s) and MD simulation $(0.07 \times 10^{-5}$ cm^2/s) values in the literature, even though no attempt has been made to match the relative viscosity of the lipid phase.

To investigate the structure of mixtures of lipid/surfactant bilayers, a series of simulations was done with 540 membrane molecules, surfactant or lipid, at a surfactant mole-fraction varying from 10 % to 100 % in steps of 10 %. Simulations have been done both with $C_{12}E_6$ (bead structure c_4e_4) and with C_9E_8 (bead structure c_3e_5). First we concentrate on the results regarding $C_{12}E_6$. The system with 90 % mole-fraction $C_{12}E_6$ displays holes that move around. They are relatively stable, the typical life-time is 20 –

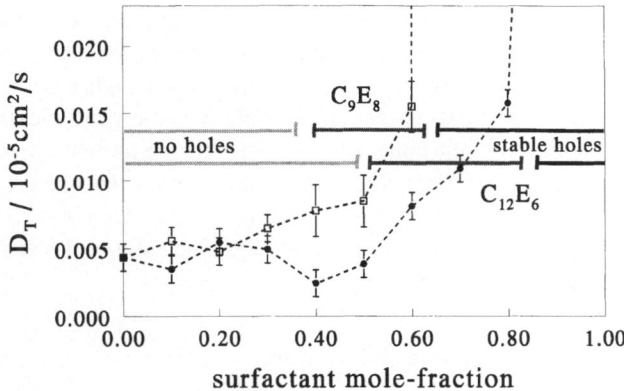

Fig. 20. Transverse water diffusion coefficient of water through the bilayer in the presence of surfactant, after [13]

40 ns, and over the course of the run there is hardly ever a conformation without a hole in the patch of $124\,\mathrm{nm}^2$. Systems with mole-fraction 80 % or less of surfactant do not show stable perforated conformations. Occasionally small holes do appear, but they disappear very quickly. At 80 % mole-fraction surfactant the typical life-time of a hole is some $0.4 - 1\,\mathrm{ns}$.

The diffusion of water perpendicular to the layers provides a useful way of testing the porosity of the bilayers. To interpret these results, we first solve the diffusion equation for the mean square displacement of water in a narrow slit of impenetrable walls at distance L. For diffusion coefficient D, the mean square displacement perpendicular to the walls follows analytically as

$$\frac{R_s^2(t)}{L^2} = \frac{1}{6} - \frac{16}{\pi^4} \sum_{n\ \mathrm{odd}} n^{-4} \exp\left(-\frac{n^2 \pi^2 D t}{L^2} \right) \approx \frac{1}{6} \left[1 - \frac{96}{\pi^4} e^{-t/\tau} - \left(1 - \frac{96}{\pi^4} \right) e^{-9t/\tau} \right]. \quad (6)$$

The first expression is exact. The second is a good approximation that can be used for curve fitting. To find a reliable value for the transverse diffusion coefficient, the mean square displacement of water normal to the bilayers is recorded, and the transverse diffusion is obtained by fitting the data to

$$R^2(t) = a R_s^2(t; \tau) + 2 D_\perp t,$$

where R_s^2 is the diffusion in a slit given by (6), and a and τ are free fit parameters. The resulting transverse diffusion of water is shown in Fig. 20 as a function of the surfactant mole-fraction in the layer.

It is found that up to, and including, mole-fraction of 50 % $C_{12}E_6$, the diffusion of water through the bilayer is independent of the amount of added surfactant. From that point onwards the water diffusion through the layers steadily increases. This increase is attributed to the formation of small holes that open and close on a time-scale of 0.5 ns or less, depending on the surfactant content. Permanent holes occur at 90 % surfactant. Consequently the transverse diffusion coefficient increases sharply above 80 % surfactant.

The transverse diffusion of water through the PE/C_9E_8 bilayer is also shown in Fig. 20. We see the same qualitative picture involving the creation and annihilation of small transient holes, followed by a phase with stable holes at higher surfactant content. However, the range of surfactant concentrations over which transient holes occur is much shorter than for $C_{12}E_6$. Also, the point where stable holes are formed is reached already at 70 % mole-fraction of surfactant. The real transition points must be between 60 and 70 % for C_9E_8 and between 80 and 90 % for $C_{12}E_6$, i.e. some 56 % and 78 % by weight respectively. The approximate ranges where no holes, fluctuating holes and stable holes occur are indicated in Fig. 20 by bars.

4.3 Biomembrane Deformation and Rupture

So far we have been concerned only with membranes at vanishing surface tension. However, in many cases it is found that dividing cells are particularly vulnerable. Dividing cells are not necessarily in a state of vanishing membrane tension. For instance, when yeast cells divide, their cell membrane buds out of the cell wall and is no longer protected by it. Instead the membrane is exposed to the solution. The osmotic pressure difference between inside and outside of the cell then leads to a finite surface tension on the membrane. The membrane will react to this osmotic pressure by expanding, which is an obvious prerequisite for cell division when the budding mechanism is pertinent. If the membrane cannot withstand this expansion, the cell will die. For these reasons it is prudent to simulate cell membranes under strain, rather than to study them at zero surface tension, as far as the mechanism for cell-death is concerned. Simulations of mixed membranes of lipid and $C_{12}E_6$ were undertaken in which the membrane is stretched over time, leading to increasing tension and ultimately to rupture. An example of this process is shown in Fig. 21, where the actual creation and expansion of holes is monitored. The successive frames are taken at time intervals of 1.2 ns and the patches are 17×17 nm^2 across. This membrane consists of 70 % PE and 30 % $C_{12}E_6$. It ruptures when its area is increased by 74 %.

The full stress history of the expanding membrane is followed in simulation. Each system is left to equilibrate over 5.3 ns after which time the y- and z-coordinates are expanded by a factor 1.03, while the x-coordinates are contracted by a factor 0.94. This cycle is repeated 12 times. This gives the yield curves shown in Fig. 22 for 10, 50 and 80 % mole fraction surfactant. Each simulation shows a clear rise in surface tension, up to a critical point where the layer fails. These simulations predict that adding surfactant to a lipid membrane significantly reduces the strength and maximum stretch of the membrane. This holds even at amounts of surfactant that have no measurable influence on the level of water diffusion through a stress-free bilayer. Without surfactant the membrane area may be increased by 100 % before it ruptures, but at a 50 % mole-fraction of surfactant this tolerance is reduced to a mere 50 % area increase. Also the maximum tension that the membrane can take reduces from 67 mN / m at 0 % surfactant, to 41 mN / m at 50 % surfactant.

The trends predicted imply that the cell will become more sensitive to the osmotic pressure difference between inside and outside, when it is exposed to a surfactant solution. For a bilayer containing 50 % mole-fraction surfactant, the pressure tolerance is reduced by some 40 %. This will have dramatic influence on the survival chances of

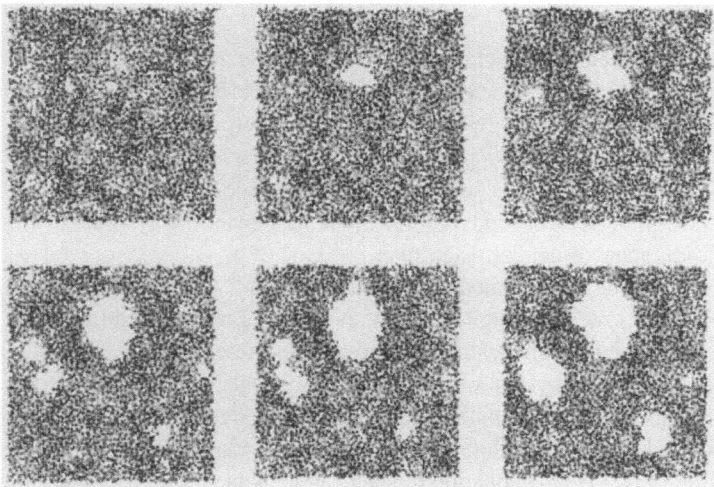

Fig. 21. Rupture process of a simulated biomembrane, after [13]

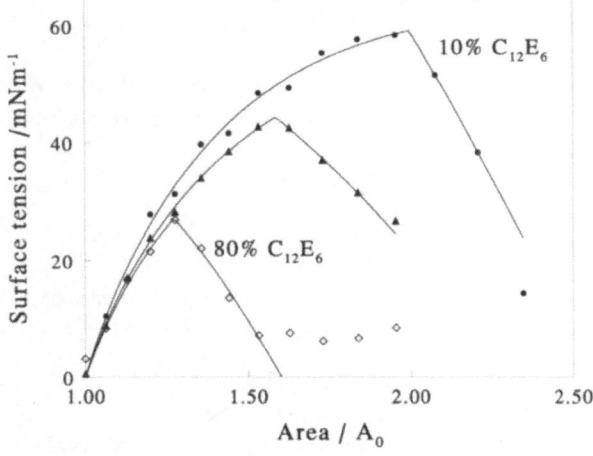

Fig. 22. Stress-strain curves for biomembranes at various amounts of surfactant, after [13]

dividing cells. Another system for which these simulations are relevant is red blood cells. These cells do not have a cell wall, but only a cell membrane. Therefore the membrane is directly exposed to the solution, and has to accommodate for all osmotic pressure differences. When the maximum pressure that a cell can withstand by incorporation of surfactant decreases below the actual osmotic pressure, the membrane ruptures. These simulations give a possible explanation why red blood cells lyse when they are exposed to a surfactant solution.

5   Conclusions

In summary, dissipative particle dynamics is a flexible method and easy to code simulation method. It has already been applied successfully to a wide variety of problems, even though we deal with a relatively new technique. The strong points of the method are: it is very competitive for hydrodynamics of polymers and mesophases, useful for multiphase flows, porous media, colloidal dispersions, etc. It is able to produce molecularly detailed simulations up to microseconds. In this mode it is faster than full atomistic Molecular Dynamics by many orders of magnitude.

The down sides of the method are the following: diffusion is too fast, the speed of sound is too low, the equation of state not always realistic, and parameterisation is a problem for detailed chemistry. With respect to these points it should be mentioned that the first is not always a problem, but actually contributes to the speed of evolution. For multiphase flow where both diffusive and hydrodynamic processes are important, this flaw can be repaired using the Andersen Monte Carlo method for the velocity randomisation [23]. Also the equation of state can be made more realistic if required [19]. Finally, the parameterisation problem for molecular simulation is a general problem in mesoscopic simulation, and not specific to dissipative particle dynamics.

Acknowledgements

S. Jury, P. Bladon, M. Cates, S. Krishna, M. Hagen, N. Ruddock, P. Warren, N. Spenley, C. Wijmans, B. Smit, T. Madden, D.J. Tildesley, and K. Rabone are kindly acknowledged for permission to reproduce their work.

References

1. D.B. Tieleman, S.J. Marrink, H.J.C. Berendsen: BBA-Rev. Biomembr. **1331**, 235 (1997)
2. H. Heller, M. Schaefer, K. Schulten: J. Phys. Chem. **97**, 8343 (1993)
3. R. Lipowsky, S. Grotehans: Europhys. Lett. **23**, 599 (1993)
4. E. Lindahl, O. Edholm: Biophys. J. **79**, 426 (2000)
5. R.D. Groot, P.B. Warren: J. Chem. Phys. **107**, 4423 (1997)
6. P.J. Hoogerbrugge, J.M.V.A. Koelman: Europhys. Lett. **19**, 155 (1992)
7. P. Espanol: Phys. Rev. E **52**, 1734 (1995)
8. P. Espanol, P. Warren: Europhys. Lett. **30**, 191 (1995)
9. I. Vattulainen, M. Karttunen, G. Besold, J.M. Polson: J. Chem. Phys. **116**, 3967, (2002)
10. M.P. Allen, D.J. Tildesley: *Computer Simulation of Liquids* (Clarendon, Oxford 1987)
11. W.K. Den Otter, J.H.R. Clarke: Int. J. mod. Phys. C **11**, 1179 (2000)
12. I. Pagonabarraga, M.H.J. Hagen, D. Frenkel: Europhys. Lett. **42**, 377 (1998)
13. R.D. Groot, K.L. Rabone: Biophys. J. **81**, 725 (2001)
14. J.R. Partington, R.F. Hudson, K.W. Bagnall: Nature **169**, 583 (1952)
15. C.M. Wijmans, B. Smit, R.D. Groot: J. Chem. Phys. **114**, 7644 (2001)
16. P. Espanol: Europhys. Lett. **40**, 631 (1997)
17. J.B. Avalos, A.D. Mackie: Europhys. Lett. **40**, 141 (1997)
18. P. Espanol: Phys. Rev. E **57**, 2930 (1998)
19. I. Pagonabarraga, D. Frenkel: J. Chem. Phys. **115**, 5015 (2001)

20. X.F. Yuan, R.C. Ball, S.F. Edwards: J. Non-Newtonian Fluid Mech. **46**, 331 (1993)
21. J.J. Monaghan: Annu. Rev. Astron. Astr. **30**, 543 (1992)
22. C.P. Lowe, M.W. Dreischor: Simulating the Dynamics of Mesoscopic Systems, Lect. Notes Phys. **640**, 35 (2004)
23. C.P. Lowe: Europhys. Lett. **47**, 145 (1999)
24. A.K. Gunstensen, D.H. Rothman, S. Zaleski, G. Zanetti: Phys. Rev. A **43**, 4320 (1991)
25. J.G.E.M. Fraaije, B.A.C. van Vlimmeren, N.M. Maurits, M. Postma, O.A. Evers, C. Hoffmann, P. Altevogt, G. Goldbeck-Wood: J. Chem. Phys. **106**, 4260 (1997)
26. E.S. Boek, P.V. Coveney, H.N.W. Lekkerkerker, P. van der Schoot: Phys. Rev. E **55**, 3124 (1997)
27. A.T. Clark, M. Lal, J.N. Ruddock, P.B. Warren: Langmuir **16**, 6342 (2000)
28. K.E. Novik, P.V. Coveney: Phys. Rev. E **61**, 435 (2000)
29. P.B. Warren: Phys. Rev. Lett. **8722**, 5702 (2001)
30. M.E. Cates, V.M. Kendon, P. Bladon, J-C. Desplat: Faraday Discuss. **112**, 1 (1999)
31. F.S. Bates, G.H. Fredrickson: Annu. Rev. Phys. Chem.**41**, 525 (1990)
32. L. Leibler: Macromolecules **13**, 1602 (1980)
33. M. Doi, S.F. Edwards: *The Theory of Polymer Dynamics* (Clarendon, Oxford 1986)
34. N.A. Spenley: Europhys. Lett. **49**, 534 (2000)
35. F.S. Rowlinson, B. Widom: *Molecular Theory of Capillarity* (Clarendon, Oxford 1982)
36. R.D. Groot, T.J. Madden: J. Chem. Phys. **108**, 8713 (1998)
37. M.W. Matsen, F.S. Bates: Macromolecules **29**, 1091 (1996)
38. R.D. Groot, T.J. Madden, D.J. Tildesley: J. Chem. Phys. **110**, 9739 (1999)
39. J. Zhao, B. Majumdar, M.F. Schulz, F.S. Bates, K. Almdal, K. Mortensen, D.A. Hajduk, S.M. Gruner: Macromolecules **29**, 1204 (1996)
40. I.W. Hamley, K.A. Koppi, J.H. Rosedale, F.S. Bates, K. Almdal, K. Mortensen: Macromolecules **26**, 5959 (1993)
41. G.H. Fredrickson, E. Helfand: J. Chem. Phys. **87**, 697 (1987)
42. R.G. Larson: J. Chem. Phys. **96**, 7904 (1992)
43. M. Schwab, B. Stühn: Colloid and Polym. Sci. **275**, 341 (1997)
44. R.D. Groot, T.J. Madden. In: *Structure and Dynamics in the Mesoscopic Domain*, ed. by Kulkami, Lal (Imperial College Press, London 1998), p. 288
45. N.P. Balsara, B.A. Garetz, M.C. Newstein, B.J. Bauer, T.J. Prosa: Macromolecules **31**, 7668 (1998)
46. M.W. Matsen: Phys. Rev. Lett. **80**, 4470 (1998)
47. R.G. Larson: Personal Communication, (1998)
48. Y. Kong, C.W. Manke, W.G. Madden, A.G. Schlijper: J. Chem. Phys. **107**, 592 (1997)
49. S. Jury, P. Bladon, M. Cates, S. Krishna, M. Hagen M, N. Ruddock, P. Warren: Phys. Chem. Chem. Phys. **1**, 2051 (1999)
50. R. Nagarajan: J. Chem. Phys. **90**, 1980 (1989)
51. E. Ruckenstein, G. Huber, H. Hoffmann: Langmuir **3**, 382 (1987)
52. E.D. Goddard, K.P. Ananthapadmanabhan: *Interactions of Surfactants with Polymers and Proteins* (CRC Press, London 1993)
53. B. Cabane: J. Phys. Chem. **81**, 1639 (1977)
54. K. Chari, B. Antalek, M.Y. Lin, S.K. Sintra: J. Phys. Chem. **100**, 5294 (1994)
55. J. Vanstam, W. Brown, J. Fundin, M. Almgren, C. Lindblad: ACS Symp. Ser. **532**, 194 (1993)
56. P. M. Claesson, M. L. Fielden, A. Dedinaite, W. Brown, J Fundin: J. Phys. Chem. B **102**, 1270 (1998)
57. S. J. Mears, T. Cosgrove, T. Obey, L. Thompson, I. Howell: Langmuir **14**, 4997 (1998)
58. R.D. Groot: Langmuir **16**, 7493 (2000)
59. R.D. Groot, A. Bot, W.G.M. Agterof: J. Chem. Phys. **104**, 9202 (1996)

60. J. de Cloizeaux, G. Jannink: *Polymers in Solution, Their Modelling and Structure* (Clarendon, Oxford 1990)
61. S.L. Moore: The Mechanisms of Antibacterial Action of Some Non-Ionic Surfactants. D. Phil. Thesis, University of Brighton (1997)
62. P. Schlieper, E. Derobertis: Arch. Biochem. Biophys. **184**, 204 (1977)
63. J. Burgoyne, M.C. Holmes, G.J.T. Tiddy: J. Phys. Chem. **99**, 6054 (1995)
64. J. Gustafsson, G. Oradd, M. Almgren: Langmuir **13**, 6956 (1997)
65. J. Gustafsson, G. Oradd, M. Nyden, P. Hansson, M. Almgren: Langmuir **14**, 4987 (1998)
66. P.A. Barneveld, J.M.H.M. Scheutjens, J. Lyklema: Langmuir **8**, 3122 (1992)
67. S. Saeki, N. Kuwahara, M. Nakata, M. Kaneko: Polymer **17**, 685 (1976)
68. J.R. Lu, Z.X. Li, R.K. Thomas, E.J. Staples, I. Tucker, J. Penfold: J. Phys. Chem. **97**, 8012 (1993)
69. M.H. Cohen, D. Turnbull: J. Phys. Chem. **31**, 1164 (1959)
70. W. Pfeiffer, T. Henkel, E. Sackmann, W. Knoll, D. Richter: Europhys. Lett. **8**, 201 (1989)

Simulating the Dynamics of Mesoscopic Systems

Christopher P. Lowe and Menno W. Dreischor

Department of Chemical Engineering, University of Amsterdam, Nieuwe Achtergracht 166, 1018 WV Amsterdam, The Netherlands

Abstract. We consider how to simulate quantitatively the dynamics of mesoscopic systems on macroscopic time-scales. In this respect we consider two relatively simple systems, colloidal suspensions of hard spheres and a simple model polymer. To simulate the behaviour of these two systems on macroscopic time-scales we have to simplify them considerably. Our emphasis here is on how we do so, what we can throw out and what we must keep. We argue that a strategy for checking the consequences of this loss of detail is essential. Particularly, we emphasize the importance of hydrodynamics and how we can include fluctuating hydrodynamic behaviour. A simple DPD-like model solvent (an ideal gas coupled to a Lowe-Andersen thermostat) more than adequately achieves this. Considering a long polymer in an external potential we show that it is possible to predict with confidence its static and dynamic properties with a renormalized model polymer consisting of only a few beads. To do so, however, we have to renormalize the dynamics in the sense that the model shows long polymer dynamic behaviour for any number of beads in the renormalized model.

1 Introduction

Mesoscopic systems possess characteristic length scales that are, by atomic standards, very long. Two examples, that we will consider here in some detail, are colloidal suspensions and polymeric solutions (see Fig. 1). The former consist of large particles dispersed in a solvent. These big, or "colloidal", particles are big in the sense that they are much larger than the solvent molecules (one micron would be typical). Polymers are very long molecules consisting of many repeating units (monomers). Again, by atomic standards,

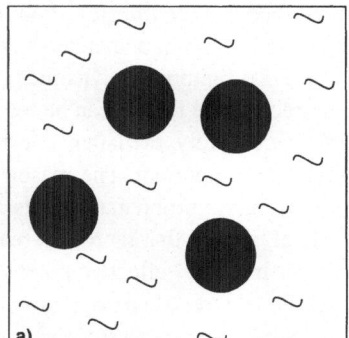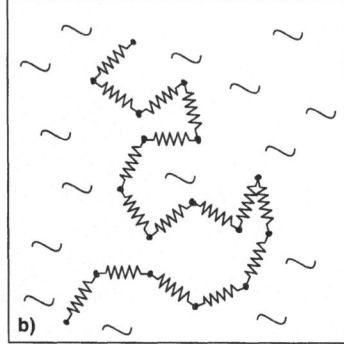

Fig. 1. A schematic representation of the two mesoscopic systems we consider in this article. Left a colloidal suspension of hard spheres. Right, an ideal polymer chain

they are very large. The fact that the typical length scales present in mesoscopic systems are atomistically long basically precludes the use of techniques that explicitly resolve the atomic scale. Notably, molecular dynamics [1]. There is a further complication. The presence of these long length scales implies that the dynamics of mesoscopic systems evolve on correspondingly longer times scales. Whereas a picosecond of time might be sufficient to capture the dynamics of a simple (small) molecular system, for a meso-scopic system it most definitely is not. This will be discussed more quantitatively later on. Relevant time scales in the range of seconds are not uncommon. It is partially the fact that the dynamics of these systems evolve so slowly that make their dynamic behaviour so rich and distinct from simple atomic systems.

We are therefore confronted with the fundamental problem that in order to access the length and time-scales necessary to predict the true macroscopic behaviour of a mesoscopic system, it is hopeless to model the system in terms of the atoms of which it is composed. The only thing we can do is simplify the system, and, as we shall see, simplify it quite dramatically. By simplifying the system it is inevitable that we will have to throw out a lot of details. The question then is, what do we throw out and what do we keep? If we want to keep something, how in practical terms do we do it? Can we *quantitatively* and with confidence describe properties of the original system? Can we be sure that in simplifying the system we have not thrown out the baby with the bath water? Although the problem of mesoscopic modelling is quite general, here our aim is to address these questions for the two systems outlined above.

1.1   Renormalizing Static Properties

In this article we are primarily interested in modelling correctly the *dynamics* of meso-scopic systems. However, an important pre-requisite for modelling the dynamics cor-rectly is that we first renormalize the static properties correctly. By "renormalize" we mean that we represent our original mesoscopic system with a much smaller number of particles. That is, we simplify our original system while maintaining some coarse grained feature of its structure. The reason that this is important is that, in equilibrium, the trans-port properties can be determined by analyzing the decay of fluctuations [2]. The decay of fluctuations will be governed by two factors. First there is a thermodynamic "driving force". If the structure of the system is slightly perturbed from equilibrium, it will want to return and this thermodynamic force drives the relaxation. Second, there is the dynamic response of the system to this thermodynamic force. To capture the dynamics correctly we need to model both the force and response correctly. The former can be regarded as the problem of renormalizing the static properties adequately, the latter renormalizing the dynamic response correctly. Thus, both effects are important. The reason that we concentrate here on the dynamic part is that it is far more problematic. This is not to say that renormalizing the static properties is trivial, simply that there is a reasonably well defined route by which it can be achieved. For instance "effective potentials" can be calculated from simulations that reproduce certain features of an original system in a model with fewer particles [3,4]. By "effective potential" we mean a somewhat fictitious potential intended solely for the purpose of reproducing some of the static properties of the original system. This of course begs the question, exactly what features should we try to keep in the renormalized model? This we address briefly later, sufficient here to say that, once this question is answered, renormalizing the static properties is in principle possible.

The reason for choosing these two particular systems is because they are probably the simplest non-trivial mesoscopic systems for which renormalizing the static properties is relatively straightforward. Since we are interested in how well we can, in principle, model the dynamics we do not want the additional complication of working out how to renormalize the statics. Specifically, a colloidal particle is solid, so we can neglect its internal structure. Because the solvent molecules are small they do not influence the static properties of the colloidal phase. For example, simple model suspensions composed of hard spheres (which we will restrict ourselves to here) have the static properties that are simply those of an equivalent hard sphere fluid, so no additional effective potential is necessary. As we will see though, this does not imply that their dynamics are similar. Thus it is possible to make considerable progress modelling the dynamics of this particular mesoscopic system and, from the point of view of understanding what makes mesoscopic dynamics tick, it is an extremely useful starting point. Polymers, on the other hand, do have internal structure. The same issues involved in modelling colloidal suspensions apply, but the internal structure adds an extra degree of complexity. Here we will further restrict ourselves to a particular model of a polymer, namely an ideal (or "Gaussian") chain [5]. In this model, "beads" intended to represent units of some real polymer, are connected by harmonic springs. The reason for choosing this particular model is that the full N-particle distribution function for the beads is known analytically [5]. This means that if we take an original ideal polymer chain consisting of $N + 1$ beads connected by springs with a potential

$$ U(\mathbf{r}_{ij}) = \frac{3}{2} \frac{k_{\mathrm{B}}T}{b^2} \mathbf{r}_{ij}^2 \, , $$

we can exactly renormalize it to a shorter chain consisting of $n + 1$ beads such that the average of the end to end vector will be the same for both systems. This simply requires using an effective spring interaction of the form

$$ U^*(\mathbf{r}_{ij}) = \frac{3}{2} \frac{k_{\mathrm{B}}T}{(N/n)b^2} \mathbf{r}_{ij}^2 \, . $$

Whether keeping the correct end to end distance is the best criterion is a question for later. For now it is sufficient to note that although we have the complication of internal structure, we at least know exactly what it is. It is a best case scenario for a system with internal structure, but still, as we will see, far from trivial.

In summary, if we cannot demonstrate that we can do *quantitative* dynamic meso-scopic simulations for these two systems, for which the static properties are relatively straightforward, one would really be forced to the depressing conclusion that the whole undertaking is doomed.

1.2 Time-Scales

As an illustration of the problem we are confronted with when trying to study, numerically, the dynamics of mesoscopic systems it is useful to consider some of the time-scales we know must be involved. There is firstly the velocity decay time, τ_{v}. This is the time it takes colloidal particles or polymer molecules (subsequently referred to simply as "particles") to forget their velocity. That is, if at some initial instant of time the particles have some velocity, $\mathbf{v}(0)$, at a time τ_v later on the velocity will not be correlated with the initial velocity. To estimate τ_v, consider the velocity autocorrelation function $C(t)$ defined, in terms of one component of the velocity v, as

$$C(t) = \langle v(t)v(0) \rangle .$$

This can be related to the diffusion coefficient by the Green–Kubo relation

$$D = \frac{k_\mathrm{B}T}{m} \int_0^\infty \mathrm{d}t\, \frac{C(t)}{C(0)}. \qquad (1)$$

Here we have made use of the fact that for a system in equilibrium $\langle v^2 \rangle = k_\mathrm{B}T/m$, where T is the temperature, m the particle mass and k_B Boltzmann's constant. The integral over all time of the normalized velocity autocorrelation function $C(t)/C(0)$ has units of time. So, following our definition of the velocity decay time, a reasonable formal definition is

$$D = \frac{k_\mathrm{B}T}{m} \tau_v . \qquad (2)$$

Given an estimate for D we can therefore estimate τ_v. For colloidal spheres of radius a in a solvent of shear viscosity η and density ρ, it can be estimated from the Stokes–Einstein relation

$$D = \frac{k_\mathrm{B}T}{6\pi\eta a} . \qquad (3)$$

If we assume that the colloidal particles have the same density as the solvent, always a good approximation in practice, then from (2) and (3) we find that

$$\tau_v \sim \frac{a^2}{\nu} ,$$

where $\nu\,(= \eta/\rho)$ is the kinematic viscosity. For a one micron colloidal particle dispersed in water at room temperature, this gives a velocity decay time of about one microsecond. This may seem short, and indeed for suspensions (or polymer solutions) it is referred to as the "short" time-scale. However, it should be born in mind that by atomic standards it is already very long. In molecular dynamics, for example, where the dynamics are simulated at the scale of the atoms, the equations of motion are typically integrated with a time-step of 10^{-15} s. Already we see that an atomic level approach is hopeless. In fact it gets worse.

There is a second time-scale we can consider, the positional time-scale τ_p. This is the time it actually takes the particles to move significantly. That is, if we look at the system then come back a time τ_p later and look again, the particles will be in quite different positions. To estimate this time-scale we can make use of the Einstein definition of the diffusion coefficient in terms of the increase with time of the root mean squared displacement in a given direction, and equate this with the size of the particles. For a colloidal particle this gives

$$\tau_{\mathrm{p}} = \frac{a^2}{2D}.$$

Again substituting the Stokes–Einstein value for the diffusion coefficient of a micron sized particle dispersed in water at room temperature we find $\tau_{\mathrm{p}} \sim 1\,\mathrm{s}$. Thus, the positional time-scale is six orders of magnitude longer than the velocity decay time. That is, on the time-scale the velocities of the particles change they go essentially nowhere. The same is true for polymer molecules. A reasonable measure for the size of a polymer coil is the radius of gyration, R_{g}, which for an ideal chain of N beads connected with a Kuhn length b, is

$$R_{\mathrm{g}} = \sqrt{\frac{N}{6}}\, b.$$

The diffusion coefficient of a long ideal polymer chain in a solvent of shear viscosity η can be calculated from the Zimm result [6]

$$D = \frac{k_{\mathrm{B}}T}{6\pi\eta R_{\mathrm{h}}}, \tag{4}$$

where R_{h} is the hydrodynamic radius of the polymer coil (equivalent to a for colloidal particles). For long ideal chains the hydrodynamic radius is

$$R_{\mathrm{h}} = \frac{3}{8}\sqrt{\pi}R_{\mathrm{g}},$$

so the positional time scale τ_{p} is accordingly given by

$$\tau_{\mathrm{p}} = \frac{27\pi^2}{1024}\sqrt{\frac{\pi}{6}}\frac{N^{3/2}b^3\eta}{k_{\mathrm{B}}T}.$$

For a chain of 10^4 segments and Kuhn length of about 1 nm, τ_{p} is in the order of 0.01 s. Again, the positional time scale is extremely long by atomic standards.

So in general, for a mesoscopic system, we have at least two well separated time-scales ($\tau_{\mathrm{p}} \gg \tau_{\mathrm{v}}$), that are both long when compared to a molecular dynamics time-step ($\Delta t \sim 10^{-15}$ s). The question that remains is what to do about this problem. The simplest answer is to throw out a lot of details and hope that this will not significantly influence the outcome of our model. But how will we know?

2   Absolutely Minimal Mesoscopic Dynamics

Moving on to simulating the dynamics of our two mesoscopic systems we begin with the absolute minimal level of description of the solvent. Any explicit solvent is in fact thrown out altogether. The effect of the solvent at this level of approximation is simply to supply the fluctuations necessary to maintain the system at the required temperature and a degree of friction (related to the diffusion coefficient of the colloidal particles or monomer beads). Even at this level of description we still have nonetheless a velocity relaxation time τ_v and a positional time τ_p. What it *does not* have is hydrodynamic interactions. That is, it does not take into account the influence the motion of one particle has on its neighbours because of the flow fields it generates in the solvent. Let us begin by explicitly examining a method of simulating the dynamics at this level of approximation.

2.1   Modelling the Solvent with an Andersen Thermostat

The Andersen thermostat [7] was originally developed as a means of carrying out molecular dynamics simulations in the canonical (constant NVT) ensemble. Molecular dynamics involves solving Newton's equations of motion for a set of particles interacting through some potential. As such, the total energy E in the system is conserved. It samples the micro-canonical (constant NVE) ensemble. The Andersen method proceeds as follows. Suppose we describe a system of N particles with mass m, whose positions are $\mathbf{r}_i$ and velocities $\mathbf{v}_i$. The system evolves in time by integrating Newton's equations of motion over a discrete time-step Δt, using for example a Verlet algorithm [1] to update the positions and velocities,

$$\mathbf{r}_i(t + \Delta t) = \mathbf{r}_i(t) + \Delta t\, \mathbf{v}_i(t) + \frac{1}{2}(\Delta t)^2 \frac{\mathbf{f}_i(t)}{m}\,;$$

$$\mathbf{v}_i(t + \Delta t) = \mathbf{v}_i(t) + \frac{\Delta t}{2m}(\mathbf{f}_i(t) + \mathbf{f}_i(t + \Delta t))\,.$$

Note that this conserves total energy. However, at this point, with a probability $\Gamma \Delta t$, the velocity of a particle may instead be taken from the thermal (Maxwellian) distribution of velocities. This violates conservation of energy and effectively kicks the system onto another constant energy surface. Taking the velocities from a Maxwellian ensures that these constant energy shells are sampled with the correct statistical weight to generate the canonical distribution. This procedure, conceptually at least, also mimics the effect of the particle undergoing a collision with the solvent. It maintains the temperature of the particles at the set (or "bath") temperature. Here Γ, the bath collision frequency, is a parameter that can in principle be chosen freely (the limiting distribution is Canonical independent of Γ). From a dynamic point of view, 'bath' collisions completely de-correlate particle velocities so that the velocity autocorrelation function takes the form:

$$C(t) = \frac{k_B T}{m}\, \mathrm{e}^{-\Gamma t}\,, \tag{5}$$

and the diffusion coefficient (from (1)) will be

$$D = \frac{k_B T}{\Gamma m}\,.$$

Following from our discussions of the relevant dynamic time-scales we therefore have a velocity decay time $\tau_v = 1/\Gamma$ and positional time $\tau_p = a^2 \Gamma m / k_B T$, where a is the particle radius. Thus the model has only one relevant parameter, the ratio of the two time-scales

$$\frac{\tau_p}{\tau_v} = \frac{a^2 \Gamma^2 m}{k_B T}\,.$$

Modelling the solvent with an Andersen thermostat is very closely related to another approach which describes the motion of particles in the solvent by means of a stochastic differential equation – the Langevin equation. In this case the equations of motion are

$$\frac{d\mathbf{r}_i(t)}{dt} = \mathbf{v}_i(t)$$

and

$$m\frac{d\mathbf{v}_i(t)}{dt} = \mathbf{f}_i(t) = -\gamma\mathbf{v}_i(t) + \mathbf{F}_R(\mathbf{r}_i(t)) + \mathbf{F}_I(\mathbf{r}_i(t)).$$

Here γ is the friction coefficient (quantifying the drag force exerted by the solvent on the particles), $\mathbf{F}_R(\mathbf{r}_i(t))$ is a random force (representing the thermal fluctuations induced by the solvent), and $\mathbf{F}_I(\mathbf{r}_i(t))$ is the sum of any other forces acting on particle i. Solving the Langevin equation yields a velocity autocorrelation function of the form

$$C(t) = \frac{k_B T}{m}\,e^{-\gamma t/m},$$

which is the same result we get from the Andersen thermostat (cf. (5)) if we interpret the friction coefficient as $\gamma = m\Gamma$. One might well ask why use the Andersen thermostat rather than solve the Langevin equation? The answer is simply that it is easier to do computationally. Andersen's method satisfies the detailed balance condition and as such is a valid Monte Carlo method. In practical terms it will always give the static properties of the system (temperature, pressure etc.) correctly. Constructing an algorithm to solve the Langevin equation that also satisfies this condition is far more involved.

2.2 Langevin Dynamics of a Gaussian Chain

The Langevin level of approximation for the dynamics of a single ideal polymer chain is known as the Rouse model [8]. It can be solved analytically for long chains so we do not at this point need to do any simulations to at least to extract the dynamics the model predicts. Simply by analyzing the dynamics we can nonetheless come to a very important conclusion about simulating mesoscopic dynamics. The question we want to address here is the following. Suppose we have a very long ideal chain and we want to renormalize it to a model chain consisting of fewer beads. How far do we need to renormalize it to make simulating it at least feasible?

For a Rouse chain of any chain length, the center of mass diffusion coefficient will be

$$D = \frac{k_B T}{N\gamma},$$

where N is the number of monomers and γ the monomer friction coefficient. This simply means the friction of the whole polymer is N times the friction of a monomer. Note that this differs fundamentally from the Zimm result (see (4)). This is because we are currently neglecting hydrodynamics. We have already discussed how we can renormalize a long ideal polymer chain down to a smaller number of beads in Sect. 1.1. Since we know the N dependence of the diffusion coefficient we can also renormalize the dynamics, simply by measuring time in units of the time it takes the polymer to diffuse a distance

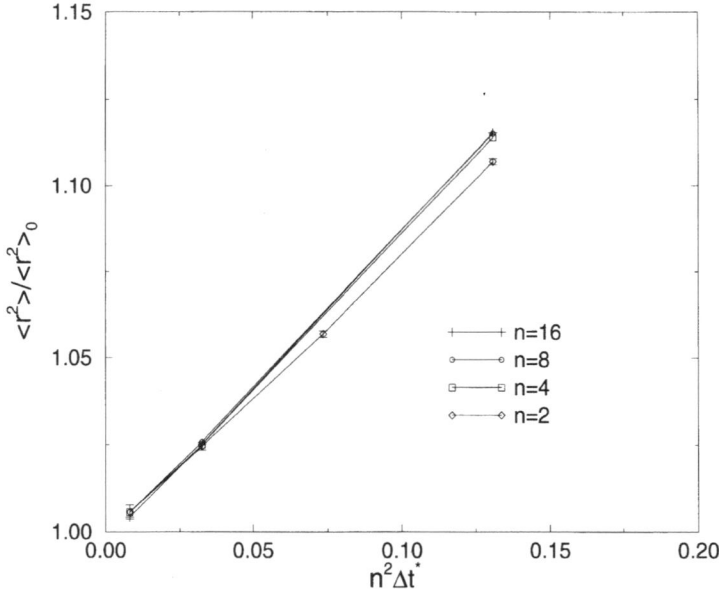

Fig. 2. Ratio of the mean bond length $\langle \Delta r^2 \rangle$ calculated using the Andersen thermostat to the correct value $\langle \Delta r^2 \rangle_0$ as a function of $n\Delta t^*$. Here n is the number of particles in the model and Δt^* the dimensionless time-step defined in the text

the order of its end to end length ($t^* = tD/nb^2$) where n is the number of beads in the renormalized model and the original polymer simply corresponds to setting $n = N$.

What happens then if we fix the numerical accuracy we require, and vary the number of beads in the renormalized model? In Fig. 2 we have plotted the relative error in the mean squared bond length as a function of the dimensionless time step $n^2\Delta t^*$. Clearly, to maintain a fixed accuracy we must have $\Delta t^* \propto 1/n^2$. This is to be expected because if we fix the Kuhn length the potential between beads is the same so the same time step should give the same accuracy independent of n. However, the centre of mass diffusion coefficient decreases proportionally with n and the mean square end to end distance increases proportionally with n. So the same time step is, in dimensionless terms, a factor n^2 smaller. Given that the amount of central processing unit (CPU) time required per time step is proportional to n, this implies that the real world time simulated per unit CPU time goes down in proportion to n^3. So if we take an original chain and renormalize from N to n beads the real world time we can simulate per unit CPU time scales as $(N/n)^3$. That is, for every factor of two we can renormalize, we gain a factor eight in the real world time we can simulate for a given amount of CPU time. Quantitatively, using the data shown in Fig. 2 and assuming we fix an accuracy of 2 %, we will illustrate what this means in practice with two examples. Say we want to model a *short* chain of polyethylene ($N = 1024$, $b = 5 \times 10^{-8}$ cm). In this case, with $n = 2$ (representing the polymer as a dimer) 1 s of CPU time represents approximately 0.1 s of real time. However, if we wanted to model the complete system ($n = N = 1024$) 1 s of CPU time would only represent 10^{-10} s. Now suppose, for our second example, we want to model a strand of DNA ($N = 1024$, $b = 5 \times 10^{-6}$ cm). If we represent this strand with only two beads, 1 s of CPU time represents approximately 10^5 s which is roughly one day. The computer far outstrips reality. Hopefully, this analysis illustrates firstly that, even for simple model polymer chains, there is much to be gained if we can renormalize their dynamics. Secondly, to simulate macroscopic time-scales in a reasonable amount of computer time, we need to be able to drastically simplify the system down to a handful of beads.

2.3 Brownian Dynamics

The Langevin description may seem like the minimum level of description of the dynamics. However, given the separation of time-scales $\tau_v \ll \tau_p$, it is possible to integrate the Langevin equation over the velocity decay time. This procedure essentially throws out τ_v, and arrives at a simplified Smoluchowski equation. This is an N particle diffusion equation for the single particle distribution function $p(\mathbf{r}_i, t)$ of interacting particles where the only relevant time-scale is τ_p,

$$\frac{dp(\mathbf{r}_i, t)}{dt} = D\nabla_i \left\{ \nabla_i + \frac{\mathbf{f}_i(\mathbf{r}_i)}{k_B T} \right\} p(\mathbf{r}_i, t) \,.$$

This equation can therefore be solved purely at the positional level (the velocity decay time now being irrelevant). A particularly simple algorithm for doing so is that of Ermak and McCammon [9] where the time evolution of the particle positions is simply

$$\mathbf{r}_i(t + \Delta t) = \mathbf{r}_i(t) + \frac{D}{k_B T} \mathbf{f}_i(\mathbf{r}_i) + \Delta \mathbf{r}_i \,,$$

where the random displacement $\Delta \mathbf{r}_i$ satisfies the condition

$$\langle \Delta \mathbf{r}_i \cdot \Delta \mathbf{r}_i \rangle = 2D\Delta t \,.$$

These equations naturally no longer involve velocities.

2.4 The Dynamics of a Colloidal Suspension

It is possible to check the consequences of going further in simplifying the dynamics (integrating over τ_v) by keeping τ_v in the problem (using the Andersen thermostat detailed above), imposing the separation of time-scales $\tau_p \gg \tau_v$ and checking to what extent the results are then independent of the velocity decay time (τ_v). According to the Smoluchowski equation it should be irrelevant. A useful property to probe in this respect is the visco-elastic response because, for colloidal suspensions of essentially hard spheres [10], it has been carefully measured experimentally [11]. There is thus a firm point of comparison. Numerically, the viscoelastic response can be calculated from the stress-stress correlation function. The xy contribution to the time dependent stress-stress correlation function is

$$\Sigma_{xy}(t) = \frac{1}{k_B T V} \langle \sigma_{xy}(0) \sigma_{xy}(t) \rangle \,,$$

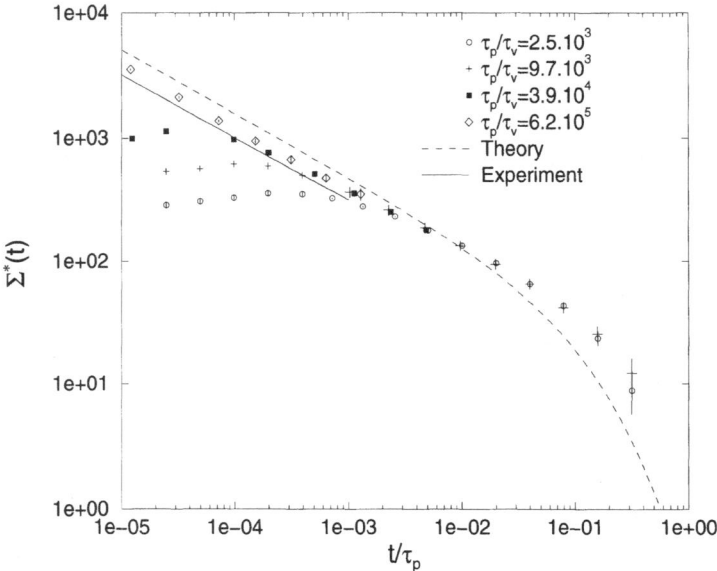

Fig. 3. Dimensionless stress-stress correlation function as a function of t/τ_p, where τ_p is the positional time-scale (see text). The different sets of simulation data correspond to different values of the ratio $\tau_\mathrm{p}/\tau_\mathrm{v}$, where τ_v is the velocity decay time

where the xy stress σ_{xy} is given in terms of the separation vector between interacting particles i and j, $\mathbf{r}_{ij}$, and the inter-particle forces $\mathbf{f}_{ij}$:

$$\sigma_{xy} = \sum_i m v_{ix} v_{iy} + \sum_{i,i\neq j} r_{ijx} F_{ijy} .$$

The first term is negligible in all cases, except in a gas. The viscosity can in turn be calculated from the time integral of the stress-stress correlation function

$$\eta = \int_0^\infty \mathrm{d}t\, \Sigma_{xy}(t) ,$$

or, more generally, the frequency dependent viscosity can be calculated from the Laplace transform of the stress-stress correlation function. Of course the other off diagonal elements of the stress tensor all behave in the same way because the system is isotropic, so in practice we average over all six.

For a suspension with a volume fraction of 45 % the results we obtain for the time dependent (dimensionless) stress-stress correlation function as a function of t/τ_p are shown in Fig. 3. Notice that we are really doing dynamic mesoscopic simulation here because in the real world τ_p is very much greater than a molecular time-scale. Following from our discussion of time-scales we know that for particles of diameter $1\,\mu$, $\tau_\mathrm{p} \sim 1\,\mathrm{s}$ in which case the time-scale covered by the data plotted in the figure extends to $0.1\,\mathrm{s}$. The simulation data is for different ratios of the positional and velocity decay times (although in all cases $\tau_\mathrm{p} \gg \tau_\mathrm{v}$ as we require for a realistic system). We see that as we increase this ratio the function follows the same curve, independent of τ_v, from progressively shorter times. From then on it only depends on τ_p and as such corresponds to the solution to the Smoluchowski equation (even though we are effectively solving a Langevin equation to get it). This behaviour is entirely consistent with the picture outlined above. At times significantly greater than τ_v only the positional time is relevant and we could just as well have thrown out the velocity decay time altogether and solved the Smoluchowski equation instead.

Another point worth noting from Fig. 3 is that at short times the results following the Smoluchowski curve (that is, where the data are independent of the ratio τ_p/τ_v) are in good agreement with the experimentally observed behaviour (obtained by performing the inverse transform of the high frequency asymptote of the frequency dependent viscosity reported in [11]). However, we can also calculate the time integral of the stress-stress correlation function which should asymptote at long times to the viscosity. This function is plotted in Fig. 4. The data shows that (to within the statistical accuracy) the viscosity approaches the same value independent of the ratio of τ_p/τ_v. That is, the velocity decay time influences the decay of the stress-stress correlation function at short times but this does not affect the value we obtain for the viscosity. More importantly, the simulations disagree significantly from the (approximate) expression derived by Brady [12]. The point about Brady's result is that it agrees with experiment, so if the simulations do not agree with this theory they do not agree with experiment either. The discrepancy is in fact almost a factor of two. We know that this cannot have anything to do with throwing out the velocity decay time, because in the simulations we explicitly resolve it. For the

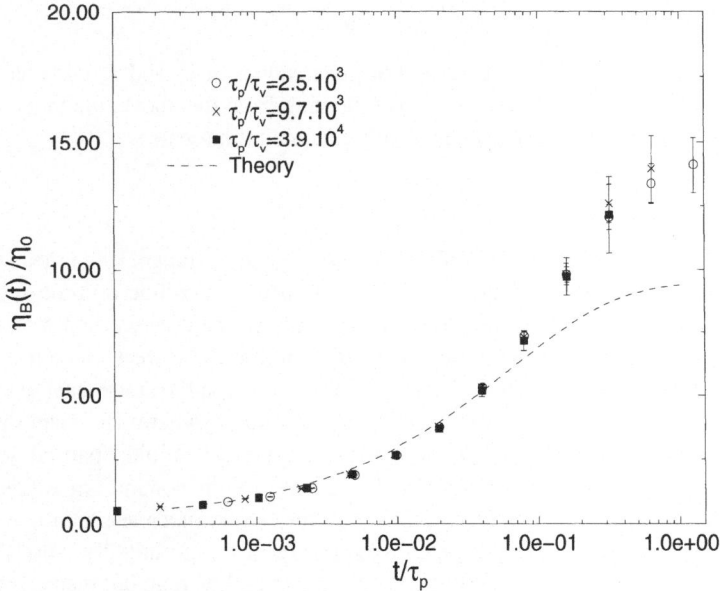

Fig. 4. The time dependent dimensionless suspension viscosity as a function of t/τ_p, where τ_p is the positional time-scale (see text). The different sets of simulation data correspond to different values of the ratio τ_p/τ_v, where τ_v is the velocity decay time

origins of this discrepancy we have to look at the consequences of using such a minimal model for the solvent.

So by looking at the dynamics of stress relaxation, we can conclude that we get essentially the same viscoelastic response if we throw out τ_v, although the actual answer is wrong. The question remains whether even this is true for everything? As the astute reader will guess, the answer of course is no. There is an additional time-scale in the problem, the collision frequency f_c for collisions between colloidal spheres. For a hard sphere fluid the collision frequency is given by

$$f_c = 4na^2 g(\sigma)\sqrt{\frac{k_B T}{m}},$$

where n is the number density of colloidal particles and $g(\sigma)$ the radial distribution function at contact. It is fairly straightforward to show, based on equilibrium arguments, that this must also be the colloid-colloid collision frequency in a suspension, independent of the solvent. This collision frequency introduces a new time scale, a colloidal particle collision time $\tau_c (= 1/f_c)$. Each colloidal particle collides on average once with a neighbour during a time τ_c. In fact, looking at our definitions of τ_v and τ_p, τ_c can be related to τ_v and τ_p by

$$\frac{\tau_c}{\tau_v} \sim \sqrt{\frac{\tau_p}{\tau_v}} \tag{6}$$

and

$$\frac{\tau_c}{\tau_p} \sim \sqrt{\frac{\tau_v}{\tau_p}},$$

so it is not an independent time-scale. For given values of τ_v and τ_p its value is fixed. We showed that in practice $\tau_p \gg \tau_v$, so following from the above, the mean collision time fits roughly in between these two well separated time scales

$$\tau_v \ll \tau_c \ll \tau_p.$$

However, if we think what this implies it is actually quite strange. If the mean collision time is very much greater than τ_v it means that particles collide on time-scales long compared to those on which they forget their velocity. The process cannot be regarded as simply ballistic. On the other hand, the collision time being very much less than the positional time means that on the time scale over which, on average, each particle has collided once, the particles have in fact hardly moved anywhere. How can this be the case? Surely the particles have to move some sort of typical colloidal particle separation before they can collide. Apparently not. The answer to this conundrum is the fact that every particle *on average* collides in a time τ_c. If this is very much less than τ_p only a few pairs of particles can possibly have collided (those that were initially close together). These few particles which do collide must therefore collide together many many times to give an average collision time of τ_c. Using the same methodology described above (in which we resolve τ_v) we tested this hypothesis by calculating a function P_N. This we define as the probability that two colliding particles subsequently re-collide with each

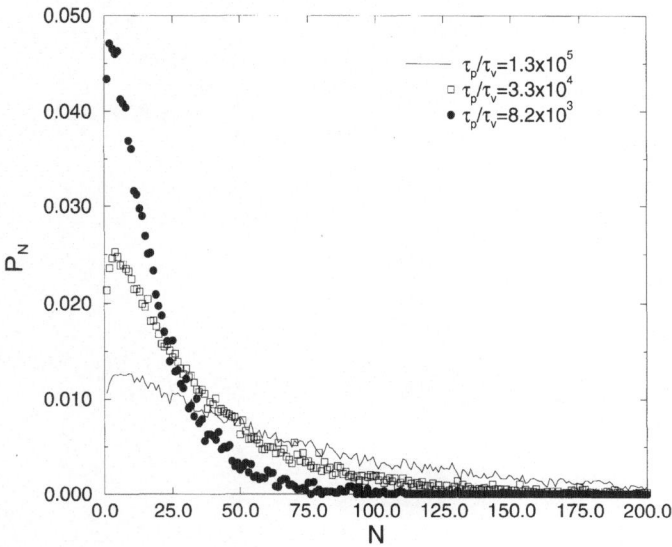

Fig. 5. The probability two colloidal particles colliding with each other subsequently re-collide N times with neither undergoing an intervening collision with another colloidal particle. Different sets of simulation data correspond to different values of the ratio τ_p/τ_v. The volume fraction of the suspension was 10 %

other N times without either particle undergoing an intervening collision with any other particle. These are N pure binary re-collisions.

The results we obtain for P_N in a model suspension with a volume fraction of 10 % are shown in Fig. 5 for different values of the ratio τ_p/τ_v. From the figure we can see that while the most probable number of these binary collisions is relatively small, there is a significant probability that there will be a very large number of re-collisions. Furthermore as we increase the ratio τ_p/τ_v the distribution shifts, large numbers of re-collisions become more probable. From these distributions we can calculate the average number $\bar{N}$ of pure binary re-collisions a pair of colliding particles subsequent undergo. For the data shown in the figure we find $\bar{N} = 17$, 34, and 68 for $\tau_p/\tau_v = 8.2 \times 10^3, 3.3 \times 10^4$ and 1.3×10^5 respectively. The latter value is still perfectly plausible for a typical colloidal suspension and the value $\bar{N} = 68$ implies that any two particles which collide will on average collide 68 times with each other before either hits another particle. This appears to confirm the hypothesis we put forward above. We can also analyze the time these re-collisions typically take by calculating the average time it takes for N re-collisions, $\tau(N)$. The value of $\tau(N)$, divided by the mean collision time is plotted in Fig. 6. Clearly, for more than a few repeated re-collisions the value is constant, with the same constant for each set of data. This means that the repeated rattling re-collisions occur on a time-scale related to the colloid-colloid collision time. However, remember that this collision time is a function of both the positional time and the velocity decay time (see (6)). That is, these re-collisional dynamics are partially determined by τ_v so any scheme that attempted to throw out the velocity decay time could never resolve correctly the collisional dynamics of the suspension.

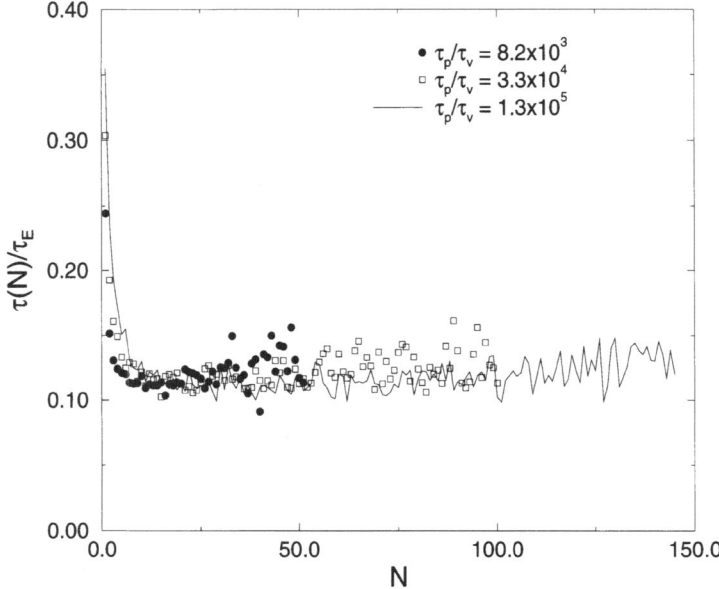

Fig. 6. The average time for a pure binary re-collision $\tau(N)$ relative to the mean collision time (here denoted τ_E). This is plotted as a function of the number of re-collisions in the sequence

2.5   What Does Minimal Dynamics Teach Us

- We can conclude that even a simple model of a long polymer (the ideal chain) needs to be drastically simplified before its dynamics can be simulated on macroscopic time-scales.
- Simplify things further by throwing out the velocity decay time τ_v is okay for some things, but wrong for others. The only way to check if it is a valid approximation is by comparing the case with and without τ_v and see if it makes a difference.
- The simplest approach simply does not work. It gives completely the wrong diffusion coefficient for a polymer and the wrong viscosity for the suspension. Like it or not, we need to introduce some new levels of complexity.

3   More Realistic Solvent Dynamics – Hydrodynamics

3.1   The Importance of Hydrodynamics

The minimal model of the solvent we described above, while conceptually and computationally simple, unfortunately does not work. We threw out hydrodynamics which is equivalent to neglecting conservation of momentum. A reasonable answer to the question, what makes a fluid behave like a fluid? is – conservation of momentum. It is therefore unsurprising that the solvent-less approach failed. To incorporate the fluid like nature of the solvent correctly we need to include fluid like, or "hydrodynamic" behaviour. The simplest answer to the question, what is hydrodynamics? is that the flow of a viscous fluid is determined by Navier–Stokes equations [13]. We will not write the equations here, simply point out that the physical content of these equations is simply; conservation of particle number and momentum, isotropy, and Galilean invariance. Any model that satisfies these criteria will reproduce hydrodynamic behaviour on long enough length and time-scales.

It is because we neglected hydrodynamics that the minimal model gave the wrong viscosity for a suspension. We know this because much more sophisticated simulations in which hydrodynamics are included give the correct answer [14]. However, here, to illustrate how important hydrodynamics is with respect to the dynamics of mesoscopic systems, we return to the case of a single polymer chain in solution. Without hydrodynamics the diffusion coefficient of a polymer coil in solution is simply given by the Rouse result [8]

$$\frac{D}{D_{\mathrm{m}}} = \frac{1}{N} \,, \tag{7}$$

where this time we have written it in terms of D_{m}, the monomer diffusion coefficient (i.e. the diffusion coefficient a single bead would have if it was detached from the chain).

Suppose we regard a bead moving in the chain as exerting a point force $\mathbf{F}$ acting on the fluid. It will then set up a velocity field in the solvent given by the Oseen tensor (the solution to the Navier–Stokes equations for a point force). If the force is applied in the α direction then we have

$$u_\alpha = \frac{F_\alpha}{8\pi\eta} \left\{ \frac{1}{r} + \frac{r_\alpha{}^2}{r^3} \right\} \tag{8}$$

and

$$u_\beta = \frac{F_\alpha}{8\pi\eta} \left\{ \frac{r_\alpha r_\beta}{r^3} \right\} \,,$$

where β is either of two directions orthogonal to α. These flow fields will of course influence the motion of the particles' neighbours, there is a "hydrodynamic interaction". Note also that this hydrodynamic interaction is long ranged, decaying as $1/r$, just like a Coulomb force. Indexhydrodynamic interaction

We can work out what the effect of these interactions will be on the centre of mass diffusion coefficient of the polymer by considering its mobility. By imagining a small force F_x/N applied to each bead in the chain and calculating the corresponding centre of mass velocity $\bar{v}$, we can define the mobility M

$$M = \frac{\bar{v}}{F_x} \,.$$

The diffusion coefficient is then given by the generalized Einstein equation

$$D = k_{\mathrm{B}} T M \,. \tag{9}$$

If we applied such an external force to our polymer then, allowing for the fluid flow (from (8)) the velocity of a particle i located at $\mathbf{r}_i$, would be

$$v_{i\alpha} = \frac{F_x}{N} \frac{1}{6\pi\eta a} + \frac{F_x}{N} \frac{1}{8\pi\eta} \sum_{i\neq j} \left\{ \frac{1}{r} + \frac{r_{ij\alpha}{}^2}{r^3} \right\} \,. \tag{10}$$

Here the first term in the equation uses an approximation for what is called the "self-term". This is the mobility M_s of a single particle were it to have no interactions (hydrodynamic or otherwise) with the rest of the polymer. It is related to the monomer diffusion coefficient, D_m

$$M_s = \frac{D_m}{k_B T} = \frac{1}{6\pi\eta a} \, . \tag{11}$$

This serves to define a hydrodynamic radius a for a bead in the chain. The second term in (10) represents the collective effect of the neighbouring monomers. The average drift velocity of the chain is obtained by averaging over all particles to give

$$\langle \bar{v} \rangle = \frac{F_x}{N6\pi\eta a} + \frac{F_x}{N^2 8\pi\eta} \left\langle \sum_i \sum_{i \neq j} \left\{ \frac{1}{r} + \frac{r_{ij\alpha}{}^2}{r^3} \right\} \right\rangle .$$

Combining this with (9) and (11) we have the following relation for the polymer diffusion coefficient:

$$\frac{D}{D_m} = \frac{1}{N} + \frac{3a}{4N^2} \left\langle \sum_i \sum_{i \neq j} \left\{ \frac{1}{r} + \frac{r_{ij\alpha}{}^2}{r^3} \right\} \right\rangle . \tag{12}$$

The average in the second term is simply a static property of the distribution of mass in the chain. Since we know the N particle distribution function for an ideal chain it can be evaluated (at least for a long chain) and the second term written as

$$\frac{3a}{4N^2} \left\langle \sum_i \sum_{i \neq j} \left\{ \frac{1}{r} + \frac{r_{ij\alpha}{}^2}{r^3} \right\} \right\rangle_{N \to \infty} = \frac{8a}{3b} \sqrt{\frac{6}{N\pi}} = \frac{a}{R_h} , \tag{13}$$

where R_h can be regarded as a hydrodynamic radius for the entire polymer chain. So the expression for the polymer diffusion coefficient becomes

$$\frac{D}{D_m} = \frac{1}{N} + \frac{8a}{3b} \sqrt{\frac{6}{N\pi}} \, .$$

The first term on the right hand side is simply the Rouse result (cf. 7) so for short chains the hydrodynamics have little effect. For large N (that is, a real polymer) it is clear that the second (hydrodynamic) term dominates. Note that the difference is quite dramatic because without hydrodynamics we have the scaling $D \propto 1/N$ whereas it should be $D \propto 1/\sqrt{N}$. Further, without hydrodynamics the diffusion coefficient depends on the hydrodynamic radius of the beads, a, (effectively the monomer mobility) whereas with hydrodynamics it does not. The diffusion coefficient is simply equivalent to that of a sphere of radius R_h. The effect of the hydrodynamics is to make the dynamics almost totally collective.

3.2   Putting Back the Hydrodynamics

So how can we put back the hydrodynamics? There are several options all of which inevitably represent a substantial increase in computational complexity relative to the solvent-less approach discussed above. However, as this gives incorrect dynamics, this is an unavoidable problem. One method for including hydrodynamics does still avoid any explicit inclusion of the solvent. Stokesian dynamics [15] solves a more complex Smoluchowski equation in which the hydrodynamic interactions between particles are included at the level of pair mobilities. However, these interactions are many bodied and long ranged and without sophisticated numerical techniques the method has a poor scaling of the CPU time with particle number. The approach also violates Galilean invariance so can only strictly be used at vanishingly low Reynolds numbers. The second approach is to explicitly include the solvent but at a simplified level. Its role is simply to generate the correct hydrodynamic behaviour. There are several of these particle models on the market. The lattice Boltzmann method [16,17] solves a discretized Boltzmann equation which yields collective hydrodynamic behaviour. However, the particles move on a lattice which is inconvenient if one is interested in complex geometries that do not easily map onto a lattice. The model, because it describes "average" fluid behaviour, has no spontaneous fluctuations. These are of course necessary if we are interested in thermal motion. They can be introduced but not in a totally satisfactory manner. The system still does not have a well defined thermodynamic equilibrium distribution. Nonetheless, this approach has been relatively successfully applied for simulating polymer dynamics [18]. Of the off-lattice model solvents a method has recently been proposed by Malevanets and Kapral [19,20] which is very similar to the direct simulation Monte Carlo method of Bird, the latter being widely used to study gas flows. Their method has the disadvantage that is is not Galilean invariant and again introduces boxes (the off-lattice equivalent of a lattice). The dynamics of the model fluid are also fundamentally gas like not liquid like (something we will consider later). Another approach that has been widely applied is dissipative particle dynamics (DPD) [21] which has the advantage of conserving momentum, being isotropic and Galilean invariant. It is however somewhat *ad hoc* in other respects, notably in that particles interact through a (soft) potential that introduces a somewhat artificial length scale into the problem. Nonetheless, while bearing in mind that there are other options, given that this method fulfills the minimal requirements for reproducing hydrodynamic behaviour, it is the one we will pursue here. We will also discuss a related approach, the Lowe–Andersen thermostat [22].

3.3   Dissipative Particle Dynamics and the Lowe–Andersen Thermostat

Dissipative Particle Dynamics (DPD) is a particle based model. The time evolution of these interacting particles is governed by Newton's equations of motion

$$\frac{d\mathbf{r}_i}{dt} = \mathbf{v}_i \,; \quad m\frac{d\mathbf{v}_i}{dt} = \frac{d\mathbf{p}_i}{dt} = \mathbf{f}_i \,.$$

The force contains three parts, each of which is pairwise additive

$$\mathbf{f}_i = \sum_{i \neq j} \mathbf{F}_{ij}^{\mathrm{D}} + \sum_{i \neq j} \mathbf{F}_{ij}^{\mathrm{R}} + \sum_{i \neq j} \mathbf{F}_{ij}^{\mathrm{C}} \,,$$

where $\mathbf{F}_{ij}^{\mathrm{D}}$ is the dissipative force and $\mathbf{F}_{ij}^{\mathrm{R}}$ is the random force. These take the form

$$\mathbf{F}_{ij}^{\mathrm{D}} = -\gamma \omega^{\mathrm{D}}(r_{ij})(\hat{\mathbf{r}}_{ij} \cdot \mathbf{v}_{ij})\,\hat{\mathbf{r}}_{ij}$$

and

$$\mathbf{F}_{ij}^{\mathrm{R}} = \sigma \omega^{\mathrm{R}}(r_{ij}) \theta_{ij} \, \hat{\mathbf{r}}_{ij} \,,$$

where $\mathbf{v}_{ij} = \mathbf{v}_i - \mathbf{v}_j$, $\omega(r_{ij})$ is a weight function, which tends to zero at some cut-off radius $r = r_c$, and θ_{ij} is a random number with zero mean and unit variance. One of the weight functions can be chosen arbitrarily, thereby fixing the other weight function. In order to have a canonical equilibrium distribution, the weight functions should obey [21]

$$[\omega^{\mathrm{R}}(r_{ij})]^2 = \omega^{\mathrm{D}}(r_{ij})$$

and

$$\sigma^2 = 2k_B T \gamma \,.$$

The remaining force, $\mathbf{F}_{ij}^{\mathrm{C}}$, is the conservative force and can in principle be chosen arbitrarily (it is usually taken as a soft repulsion with the same range as the dissipative and random forces). The method has the following properties.

- The dissipative and random forces act as a thermostat (give a limiting canonical distribution).
- It is stochastic but conserves particle number and momentum (all forces act in equal and opposite pairs).
- It is Galilean invariant because the thermostat acts on relative velocities.
- It is isotropic because all the interactions are spherically symmetric.

In other words it has all the properties needed to reproduce hydrodynamics. However, it is not easy to implement numerically, due to the velocity dependent forces. In the limit of an infinitely small time-step it will give the correct (canonical) distribution but for larger time-steps, unless an iterative self-consistent algorithm is used, there can be serious spurious effects [23]. For a numerical investigation of these issues see [24,25].

An alternative approach is to use a Lowe–Andersen thermostat [22]. As we have seen, with the Andersen thermostat the equations of motion are integrated as normal and periodically a new velocity is taken from a Maxwellian (a "bath" collision). Bath collisions do not conserve momentum. The Lowe–Andersen thermostat, on the other hand, like DPD uses the relative velocities of pairs of particles located within a cut-off radius r_c of each other. For each pair of particles whose velocities are to be "thermalized" we work on the component of the velocity parallel to the line of centres (to conserve angular momentum) and generate a new relative velocity $[\mathbf{v}_{12}]' \cdot \hat{r}_{ij}$ from a distribution $\xi_{ij}\sqrt{2k_B T/m}$. Here ξ_{ij} is a Gaussian distributed random number with zero mean and unit variance. The factor of $\sqrt{2}$ reflects the fact that we are using the Maxwellian for relative velocities. To impose this new relative velocity on the particles, and conserve momentum, we write the new particle velocities

$$[\mathbf{v}_i]' = \mathbf{v}_i + \Delta_{ij} \,; \quad [\mathbf{v}_j]' = \mathbf{v}_j - \Delta_{ij} \,,$$

where

$$2\Delta_{ij} = \hat{r}_{ij}([\mathbf{v}_{ij}]' - \mathbf{v}_{ij}) \cdot \hat{r}_{ij} \,.$$

Note that this method is also Galilean invariant (because it works on relative velocities), conserves momentum and is isotropic by construction. It could also be used with any kind of conservative force. What it does not have is weight functions, but no one has ever shown these fulfill any useful function in DPD. If something is not demonstrably useful it is best to throw it out. Basically then, this method has all the positive features of DPD but is numerically simpler and allows one to use a longer time-step without introducing any spurious effects [25].

What about the conservative force? So far all we have said is that in principle it can be anything. For the ideal chain we know how to derive the effective potential between particles in the renormalized model. What we do not now want is the solvent to interfere with this, which it would if we put in a solvent polymer conservative interaction. Thus, the best conservative force here is no conservative force. That is we model the solvent simply as a thermostatted ideal gas. But a gas is not a liquid, so can this be justified?

3.4 Parametrically Correct Solvent Modelling

We now have a way to put in the hydrodynamics of our model polymer system. The question now is, can we construct it in such a way that it is parametrically correct? That is, such that transport processes in the model solvent occur on realistic time-scales. This is something that is often neglected. It is not however sufficient simply to reproduce hydrodynamics, the hydrodynamic variables must be coupled in a realistic way to the dynamics of the rest of the system. Here the coupling between the polymer and the solvent is carried out by using the same thermostat procedure operating between the polymer beads and the solvent as that acting between the solvent particles themselves. Thus a polymer bead is hydrodynamically equivalent to a solvent particle. Its diffusion coefficient (and also its self mobility) will be the same as that of a solvent particle. Note that there is no thermostat interaction between polymer beads. This would lead to an artificial contribution to the polymer dynamics that would be influenced by the distribution of mass within the polymer itself. All the hydrodynamic information should go through the model solvent. Using this approach it does.

Basically we now have three parameters we can play with. The interaction cut-off radius, r_c, the Kuhn length b, and one parameter characterizing the dynamics of the solvent, $\lambda = \sqrt{k_B T / \Gamma^2 r_c^2 m^2}$. This basically characterizes the relative importance of diffusive and ballistic transport of momentum in the model solvent. To minimize the latter (which is inconsistent with hydrodynamics) we need to satisfy the condition $\lambda < 1$. To simulate a realistic liquid solvent what other conditions do we need to satisfy? In a liquid the Schmidt number Sc $= \nu/D$ (where ν is the kinematic viscosity) is very high, typically of the order of 10^3. In a gas it is order unity. The kinematic viscosity characterizes the rate at which momentum is transported whereas the diffusion coefficient characterizes the rate at which particles actually move. So if the Schmidt number is high, the hydrodynamic interactions propagate rapidly when compared to the time it takes anything to move. Almost all particle based solvent simulations to date do not satisfy this requirement and are therefore dynamically unrealistic [26]. In some cases it may not matter but in general it will. It can lead to artificial screening of the hydrodynamic interactions. This would mean that considerable effort was being put into including hydrodynamics but then, by running with unrealistic parameters, the hydrodynamics was actually rendered largely irrelevant. And there are other problems. With the model we describe we can only expect hydrodynamics (a continuum theory) to be recovered on length scales of the order of $l \geq r_c$, or at best a few solvent inter-particle separations. So, to resolve the hydrodynamic interactions between adjacent beads in the chain we need $b \geq r_c$. Furthermore we would like to maximize r_c, since it will increase the Schmidt number [22]. However a small value for r_c is computationally more convenient (there are less interactions to evaluate). A compromise has to be sought. We use a value $(4/3)\pi r_c{}^3 n = 8$, where n is the number density of the solvent. This means one particle interacts, on average, with eight other particles.

Once the cut-off radius is fixed, the Schmidt number can in fact be made arbitrarily high by reducing λ. This is because the diffusion coefficient of the particles decreases with decreasing λ while the viscosity becomes independent of λ because all the momentum is basically transported through the thermostat interaction. However we then encounter another problem. The solvent is of course relatively compressible (even if it has a liquid-like Schmidt number). How compressible a real fluid is, is actually a matter of length scales. Momentum can propagate by sound propagation on a time-scale τ_s, the time it takes sound to propagate a distance l ($\tau_s = l/c_s$). However the time for viscous transport of momentum over the same length l is $\tau_\nu = l^2/\nu$. If we let α denote the ratio between the two ($\alpha = \tau_p/\tau_s = lc_s/\nu$) we can now see that this ratio is length scale dependent. Using the sound speed and compressibility of water, if the length scale of our simulations is $1\,\mu m$, we should have $\alpha = 1000$. If the length scale of our simulations is $1\,nm$, $\alpha = 1$.

In a system with an ideal gas equation of state, the speed of sound is given by:

$$c_s = \sqrt{\frac{k_B T}{m}} \, .$$

If we choose our units such that Γ (thermostat frequency), m (particle mass), and r_c are all unity, we then have $\lambda = \sqrt{k_B T}$. Reducing λ is therefore a bad thing, because it reduces α. However reducing λ is also a good thing because it increases the Schmidt number. Plotting the λ dependence of both the Schmidt number and α we find an optimum value of λ, $\lambda \sim 0.02$, for which the fluid is reasonably incompressible ($\alpha \sim 1$) while the Schmidt number is still quite high (Sc ~ 100). In actual fact there are reasonable grounds to believe that while this value of α is ideally somewhat too low this might not matter. This reason is simply that sound propagation has no net effect on hydrodynamics. Careful analysis of the wave-vector dependent transverse current-current correlation function shows that sound propagation only induces wiggles about the correct result [27]. Nonetheless, it remains true that we could not trust the dynamics of our results on times $t < \tau_s$ even if we still get the correct transport coefficients. In other words τ_s should certainly be minimized even if cannot be completely parametrically correct.

3.5   Concluding Remarks

- It is possible to put in hydrodynamics by several routes, but there is a significant computational price to be paid.
- If one uses a particle model for the solvent, the parameters of any particle model are very tightly controlled if the model fluid is to be realistic.

- For polymer chains it is not sufficient just to put in the hydrodynamics because, even with hydrodynamics, the dynamics of long polymer chains are very different from short ones.

4 An Example Problem – A Long Polymer Chain in an External Potential

In this section we will address the following problem: *If we put a long ideal polymer chain in an external potential*

$$U^X = \frac{\alpha}{2} r^2 ,$$

how will its radius of gyration and diffusion coefficient change?

This may seem a somewhat artificial example, and indeed we are primarily interested in it from the point of view of *quantitatively* predicting the dynamics from a renormalized model. However, one might imagine a polymer held in an optical trap as being an analogous experimental system. We should also point out that the Gaussian chain is a rather artificial model for a real polymer chain. It neglects excluded volume effects and polymer solvent interactions for example. However, we are using it here because it is a best case scenario. We have a very good idea how to reproduce its static (conformational) properties in a renormalized model. If we cannot make any progress with this problem we would be forced to conclude the whole undertaking is doomed.

We do however still have some obstacles to overcome before we can tackle our model problem and assess how well we can do in practice. Firstly, in Sect. 1.1 we said that the ideal chain was an ideal test case for dynamic mesoscopic simulation because it was straightforward to reproduce the static (configurational) properties. As to what properties we should reproduce, this was left as a bit of an open question. Secondly, we now have options for putting hydrodynamics into our simulations. However, the hydrodynamics of long polymer chains are not the same as long ones.

4.1 Renormalizing the Static Properties of the Ideal Chain

The reason that renormalizing the static properties of an ideal chain is relatively straightforward is that, if the potential between connected beads in an original polymer is

$$U(r_i - r_{i+1}) = \frac{3}{2} \frac{k_B T}{j b^2} (r_i - r_{i+1})^2 ,$$

then the correct statistical distribution between beads that are not adjacent can always be written as

$$\langle (r_i - r_{i+j})^2 \rangle = j b^2 .$$

What does this imply for the bead-bead potential in the renormalized model? Clearly the stretching potential becomes an effective potential aimed at reproducing some property of the original system with fewer beads. So what should it reproduce? The answer, we would suggest, is the structure factor $S(\mathbf{k})$ at small wave-vectors $\mathbf{k}$. The reason for this is that linear response theory tells us [28] that if a system is subject to a wave-vector dependent external potential $U(\mathbf{k})$ the corresponding change in the density $\rho(\mathbf{k})$ will be

$$\frac{\Delta\rho(\mathbf{k})}{\rho} = S(\mathbf{k})\frac{U(\mathbf{k})}{k_\mathrm{B}T},\qquad(14)$$

where $S(\mathbf{k})$ is the structure factor

$$S(\mathbf{k}) = \frac{1}{N}\sum_i\sum_j e^{-i\mathbf{k}\cdot(\mathbf{r}_i-\mathbf{r}_j)}.\qquad(15)$$

That is, the structure factor is the response function for density perturbations. In other words, (14) tells us that if we reproduce the long wave-length structure factor in a renormalized model of the original polymer the configurational response of the model system should be the same as the original system for an external perturbation of long-wavelength. For higher wavelength perturbations (that is external potentials that vary significantly over, say, the length of the polymer) we will have to resolve the structure factor to correspondingly shorter wave-lengths.

Having decided the structure factor is the key quantity we should reproduce in our renormalized model, (15) immediately suggests a problem. In the limit of long wave-lengths, $\mathbf{k}\to 0$ and $S(\mathbf{k})\to N$. So for our original polymer the long wave-length structure factor will always approach N, whereas in any renormalized model where we try to represent the polymer with n ($n < N$) beads $S(\mathbf{k})$ will always approach n. They cannot possibly be matched. What we can do is study $s(\mathbf{k}) = S(\mathbf{k})/N$ and $U(\mathbf{k})\to NU(\mathbf{k})$ in which case (14) becomes

$$\frac{\Delta\rho(\mathbf{k})}{\rho} = s(\mathbf{k})\frac{NU(\mathbf{k})}{k_\mathrm{B}T}.$$

That is, we have to renormalize the external potential as well as the polymer.

Now all we have to do is to construct a simplified model ideal chain which reproduces the long-wavelength $s(\mathbf{k})$ of the original polymer. This is relatively straightforward because, if we consider the α component of the wavevector, for an ideal chain we have

$$s(k_\alpha) = 1 - \frac{k^2}{N^2}\sum_{i=1}^N\sum_{j=1}^N\left\langle(r_{i\alpha}-r_{j\alpha})^2\right\rangle = 1 - \frac{k^2b^2}{18}\frac{(N-1)(N+1)}{N}.$$

So in the renormalized model, measuring lengths in units of $b\sqrt{N}$, the renormalized Kuhn length $\bar{b}$ is related to the original Kuhn length b according to

$$\bar{b}^2\frac{(n-1)(n+1)}{n} = b^2\frac{(N-1)(N+1)}{N},$$

so for large N

$$\bar{b} = \sqrt{\frac{n}{(n-1)(n+1)}}$$

(which actually implies that you keep the original radius of gyration).

This then is how we renormalize the static properties of the ideal chain. It is worth here asking the question, what do the 'renormalized' particles actually represent. Suppose we renormalize to $n = 2$, a dimer. Do the particles in the renormalized model represent the centres of mass of the renormalized segments of the chain? Actually no, because $\bar{b} = \sqrt{2/3}$ and if this were to be the case it would be $\bar{b} = \sqrt{1/3}$. Do they represent the distribution between the centres of the segments in the original chain? No again, because in this case we would have $\bar{b} = \sqrt{1/2}$. The only answer we can come to is that they are just constructs required to make the renormalized system behave like the original one.

4.2 Long Polymer Dynamics from Short Model Polymers

In Sect. 3.1 we derived an expression for the center of mass diffusion coefficient of an ideal polymer chain (see (13)) which we saw implied that the diffusion coefficient of a long polymer scaled with the number of monomers N as $D \propto 1/\sqrt{N}$. This basically reflects the fact that for a long polymer chain the dynamics are completely determined by collective hydrodynamic motion. However, we also saw that, even with the full hydrodynamic analysis, this behaviour was only recovered for long chains and one would not expect it for short chains. That is, for short chains the effect of the hydrodynamics is underestimated and this is simply because there are insufficient beads to generate enough collective motion. This seems like an insurmountable problem if we want to simulate the dynamics of long polymers with short model ones because, this analysis shows, they simply do not behave in the same way.

One thing that we might attempt to do, is to construct short model polymers that at least display the asymptotic scaling with the minimum number of beads. This we could argue would at least be realistic in the sense that if we imagined cutting our model polymer in half, the two smaller polymers would diffuse with a diffusion coefficient $\sqrt{2}D$, where D was the diffusion coefficient of the original polymer. This at least would be the correct long polymer behaviour. Achieving such an aim would also be useful in the following respects:

- If we were studying polydisperse systems, at least the ratios of the diffusion coefficients of the various polymer lengths would be correct.
- For systems that are not dilute, it would also be important for polymer-polymer hydrodynamic interactions. Since D sets the hydrodynamic radius the ratio of the hydrodynamic radius to the actual radius could be kept correct.
- Perhaps most importantly, if we know $D(n)$, where n is the number of beads in the renormalized model we also know $D(N)$, where N is the number in the actual polymer, simply through dimensional analysis.

But is it possible to achieve this? One thing that suggests itself is that there are significant corrections to the asymptotic (large N) result for the static average. The first correction has been calculated by Dünweg et al. [29] to be

$$\frac{D}{D_{\mathrm{m}}} = \frac{1}{N} + \frac{a}{b}\left(\frac{3.685}{\sqrt{N}} - \frac{4.036}{N}\right).$$

Suppose we were to set the ratio of the hydrodynamic radius, a to the Kuhn length, b, to be $a/b = 1/4.036$. This results in the N^{-1} terms canceling out, at least to this degree

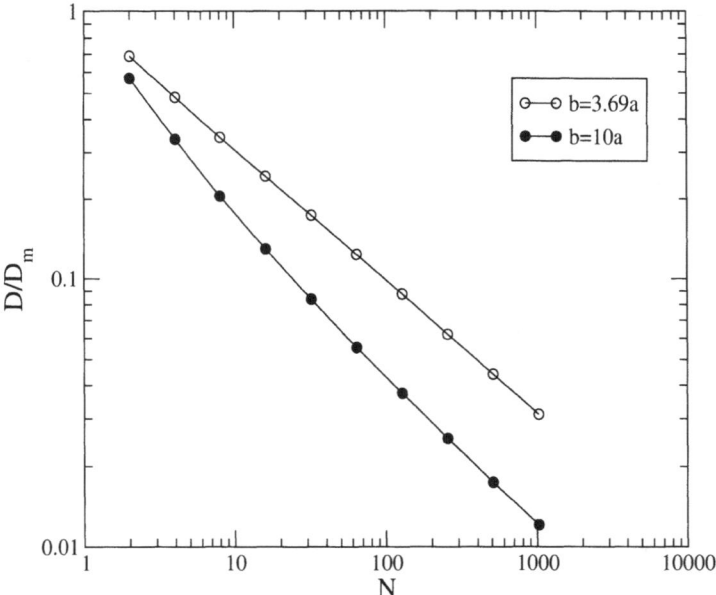

Fig. 7. The diffusion coefficient of an ideal chain relative to the monomer diffusion coefficient D_m as a function of the number of monomers. The different data correspond to different ratios of the Kuhn length b to the monomer hydrodynamic radius a

of approximation, and $N^{-1/2}$ scaling for all N. A second approach is to set a/b to a value $a/b = 1/3.685$ (similar to the value we had before). In this case the asymptotic expression for polymer diffusion coefficient is

$$\frac{D}{D_m} \lim_{N \to \infty} = \frac{1}{\sqrt{N}},$$

that is, the constant of proportionality is unity. We now have a function that by definition starts at unity and decays asymptotically as $1/\sqrt{N}$ so we might expect that in between the asymptotic scaling works pretty well because it extrapolates exactly to the correct $N = 1$ value.

We can check how well this works, simply by evaluating D/D_m by calculating the static average in (12) numerically. This we do by Monte Carlo simulations of the ideal chain. In Fig. 7 we have plotted logarithmically D/D_m calculated in this way for ideal chains with varying numbers of beads. As we can see, with $a/b = 1/3.685$ the function is essentially linear with slope $-1/2$ (corresponding to the asymptotic behaviour). That is, we do, to a very good approximation, get $D/D_m = 1/\sqrt{N}$ for all N. This was what we required. The other value we suggested $a/b = 1/4.036$ gives similar results but for small N does not work quite so well. We thus prefer the former. To emphasize the importance of setting this parameter correctly, we have also plotted the function for $a/b = 1/10$ for which it is obvious that the asymptotic scaling only sets in for very large N and for small N the scaling is essentially the non-hydrodynamic $D/D_m = 1/N$ behaviour.

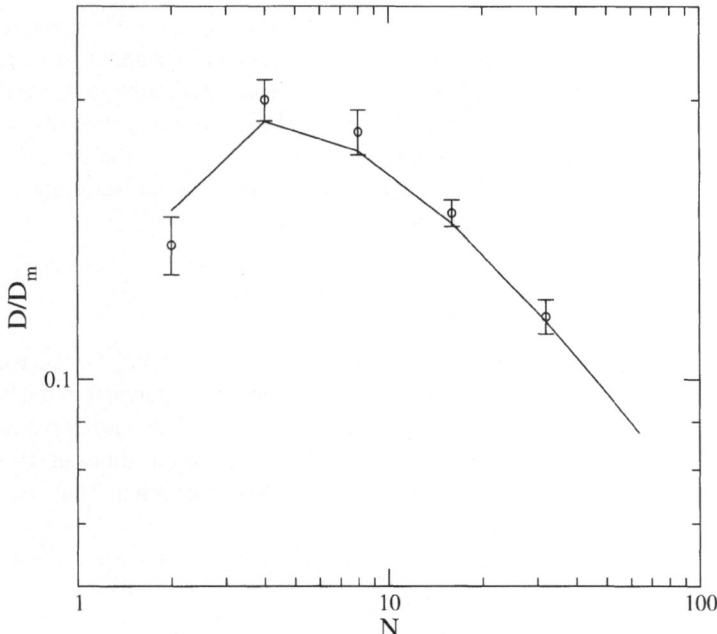

Fig. 8. The hydrodynamic contribution to the diffusion coefficient of an ideal chain relative to the monomer diffusion coefficient D_m as a function of the number of monomers. These results were obtained using the Lowe–Andersen thermostat to model the solvent. The ratio of the Kuhn length to the monomer hydrodynamic radius is the "magic" value of 3.68

In our discussions of setting the relevant parameters for a particle model solvent, we concluded that, to carry out the simulations realistically, there was very little freedom to vary the parameters. Up until now though the Kuhn length b could have been anything. Now, if we have to set this Kuhn length to the "magic" value we no longer have any freedom to vary this parameter. This is because the hydrodynamic radius of the solvent (and hence the hydrodynamic radius of the polymer beads) has a constant value independent of the parameter λ, when λ is very much less than unity as we require. Its value is to a very good approximation $1/5\,r_c$, which is actually about $1/4$ of a typical solvent interparticle separation. This means that b itself must be about one interparticle separation. But in our discussion of what parameters we should use we stated that b should be preferably at least the length of the cut-off radius and minimally a typical interparticle separation. How well can we possibly resolve the hydrodynamic interactions between beads so closely separated in terms of the separation of solvent particles. To test this, we carried out simulations of ideal polymer chains using the magic Kuhn length, all other parameters fixed by values arrived at from the discussion in Sect. 3.4. The solvent was modelled using the Lowe–Andersen thermostat coupled to an ideal gas, again as discussed in Sect. 3.4. To test how well the model reproduces the hydrodynamics we have plotted in Fig. 8 the hydrodynamic contribution to the polymer diffusion coefficient

$$\frac{D_h}{D_m} = \frac{D}{D_m} - \frac{1}{N},$$

that is, the contribution that comes purely from the hydrodynamic interactions. The reason for doing this is that the model will get the none-hydrodynamic $1/N$ contribution correctly by construction. For short chains the hydrodynamic contribution to the diffusion coefficient is relatively small, so just calculating D/D_m is not a very strict test. However, as we see from the figure, the model does a remarkably good job of resolving the hydrodynamic interactions between beads, even though adjacent beads are separated, on average, by just a typical solvent interparticle separation.

4.3   Results for the Test Problem

Putting all this together how do we do for our test problem? In Fig. 9 we have plotted simulation results for the ratio of the radius of gyration for the polymer in the harmonic potential to its unperturbed radius of gyration. As we pointed out earlier we can either carry out the renormalization explicitly or simply measure all quantities in units related to the original long polymer. We have chosen here to do the latter. Thus we define a dimensionless external potential parameter

$$\alpha^* = \alpha \frac{b^2}{N k_\mathrm{B} T} \,,$$

where N is the number of beads in the original polymer. Allowing for the fact that we also need to renormalize the external potential itself what we expect is that if we have successfully renormalized the model the radius of gyration for a given value of $n\alpha^*$ will not depend on n. If we look at Fig. 9 we see that this is indeed the case. Only the $n = 2$ model (dimer) does not superimpose at some point to a function independent of n. This simply tells us that two particles is not enough. For larger numbers of beads in the renormalized model the data is independent of n up to some value of $n\alpha^*$ and the value for which this is the case increases with the number of beads. This just means that more beads are required to model the behaviour of a long polymer in an increasingly strong external potential. We need, in other words, to resolve more detail. Because the $n = 8$ curve is indistinguishable from the $n = 16$ curve up to $n\alpha^* \sim 20$ we can conclude that the $n = 8$ model reproduces the behaviour of a long chain up to this magnitude of external potential. We can also conclude that over this range we have a function independent of n so the curve is the same for any n. We can work out data for the original long polymer simply by substituting $n = N$ and reading off the appropriate value for r_g/r_g^0. For stronger external potentials we cannot be sure that we have data independent of n without simulating longer renormalized chains. This is of course perfectly possible and at least we know with confidence where we exactly know the answer for a long polymer and where we still do not. From the point of view of this static property we have therefore succeeded in solving the problem we set out to solve. What about the dynamics?

In Fig. 10 we have plotted $\sqrt{n}D/D_\mathrm{m}$ as a function of the same dimensionless potential parameter. We see that, as for the radius of gyration, with increasing numbers of particles in the renormalized model the data collapse to a single function. Again, the stronger the potential the more particles we need in the model to reproduce this function, but for the range of $n\alpha^*$ where we do know it we can get the answer for any length of chain (the function does not depend on n so we can substitute $n = N$ to get the answer

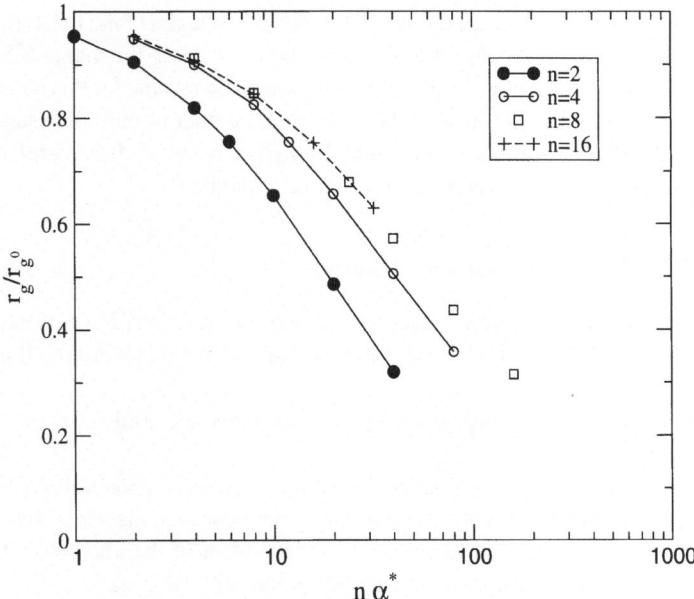

Fig. 9. The radius of gyration R_g, relative to the unperturbed value R_g^0, for a model polymer in an external harmonic potential as a function of the dimensionless potential parameter α^*. The different data correspond to model chains with different numbers of beads n

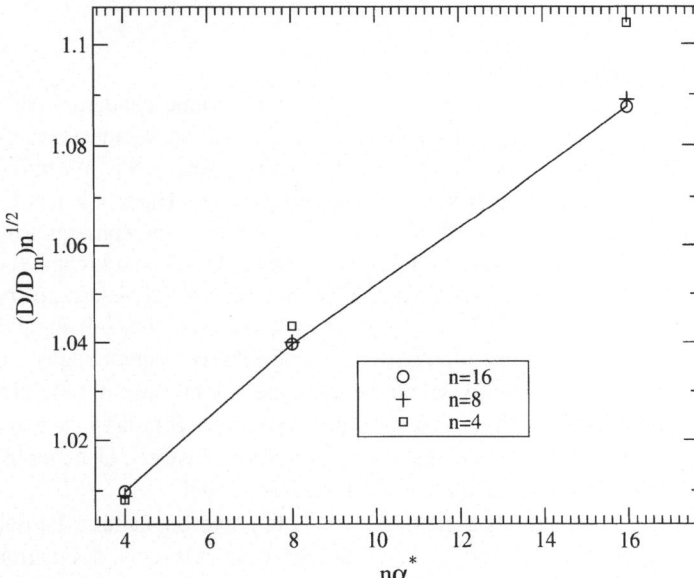

Fig. 10. The diffusion coefficient D, relative to the monomer diffusion coefficient D_m, for a model polymer in an external harmonic potential as a function of the dimensionless potential parameter α^*. The different data correspond to model chains with different numbers of beads n

for any original polymer). Note that this is not a trivial extension of the behaviour we get for the radius of gyration because it depends crucially on using the magic Kuhn length which gives $D \propto 1/\sqrt{n}$ scaling for all n. If we took any other value for this parameter the functions would not (at least for this relatively small number of particles) superimpose and we would still not know how our original polymer behaved. It is therefore by this means that we have successfully renormalized the dynamics.

4.4   What Do We Learn from This Example

- We have at least set out to achieve the goal we set ourselves. For this example we can predict the behaviour of a long polymer chain with a relatively small number of beads.
- This small number of beads cannot be two, this does not resolve the long polymer well enough.
- We can establish how many beads we need for a given situation only by being able to put more information about the original system back (in the sense that we put in more beads). Only by finding behaviour independent of the number of beads can we, with confidence, predict the behaviour of the original system.
- In order to get the correct dynamic behaviour we need to be able to mimic the dynamic behaviour of a long polymer with a short one in the sense that long polymer scaling holds for all numbers of beads in the polymer.

5   General Conclusions

The dynamics of mesoscopic systems are very slow by atomic standards. Thus, in order to capture correctly the macroscopic behaviour in a simulation, we must construct a very simplified model of the system. A lot of detail must be thrown away. We have to decide what we are going to throw out and what we are going to keep. The problem is, if we throw out some information it is difficult to know with confidence the consequences without having a plan B available to put it back in. So, for example, we could ignore the velocity decay time for a colloidal suspension and, if we look at the visco-elastic response, this is a reasonable approximation. However, if we look instead at the collisional dynamics, we find that the velocity decay time plays a crucial role. We cannot neglect it. This we only know by using a model in which we keep the velocity decay time. That is, we had the option to put this information back into our system. Similarly we can throw out hydrodynamics, but again we find that we get the wrong answers. Once more, we need an option for putting this level of detail back into our model.

There are several methods available for including hydrodynamics. Particle models that do not involve a grid or lattice, and that have correct thermal fluctuations, are an attractive option. However, if we use these type of models and we want to do the dynamics realistically, we must pay careful attention to making sure that the transport properties of the model solvent are parametrically sensible for a fluid. It is possible to do so (with the Lowe–Andersen thermostat at least) but one ends up with quite a limited set of values for the parameters describing the solvent which can plausibly be described as realistic. Having done so, we demonstrated that at least the model does a remarkably good job of capturing hydrodynamic interactions, even on length scales the order of a typical solvent particle separation.

For polymeric systems, where the particles making up the mesoscopic system have internal structure, we suggest that a renormalized model (in which the original system is represented by fewer particles) should reproduce the long wave-length behaviour of the structure factor of the original system. However, this cannot possibly be achieved without also renormalizing any external perturbation to which the system is exposed. Renormalizing the dynamics is more problematic. Our approach was to construct a model for which long polymer scaling for the diffusion coefficient held for all numbers of beads in the chain. This approach is not as yet particularly general. It relies on a reasonable understanding of the theoretical background to polymer diffusion. Nonetheless we were able to pose a simple question as to the behaviour of a very long polymer and demonstrate that with a relatively simple renormalized description, using only a handful of particles to describe the chain, we could demonstrably produce results valid for the long polymer. However, to demonstrate that this was the case we again need the ability to put back information we would otherwise have thrown away. In this case it was the degree to which we had simplified the original polymer i.e. the number of beads in the model.

If we had not been able to solve this problem that we set ourselves we would simply have given up and concluded that quantitative simulation of the dynamics of mesoscopic systems is impossible. While we still do not have a general framework for doing so we are at least encouraged by this observation to continue. It is often said that mesoscopic simulation is difficult. It is not. It is easy to cook up some model that is not atomistic and then claim that *de facto* it is a mesoscopic model. The difficult part, we hope to have illustrated, is demonstrating that a mesoscopic model actually bares some quantitative relation to reality.

Acknowledgements

We would like to acknowledge the contribution of Andrew Masters and Daan Frenkel.

References

1. M.P. Allen, D.J. Tildesley: *Computer Simulation of Liquids* (Oxford University Press, Oxford 1987)
2. J.-P. Boon, S. Yip: *Molecular Hydrodynamics* (McGraw Hill, New York 1980)
3. R.L.C. Akkermans, W.J. Briels: J. Chem. Phys. **113**, 6409 (2000)
4. K. Kremer, F. Müller-Plathe: Mol. Sim. **28**, 729 (2002)
5. M. Doi, S.F. Edwards: *The Theory of Polymer Dynamics* (Clarendon, Oxford 1986)
6. B.H. Zimm: J. Chem. Phys. **24**, 269 (1956)
7. H.C. Andersen: J. Chem. Phys. **72**, 2384 (1980)
8. P.E. Rouse: J. Chem. Phys. **21**, 1272 (1953)
9. D.L. Ermak, J.A. McCammon: J. Chem. Phys. **69**, 1352 (1978)
10. C.P. Lowe, A.J. Masters: J. Chem. Phys. **111**, 8708 (1999)
11. J.C. van der Werff, C.B. de Kruif, C. Blom, J. Mellema: Phys. Rev. A**39**, 795 (1989)
12. J.F. Brady: J. Chem. Phys. **99**, 567 (1993)
13. L.D. Landau, E.M. Lifshitz: *Fluid Mechanics* (Pergamon Press, Oxford 1987)

14. D.R. Foss, J.F. Brady: J. Fluid Mech. **407**, 167 (2000)
15. A. Sierou, J.F. Brady: J. Fluid Mech. **448**, 115 (2001)
16. A.J.C. Ladd: J. Fluid. Mech. **271**, 285 (1994)
17. I. Pagonabarraga: Lattice Boltzmann Modeling of Complex Fluids: Colloidal Suspensions and Fluid Mixtures, Lect. Notes Phys. **640**, 275 (2004)
18. P. Ahlrichs, B. Dünweg: Int. J. Mod. Phys C **9**, 1429 (1998)
19. A. Malevanets, R. Kapral: J. Chem. Phys. **112**, 7260 (2000)
20. A. Malevanets, R. Kapral: Mesoscopic Multi-particle Collision Model for Fluid Flow and Molecular Dynamics, Lect. Notes Phys. **640**, 112 (2004)
21. P. Español, P.B. Warren: Europhys. Lett. **30**, 191 (1995)
22. C.P. Lowe: Europhys. Lett. **47**, 145 (1999)
23. I. Pagonabarraga, M.H.J. Hagen, D. Frenkel: Europhys. Lett. **42**, 377 (1998)
24. I. Vattulainen, M. Karttunen, G. Besold, J. M. Polson: J. Chem. Phys. **116**, 3967 (2002)
25. P. Nikunen, M. Karttunen, I. Vattulainen: Comput. Phys. Commun. **153**, 407 (2003)
26. N. A. Spenley: Europhys. Lett. **49**, 534 (2000)
27. A. F. Bakker, C. P. Lowe: J. Chem. Phys. **116**, 5867 (2002)
28. J.-P. Hansen, I. R. McDonald: *The Theory of Simple Liquids* (Academic Press, London 1986)
29. B. Dunweg, D. Reith, M. Steinhauser, K. Kremer: J. Chem. Phys. **117**, 914 (2002)

Statistical Mechanics of Coarse-Graining

Pep Español

Departamento de Física Fundamental, UNED, Apartado 60141, 28080 Madrid, Spain

Abstract. We review the concepts of the theory of coarse-graining and its mathematical background based on the projection operator technique. The objective of the theory is to derive the Fokker–Planck equation that governs the probability distribution of the coarse-grained variables. The essential practical problem of an explicit coarse-graining procedure from the microscopic dynamics, which is the high dimensionality of state space, is pinpointed. This problem enforces one to model the objects appearing in the Fokker–Planck equation. In this case, the program of coarse-graining helps in producing strong conditions on the possible forms of these objects. In particular, we review the stringent GENERIC structure that emerges when the dynamical invariants can be expressed as functions of the coarse-grained variables. Finally, we illustrate how the GENERIC framework may help in the task of inventing new *discrete* models for the simulation of complex fluids that are thermodynamically consistent by construction.

1 Introduction

Complex fluids are fluids which despite their continuum appearance at macroscopic scales present structure at mesoscopic scales. Suspended particles or macromolecules, bubbles, droplets, vesicles, micelles, and lamellae are examples of such structures. Many fluids of technological importance, such as lubricants, paints, surfactant-aided oil recovery fluids, liquid crystals, plastics, shampoos, etc., are complex fluids. What makes these fluids "complex" is the conspicuous coupling between the mesoscopic structures of the fluids with their flow properties. Such highly non-linear coupling defies analytical approaches and it is usually necessary to resort to computer simulations in order to have a better understanding of the physical behaviour of these systems. Of course, there is a great interest from an economic perspective to simulate complex fluid systems of industrial and technological importance. A simulation allows one to gain insight in the processing of materials and propose new directions for design without expensive and time consuming experimentation in a laboratory. However, computer equipment is also a limited resource and there is need for developing simulation models that allow to capture the essential features of the materials with the minimum of computational units and computer time.

The process of representing a system with fewer degrees of freedom than those actually present in the system is called coarse-graining. By coarse-graining, one eliminates the "uninteresting" fast variables and keeps the coarse-grained variables with time scales much larger than typical molecular scales. Therefore, by coarse-graining we not only gain in terms of a reduction of computational units, but also on the possibility of

P. Español, Statistical Mechanics of Coarse-Graining, Lect. Notes Phys. **640**, 69–115 (2004)
http://www.springerlink.com/   © Springer-Verlag Berlin Heidelberg 2004

exploring a much larger time span. In the field of complex fluids, coarse-grained models are usually constructed with a judicious balance of physical intuition, simplicity, and respect of symmetries. The validity of such coarse grained models is inferred a posteriori, by confronting its predictions with experiments. However, molecular specificity is usually lost in these phenomenological approaches. Only by appealing to a microscopic foundation of coarse-graining one can expect to relate a coarse grained model with the underlying molecular system. Of course, several question may spring to the mind: Is there a general method for coarse-graining? Is it always possible to coarse-grain a system? How to ensure thermodynamic consistency in a coarse-grained model? We will try to answer these questions later in this chapter, but for the time being we simply state that a microscopic basis for coarse-graining exists. Actually, the general program for coarse-graining was put on a rigorous basis in middle of last century by people like Kirkwood [1], Green [2], Zwanzig [3], Mori [4], and many, many others. In particular, starting from the microscopic equations governing atomic variables, projection operator techniques allowed to derive the equations of motion of the coarse-grained variables that evolve in distinctly slow time scales from the rest of variables in the system [5]. It is our belief that, despite of its "too formal" appearance, a clear understanding of the theory behind coarse-graining is necessary for anyone willing to simulate the behaviour of complex fluids.

This chapter is structured as follows. Sections 2 and 3 introduce the general concepts of the theory and illustrate them for the case of a colloidal suspension, respectively. The mathematical details of the theory are given in Sect. 4, where the projection operator derivation of the Fokker–Planck equation is presented [3,5]. We return to colloidal suspensions in Sect. 5, where we show how the different objects of the Fokker–Planck equation can be explicitly computed in principle, giving the microscopic expressions for them. In Sect. 6 we present a general strategy for computing the objects in the Fokker–Planck equation by running a molecular dynamics of the system and highlight a basic problem concerning the high dimensionality of the coarse state space. Nevertheless, the coarse-graining theory still provides for a large number of restrictions that help in the modelling of the objects in the Fokker–Planck equation. We review these restrictions in Sect. 7 for the case in which the dynamical invariants of the system can be expressed in terms of the coarse-grained variables. In this case, a powerful thermodynamically consistent structure called GENERIC emerges [6]. In Sect. 8 we illustrate how to use the GENERIC formalism to design discrete models for complex fluids. As particular examples, the Smoothed Particle Hydrodynamics and Dissipative Particle Dynamics models will be presented. Finally, we will suggest how these particle models can be enriched by using additional internal variables in order to simulate viscoelastic fluids, transport of pollutants, and chemically reacting mixtures.

2   The Theory of Coarse-Graining in a Nutshell

In this section we first enumerate the fundamental concepts on which the theory of coarse graining is based on. In the next section these concepts will be illustrated in a specific example of a complex fluid, a colloidal suspension. The mathematical aspects of the theory are deferred to Sect. 4.

A given system of many degrees of freedom, such as a complex fluid, may be described at different levels of description. Each level of description is characterized by a set of relevant variables that specify the state of the system at that level. The word "level" suggests a hierarchical structure and, in fact, the levels of description for a given system order themselves in terms of the amount of information captured by the relevant variables. Less detailed levels (coarser levels) have a smaller number of variables and capture less information. For each level of description there is a dynamic equation for the relevant variables. The evolution of the relevant variables occurs with a time scale characteristic of the level of description. A coarse grained level is valid for describing phenomena that occur at time scales equal or larger than the typical time scale of the level. It cannot reproduce the behaviour at shorter time scales. In any system, among all the possible levels of description two of them are particularly important because they lay at the extremes of the hierarchy. They are the microscopic and macroscopic levels. The microscopic level has the position and momenta of all the atoms of the system as set of relevant variables, the dynamic equations are Hamilton's equations, and the time scale is a typical collision or vibration time. The macroscopic level is the level of Thermodynamics where the relevant variables are the dynamical invariants of the system (mass, momentum, energy). At the macroscopic level, there is no equation of motion, because the relevant variables are constant in time and the time scale is infinite. Any other level of description is in between these two levels and could be named as a mesoscopic level. It is apparent that a given system may have many mesoscopic levels, each one of different detail. The connection between these levels of description is made through a coarse graining procedure in which some degrees of freedom are eliminated from the description. It may happen that there exists a clear separation of the time scales of the relevant variables and the time scales of the eliminated degrees of freedom. When this fortunate case occurs, the coarse-grained level of description has evolution equations for the relevant variables that are Markovian, that is, the future state of the system is determined by the present, but not past, values of the relevant variables. This is fortunate because, from a mathematical point of view, non-Markovian dynamic equations are integro-differential equations which are much more difficult to treat than the usual differential equations that result from a Markovian description. The loss of information that occurs when going from one level of description to a more coarser one is reflected on a stochastic description of coarse-grained levels. The idea is that even if the description is Markovian, the knowledge of the coarse state of the system at a given time is not sufficient to predict the future coarse state with certainty. This is because many microscopic states are compatible with the initial coarse state. The uncertainty in the actual initial microscopic state amounts to an uncertainty to the final coarse state. These uncertainties are ultimately responsible for the need of a statistical description based on the probability distribution function for the relevant variables. The dynamic equation is therefore an equation for the probability distribution function and has, under the Markovian approximation, the form of a (FPE). When the evolution equation is of the Fokker–Planck type, the eliminated degrees of freedom show up in a systematic dissipative effect on the relevant variables and also in the form of thermal fluctuations. Fluctuations and dissipation have the same essential origin, the elimination of degrees of freedom. They are related through the fluctuation-dissipation theorem.

The main objective of the theory of coarse-graining is to derive the FPE that governs the probability distribution of the coarse-grained variables. This was achieved in a landmark paper by Green [2], and the FPE was subsequently re-derived by using a projection operator technique by Zwanzig [3]. In this way, explicit molecular expressions for the coefficients appearing in the FPE were obtained in terms of the microscopic dynamics. The technique can be generalized for jumping not only from the microscopic level to a mesoscopic level but also from any mesoscopic level to any other more coarse-grained mesoscopic level [7].

The FPE can only be obtained under the assumption of clear separation of time scales between the selected variables and the rest of variables of the system. In particular, the coarse-grained variables should be slow. But, how do we know if such separation of time scale exists? The sad answer is that we do not have any a priori way to know in advance the time scales of evolution of our selected relevant variables. A large degree of physical intuition is required in order to select a proper set of relevant variables. If the theory does not predict correctly the results of experiments or simulations, we should think what other variables might evolve in the time scale of the description and include them also as relevant variables with the hope that we now capture all the slow dynamics of the system.

Before closing this introductory section, it is worth to compare the coarse-graining strategy based on projection operators with the older and venerable approach of kinetic theory [8]. Kinetic theory is also a theory of coarse graining in which the reduction of information is achieved by considering the probability distribution functions of small subsets of particles. Starting from the Liouville equation, one constructs a hierarchy of equations (the BBGKY hierarchy) for the probability distribution functions at a given time in which the equation for a given n-particle distribution depends on higher particle distributions. Therefore, the equation for the lowest distribution, although being a differential equation involving only the present time, is not closed. This is in contrast with the projection operator approach in which one obtains a closed equation for the distribution function, which, however, involves past times. Of course, one has to be as careful in truncating the kinetic hierarchy in order to obtain a closed equation as in approximating the memory equation in the projection operator approach in order to have a memoryless equation.

3 Example: A Colloidal Suspension

In this section we will illustrate the fundamental concepts of the theory by studying several different levels of description that are used to describe colloidal suspensions. A colloidal suspension is made of a collection of small solid objects (like spherical latex particles) of the size of, say, a micron suspended in a fluid such as, for example, water. A roadmap of this section is shown in Fig. 1.

3.1 Microscopic Level: Classical Mechanics

At the most microscopic level, we can model a colloidal suspension by assuming that the solid suspended objects are spherical and we need only 6 degrees of freedom for

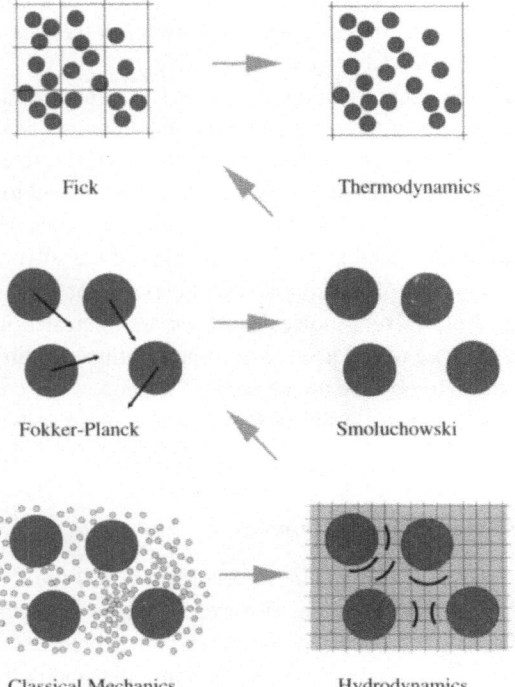

Fick   Thermodynamics

Fokker-Planck   Smoluchowski

Classical Mechanics   Hydrodynamics

Fig. 1. Different levels of description in a colloidal suspension. Arrows denote the direction of coarse graining from the Classical Mechanics level at the lower left hand corner to Thermodynamics level at the top right hand corner

describing the state of the object, the position $\mathbf{Q}_i$ and the momentum $\mathbf{P}_i$ of its center of mass. For irregular objects we would need also to consider orientation, angular velocities, etc. The fluid in which these solid colloidal particles are suspended will be described at the most microscopic level by the positions $\mathbf{q}_i$ and momenta $\mathbf{p}_i$ of the centers of mass of the molecules constituting the fluid. Again, we assume spherical molecules for simplicity. The microscopic state will be denoted by $z = \{\mathbf{q}_i, \mathbf{p}_i, \mathbf{Q}_i, \mathbf{P}_i\}$. The evolution of the microstate is governed by Hamilton's equations,

$$\dot{\mathbf{q}}_i = \frac{\partial H(z)}{\partial \mathbf{p}_i}, \qquad \dot{\mathbf{Q}}_i = \frac{\partial H(z)}{\partial \mathbf{P}_i},$$

$$\dot{\mathbf{p}}_i = -\frac{\partial H(z)}{\partial \mathbf{q}_i}, \qquad \dot{\mathbf{P}}_i = -\frac{\partial H(z)}{\partial \mathbf{Q}_i}, \tag{1}$$

where the Hamiltonian is given by

$$H(z) = \sum_i \left(\frac{p_i^2}{2m_i} + \frac{P_i^2}{2M_i} \right) + \frac{1}{2} \sum_{ij} \left(V_{ij}^{\mathrm{SS}}(q) + V_{ij}^{\mathrm{SC}}(q, Q) + V_{ij}^{\mathrm{CC}}(Q) \right). \tag{2}$$

Here, m_i is the mass of a solvent molecule, M_i the mass of a colloidal particle, and V^{SS}, V^{SC}, V^{CC} are the potentials of the forces between solvent molecules, solvent and colloidal particles, and colloidal particles, respectively.

In principle, the differential equations (1) can be solved numerically with a computer. The technique is known as molecular dynamics and allows us to keep track of all the microscopic dynamics of the system [9]. The smallest typical time scale is a collision time in the range of picoseconds and, consistently, we will need to use a time step for the numerical solution which is much smaller than this time scale. However, if the mass of the colloidal particles is much larger than the mass of the solvent particles, as it occurs in reality, the evolution of the colloidal particles will be very slow in comparison with the evolution of the solvent molecules. If we are interested in the motion of the colloidal particles, then we would need an enormous number of time steps (and, therefore, of computer time) to observe an appreciable motion of the colloidal particles. To study these large time scales in a colloidal suspension, molecular dynamics is absolutely impracticable.

3.2   Mesoscopic Level 1: Hydrodynamics

If we look at the motion of the solvent molecules, we will see that they collide with each other resulting in a rapid motion. However, if we look "from a distant point" to the multitude of molecules, a collective motion will be appreciated in which molecules in a region of space move coherently (overwhelming the small erratic motions due to collision). It will be possible to appreciate slowly evolving waves, vortices and other sort of collective motion. The variables that capture these collective motions are the hydrodynamic variables. These variables are the mass density field $\rho_\mathbf{r}(z)$, the momentum density field $\mathbf{g}_\mathbf{r}(z)$, and the energy density field $e_\mathbf{r}(z)$, defined by

$$\rho_\mathbf{r}(z) = \sum_i m \bar{\delta}(\mathbf{r} - \mathbf{q}_i),$$

$$\mathbf{g}_\mathbf{r}(z) = \sum_i \mathbf{p}_i \bar{\delta}(\mathbf{r} - \mathbf{q}_i),$$

$$e_\mathbf{r}(z) = \sum_i e_i \bar{\delta}(\mathbf{r} - \mathbf{q}_i), \tag{3}$$

where $\bar{\delta}(\mathbf{r} - \mathbf{q}_i)$ is a coarse-grained delta function (a function with a support over a finite small region and normalized to unity, see Fig. 2). In the above equations, e_i is the energy of particle i (the sum of its kinetic energy plus half the potential energy due to the interaction with its neighbours). It may appear as a contradiction to coarse-grain through a set of field variables (which have, in principle an infinite number of degrees of freedom). However, we should note that the above fields involving the coarse-grained delta functions are "smooth" fields (which have a small number of Fourier components with large wavelengths). Implicit in the definition of the hydrodynamic variables in (3) there is a partition of physical space in little cells that contain many solvent molecules. The hydrodynamic variables tell us about how many particles have a given average velocity and energy in a certain region of space, rather than providing the exact location and velocity of each solvent molecule. The reduction of information in passing from

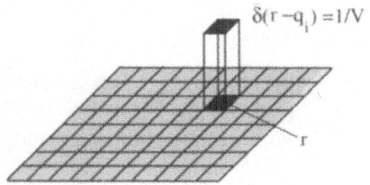

Fig. 2. Coarse-grained delta function $\bar{\delta}(\mathbf{r} - \mathbf{q}_i)$ takes the value $1/V$, V being the volume of the cell, if $\mathbf{q}_i$ is within the cell labeled with r, and zero otherwise. The "point" r actually represents a portion of space

solvent atoms to smooth hydrodynamic fields is then apparent. Therefore, at a coarse-grained level we can describe the colloidal suspension by the set of hydrodynamic fields and the position and momenta of the colloidal particles.

The equations that govern the hydrodynamic variables are the well-known hydrodynamic equations [10]. They include the microscopic details of the solvent (what particular molecules constitute the solvent) through the pressure and temperature equations of state, and through the particular values of the transport coefficients (shear and bulk viscosities, and thermal conductivity). The time scales of the hydrodynamic variables depend on the length scale being considered. Typically, perturbations with large wavelengths evolve slowly. In a colloidal suspension these length scales are either the radius of the particles or the typical distance between colloidal particles. Roughly speaking, there is a hydrodynamic time scale for mass evolution which is a sound traversal time, a time scale for momentum evolution given by a a viscous time, and a time scale for energy transport, given by a thermal diffusion time. The hydrodynamic equations for the solvent fields are coupled with the Newton's equation for the colloidal particles through boundary conditions, whereas the forces on the colloidal particles are given through the integral of the stress tensor of the fluid over the surface of the colloidal particle. We observe that the motion of a given colloidal particle depends on the motion of the rest through the coupling with the hydrodynamic fields of the solvent. These solvent mediated interactions between colloidal particles are known as hydrodynamic interactions.

As a result of the coarse-graining of the solvent, the hydrodynamic variables are subject to fluctuations. Physically, these fluctuations came from the fact that solvent molecules can enter and go out of the little cells that are used to defined the hydrodynamic variables. Mathematically, the fluctuations are described as stochastic contributions to the stress tensor and heat flux in the Navier–Stokes equations, as proposed by Landau and Lifshitz in their theory of fluctuating hydrodynamics [11]. A very remarkable property of these fluctuations is that their variance is proportional to the transport coefficients of the solvent, this is, to the shear and bulk viscosities and to the thermal conductivity. This is the fluctuation-dissipation theorem for this level of description. The hydrodynamic fluctuations are transmitted to the colloidal particles through the boundary conditions on the surface of the particles and are ultimately responsible for the Brownian motion of the particles [12]. In the highly dilute limit (that is, for an isolated colloidal particle) and for neutral buoyant particles (with density equal to the solvent density) it is possible to compute explicitly the velocity autocorrelation of the colloidal particle [13], which exhibits the famous long-time tail, an algebraic decay as $t^{-3/2}$ [14]. The physical origin of this slow algebraic decay is that, due to a spontaneous fluctuation, the colloidal particle pushes the fluid in front of it. The push generates a vortex centered on the particle that, in turn, pushes the particle from its rear, thus maintaining the initial state of motion of the particle. These long-time tails have been observed experimentally for colloidal systems [15].

There is a large number of simulation techniques for the hydrodynamic level of description. In general, any Navier–Stokes solver for the solvent that allows to describe thermal fluctuations and Molecular Dynamics for the colloidal particles will do. Lattice Boltzmann [16,17] or Dissipative Particle Dynamics [18] have been used for this purpose. The Malevanets–Kapral method can also be used in this problem [19,20].

3.3 Mesoscopic Level 2: Fokker–Planck

We have already noted that if the colloidal particles are massive their motion will be slow. In that case, we could expect that the propagation of the hydrodynamic interactions is very fast, instantaneous, compared with the time scale of variation of the colloidal position and momenta. If this is the case, we can eliminate the solvent hydrodynamic variables from the description. The state of the system at this mesoscopic level of description is given by $x = \{\mathbf{Q}_i, \mathbf{P}_i\}$. The FPE that corresponds to the mesoscopic level 2 is given by [22]

$$\frac{\partial}{\partial t} P(x,t) = -\sum_i \left[\mathbf{V}_i \cdot \frac{\partial}{\partial \mathbf{Q}_i} + \mathbf{F}_i^{\mathrm{CC}} \cdot \frac{\partial}{\partial \mathbf{P}_i} \right] P(x,t)$$

$$+ k_{\mathrm{B}} T \sum_{ij} \frac{\partial}{\partial \mathbf{P}_i} \cdot \boldsymbol{\zeta}_{ij}(Q) \cdot \left[\frac{\partial}{\partial \mathbf{P}_j} + \frac{\mathbf{P}_i}{M_i k_{\mathrm{B}} T} \right] P(x,t). \tag{4}$$

Here, $\mathbf{V}_i = \mathbf{P}_i/M_i$, $\mathbf{F}_i^{\mathrm{CC}}$ is the effective force due to the rest of colloidal particles exerted on particle i and $\boldsymbol{\zeta}_{ij}(Q)$ is a friction tensor which depends on the position of the colloidal particles. The physical picture behind (4) can be best appreciated in the mathematically equivalent stochastic differential equations (SDE)

$$d\mathbf{Q}_i = \mathbf{V}_i dt \quad \text{and} \quad d\mathbf{P}_i = \mathbf{F}_i^{\mathrm{CC}} dt - \sum_j \boldsymbol{\zeta}_{ij}(Q) \cdot \mathbf{V}_j dt + d\tilde{\mathbf{F}}_i.$$

We observe that the particles evolve according to their velocities and that they are subjected to forces due to the other colloidal particles that depend on their positions, $\mathbf{F}_i^{\mathrm{CC}}$, and velocities, $-\boldsymbol{\zeta}_{ij}(Q) \cdot \mathbf{V}_j$. Note that if a colloidal particle j is moving, it will exert forces on the colloidal particle i. These forces are the result of the hydrodynamic interactions that are captured at this level of description through the friction tensor $\boldsymbol{\zeta}_{ij}(Q)$. Finally, the particles are also subject to stochastic forces $d\tilde{\mathbf{F}}_i$ that are mathematically described in terms of Wiener processes. The variance of these forces is given by the Fluctuation-Dissipation theorem which, at this level of description, takes the form $d\tilde{\mathbf{F}}_i d\tilde{\mathbf{F}}_j = 2k_{\mathrm{B}} T \boldsymbol{\zeta}_{ij}(Q) dt$.

If the colloidal particles are very far from each other, as it happens when the suspension is dilute, we may expect that the mutual influence among colloidal particles is negligible and that the friction tensor is diagonal, this is $\zeta_{ij} = \delta_{ij}\mathbf{1}\zeta$, where ζ is called the friction coefficient. In this case, the SDE equivalent to the FPE (4) decouple into a set of independent equations, called Langevin equations. The Langevin equations for a dilute suspension predict that the velocity autocorrelation function of a colloidal particle decays exponentially. As we have seen, this is at variance with experiments that show a clear long-time tail for neutrally buoyant particles. The reason for this discrepancy between theory and experiments for neutrally buoyant particles can only be attributed to the fact that the time scales of evolution of the colloidal variables $\mathbf{Q}_i, \mathbf{P}_i$ are not neatly separated from the time scales of hydrodynamic variables. We have here one example of the hazards of selecting a too low number of variables. Even though $\mathbf{Q}_i, \mathbf{P}_i$ look as a reasonable set of relevant variables, they are not. If we are interested in the time scale in which the momentum of the particles evolves, we have to include, necessarily, the hydrodynamic modes as a part of the description. This time scale is typically $10^{-6}s$. It turns out, however, that the density ratio between the colloidal particle density and the solvent density governs the separation of time scales between colloidal momenta and hydrodynamic variables [12]. For colloidal suspensions of particles which are much denser than the solvent, these time scales are well-separated and the Fokker–Planck level of description is a good level of description. Finally, we note that the numerical solution of the SDE (5) is called Brownian Dynamics [9,21].

3.4   Mesoscopic Level 3: Smoluchowski

Even though the coarse-grained variables $\mathbf{Q}_i, \mathbf{P}_i$ are not a good set of coarse variables for neutrally buoyant suspensions, it turns out that the $\mathbf{Q}_i$ alone do actually define a proper level of description. The reason is that the time scale of evolution of the positions of the colloidal particles is much longer than the time scale of evolution of the momentum and hydrodynamic modes. A typical time scale of evolution of the position variables for a colloidal suspension is 10^{-3} s. Therefore, by using $\mathbf{Q}_i$ alone, we will be able to describe correctly the phenomena that occur in time scales above 10^{-3} s.

The FPE that now governs the probability density $P(\mathbf{Q}, t)$ is called the Smoluchowski equation. It has the form [22]

$$\frac{\partial}{\partial t}P(\mathbf{Q}, t) = -\sum_i \frac{\partial}{\partial \mathbf{Q}_i} \cdot \left[\mathbf{D}_{ij} \cdot \frac{\mathbf{F}_j^{\mathrm{CC}}}{k_{\mathrm{B}}T} P(\mathbf{Q}, t) \right] + \sum_{ij} \frac{\partial}{\partial \mathbf{Q}_i} \mathbf{D}_{ij} \frac{\partial}{\partial \mathbf{Q}_j} P(\mathbf{Q}, t), \quad (5)$$

where D_{ij} is called the diffusion tensor, and it captures the mutual influence between colloidal particles due to the elimination of the solvent and momentum variables. The mathematically equivalent set of SDE corresponding to (5) is

$$d\mathbf{Q}_i = \sum_j \boldsymbol{D}_{ij}(Q)\mathbf{F}_i^{\mathrm{CC}}(Q)\mathrm{d}t + k_{\mathrm{B}}T \sum_j \frac{\partial}{\partial Q_j} \boldsymbol{D}_{ij}(Q)\mathrm{d}t + d\tilde{\mathbf{Q}}_i,$$

and the Fluctuation-Dissipation theorem at this level of description takes the form $d\tilde{\mathbf{Q}}_i d\tilde{\mathbf{Q}}_j = 2k_{\mathrm{B}}\boldsymbol{D}_{ij}(Q)\mathrm{d}t$. Again, for dilute suspensions $\mathbf{D}_{ij} = \delta_{ij}\mathbf{1}D$, where D is the self-diffusion coefficient of the colloidal particles.

3.5   Mesoscopic Level 4: Fick

Further coarse graining can be performed if we are interested not in the actual positions $\mathbf{Q}_i$ of the colloidal particles but on the number of colloidal particles that are in a region of the space located around $\mathbf{r}$. This amounts to the introduction of the number of colloidal particles per unit volume and to the use of a concentration field $c(\mathbf{r})$ as the variables x describing the state of the system. The concentration variable contains a much reduced amount of information because it tells us how many particles are in a region, but not which one is exactly where. We expect that the concentration field evolves in a time scale much larger than the time scale of evolution of the positions of the colloidal particles because in order to have an appreciable change in the number of particles in a given sufficiently large region, these particles must move (diffuse) for a long time in order to traverse this region.

Due to the field character of the concentration variable, we have now a probability functional $P[c(\mathbf{r}), t]$ which will obey a functional FPE. The mathematically equivalent SDE takes the form of a continuity equation

$$\partial_t c = -\nabla \mathbf{J} \qquad \text{with} \qquad \mathbf{J} = -\frac{c}{\zeta}\nabla\mu + \tilde{\mathbf{J}}, \tag{6}$$

where the mass flux $\mathbf{J}$ has a systematic contribution proportional to the chemical potential of the colloidal particles plus a stochastic part $\tilde{\mathbf{J}}$ with a variance proportional to the transport coefficient c/ζ. For the sake of simplicity, we have assumed that the suspension is dilute, in such a way that hydrodynamic interactions can be neglected. Otherwise, one obtains non-local in space equations [7]. When the suspension is dilute, we may use the ideal gas expression for the chemical potential $\mu = k_B T \ln c$ that leads to the usual diffusion equation $\partial_t c = D\nabla^2 c - \nabla \tilde{\mathbf{J}}$ where $D = k_B T/\zeta$ is the Einstein expression for the diffusion coefficient. Equation (6) can be easily simulated by discretizing the resulting stochastic diffusion equation with finite differences or any other discretization technique for stochastic partial differential equations [23].

3.6   Macroscopic Level: Thermodynamics

Finally, we might be interested in very long time scales in which the system has arrived at equilibrium. In this case, the only relevant variables are the dynamical invariants, i.e., those coarse-grained variables that are independent of time due to particular symmetries of the microscopic Hamiltonian, like total energy, or those coarse-grained variables like mass and volume, that are constant parameters in the Hamiltonian. Note that the volume of the container of the colloidal suspension can be understood as a parameter of a confining potential in the Hamiltonian. Obviously, there is no equation of motion for this level of description because we are interested in the long time, equilibrium state of the system.

In summary, in this section we have illustrated several different levels of description that can be defined for the study of a colloidal suspension. The general theme is that every level captures less information than its predecessor and allows one to describe time scales at and above the typical time scale of evolution of the coarse-grained variables.

The levels of description presented above for a colloidal suspension do not exhaust the possibilities and they have been selected as providing a pedagogical presentation of the main ideas involved in coarse-graining. Of course, other levels of intermediate complexity can and must be formulated depending on the physical situation that one wants to describe. For example, one can define a level of description where the relevant variables are the hydrodynamic solvent variables plus the concentration field for the colloidal particles. This purely hydrodynamic level of description is particularly useful for the description of transport of dilute pollutants by fluid flows. In this case, the equations are the hydrodynamic equations (that contain the effect of the dilute suspension through an osmotic pressure term) plus an advection-diffusion equation for the concentration field. If the colloidal suspension is more concentrated and anisotropic effects due to flow start to play a role, then it is necessary to enrich the level of description by including a vectorial or tensorial quantity representing the degree of local anisotropy in the system [24].

4   The Mathematics of the Theory of Coarse-Graining

In the previous section we have presented the evolution equations for the relevant variables of a colloidal suspension at different levels of description without explaining the origin of these equations. These evolution equations are Fokker–Planck equations or its equivalent stochastic differential equations. Actually, the objective of a theory for coarse-graining is to provide for the particular form of the different objects that appear in the FPE corresponding to a particular level of description. More precisely, one would like to relate the drift and diffusion terms of the FPE with molecular details or with the drift and diffusion of a more detailed mesoscopic level. In this section, we will present a method based on a projection operator that allows one to derive the FPE at a given level of description from the Liouville equation that governs the microscopic level. In this procedure, one obtains explicit expressions for the objects that appear in the FPE. It is also possible to relate different mesoscopic levels, obtaining expressions for the coefficients of the FPE at a mesoscopic level in terms of the dynamics at a more detailed but still mesoscopic level [7]. However, and for the sake of simplicity, we will consider here only the coarse graining from the most detailed, microscopic level.

Apparently, there are many different projection operators [5] and one can be bewildered about which one is "the correct one". Here we select the Zwanzig projection because: (1) it leads to non-linear dynamic equations describing arbitrarily far from equilibrium situations (the Mori projection leads to linear equations valid near equilibrium and it is a particular case of the Zwanzig projection) (2) it allows to obtain the FPE which captures all the statistics of the stochastic process. Projection onto relevant variables instead of onto distribution of relevant variables leads to an incomplete description of the stochastic process [25], and (3) it is time-independent, which is much simple to handle than time-dependent projections [5].

The scope of this chapter has some restrictions. In particular, we will consider only isolated systems. We will assume also that at the most microscopic level the system can be well described by classical mechanics or, to be more specific, by Hamilton's equations. No reference to quantum mechanics will be made. Another crucial assumption is that the Hamiltonian dynamics of the system has a well-defined equilibrium state that is reached by the system as the time proceeds. The assumption of isolated system implies that we will look at the relaxational dynamics of the system towards its equilibrium state. This might seem a strong restriction from an experimental point of view. Experiments often deal with situations in which a system is subject to the action of "external influences", usually through the boundary of the system. Nevertheless, the theory for isolated systems already provides the basic model equations to which boundary conditions can be applied in a latter stage.

4.1 The Microscopic Level

All the complex fluid systems we are interested in can be described with classical mechanics. For this reason, we stop at reviewing the basic concepts of classical mechanics for a Hamiltonian system. Let us denote the microstate of the system by $z = \{\mathbf{q}_i, \mathbf{p}_i\}$. The set of all z constitutes the phase space Γ of the system. Hamilton's equations are

$$\dot{\mathbf{q}}_i = \frac{\partial H(z)}{\partial \mathbf{p}_i}, \qquad \dot{z} = L_0 \frac{\partial H(z)}{\partial z}, \qquad (7)$$
$$\dot{\mathbf{p}}_i = -\frac{\partial H(z)}{\partial \mathbf{q}_i},$$

where the right hand side is a condensed way of writing Hamilton's equations, with the matrix L_0 having a block diagonal matrix form with the blocks given by

$$\begin{pmatrix} 0 & 1 \\ -1 & 0 \end{pmatrix}. \qquad (8)$$

Note that Hamilton's equations are first order differential equations that require the knowledge of an initial condition. Given an initial microstate z_0, the solution of Hamilton's equations can be denoted by $z(z_0, t) = T_t z_0$, where we have introduced formally the time evolution operator T_t. This operator satisfies $T_0 = 1$ and $T_t T_{t'} = T_{t+t'}$. In addition, the coordinate transformation $z \to T_t z$ has unit Jacobian for all t (this is the theorem of the integral invariants of Poincaré [26]).

4.2 Liouville Theorem

Even though the evolution of $z(t)$ is deterministic, it is usually difficult to know with precision the initial state z_0 of the system. All what we know is the probability distribution function of the initial state $\rho_0(z)$. The probability distribution function at a subsequent time is denoted by $\rho_t(z)$ and it obeys the Liouville equation which we derive here for completeness. Let M be a region of not vanishing measure of Γ and $T_t M$ the region resulting from the evolution of each point of M according to Hamilton's equations. It is obvious that the probability that the system is in the region M at $t = 0$ is identical to the probability of being at $T_t M$ at $t = t$. For this reason,

$$\int_M \rho(z, 0) \mathrm{d}z = \int_{T_t M} \rho(z, t) \mathrm{d}z. \qquad (9)$$

By performing the change of variables $z' = T_{-t} z$ (with unit Jacobian) the integral in the left hand side becomes

$$\int_{T_t M} \rho(z, t) \mathrm{d}z = \int_M \rho(T_t z, t) \mathrm{d}z \qquad (10)$$

This is true for any region M and, therefore, the integrand of the right hand side of (9) and the left hand side of (10) should be equal, i.e.

$$\rho(z,0) = \rho(T_t z, t),\tag{11}$$

or, by a simple change of variables,

$$\rho(z,t) = \rho(T_{-t} z, 0).\tag{12}$$

By taking the time derivative on both sides of (11) we obtain the Liouville theorem,

$$\frac{d}{dt}\rho(T_t z, t) = \frac{d}{dt}\rho(z,0) = 0.\tag{13}$$

Further application of the chain rule leads to the Liouville equation for the probability density in phase space,

$$\frac{\partial}{\partial t}\rho(z,t) = -iL\rho(z,t),\tag{14}$$

where the Liouville operator is defined by

$$iL \equiv \sum_i \left(\frac{\partial H}{\partial \mathbf{p}_i} \frac{\partial}{\partial \mathbf{q}_i} - \frac{\partial H}{\partial \mathbf{q}_i} \frac{\partial}{\partial \mathbf{p}_i} \right) = \frac{\partial H}{\partial z} L_0 \frac{\partial}{\partial z}.\tag{15}$$

This operator satisfies

$$\int dz \rho^{eq}(z) A(z) iL B(z) = -\int dz \rho^{eq}(z) B(z) iL A(z),\tag{16}$$

for arbitrary functions $A(z), B(z)$.

4.3 Equilibrium at the Microscopic Level

Let us consider now the final state predicted by the Liouville equation. A basic mathematical question that has not been solved in its full generality is under which conditions the Liouville equation (14), which is a first order partial differential equation, leads to a stationary solution with $\partial_t \rho(z,t) = 0$. That this is not generally the case can be seen by considering a delta like initial distribution $\rho(z,0) = \delta(z - z_0)$, where we know exactly the initial state z_0. In this case, we know that the solution is given by $\rho(z,t) = \delta(T_t z_0 - z)$, this is, the distribution function remains peaked at the solution of Hamilton's equations. There is no broadening of the distribution function and the system does not reach a stationary state. However, if the dynamics generated by the Hamiltonian is highly unstable (i.e. chaotic), we may expect that any non-delta initial distribution will evolve with a sort of broadening. To be more specific, if the dynamics of the system is of the mixing type, then the system reaches an effective stationary state [27]. We will assume, without rigorous proof, that the most general stationary distribution function of (14) is a function of the dynamical invariants of the system. A dynamical invariant $I(z)$ is a dynamical function that does not change in time, this is,

$$\frac{d}{dt}I(z(t)) = 0.$$

The Hamiltonian $H(z)$ is a dynamical invariant (and we say that energy is conserved). As it is well-known, if the Hamiltonian is translationally and rotationally invariant, the total momentum and angular momentum are also dynamical invariants. The number of particles of the system and the parameters of the Hamiltonian (the mass of the particles, the range of potential, etc.) are also trivial dynamical invariants.

Any distribution function $\rho(z)$ which is a function $g(I(z))$ will be, therefore, a stationary solution of the Liouville equation and, as stated, we will assume that any stationary solution is of this type. This stationary distribution is called the equilibrium ensemble $\rho^{\text{eq}}(z)$. Therefore,

$$\lim_{t \to \infty} \rho(z, t) = \rho^{\text{eq}}(z) = g(I(z)). \tag{17}$$

Let us investigate the meaning of the function $g(I)$ by considering the probability distribution $P^{\text{eq}}(I)$ of dynamical invariants at equilibrium. By definition,

$$P^{\text{eq}}(I) = \int dz \rho^{\text{eq}}(z)\delta(I(z) - I) = \int dz g(I(z))\delta(I(z) - I) = g(I)\Omega^{\text{eq}}(I), \tag{18}$$

where we have introduced the measure $\Omega^{\text{eq}}(I)$ of the region of phase space compatible with a given set of dynamical invariants

$$\Omega^{\text{eq}}(I) = \int dz \delta(I(z) - I). \tag{19}$$

Equation (18) allows to identify $g(I)$ and (17) becomes

$$\rho^{\text{eq}}(z) = \frac{P^{\text{eq}}(I(z))}{\Omega^{\text{eq}}(I(z))}. \tag{20}$$

Therefore, the equilibrium ensemble is fully determined by the probability distribution of dynamical invariants at equilibrium. It is obvious that the distribution of dynamical invariants at any time is itself invariant (use $\rho(z, t)$ instead of $\rho^{\text{eq}}(z)$ in (18), take the time derivative and use the Liouville equation). This means that the equilibrium ensemble is fully determined once the initial distribution of dynamical invariants is known at the initial time. As a particular example of (20), let us assume that at the initial time we know with absolute precision the values of the invariants I_0. In this case, the equilibrium probability density (20) becomes the microcanonical ensemble

$$\rho^{\text{mic}}(z) = \frac{\delta(I(z) - I_0)}{\Omega^{\text{eq}}(I_0)}. \tag{21}$$

4.4 The Mesoscopic Level

At a given mesoscopic level, the state of the system is described by certain set of dynamical functions $X(z)$ which may take numerical values x. These functions are actually the coarse-grained variables or relevant variables that define the level of description. For the case of a colloidal suspension, the above functions can be the hydrodynamic fields, or the positions and momenta of the colloidal particles, or the concentration field, depending on the level of description selected. The coarse grained variables evolve in time due to the implicit evolution of z. Actually, we can compute the time derivative of $X(T_t z_0)$ through the chain rule

$$\frac{\mathrm{d}}{\mathrm{d}t} X(T_t z_0) = (\mathrm{i}LX)(T_t z_0). \tag{22}$$

A formal solution of this equation is

$$X(T_t z_0) = (\exp\{\mathrm{i}Lt\}X)(z_0), \tag{23}$$

where the exponential operator is defined formally through the Taylor series

$$\exp\{\mathrm{i}Lt\} \equiv 1 + \mathrm{i}Lt + \frac{1}{2!}(\mathrm{i}Lt)^2 + \frac{1}{3!}(\mathrm{i}Lt)^3 + \cdots . \tag{24}$$

When $X(z) = z$, we see that the evolution operator has the form $T_t = \exp\{\mathrm{i}Lt\}$.

We discuss now the essential source of stochasticity that appears in any coarse-graining process. Let us assume that we know the initial numerical values of the coarse grained variables x_0 with precision and consider two different microstates z_0, z_0' that are both compatible with x_0, this is, $X(z_0) = X(z_0') = x_0$. As the initial microstates evolve, the functions $X(T_t z_0), X(T_t z_0')$ will take different numerical values in general, i.e. $X(T_t z_0) \neq X(T_t z_0')$. Therefore, even if we know with certainty the actual initial state x_0 different outcomes are possible for the value of the coarse-grained variables at later times, just because there is no control over the actual microscopic state that produces the given macroscopic initial state x_0. In order to describe our ignorance about the values of the coarse-grained variables at later times, we will use a distribution function $P(x, t)$. Of course, the distribution function at the mesoscopic level is related to the distribution function at the microscopic level through

$$P(x, t) = \int \delta(X(z) - x)\rho(z, t)\mathrm{d}z. \tag{25}$$

Note that (25) is the general way of relating the probability distribution of a function $X(z)$ of a stochastic variable with the probability distribution of the stochastic variable z itself. By using the solution (12) and a change of variables we can rewrite (25) also as

$$P(x, t) = \int \delta(X(T_t z) - x)\rho(z, 0)\mathrm{d}z, \tag{26}$$

where it is made apparent that all the stochasticity at a mesoscopic level of description comes essentially from our ignorance about the initial microscopic state, represented by the distribution function $\rho(z, 0)$.

A basic question that arises now is, what is the actual functional form of $\rho(z, 0)$? As mentioned, in principle we cannot measure the initial microscopic state z_0 exactly. If we are going to describe a system at a given coarse-grained level, we must assume that we have access to the measurement of the coarse-grained variables $X(z)$. In general, all the information we have about our system at the initial time is a particular distribution $P(x, 0)$, which is the outcome of a repeated set of measurements of the functions $X(z)$

with numerical outcomes x over the system prepared in an identical manner at the initial time. Therefore, we have to determine the distribution function $\rho_0(z)$ with the sole information that it should provide precisely the distribution $P(x,0)$, according to (25), this is

$$P(x,0) = \int dz\delta(X(z) - x)\rho_0(z). \tag{27}$$

However, there are many different $\rho_0(z)$ that can produce the same $P(x,0)$. Which is the correct one? Perhaps there is no answer to this question, but there is an answer to the best selection. According to information theory [28], the least biased distribution which is compatible with the restriction (27) is the one that maximizes the entropy functional

$$S[\rho_0] \equiv -k_B \int_\Gamma \rho_0(z) \ln \frac{\rho_0(z)}{\rho^{eq}(z)} dz. \tag{28}$$

We encounter here a problem of Lagrange multipliers. By introducing the multipliers $\lambda(x)$ for the continuum set of restrictions (27) (one for each x), we maximize the functional $I[\rho_0] = S[\rho_0] + \int dx\lambda(x)P(x,0) + \mu \int dxP(x,0)$, where the μ Lagrange multiplier stands for the normalization to unity restriction of $\rho_0(z)$. The Lagrange multipliers are obtained by substituting the maximum value into the restriction (27). The following final result is obtained [29]

$$\rho_0(z) = \frac{P(X(z),0)}{\Omega(X(z))}, \tag{29}$$

where $\Omega(x)$ is given by

$$\Omega(x) = \int dz\delta(X(z) - x)\rho^{eq}(z). \tag{30}$$

$\Omega(x)$ can be interpreted as a measure of the region of phase space which is compatible with the state x. Note also that $\Omega(x)$ must be the equilibrium distribution function at the coarse-grained level of description, because (30) is just the way in which the probability at the coarse-grained level relates with the microscopic probability.

4.5 Exact Equation for $P(x,t)$

In (25) we observe that the evolution of $\rho(z,t)$ according to the Liouville equation induces an evolution of $P(x,t)$. Of course, we would like to have a closed dynamical equation for $P(x,t)$ that makes no reference to the underlying dynamics given by $\rho(z,t)$. This closed equation can be obtained with the help of a projection operator technique. Following Zwanzig [3], we introduce a projection operator $\mathcal{P}$ that applies to any function $F(z)$ of phase space Γ

$$\mathcal{P}F(z) = \langle F\rangle^{X(z)}, \tag{31}$$

where we have introduced the constrained average $\langle F\rangle^x$ by

$$\langle F\rangle^x = \frac{1}{\Omega(x)} \int dz\rho^{eq}(z)\delta(X(z) - x)F(z), \tag{32}$$

Note that the effect of the operator $\mathcal{P}$ on an arbitrary function of phase space is to transform it into a function of the relevant variables $X(z)$. The operator $\mathcal{P}$ satisfies the projection property $\mathcal{P}^2 = \mathcal{P}$. We introduce also the complementary projection operator $\mathcal{Q} = 1 - \mathcal{P}$ which satisfies $\mathcal{P}\mathcal{Q} = 0$ and $\mathcal{Q}^2 = \mathcal{Q}$. The operators $\mathcal{P}, \mathcal{Q}$ satisfy

$$\int dz \rho^{\mathrm{eq}}(z) A(z) \mathcal{P} B(z) = \int dz \rho^{\mathrm{eq}}(z) B(z) \mathcal{P} A(z), \tag{33}$$

for arbitrary functions $A(z), B(z)$. It is convenient to introduce the following notation

$$\Psi_x(z) = \delta(X(z) - x), \tag{34}$$

and consider the Dirac's delta function as an ordinary phase function with a continuum index x. According to the formal solution (23) this phase function will evolve according to

$$\Psi_x(T_t z) = \exp\{iLt\} \Psi_x(z), \tag{35}$$

and, therefore,

$$\partial_t \Psi_x(T_t z) = \exp\{iLt\} iL \Psi_x(z). \tag{36}$$

Now we introduce a mathematical identity between operators

$$\exp\{iLt\} = \exp[iLt]\mathcal{P} + \int_0^t dt' \exp[iLt']\mathcal{P}iL\mathcal{Q}\exp[iL\mathcal{Q}(t - t')] + \mathcal{Q}\exp[iL\mathcal{Q}t]. \tag{37}$$

This identity can be proved by taking the time derivative on both sides. If two operators have the same derivative and coincide at $t = 0$ then they are the same operator. We now apply this identity (37) to the left hand side of (36). After some algebra, which uses the explicit form of the operators $\mathcal{P}, \mathcal{Q}$, the properties (16), (33), and the chain rule in the form

$$iL\Psi_x(z) = iLX_i \frac{\partial}{\partial x_i} \Psi_x(z),$$

where summation over repeated indices is implied, one obtains,

$$\partial_t \Psi_x(T_t z) = -\frac{\partial}{\partial x_i} \cdot v_i(x) \Psi_x(T_t z)$$
$$+ \int_0^t dt' \int dx' \Omega(x') \frac{\partial}{\partial x_i} \cdot D_{ij}(x, x', t - t') \cdot \frac{\partial}{\partial x'_j} \frac{\Psi_{x'}(T_{t'} z)}{\Omega(x')}$$
$$+ \mathcal{Q}\exp\{iL\mathcal{Q}t\} \mathcal{Q}iL\Psi_x(T_t z). \tag{38}$$

We have defined the drift $v_i(x)$ and the diffusion tensor $D_{ij}(x, x', t)$ through

$$v_i(x) \quad = \langle iLX_i \rangle^x,$$
$$D_{ij}(x, x', t) = \langle (iLX_j - \langle iLX_j \rangle^{x'}) \exp\{iL\mathcal{Q}t\} \Psi_x(iLX_i - \langle iLX_i \rangle^x) \rangle^{x'}. \tag{39}$$

If we multiply (38) by $\rho(z, 0)$, integrate over z, and use (26), we obtain a final exact and closed equation for $P(x, t)$

$$\partial_t P(x, t) = -\frac{\partial}{\partial x_i} \cdot v_i(x) P(x, t)$$

$$+ \int_0^t dt' \int dx' \, \Omega(x') \frac{\partial}{\partial x_i} \cdot D_{ij}(x, x', t - t') \cdot \frac{\partial}{\partial x'_j} \frac{P(x', t')}{\Omega(x')}, \qquad (40)$$

where we have used that the initial ensemble (29) is a function of $X(z)$ and, therefore,

$$\int dz \rho(z, 0) \mathcal{Q} \exp\{iL\mathcal{Q}t\} \mathcal{Q} iL\Psi_x(T_t z) = 0,$$

where (33) has been used along with the property $\mathcal{Q}f(A) = 0$.

4.6 The Fokker–Planck Equation

Equation (40) is an exact an rigorous closed equation governing the distribution function $P(x, t)$. No approximations have been made and, essentially, it is another way of rewriting the Liouville equation. In principle, it is as difficult to solve as the original Liouville equation. However, as it happens often in Physics, just by rewriting the same thing in a different form, it is possible to perform suitable approximations that allow for an advance in the understanding of the problem. In the case of (40), the approximation is called the Markovian approximation and transforms the integro-differential equation into a simple Fokker–Planck equation.

The Markovian assumption is one about separation of time scales between the time scale of evolution of the phase function $X(z)$ and the rest of variables of the system. If this separation of time scales exists then, in the time scale in which the tensor $D_{ij}(x, x', t - t')$ decays, the probability $P(x, t')$ has not changed appreciably. Schematically, we write the memory term in (40) as

$$\int_0^t dt' D(t - t') P(t') \approx P(t) \int_0^\infty D(t') dt'. \qquad (41)$$

This approximation is depicted in Fig. 3. We have extended in (41) the upper limit of integration to infinity, due to the fast decay of $D(t)$. Note that the tensor $D_{ij}(x, x', t - t')$ is a quantity of order $(iLX)^2$, i.e. second order in the time derivatives of the relevant variables. The time scale of evolution of $P(x, t)$ is the same as the time scale of the variables $X(z)$. The approximation (41) amounts, therefore, to neglect third order time derivatives of the relevant variables in front of second order terms. We, therefore, consistently perform a formal expansion of the tensor $D_{ij}(x, x', t - t')$ in (39) in terms of iLX and keep only second order terms. Then,

$$\exp\{iL\mathcal{Q}t\}\Psi_x \mathcal{Q} iLX = \Psi_x \exp\{iL\mathcal{Q}t\} \mathcal{Q} iLX + \mathcal{O}(iLX)^2. \qquad (42)$$

Therefore, up to terms of order $\mathcal{O}(iLX^3)$ we have

$$D_{ij}(x, x', t) = \delta(x - x') \langle (iLX_j - \langle iLX_j \rangle^x) \exp\{iL\mathcal{Q}t\}(iLX_i - \langle iLX_i \rangle^x) \rangle^x, \qquad (43)$$

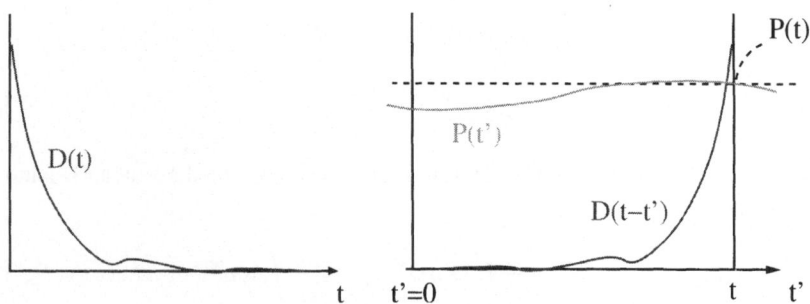

Fig. 3. The Markovian approximation

and the tensor becomes diagonal in x, x'. By substitution of the approximate form (43) into the exact equation (40) and using (41) we obtain

$$\partial_t P(x,t) = -\frac{\partial}{\partial x_i} \cdot v_i(x) P(x,t) + \frac{\partial}{\partial x_i} \cdot \Omega(x) D_{ij}(x) \cdot \frac{\partial}{\partial x_j} \frac{P(x,t)}{\Omega(x)}, \qquad (44)$$

where we have defined the diffusion tensor

$$D_{ij}(x) = \int_0^\infty dt' \langle (iLX_j - \langle iLX_j \rangle^x) \exp\{iL\mathcal{Q}t'\} (iLX_i - \langle iLX_i \rangle^x) \rangle^x, \qquad (45)$$

and we summarize for completeness the rest of quantities appearing in (44)

$$v_i(x) = \langle iLX_i \rangle^x,$$

$$\langle \ldots \rangle^x = \frac{1}{\Omega(x)} \int dz \rho^{eq}(z) \delta(X(z) - x) \ldots,$$

$$\Omega(x) = \int dz \rho^{eq}(z) \delta(X(z) - x). \qquad (46)$$

Equation (44) is the desired Fokker–Planck equation for the dynamics at the mesoscopic coarse-grained level of description. This equation is one of the cornerstones of non-equilibrium statistical mechanics and was obtained by Zwanzig in 1961 [3] following the path pioneered by Green in [2]. In this equation, all the objects $v_i(x)$, $D_{ij}(x,t)$ and $\Omega(x)$ have a definite microscopic definition. In particular, (45) is a general form of the well-known Green–Kubo formulae that relates the transport coefficients $D_{ij}(x)$ with a time integral of a correlation function of microscopic variables. The FPE (44) is valid whenever there is a clear separation of time scales such that the Markovian approximation is valid. Clearly, the FPE will describe correctly the evolution of $P(x,t)$ only for times which are larger than the typical time scales of decay of the correlation involved in $D_{ij}(x)$. We cannot investigate with this equation shorter time scales. For this short time scales, the transport coefficients start to depend on time.

As a final remark, we can check that, despite the approximations taken to derive the FPE (44), $\Omega(x)$ is still the equilibrium solution. By substituting $\Omega(x)$ into (44) we have that $\Omega(x)$ will be a stationary solution of the FPE (44) if and only if

$$\sum_i \frac{\partial}{\partial x_i} v_i(x)\Omega(x) = 0. \tag{47}$$

This is easily proved by using the definition (39), the constrained average (32) and the chain rule, i.e.,

$$\frac{\partial}{\partial x_i} v_i(x)\Omega(x) = \int dz \rho^{\mathrm{eq}}(z) \frac{\partial}{\partial x_i} \Psi_x(z) \mathrm{i}LX_i(z) = - \int dz \rho^{\mathrm{eq}}(z) \mathrm{i}L\Psi_x(z) = 0, \tag{48}$$

where we have used (16) and $\mathrm{i}L\rho^{\mathrm{eq}} = 0$.

5 Example: Smoluchowski Level

We illustrate now how the general formalism developed in the previous section can be applied to the case of a colloidal suspension in order to derive the FPE at the Smoluchowski level discussed in Sect. 3.4. The idea is simply to translate to our system the different objects defined in (45), (46) that appear in the FPE (44). The microscopic Hamiltonian is given in (2). The equilibrium ensemble is given by, for example, the canonical ensemble

$$\rho^{\mathrm{eq}}(z) = \frac{1}{Z} \exp\{-\beta H(z)\},$$

where $\beta = (k_{\mathrm{B}}T)^{-1}$ is proportional to the inverse of the temperature T. The relevant variables $X(z) = x$ at the Smoluchowski level are the positions of the colloidal particles $\mathbf{Q}_i$ which take numerical values $\overline{\mathbf{Q}}_i$. The equilibrium solution $\Omega(x)$ of the FPE in (46) is given by

$$\Omega(\overline{Q}) = \int dz \rho^{\mathrm{eq}}(z) \prod_i \delta(\mathbf{Q}_i - \overline{\mathbf{Q}}_i)$$

$$\propto \exp\{-\beta V^{\mathrm{CC}}(\overline{Q})\} \int dq \exp\{-\beta \left(V^{\mathrm{CS}}(\overline{Q}, q) + V^{\mathrm{SS}}(q)\right)\}$$

$$\equiv \exp\{-\beta \overline{V}^{\mathrm{eff}}(\overline{Q})\}, \tag{49}$$

where $\int dq$ is a condensed notation for the integral over solvent positions. We have introduced the effective potential as

$$\overline{V}^{\mathrm{eff}}(\overline{Q}) = V^{\mathrm{CC}}(\overline{Q}) - k_{\mathrm{B}}T \ln \int dq \exp\{-\beta \left(V^{\mathrm{CS}}(\overline{Q}, q) + V^{\mathrm{SS}}(q)\right)\}. \tag{50}$$

The effective potential has a contribution $V^{\mathrm{CC}}(\overline{Q})$ coming from the direct interaction potential and an additional contribution that represents the effect of the static and equilibrium averaged solvent mediated interaction between colloidal particles.

After performing the integrals over the Dirac delta functions, the constrained average in (46) now takes the form

$$\langle \cdots \rangle^{\overline{Q}} = \frac{1}{\Omega(\overline{Q})} \int dq \rho^{\mathrm{eq}}(q, \overline{Q}) \cdots .$$

Note that this constrained average is simply an equilibrium average over the solvent degrees of freedom, in which the colloidal particles are assumed to be fixed at the values $\overline{Q}$. It is, therefore, an equilibrium ensemble average in which the colloidal particles act as external static force fields.

Because the time derivatives iLX are simply P_i/M_i, the drift term $v(x) = \langle iLX \rangle^x$ defined in (46) is now the constrained equilibrium average of the momentum of the colloidal particles, which is zero by isotropy of the equilibrium ensemble. The diffusion tensor (45) becomes

$$\mathbf{D}_{ij}(\overline{Q}) = \int_0^\infty dt' \langle \mathbf{V}_j \exp\{iL\mathcal{Q}t'\} \mathbf{V}_i \rangle^{\overline{Q}}.$$

The final FPE (44) takes now the form

$$\partial_t P(Q,t) = \frac{\partial}{\partial \mathbf{Q}_i} \left[\beta \mathbf{D}_{ij}(Q) \frac{\partial \overline{V}^{\mathrm{eff}}(Q)}{\partial \mathbf{Q}_j} \right] P(Q,t) + \frac{\partial}{\partial \mathbf{Q}_i} \mathbf{D}_{ij}(Q) \cdot \frac{\partial}{\partial \mathbf{Q}_j} P(Q,t),$$

which, has been presented already in (5).

In this example, we observe how the general Fokker–Planck description can be applied to a specific level of description of a given system. The essential benefits of this approach are that it is very simple to get the structure of the coarse-grained equation. Also, we obtain explicit microscopic expressions for all the objects in the FPE. In particular, the diffusion tensor which describes the mutual, solvent-mediated influence of the colloidal particles is given in terms of the auto and cross-correlations of the velocities of the colloidal particles, where the averages are taken over the solvent degrees of freedom which are distributed according to an equilibrium ensemble in the presence of the external fields due to the static colloidal particles.

6   How to Compute the Objects in the FPE from a MD Simulation?

The FPE (44) governs the full dynamics at a coarse-grained level of description. In the equation there appear three objects, the drift $v_i(x)$, the diffusion tensor $D_{ij}(x)$ and the equilibrium distribution function $\Omega(x)$ which are defined in terms of the microscopic dynamics of the system. It would be very useful to have a definite way to compute these objects through a numerical solution of the microscopic equations, that is, through a molecular dynamics (MD) simulation. In the previous example of the colloidal system at the Smoluchowski level, one could run a molecular dynamics simulation in which the colloidal particles are at fixed positions and compute the velocity correlation functions in order to extract the corresponding diffusion tensors. Note that this is perhaps the only way of using a molecular dynamics in order to study a colloidal suspension. In this example, it has been possible to deal with the Dirac delta functions over the relevant variables that appear in the constrained averages (46) because the relevant variables are very simple functions of the microscopic state. In a general case, it is not possible to perform explicitly the integral over the Dirac delta functions appearing in the definitions (45), (46) and it is necessary to devise a new method for dealing with the constrained averages. In this section, we briefly sketch a general route to compute the quantities that appear in the general FPE and highlight the problems that usually appear.

First, we have to compute $\Omega(x)$ which is the equilibrium distribution function of our state variables. One route for getting this distribution function is to simply bin the x state space and get an histogram of occurrences of the different values of $x = X(z)$ as the microscopic state z evolves in our equilibrium MD simulation. However, we will need also to compute constrained averages in the drift and diffusion tensors, and the presence of a Dirac delta function complicates such computation enormously. For this reason, it is convenient to find another route for the calculations which is not microcanonical but canonical, in the following sense. Consider the following partition function $Z(\lambda)$ defined as

$$Z(\lambda) = \int dz \rho^{\text{eq}}(z) \exp\{-\lambda X(z)\} = \langle \exp\{-\lambda X\} \rangle^{\text{eq}}. \tag{51}$$

This quantity can be understood as the normalization factor of an equilibrium ensemble in which there is a coupling between external fields λ and the relevant variables. Strictly speaking, it is the equilibrium average of the phase function $\exp\{-\lambda X(z)\}$, which is very easy to compute during a MD run. Note that, in principle, a single run allows to compute $Z(\lambda)$ for all λ simultaneously. By introducing the following identity

$$1 = \int dx \delta(X(z) - x), \tag{52}$$

within the integral sign of (51) we can relate $Z(\lambda)$ with $\Omega(x)$

$$Z(\lambda) = \int dx \Omega(x) \exp\{-\lambda x\}.$$

In most applications, we will have that the argument of the integral is highly peaked at a particular maximum value x^* and, by the usual steepest descent argument, we can approximate

$$Z(\lambda) \approx \Omega(x^*) \exp\{-\lambda x^*\}, \tag{53}$$

where the maximum occurs at x^*, which is the solution of

$$\frac{\partial \ln \Omega(x^*)}{\partial x} = \lambda. \tag{54}$$

By taking the derivatives with respect to λ in (53) and using (54) we have the expression conjugate to (54)

$$-\frac{\partial \ln Z(\lambda)}{\partial \lambda} = x^*.$$

This partial derivative can be easily computed from MD, because from (51),

$$-\frac{\partial \ln Z(\lambda)}{\partial \lambda} = \frac{\langle X \exp\{-\lambda X\} \rangle^{\text{eq}}}{\langle \exp\{-\lambda X\} \rangle^{\text{eq}}}.$$

In this way, we get the functional relation $x^*(\lambda)$ that can be inverted to give $\lambda(x)$ [1] and, from (53) one can recover the equilibrium distribution function in the form

$$\Omega(x) \approx Z(\lambda(x)) \exp\{\lambda(x)x\}.$$

Note that although this is a quite involved route to get $\Omega(x)$, now we can compute the constrained averages as "canonical" averages which are defined as

$$\langle \cdots \rangle^\lambda = \int \mathrm{d}z \rho^{\mathrm{eq}}(z) \exp\{-\lambda X(z)\} \cdots .\qquad(55)$$

The idea, of course, is to introduce again (52) and resort to the sharpness of the function involved to approximate the constrained average with the canonical one, this is,

$$\langle \cdots \rangle^x \approx \langle \cdots \rangle^{\lambda(x)},$$

where the functional relation $\lambda(x)$ has already been obtained. In this way, the constrained average involved in the definition of $v_i(x)$ is now easily computed from an equilibrium MD simulation.

The term that remains to be computed is the diffusion tensor, which is given by a correlation function. Note, however, that the evolution operator that appears in the definition of the diffusion tensor corresponds to the projected dynamics, i.e. $\exp\{\mathrm{i}L\mathcal{Q}\}$, instead of being the actual dynamics $\exp\{\mathrm{i}L\}$ that is reproduced in a MD simulation. The question thus arises about whether we can approximate the projected dynamics by the real one. This can be done, actually, because under the approximation where we have neglected terms of order $(\mathrm{i}LX)^3$, both dynamics coincide. Note that $\exp\{\mathrm{i}L\mathcal{Q}t\} = \exp\{\mathrm{i}Lt\}+\mathcal{O}(\mathrm{i}LX)$, as can be seen from the Taylor series (24). However, this apparently innocent approximation has terrible consequences. To appreciate them, let us write the diffusion tensor (45) with the approximation $\exp\{\mathrm{i}L\mathcal{Q}t\} \approx \exp\{\mathrm{i}Lt\}$ in the form

$$D_{ij}(x) = \lim_{\tau\to\infty} \int_0^\tau \mathrm{d}t' \langle (\mathrm{i}LX_j - \langle \mathrm{i}LX_j \rangle^x) \exp\{\mathrm{i}Lt'\}(\mathrm{i}LX_i - \langle \mathrm{i}LX_i \rangle^x) \rangle^x,\qquad(56)$$

that leads to

$$D_{ij}(x) = \lim_{\tau\to\infty} \left[\langle \mathrm{i}LX_j X_i(\tau) \rangle^x - \langle \mathrm{i}LX_i \rangle^x \langle X_i(\tau) \rangle^x \right],\qquad(57)$$

where we have used that $\mathrm{i}LX_i = \dot{X}_i$ and we have performed the time integral explicitly. If we take $\tau \to \infty$ in this expression, we note that

$$\lim_{\tau\to\infty} \langle \mathrm{i}LX_j X_i(\tau) \rangle^x = \langle \mathrm{i}LX_i \rangle^x \langle X_i(\tau) \rangle^x,\qquad(58)$$

because as time proceeds, any two dynamical variables become uncorrelated and statistically independent. We arrive, therefore at the very surprising result that the diffusion

---

[1] Note that $\partial x^*/\partial\lambda = \langle \delta x \delta x \rangle^{\mathrm{eq}}$ which, being a second moment of a distribution, is a positive definite quantity. This is the main hypothesis of the inverse function theorem [30] that ensures that the function $x(\lambda)$ can be inverted.

tensor is zero! This is known as the plateau problem and has its origin in the uncontrol-lable step of approximating the projected dynamics $\exp\{iL\mathcal{Q}t\}$ with the actual dynamics $\exp\{iLt\}$. The solution of this problem is to not take the limit $\tau \to \infty$. Due to the clear separation of time scales, by choosing τ large enough for the correlation to have decayed sufficiently, but short in the time scale of the relevant variables, the diffusion tensor takes a constant value that does not depend appreciably on the actual value of τ. For a clear discussion of the plateau problem for the simple case of Brownian motion see [31].

In this section, we have presented a general strategy for the microscopic calculation of the different objects of the FPE from equilibrium molecular dynamics simulations. Of course, in practice there is a big problem in pursuing the above program due to the fact that the state vector x may belong to a highly dimensional space. It is very difficult to deal with multidimensional functions like $\Omega(x)$, for example. Think, for example, on the level of description of hydrodynamics where, in principle, we have several variables (mass, momentum, and energy) for every little cell in which we have divided the space occupied by our fluid. Fortunately enough, for the case of hydrodynamic variables one can focus on one cell only by resorting to the local equilibrium assumption. In this case one can reduce the problem to two variables (mass and energy densities) because the momentum variables can be easily eliminated with a Galilean transformation. In this way, the usual Green–Kubo formulae for transport coefficients can be obtained [5]. Other systems and levels of description usually lead to multivariate problems that are in general difficult, if not impossible, to handle. Even in cases where the local equilibrium assumption can be taken, if the hydrodynamic description requires further structural variables of tensorial nature, we readily obtain a difficult multivariate problem [32]. In part due to this reason, there have been not so many actual explicit implementations of the program of coarse-graining from molecular dynamics.

The alternative route is that of modeling, that is, to assume particular functional forms for the diffusion tensors, drift terms and equilibrium probability function, inspired from a good knowledge of the physical processes at hand, and check that the resulting pre-dictions agree with experiments. Of course, one would like to use as much information as possible from the microscopic derivation in order to formulate models with the min-imum of arbitrariness. In the next section, we will present the GENERIC structure for which a large number of restriction on the possible form of the objects appearing in the FPE exist. These restrictions help enormously in the task of formulating new models for complex fluids [6,33].

7   GENERIC Structure of the Fokker–Planck Equation

In this section, we study those levels of description that satisfy a very strong property: the microscopic dynamical invariants $I(z)$ of the system can be expressed as functions of the relevant variables $X(z)$ of this level of description, this is

$$I(z) = \mathcal{I}(X(z)), \tag{59}$$

where $\mathcal{I}(x)$ is a suitable function. When our level of description has this property a very powerful structure, named GENERIC, emerges [34]. As a particular case of (59), the

Hamiltonian must be expressible in terms of the relevant variables, this is

$$H(z) = E(X(z)). \tag{60}$$

Therefore, in order to have a GENERIC structure, our level of description must allow to define an energy function $E(x)$ that coincides with the actual value of the energy of the system. Note that not every level of description can have the GENERIC structure. Even when the relevant variables are slow and well time separated, if the total energy cannot be written as a function of the relevant variables, there will be no GENERIC structure. For example, in the colloidal suspension discussed in Sect. 5, the state at the Smoluchowski level is described by the set of coordinates $\mathbf{Q}_i$ of the colloidal particles. The energy of the whole system contains a contribution that comes from the solvent degrees of freedom that cannot be expressed in terms of the colloidal variables alone. Therefore, this level of description has no GENERIC structure.

In order to appreciate the beauty of the GENERIC structure we have first to exploit the consequences of (59) on the FPE (44). We first consider which form takes the average $\langle \dots \rangle^x$ when (59) holds. For this reason, we first note that the equilibrium ensemble $\rho^{eq}(z)$ is a function of dynamical invariants (see (20)) and, therefore,

$$\Omega(x) = \int dz \rho^{eq}(z) \delta(X(z) - x) = \frac{P^{eq}(\mathcal{I}(x))}{\Omega^{eq}(\mathcal{I}(x))} \overline{\Omega}(x), \tag{61}$$

where we have introduced

$$\overline{\Omega}(x) = \int dz \delta(X(z) - x). \tag{62}$$

As a result,

$$\langle \dots \rangle^x = \frac{1}{\Omega(x)} \int dz \rho^{eq}(z) \delta(X(z) - x) \dots = \frac{1}{\overline{\Omega}(x)} \int dz \delta(X(z) - x) \dots, \tag{63}$$

and we see that the constrained average does not depend on the equilibrium ensemble.

Note that we can write the equilibrium distribution function $\Omega(x)$ in the form

$$P^{eq}(x) = \Omega(x) = \frac{P^{eq}(\mathcal{I}(x))}{\Omega^{eq}(\mathcal{I}(x))} \exp\{S(x)/k_B\}, \tag{64}$$

where the entropy is defined by

$$S(x) = k_B \ln \overline{\Omega}(x). \tag{65}$$

It is apparent that the concept of entropy as defined in (65) is level dependent. It makes no sense to speak about "the" entropy of a system. Different levels of description of the same system have different entropies which depend on different sets of variables.

Equation (64) has the form of the so-called in the presence of dynamical invariants [27]. Note that if we know with precision the initial values $\mathcal{I}_0$ of the dynamical invariants, we have

$$P^{eq}(x) = \frac{\delta(\mathcal{I}(x) - \mathcal{I}_0)}{\Omega^{eq}(\mathcal{I}_0)} \exp\{S(x)/k_B\}. \tag{66}$$

An alternative to (65) is to define the entropy as $S(x) = k_\mathrm{B} \ln \Omega(x)$. However, this definition of the entropy makes problematic the treatment of situations like the one represented in (66) where we have Dirac delta functions describing exact knowledge about the dynamical invariants.

Let us continue with the consequences of (59). We may algebraically manipulate the reversible term $v_i(x) = \langle iLX_i \rangle^x$ as follows,

$$iLX_i = \frac{\partial X_i}{\partial z} iLz = \frac{\partial X_i}{\partial z} L_0 \frac{\partial H}{\partial z} = \frac{\partial X_i}{\partial z} L_0 \frac{\partial X_j}{\partial z} \frac{\partial E}{\partial x_j}(X(z)). \tag{67}$$

By using this result into the definition of the constrained average (32) we obtain

$$v_i(x) = \langle iLX_i \rangle^x = L_{ij}(x) \frac{\partial E}{\partial x_j}(x), \tag{68}$$

where the reversible operator L_{ij} is defined by

$$L_{ij}(x) = \left\langle \frac{\partial X_i}{\partial z} \cdot L_0 \cdot \frac{\partial X_j}{\partial z} \right\rangle^x. \tag{69}$$

Here, the microscopic reversible operator L_0 is the block diagonal matrix introduced in (8).

For convenience, we introduce a dissipative matrix M as $M_{ij} = D_{ij}/k_\mathrm{B}$. By using (67) and (68) in the definition (45) we easily arrive at the following degeneracy property of M

$$M \cdot \frac{\partial \mathcal{I}}{\partial x} = 0. \tag{70}$$

By collecting (64), (65), (68) and (70) we can finally write the FPE (44) in the GENERIC form

$$\partial_t P(x,t) = -\frac{\partial}{\partial x} \left[L(x) \frac{\partial E}{\partial x} + M(x) \frac{\partial S}{\partial x} \right] P(x,t) + k_\mathrm{B} \frac{\partial}{\partial x} M^S(x) \frac{\partial}{\partial x} P(x,t). \tag{71}$$

Note that only the symmetric part M^S of M appears in the last term involving the second derivatives. For reasons that will become clear later the term $L \frac{\partial E}{\partial x}$ is named the reversible part of the dynamics, and $M \frac{\partial S}{\partial x}$ the irreversible or dissipative part of the dynamics. The last term proportional to k_B is a fluctuation term that describes the broadening of the distribution function.

7.1 Properties of L and M

The first property of the matrix L is its antisymmetry, which derives from the definition in (69) and the fact that the matrix L_0 is antisymmetric. On the other hand, the matrix $M_{ij}^S = \frac{1}{2}[M_{ij} + M_{ji}]$ is, by definition, symmetric. By using (45) we can write M_{ij}^S in the form

$$M_{ij}^S = \frac{1}{2k_\mathrm{B}} \int_0^\infty dt' \langle (iLX_j - \langle iLX_j \rangle^x) \exp\{iLQt'\}(iLX_i - \langle iLX_i \rangle^x) \rangle^x$$

$$+ \frac{1}{2k_\mathrm{B}} \int_0^\infty dt' \langle (iLX_i - \langle iLX_i \rangle^x) \exp\{iLQt'\}(iLX_j - \langle iLX_j \rangle^x) \rangle^x. \tag{72}$$

The second term in the right hand side of (72) can be arranged by using the properties (16) and (33) of L and Q, which are inherited by $\exp\{iLQt'\}$ in the form

$$\langle A\exp\{iLQt'\}B\rangle^x = \langle B\exp\{-iLt'Q\}A\rangle^x.$$

In this way, (72) becomes

$$M_{ij}^S = \frac{1}{2k_B}\int_{-\infty}^{\infty}dt'\langle(iLX_j - \langle iLX_j\rangle^x)\exp\{iLQt'\}(iLX_i - \langle iLX_i\rangle^x)\rangle^x. \quad (73)$$

It is possible to show that M^S is semi definite positive. If we double contract M^S with an arbitrary vector a we obtain

$$\sum_{ij}a_i M_{ij}^S a_j = \frac{1}{2k_B}\int_{-\infty}^{\infty}\langle(A\exp\{iLQt\}A)^x dt, \quad (74)$$

where $A = \sum_i(iLX_i - \langle iLX_i\rangle^x)a_i$. Equation (74) is the time integral of an autocorrelation function. By recalling the Wiener–Knichine theorem that states that the Fourier transform of a stationary autocorrelation function is positive [10], we conclude that this double contraction is positive or, in other words, that M^S is a positive matrix. That M^S is not positive definite but semidefinite follows from (70) which shows that there are non-null eigenvectors with null eigenvalue.

A property similar to (70) but now for the matrix L takes the following form

$$\frac{\partial E}{\partial x}\cdot L\cdot\frac{\partial \mathcal{I}}{\partial x} = \frac{1}{\Omega(x)}\frac{\partial E}{\partial x_i}\int dz\rho^{eq}(z)\Psi_x(z)\frac{\partial X_i}{\partial z}\cdot L_0\cdot\frac{\partial X_j}{\partial z}\frac{\partial \mathcal{I}}{\partial x_j}$$

$$= \frac{1}{\Omega(x)}\int dz\rho^{eq}(z)\Psi_x(z)\frac{\partial H}{\partial z}L_0\frac{\partial I}{\partial z}$$

$$= \frac{1}{\Omega(x)}\int dz\rho^{eq}(z)\Psi_x(z)iLI = 0. \quad (75)$$

Another important property of the matrix L is obtained from (47) which becomes,

$$\frac{\partial}{\partial x_i}\left[L_{ij}\frac{\partial E}{\partial x_j} + M_{ij}^A\frac{\partial E}{\partial x_j}\right]\Omega(x) = 0, \quad (76)$$

where $M_{ij}^A = (M_{ij} - M_{ji})/2$ is the antisymmetric part of M. Further use of the chain rule, the (anti)-symmetries of L_{ij}, M_{ij}^A and the definition (65) leads to

$$\frac{\partial E}{\partial x_i}L_{ij}\frac{\partial S}{\partial x_j} = k_B\left[\frac{\partial L_{ij}}{\partial x_j}\frac{\partial E}{\partial x_i} + \frac{\partial M_{ij}^A}{\partial x_i}\frac{\partial S}{\partial x_j}\right]. \quad (77)$$

This seems to be a complicated differential equation to be satisfied by the different objects appearing in the FPE (71). The interest of this equation (77) is that it tells us that the left hand side is of order k_B. As we will see in Sect. 7.3, terms of this order can be neglected whenever fluctuations effects are small.

Further properties for the blocks E, S, L, M can be derived by resorting to the time reversibility of the microscopic dynamics [5], which lead to generalized Onsager reciprocity relations. Also, it is possible to study how the symmetries of the microscopic dynamics reflect on symmetries of the blocks E, S, L, M [35].

7.2   GENERIC Stochastic Differential Equation

The Fokker–Planck equation governs the probability distribution function $P(x,t)$ of a stochastic variable x and, therefore, characterizes the dynamics of these variables. Another, mathematically equivalent, way of describing the dynamics of the stochastic variables is through a stochastic differential equation (SDE) [36]. The SDE associated to the FPE (44) is, within the Ito interpretation

$$dx = \left[L(x)\frac{\partial E}{\partial x} + M(x)\frac{\partial S}{\partial x} + k_{\mathrm{B}}\nabla M^{S}(x) \right] dt + d\tilde{x}, \qquad (78)$$

where $d\tilde{x}$ is a linear combination of independent increments of the Wiener process of the form

$$d\tilde{x}_i = \sum_j B_{ij}(x)dW_j(t), \qquad (79)$$

where the (generally non-square) matrix $B(x)$ satisfies

$$B(x)B^T(x) = 2k_{\mathrm{B}}M^{S}(x). \qquad (80)$$

The term $k_{\mathrm{B}}\nabla M(x)$ in (78) is a consequence of both, the particular form in which the matrix M appears in the FPE (44), that is, in between the two partial derivatives in the last term, and the use of the Ito interpretation of the SDE.

By recalling that the independent increments of the Wiener process have a variance proportional to dt, we can write (79) and (80) in the form

$$d\tilde{x}(t)d\tilde{x}^T(t) = 2k_{\mathrm{B}}M^{S}(x(t))dt, \qquad (81)$$

which is a formal and compact statement of the Fluctuation–Dissipation theorem. This theorem, which is nothing else than the rule for obtaining the SDE from the FPE, states that the amplitude of the fluctuations is related to the matrix of transport coefficients M^{S}. Following a common practice in some stochastic physics literature, we can introduce the white noise terms $\zeta(t)$ as the time derivative of the random terms $d\tilde{x}$, although, strictly speaking, the Wiener process is not differentiable [36]. The white noise terms have zero mean and the variance is given by (81) in the form

$$\langle \zeta(t)\zeta(t')\rangle = 2k_{\mathrm{B}}M(x)\delta(t-t'). \qquad (82)$$

It is very tempting to make the identification

$$\zeta(t) \leftrightarrow \exp\{iLQt\}\left[iLX_i - \langle iLX_i\rangle^x \right], \qquad (83)$$

because, then, we can relate the Fluctuation-Dissipation (81) theorem with the Green–Kubo formula (45). Even thought this identification is certainly not rigorous, it shows that the rapidly varying functions that appear in the microscopic derivation, are modeled at the mesoscopic level as white noise, the prime example of a rapidly varying function. The identification (83) is useful in practice because, in general, it is simple to compute the time derivatives of the relevant variables, and this provides for the structure of the random terms. In turn, by having a raw understanding of the structure of the noise, one can, through the Fluctuation-Dissipation theorem (81) obtain the structure of the matrix M.

As an example, consider the following scalar product $\zeta \cdot \frac{\partial E}{\partial x}$. According to the identification (83) and by using (67) we easily arrive at the conclusion that this scalar product is zero. In terms of the properly defined noise terms $d\tilde{x}$ we have

$$d\tilde{x} \cdot \frac{\partial E}{\partial x} = 0. \tag{84}$$

In geometrical terms this expression has a lot of sense. It tells us that the stochastic kicks $d\tilde{x}$ in (78) are perpendicular to the gradient of the energy, which is a vector that is normal to the surface of constant energy. Therefore, these kicks are tangential to the energy shell, and the noise lets the state at constant energy. In other words, the noise conserves energy. This can be explicitly shown by computing the stochastic differential of the energy function $E(x)$. According to Ito calculus, one has

$$dE(t) = \frac{\partial E}{\partial x} \cdot dx + \frac{1}{2} \frac{\partial^2 E}{\partial x \partial x} : dx dx. \tag{85}$$

Note the usual occurrence of higher order terms, which is a consequence of the peculiarities of Ito calculus. By substitution of the SDE (78) in (85), and use of (81) we arrive at

$$dE(t) = \frac{\partial}{\partial x} \left[M(x) \frac{\partial E}{\partial x} \right] = 0,$$

as can be seen from the form of M in (81) and the property (84). A similar orthogonality condition holds for the rest of dynamical invariants, this is,

$$d\tilde{x} \cdot \frac{\partial \mathcal{I}}{\partial x} = 0.$$

7.3   The Size of the Fluctuations and the Deterministic Equations

The size of the thermal fluctuations is formally controlled by the Boltzmann constant k_B, which has dimensions of entropy and which is small compared with macroscopic entropies [37]. In practical situations, what controls the effect of thermal fluctuations is a dimensionless ratio k_B/S_0 where S_0 is a characteristic entropy that scales with the size of the system. With this understanding, we can take the formal limit $k_B \to 0$ in the FPE (under the assumption that all the objects E, S, L, M do not depend on k_B). This amounts to neglect the last diffusive term that contains second derivatives with respect to the state variables. The resulting equation, which has only first order derivatives, describes deterministic motion, in much the same way as the Liouville equation does. This is even more clear at the level of the SDE (78) which, in the limit $k_B \to 0$, becomes an ordinary differential equation (note that $d\tilde{x}$ scales like $k_B^{1/2}$)

$$\dot{x} = L(x) \frac{\partial E}{\partial x} + M(x) \frac{\partial S}{\partial x}. \tag{86}$$

At the same time, in the limit $k_B \to 0$ we have from (77) the following degeneracy

$$L(x)\frac{\partial S}{\partial x} = 0, \tag{87}$$

which, together with the degeneracy (70) for the case of the energy, i.e.

$$M(x)\frac{\partial E}{\partial x} = 0, \tag{88}$$

gave the name to the GENERIC structure, which is an acronym for General Equation for Non-Equilibrium Reversible Irreversible Coupling [6]. Actually, (87) and (88) are a particular way of expressing the First and Second Laws of Thermodynamics [38]. For example, one can compute the time derivative of the energy function, with the result

$$\dot{E}(x) = \frac{\partial E}{\partial x} \cdot \dot{x} = 0,$$

as a consequence of the antisymmetry of L and (87). On the other hand,

$$\dot{S}(x) = \frac{\partial S}{\partial x} \cdot \dot{x} = \frac{\partial S}{\partial x} M^S(x)\frac{\partial S}{\partial x} \geq 0, \tag{89}$$

where, again, the antisymmetry of L and M^A have been used, and the inequality stands because M^S is positive semidefinite. Note that the term $L\frac{\partial E}{\partial x}$ does not contribute to the production of entropy, which comes entirely from $M\frac{\partial S}{\partial x}$. It is for this reason and for their transformation properties under time reversibility that these terms receive their names (reversible and irreversible or dissipative terms, respectively).

The reversible dynamics $\dot{x} = L(x)\partial E/\partial x$ induces a Poisson bracket structure. For example, by using the chain rule, we have for an arbitrary function $F(x)$ that

$$\dot{F}(x) = \frac{\partial F}{\partial x}L(x)\frac{\partial E}{\partial x} \equiv \{F, E\}, \tag{90}$$

where the last equality defines the bracket. An important property of a Poisson bracket is the Jacobi identity $\{A, \{B, C\}\} + \{B, \{C, A\}\} + \{C, \{A, B\}\} = 0$, that implies the time invariance of the Poisson bracket structure, i.e.

$$\frac{\mathrm{d}}{\mathrm{d}t}\{F, G\} = \{\dot{F}, G\} + \{F, \dot{G}\}. \tag{91}$$

The Jacobi identity has not yet been derived from the microscopic definition of the matrix L in (69), but it is a generally valid property that restricts the possible forms of the L matrix [39].

It should be remarked that the Second Law is a macroscopic law. If thermal fluctuations are important, the entropy function may be a decreasing function of time due to random fluctuations. Nevertheless, there is a strong result, which applies to any Fokker–Planck equation that states that the entropy functional

$$S[P(x, t)] = -k_B \int \mathrm{d}x P(x, t) \ln \frac{P(x, t)}{P^{\mathrm{eq}}(x)}, \tag{92}$$

is a non-decreasing function of time. This result is known as the H-theorem [36]. Note that in the limit $k_B \to 0$, the distribution function $P(x, t)$ is very peaked and the numerical value of the entropy functional coincides with that of the entropy function $S(x)$ evaluated at the most probable value of $P(x, t)$, thus recovering the Second Law in the form (89).

A final word is in order about the definition of the entropy in (65). Note that $\Omega(x)$ is a probability distribution, and under a change of variables $x \to y$ it transforms as

$$\Omega'(y) = \int \mathrm{d}x \delta(Y(x) - y)\Omega(x) = \Omega(X(y)) \det \left| \frac{\partial X(y)}{\partial(y)} \right|,$$

where the last term is the Jacobian of the transformation. This implies that the entropy as defined in (65) does not transform as a scalar in a change of variables, but rather as

$$S'(y) = S(X(y)) + k_B \ln \det \left| \frac{\partial X(y)}{\partial(y)} \right|.$$

The last term is of order k_B, which is small in front of the overall entropy, for physically sensible changes of variables. This means that the entropy can be assumed to transform as an scalar whenever fluctuations can be neglected. An alternative is to define the entropy in terms of a privileged set of variables x^{ref}, in such a way that in an arbitrary set of variables, the entropy is defined as [35]

$$S'(y) = S(X(y)) - k_B \ln \det \left| \frac{\partial X^{\mathrm{ref}}(y)}{\partial(y)} \right|.$$

In this way, the entropy always transforms as an scalar under change of variables. However, there is no a priori criteria for knowing what set of variables are to be taken as privileged. As mentioned, this is not a serious problem when fluctuations are small.

8 Fluid Particle Models for Simulating Complex Fluids

Computer simulation of a fluid requires the formulation of discrete models for describing its hydrodynamic behaviour. As we have discussed in Sect. 3.2 not only a discrete representation is necessary but it may even be more adequate from a conceptual point of view. Understood as the result of a coarse-graining procedure, the continuum hydrodynamic equations emerge only as an approximation, valid when the variation of the discrete quantities defined in each little cell is small from cell to cell. This approximation is very useful because it allows one to introduce the concept of spatial derivative and to obtain analytical solutions for particularly simple problems. Yet, from a computational point of view we are interested in discrete models. The question, of course, is whether it is possible to apply the coarse-graining strategy delineated in the previous sections for the derivation of discrete models for hydrodynamics. To our knowledge this has not been done yet, although attempts in that direction have been taken [40]. What has been successfully accomplished is the derivation of the continuum hydrodynamic equations starting from the microscopic dynamics of the system, see, for example, [5], or more recently [41] for a simple fluid and [42] for a fluid mixture.

We briefly sketch how one could derive the discrete hydrodynamic equations following the strategy of the previous sections. As in any coarse-graining procedure, we have
to first define the relevant variables as functions of the microscopic degrees of freedom.
In the case of discrete hydrodynamics of a simple fluid, these variables are the extensive
variables corresponding to the densities (3)

$$M_\mu(z) = m \sum_i \chi_\mu(\mathbf{q}_i),$$

$$\mathbf{P}_\mu(z) = \sum_i \mathbf{p}_i \chi_\mu(\mathbf{q}_i),$$

$$E_\mu(z) = \sum_i \left[\frac{m}{2}(\mathbf{v}_i - \mathbf{V}_\mu)^2 + \phi_i \right] \chi_\mu(\mathbf{q}_i), \tag{93}$$

where $\mathbf{v}_i = \mathbf{p}_i/m$, $\mathbf{V}_\mu = \mathbf{P}_\mu/M_\mu$, and ϕ_i is half the potential energy of interaction
between molecule i and the rest of molecules. Here $\chi_\mu(\mathbf{r})$ is the characteristic function
of the cell μ, that takes the value 1 if $\mathbf{r}$ belongs to this cell and zero otherwise. Therefore,
M_μ is the total mass of the molecules that are in cell μ, $\mathbf{P}_\mu$ its total momentum and
E_μ its total internal energy. One recognizes the coarse-grained delta function in these
equations as $\bar{\delta}(\mathbf{r}_\mu - \mathbf{r}) = \chi_\mu(\mathbf{r})/\mathcal{V}_\mu$, where $\mathcal{V}_\mu$ is the volume of the cell labeled μ. Note
that the total Hamiltonian can actually be written in terms of the above variables because
we have the partition of unity property for the characteristic function

$$\sum_\mu \chi_\mu(\mathbf{r}) = 1,$$

that reflects that the little cells are non-overlapping and cover the whole physical space.
In this way,

$$H(z) = \sum_\mu \frac{\mathbf{P}_\mu^2(z)}{2M_\mu(z)} + E_\mu(z). \tag{94}$$

The total momentum $\mathbf{P}(z) = \sum_i \mathbf{p}_i$ can also be written in terms of the relevant variables
because $\mathbf{P}(z) = \sum_\mu \mathbf{P}_\mu(z)$. Because the dynamical invariants can be written in terms
of the relevant variables, we will have the GENERIC structure, with the energy function
given by

$$E(x) = \sum_\mu \frac{\mathbf{P}_\mu^2}{2M_\mu} + E_\mu.$$

Next, we have to compute the equilibrium distribution function $\Omega(x)$ at this level of
description. That is, we need to compute the probability density that a fluid at equilibrium
has a particular realization of the mass, momentum and internal energy in each of the little
cells in which the space has been divided. We have computed explicitly this distribution
function in [43], which turns out to have the form

$$P^{\mathrm{eq}}(M, P, E) = \frac{1}{\Omega_0} \delta(E(x) - E_0) \delta(\mathbf{P}(x) - \mathbf{P}_0) \exp\left\{ \frac{1}{k_\mathrm{B}} \sum_\mu S(M_\mu, E_\mu, \mathcal{V}_\mu) \right\}.$$

$$\tag{95}$$

Here E_0 is the total energy of the system and $\mathbf{P}_0$ is the total momentum that can be set to zero without lose of generality. The function $S(M, E, \mathcal{V})$ is the usual thermodynamic entropy of a system with mass M, energy E, and volume $\mathcal{V}$. Implicit in the derivation of (95) is the assumption that the cells in which the physical space is divided are sufficiently large to be in the thermodynamic limit for the entropy of the cell $S(M_\mu, E_\mu, \mathcal{V}_\mu)$ to be a well-defined first order function. We have also neglected the interaction energy between neighbour cells in front of the energy content of each cell (this assumption must be revised when surface tension effects need to be included [43]). Equation (95) has the form of the Einstein distribution function (64), and we can identify the entropy at our level of description as

$$S(x) = \sum_\mu S(M_\mu, E_\mu, \mathcal{V}_\mu). \tag{96}$$

Note that the entropy $S(x)$ that appears in the left hand side of (96), is a conceptually different quantity from the entropy appearing in the right hand side. In particular, they are functions of a different number of variables.

The next step would be to compute the time derivative iLX of the relevant variables needed in the two other quantities appearing in the FPE $v(x)$, $D(x)$, or, in the GENERIC framework $L(x)$, $M(x)$. Once iLX is computed we need to perform the constrained averages in $v(x)$, $D(x)$. These averages can be computed by following essentially the same approximations involved in the calculation of $P^{\mathrm{eq}}(x)$ in (95).

For reasons of space we do not dwell further into this microscopic derivation of the discrete hydrodynamic equations. Instead, we follow a very different route for the formulation of the equations governing the dynamics of the discrete hydrodynamic variables [44]. First, note that we can write the relevant variables (93) in the following form

$$M_\mu(z) = \int d\mathbf{r}\chi_\mu(\mathbf{r})\rho_\mathbf{r}(z),$$

$$\mathbf{P}_\mu(z) = \int d\mathbf{r}\chi_\mu(\mathbf{r})\mathbf{g}_\mathbf{r}(z),$$

$$E_\mu(z) = \int d\mathbf{r}\chi_\mu(\mathbf{r})\epsilon_\mathbf{r}(z), \tag{97}$$

where we have used (3). Therefore, the discrete, extensive variables (93) are just the space integral over each cell of the continuum densities (3). As we know that the continuum variables obey the continuum hydrodynamic equations, we can simply integrate the continuum equations over every little cell with the hope of obtaining a closed set of equations for the extensive variables (93). This is, actually, the well-known technique of finite volumes used to discretize the continuum equations [45]. In [44] and inspired by the work in [40], we have proposed a finite volume discretization of the Navier–Stokes continuum equations. The particularity of our approach lays in the fact that the partition of physical space is done with the help of a Voronoi tessellation that follows the flow field and thermodynamic consistency. The Voronoi tessellation is a way of partitioning the physical space by associating to a set of points named cell centers the region of space that is closer to each center. The result is a cellular structure like the one

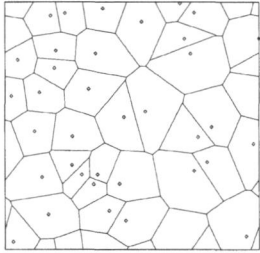

Fig. 4. The Voronoi tessellation for a randomly distributed set of points in a periodic box

in Fig. 4. One of the basic requirements of our discrete model formulation is that the model must have the GENERIC structure, thus ensuring a fulfillment of the First and Second Laws of Thermodynamics in an exact way. Of course, the continuum equations do satisfy these laws, and therefore, in the continuum limit any sensible discretization of the continuum equations will also respect them, up to certain errors related to the way in which we have performed the discretization. Our approach produces a model in which these essential laws are satisfied exactly. The physical picture is that the cells are understood as real portions of the material rather than the mathematical nodes where the continuum equations are discretized. In a way, we are pouring physics into the numerics.

The model has been tested in its basic aspects, showing that it reproduces the desired hydrodynamic and thermodynamic behaviour [44,46] in two space dimensions. However, the implementation of the Voronoi tessellation in three dimensions is rather involved and one has to rely on geometrical libraries where this implementation is done efficiently [47]. For this reason, we have devised a model that, although still based on the idea of fluid particles that represent physical portions of material, does not require the use of the Voronoi tessellation.

8.1 Soft Fluid Particles

We would like to illustrate in this section how the GENERIC structure can help in order to formulate new discrete models for the simulation of complex fluids. As a start we will construct step by step a model of fluid particles for the simulation of a simple Newtonian fluid. As a reference we have the continuum hydrodynamic equations [10]

$$\frac{d\rho}{dt} = -\rho\nabla\cdot\mathbf{v},$$

$$\rho\frac{d\mathbf{v}}{dt} = -\nabla P + \eta\nabla^2\mathbf{v} + \frac{\eta}{3}\nabla\nabla\cdot\mathbf{v},$$

$$\rho\frac{de}{dt} = -P\nabla\cdot\mathbf{v} + 2\eta\overline{\nabla\mathbf{v}} : \overline{\nabla\mathbf{v}} + \kappa\nabla^2 T. \tag{98}$$

These equations are written in Lagrangian form by introducing the well-known substantial derivative $\frac{d}{dt} = \frac{\partial}{\partial t} + \mathbf{v}\cdot\nabla$. The pressure P and temperature T are functions of the mass density ρ and the specific internal energy e (internal energy per unit mass) through the equilibrium equations of state. The transport coefficients are the shear viscosity η and the thermal conductivity κ (we assume for simplicity a zero bulk viscosity). Our aim is to present a discrete model that captures the essential physics described by the continuum equations (98).

The discrete elements of the model are the fluid particles, a concept that is actually used in order to motivate the continuum equations (98) [11]. We understand a fluid parti- cle as a small moving thermodynamic subsystem of the whole system characterized by its position $\mathbf{r}_i$, velocity $\mathbf{v}_i$, volume $\mathcal{V}_i$, mass m_i, internal energy E_i, entropy S_i, temperature T_i, and pressure P_i. Not all these variables are independent. We choose as independent variables that fully characterize the state of the system the set $x = \{\mathbf{r}_i, \mathbf{v}_i, E_i\}$. In this model we will assume for simplicity that the mass of the fluid particles is constant and equal for all particles $m_i = m$. The dependent variables are the volume $\mathcal{V}_i$, that we will assume completely determined by the relative positions between fluid particles, and the entropy $S_i = S(\mathcal{V}_i, E_i, m)$, which is a function of the extensive variables of the fluid particles given by the equilibrium entropy. The pressure and temperature are given by the thermodynamic relationships

$$\frac{1}{T_i} = \frac{\partial S_i}{\partial E_i} \quad \text{and} \quad -\frac{P_i}{T_i} = \frac{\partial S_i}{\partial \mathcal{V}_i}.$$

At this point we must specify how we assign a volume to each fluid particle. For the case of the Voronoi tessellation, there is an obvious way of giving a volume to the fluid particle, precisely the volume of the corresponding cell. If we do not want to use the Voronoi tessellation due to its difficult implementation, we must find an alternative way. Basically, we would like to assign a small volume to those particles that have many neighbours, because then the physical volume will be shared by these neighbours. One way to achieve this is by introducing a density d_i associated to each fluid particle as

$$d_i = \sum_j W(r_{ij}), \tag{99}$$

where $W(r)$ is a weight function of finite support h and normalized to unity $\int \mathrm{d}r W(r) = 1$ of the kind plotted in Fig. 5. The volume of particle i is defined as $\mathcal{V}_i = d_i^{-1}$. Note that when the particle i has many close neighbours, the sum in (99) is large, and, therefore, its volume will be correspondingly small.

The energy and entropy of the model are given by (94) and (96) which we write as

$$E(x) = \sum_i \left(\frac{m}{2}\mathbf{v}_i^2 + E_i \right) \quad \text{and} \quad S(x) = \sum_i S(E_i, \mathcal{V}_i).$$

Its derivatives with respect to the state variables are

$$\frac{\partial E}{\partial x} = \begin{pmatrix} \frac{\partial E}{\partial \mathbf{r}_i} \\ \frac{\partial E}{\partial \mathbf{v}_i} \\ \frac{\partial E}{\partial E_i} \end{pmatrix} = \begin{pmatrix} 0 \\ m\mathbf{v}_i \\ 1 \end{pmatrix}, \quad \frac{\partial S}{\partial x} = \begin{pmatrix} \frac{\partial S}{\partial \mathbf{r}_i} \\ \frac{\partial S}{\partial \mathbf{v}_i} \\ \frac{\partial S}{\partial E_i} \end{pmatrix} = \begin{pmatrix} \sum_j \frac{P_j}{T_j}\frac{\partial \mathcal{V}_j}{\partial \mathbf{r}_i} \\ 0 \\ \frac{1}{T_j} \end{pmatrix}. \tag{100}$$

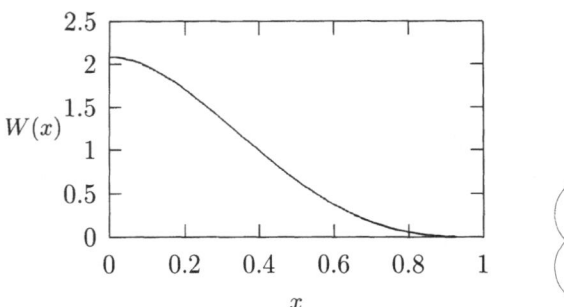

Fig. 5. The weight function $W(r/h) = \frac{105}{16\pi h^3}\left(1 + 3\frac{r}{h}\right)\left(1 - \frac{r}{h}\right)^3$, known as the Lucy function, and a pictorial view of a set of soft fluid particles, to be compared with the Voronoi tessellation in Fig. 4

We now construct the reversible part of the dynamics $\dot{x}|_{\text{rev}} = L(x)\nabla E(x)$ by defining the antisymmetric matrix L

$$
\begin{pmatrix} \dot{\mathbf{r}}_i \\ \dot{\mathbf{v}}_i \\ \dot{E}_i \end{pmatrix} = \sum_j \begin{pmatrix} 0 & \frac{\delta_{ij}}{m}\mathbf{1} & 0 \\ -\frac{\delta_{ij}}{m}\mathbf{1} & \Lambda_{ij} & \Omega_{ij} \\ 0 & -\Omega_{ji} & \Delta_{ij} \end{pmatrix} \begin{pmatrix} 0 \\ m\mathbf{v}_j \\ 1 \end{pmatrix}.
\tag{101}
$$

The first row is postulated by our desire of having the reasonable equation $\dot{\mathbf{r}}_i = \mathbf{v}_i$. The first column is then fixed by antisymmetry. The matrix Λ_{ij} should satisfy the antisymmetry $\Lambda_{ij} = -\Lambda_{ji}^T$ and the scalar should satisfy $\Delta_{ij} = -\Delta_{ji}$ in order to have L antisymmetric. Next, we make use of the degeneracy condition $0 = L(x)\nabla S(x)$ as it appears in (87)

$$
\begin{pmatrix} 0 \\ 0 \\ 0 \end{pmatrix} = \sum_j \begin{pmatrix} 0 & \frac{\delta_{ij}}{m}\mathbf{1} & 0 \\ -\frac{\delta_{ij}}{m}\mathbf{1} & \Lambda_{ij} & \Omega_{ij} \\ 0 & -\Omega_{ji} & \Delta_{ij} \end{pmatrix} \begin{pmatrix} \sum_k \frac{P_k}{T_k}\frac{\partial \mathcal{V}_k}{\partial \mathbf{r}_j} \\ 0 \\ \frac{1}{T_j} \end{pmatrix}.
\tag{102}
$$

We obtain, therefore, the conditions

$$
\sum_j \left[-\frac{P_j}{mT_j}\frac{\partial \mathcal{V}_j}{\partial \mathbf{r}_i} + \Omega_{ij}\frac{1}{T_j} \right] = 0 \quad \text{and} \quad \sum_j \Delta_{ij}\frac{1}{T_j} = 0.
\tag{103}
$$

Apparently, the simplest way of satisfying these conditions is by assuming

$$
\Omega_{ij} = \frac{P_j}{m}\frac{\partial \mathcal{V}_j}{\partial \mathbf{r}_i}, \quad \text{and} \quad \Delta_{ij} = 0.
\tag{104}
$$

Although we do not have any restriction for Λ_{ij}, we will assume for simplicity that $\Lambda_{ij} = 0$. In this way, the final reversible equations are

$$\dot{\mathbf{r}}_i = \mathbf{v}_i,$$

$$m\dot{\mathbf{v}}_i = \sum_j \frac{\partial \mathcal{V}_j}{\partial \mathbf{r}_i} P_j,$$

$$\dot{E}_i = -P_i \sum_j \frac{\partial \mathcal{V}_i}{\partial \mathbf{r}_j} \mathbf{v}_j = -P_i \dot{\mathcal{V}}_i. \tag{105}$$

The last equation makes a lot of sense if we think on the First Law expressed as $dE = -Pd\mathcal{V} + TdS + \mu dM$. In a reversible dynamics where $dS = 0$, the changes in internal energy are due only to changes in the volume (we assumed that the fluid particles have constant mass, so $dM = 0$). Note that this interpretation would break down if $\Delta_{ij} \neq 0$, so the last equation in (104) was a sensible requirement. We can use now the definition of the volume in terms of the density (99) in order to compute

$$\frac{\partial \mathcal{V}_j}{\partial \mathbf{r}_i} = -\frac{1}{d_j^2}\left[\delta_{ij} \sum_k W'(r_{ik})\mathbf{e}_{ik} + W'(r_{ij})\mathbf{e}_{ij}\right]. \tag{106}$$

Substitution of (106) into (105) leads to

$$\dot{\mathbf{r}}_i = \mathbf{v}_i,$$

$$m\dot{\mathbf{v}}_i = -\sum_j W'(r_{ij})\left[\frac{P_i}{d_i^2} + \frac{P_j}{d_j^2}\right]\mathbf{e}_{ij},$$

$$\dot{E}_i = \frac{P_i}{d_i^2}\sum_j W'(r_{ij})\mathbf{e}_{ij}\cdot\mathbf{v}_{ij}. \tag{107}$$

At this point, we would like to compare these equations with the reversible part of the continuum equations (98) (obtained by setting η, κ to zero). First, let us introduce the discrete counterpart of the mass density, $\rho_i = md_i$, and of the internal energy per unit mass $e_i = E_i/m$. If we take the time derivative of ρ_i with the help of $\dot{\mathbf{r}}_i = \mathbf{v}_i$ we obtain

$$\dot{\rho}_i = m\sum_j W'(r_{ij})\mathbf{e}_{ij}\cdot\mathbf{v}_{ij}, \tag{108}$$

which, after comparison with the continuity equation in (98), suggest that a discrete version of the divergence of the velocity field is given by

$$(\nabla\cdot\mathbf{v})_i = -\frac{1}{d_i}\sum_j W'(r_{ij})\mathbf{e}_{ij}\cdot\mathbf{v}_{ij}.$$

We can now compute $\rho_i\dot{e}_i = d_i\dot{E}_i$ which, after use of $\dot{E}_i$ in (107), is given by

$$\rho_i\dot{e}_i = \frac{P_i}{d_i}\sum_j W'(r_{ij})\mathbf{e}_{ij}\cdot\mathbf{v}_{ij} = -P_i(\nabla\cdot\mathbf{v})_i.$$

This shows that we have made a good choice by selecting $\Lambda_{ij} = 0$ and $\Delta_{ij} = 0$, because the resulting equations (107) are, term by term, very close in form to the continuum equations. We can actually understand (107) as a particularly nice discretization of the reversible part of the continuum equations (98) that conserve exactly mass, momentum, energy, and entropy.

We have now to consider the irreversible part of the dynamics. The route that we will follow consists on formulating first the stochastic noise terms $d\tilde{x} = (d\tilde{\mathbf{r}}_i, d\tilde{\mathbf{v}}_i, d\tilde{E}_i)$. As we have already noticed, we have some restrictions on the possible form of these noise, in particular (84). The basic benefit of formulating the noises is that we can construct in a straightforward manner the matrix $M(x)$ through the Fluctuation-Dissipation theorem (81), which, by construction, ensures that $M(x)$ will be symmetric and positive semi-definite. The first requirement about the noise terms is that $d\tilde{\mathbf{r}}_i = 0$. The basic reason is that we want to retain the equation $\dot{\mathbf{r}}_i = \mathbf{v}_i$. If $d\tilde{\mathbf{r}}_i \neq 0$ we would obtain additional terms in this equation. The next requirement is (84) which, after using (100), takes the form

$$\sum_i m\mathbf{v}_i \cdot d\tilde{\mathbf{v}}_i + d\tilde{E}_i = 0. \tag{109}$$

Physically, this equation states that the momentum and energy noises are not independent, because the power dissipated by the momentum forces must be canceled by (a term in) the energy noise in order to conserve total energy. We have to specify now the explicit forms of $md\tilde{\mathbf{v}}_i$, $d\tilde{E}_i$, that is, how they are expressed in terms of independent increments of the Wiener process. As a rule of thumb to follow when formulating stochastic noises, we may say that every noise term reflects an elementary transport process. For example, we expect that the stochastic change of the momentum of a fluid particle is due to a random interchange of momentum between neighbouring particles, i.e. $md\tilde{\mathbf{v}}_i = \sum_j d\mathbf{P}_{ij}$. The elementary momentum interchange $d\mathbf{P}_{ij}$ is a stochastic vector. We can construct many different stochastic vectors in terms of the Wiener process. For reasons that will became clear later, we assume

$$md\tilde{\mathbf{v}}_i = \sum_j A_{ij} d\mathbf{W}_{ij} \cdot \mathbf{e}_{ij}, \tag{110}$$

where $d\mathbf{W}_{ij} = d\mathbf{W}_{ji}$ is a matrix of Wiener processes and $\mathbf{e}_{ij}$ is the unit vector pointing from particle j to particle i. We postulate $A_{ij} = A_{ji}$ in such a way that $\sum_i md\tilde{\mathbf{v}}_i = 0$ and, therefore, momentum will be exactly conserved not only by the stochastic terms but also for the full dissipative terms. Concerning the elementary process involved in the energy transport, we assume it is of the form $C_{ij} dV_{ij}$ where the Wiener process $dV_{ij} = -dV_{ji}$ and $C_{ij} = C_{ji}$. In this way, in every stochastic elementary process, a small amount of energy $C_{ij} dV_{ij}$ is transferred from particle j to i (and particle j loses this energy, as it apparent from the symmetry properties assumed for C_{ij} and dV_{ij}). Therefore, we will assume that the stochastic energy term has the form

$$d\tilde{E}_i = \sum_j C_{ij} dV_{ij} - \frac{1}{2} \sum_j A_{ij} d\mathbf{W}_{ij} : \mathbf{e}_{ij} \mathbf{v}_{ij}. \tag{111}$$

The last term is included in such a way that (109) is exactly satisfied and represents the stochastic energy transfer that occurs as a consequence of the stochastic transfer of momentum among the fluid particles. Of course, we still have to specify the stochastic properties of the Wiener processes $\mathrm{d}\mathbf{W}_{ij}, \mathrm{d}V_{ij}$ according to its symmetry properties. They are

$$
\begin{aligned}
\mathrm{d}\mathbf{W}_{ii'}^{\alpha\alpha'}\,\mathrm{d}\mathbf{W}_{jj'}^{\beta\beta'} &= [\delta_{ij}\delta_{i'j'} + \delta_{ij'}\delta_{i'j}]\delta^{\alpha\beta}\delta^{\alpha'\beta'}\,\mathrm{d}t, \\
\mathrm{d}V_{ii'}\mathrm{d}V_{jj'} &= [\delta_{ij}\delta_{i'j'} - \delta_{ij'}\delta_{i'j}]\mathrm{d}t, \\
\mathrm{d}\mathbf{W}_{ii'}^{\alpha\alpha'}\,\mathrm{d}V_{ii'} &= 0.
\end{aligned}
\tag{112}
$$

Superscripts denote tensorial components. These properties imply the following properties of the stochastic terms (110) and (111)

$$
m^2 \frac{\mathrm{d}\tilde{\mathbf{v}}_i^\alpha \mathrm{d}\tilde{\mathbf{v}}_j^\beta}{\mathrm{d}t} = \delta_{ij}\left[\sum_k \frac{A_{ik}^2}{2}\left(\delta^{\alpha\beta} + \mathbf{e}_{ik}^\alpha \mathbf{e}_{ik}^\beta\right)\right] - \frac{A_{ij}^2}{2}\left(\delta^{\alpha\beta} + \mathbf{e}_{ij}^\alpha \mathbf{e}_{ij}^\beta\right),
$$

$$
-m\frac{\mathrm{d}\tilde{\mathbf{v}}_i^\alpha \mathrm{d}\tilde{E}_j}{\mathrm{d}t} = \delta_{ij}\left[\sum_k \frac{A_{ik}^2}{2}\left(\frac{\mathbf{v}_{ik}^\alpha}{2} + \mathbf{e}_{ik}\cdot\frac{\mathbf{v}_{ik}}{2}\mathbf{e}_{ik}^\alpha\right)\right] + \frac{A_{ij}^2}{2}\left(\frac{\mathbf{v}_{ij}^\alpha}{2} + \mathbf{e}_{ij}\cdot\frac{\mathbf{v}_{ij}}{2}\mathbf{e}_{ij}^\alpha\right),
$$

$$
-m\frac{\mathrm{d}\tilde{E}_i \mathrm{d}\tilde{\mathbf{v}}_j^\alpha}{\mathrm{d}t} = \delta_{ij}\left[\sum_k \frac{A_{ik}^2}{2}\left(\frac{\mathbf{v}_{ik}^\alpha}{2} + \mathbf{e}_{ik}\cdot\frac{\mathbf{v}_{ik}}{2}\mathbf{e}_{ik}^\alpha\right)\right] - \frac{A_{ij}^2}{2}\left(\frac{\mathbf{v}_{ij}^\alpha}{2} + \mathbf{e}_{ij}\cdot\frac{\mathbf{v}_{ij}}{2}\mathbf{e}_{ij}^\alpha\right),
$$

$$
\frac{\mathrm{d}\tilde{E}_i \mathrm{d}\tilde{E}_j}{\mathrm{d}t} = \delta_{ij}\left[\sum_k \frac{A_{ik}^2}{2}\left(\left(\frac{\mathbf{v}_{ik}}{2}\right)^2 + \left(\mathbf{e}_{ik}\cdot\frac{\mathbf{v}_{ik}}{2}\right)^2\right)\right]
$$

$$
+ \frac{A_{ij}^2}{2}\left(\left(\frac{\mathbf{v}_{ij}}{2}\right)^2 + \left(\mathbf{e}_{ij}\cdot\frac{\mathbf{v}_{ji}}{2}\right)^2\right) + \delta_{ij}\sum_k C_{ik}^2 - C_{ij}^2.
\tag{113}
$$

Equations (101) and (113) provide all the elements to formulate the deterministic equations $\dot{x} = L(x)\nabla E(x) + M(x)\nabla S(x)$, i.e.,

$$
\begin{pmatrix} \dot{\mathbf{r}}_i \\[2mm] \dot{\mathbf{v}}_i \\[2mm] \dot{E}_i \end{pmatrix} = \sum_j \begin{pmatrix} 0 & \frac{\delta_{ij}}{m}\mathbf{1} & 0 \\[2mm] -\frac{\delta_{ij}}{m}\mathbf{1} & 0 & -\frac{P_j}{m}\frac{\partial \mathcal{V}_j}{\partial \mathbf{r}_i} \\[2mm] 0 & \frac{P_i}{m}\frac{\partial \mathcal{V}_i}{\partial \mathbf{r}_j} & 0 \end{pmatrix} \begin{pmatrix} 0 \\[2mm] m\mathbf{v}_j \\[2mm] 1 \end{pmatrix}
$$

$$
+ \sum_j \begin{pmatrix} 0 & 0 & 0 \\[2mm] 0 & \frac{\mathrm{d}\tilde{\mathbf{v}}_i \mathrm{d}\tilde{\mathbf{v}}_j^T}{2k_{\mathrm{B}}\mathrm{d}t} & \frac{\mathrm{d}\tilde{\mathbf{v}}_i \mathrm{d}\tilde{E}_j}{2k_{\mathrm{B}}\mathrm{d}t} \\[2mm] 0 & \frac{\mathrm{d}\tilde{E}_i \mathrm{d}\tilde{\mathbf{v}}_j^T}{2k_{\mathrm{B}}\mathrm{d}t} & \frac{\mathrm{d}\tilde{E}_i \mathrm{d}\tilde{E}_j}{2k_{\mathrm{B}}\mathrm{d}t} \end{pmatrix} \begin{pmatrix} -\sum_k \frac{P_k}{T_k}\frac{\partial \mathcal{V}_k}{\partial \mathbf{r}_j} \\[2mm] 0 \\[2mm] \frac{1}{T_j} \end{pmatrix},
\tag{114}
$$

which, by using (113) lead to the following set of equations

$$\dot{\mathbf{r}}_i = \mathbf{v}_i,$$

$$m\dot{\mathbf{v}}_i = -\sum_j W'(r_{ij}) \left[\frac{P_i}{d_i^2} + \frac{P_j}{d_j^2} \right] \mathbf{e}_{ij} - \sum_j a_{ij} \left(\mathbf{v}_{ij} + \mathbf{e}_{ij} \cdot \mathbf{v}_{ij} \mathbf{e}_{ij} \right),$$

$$\dot{E}_i = P_i \sum_j W'(r_{ij})\mathbf{e}_{ij}\cdot\mathbf{v}_{ij} + \frac{1}{2}\sum_j a_{ij}\left(\mathbf{v}_{ij}^2 + (\mathbf{v}_{ij}\cdot\mathbf{e}_{ij})^2\right) - \sum_j c_{ij}(T_i - T_j), \quad (115)$$

where we have introduced

$$a_{ij} = \frac{A_{ij}^2}{8k_{\rm B}} \left(\frac{1}{T_i} + \frac{1}{T_j} \right) \qquad \text{and} \qquad c_{ij} = \frac{C_{ij}^2}{2k_{\rm B}T_iT_j}.$$

Equations (115) are our final set of dynamical equation for the model of soft fluid particles. Everything is known in these equations except the factors a_{ij} and c_{ij} and we have to propose reasonable expressions for these objects. In principle, these factors govern the overall amplitude of the dissipation and, therefore, they govern the dissipative transport of momentum and energy. By comparing the structure of (115) with the continuum equations (98) we observe a nice correspondence between each term in both sets of equations. We expect that a_{ij} will be related to the viscosity and c_{ij} to the thermal conductivity, and these terms will correspond to a sort of discrete version of second spatial derivatives. In order to make the connection even more explicit, consider the following integral version of the matrix of second derivatives of an arbitrary function $A(\mathbf{r})$,

$$\nabla\nabla A(\mathbf{r}_i) = \int d\mathbf{r}' [A(\mathbf{r}') - A(\mathbf{r}_i)] \frac{W'(|\mathbf{r}'-\mathbf{r}_i|)}{|\mathbf{r}'-\mathbf{r}_i|} \left[1 - 5\frac{(\mathbf{r}'-\mathbf{r}_i)(\mathbf{r}'-\mathbf{r}_i)}{(\mathbf{r}'-\mathbf{r}_i)^2} \right] + \mathcal{O}(\nabla^4 A h^2).$$

$$(116)$$

This equation is demonstrated by Taylor expanding $A(\mathbf{r}')$ around $\mathbf{r}_i$ and making use of the isotropy and normalization to unity of the weight function $W(r)$. The integral $\int d\mathbf{r}$ can be further approximated with a discrete sum $\sum_i d_i^{-1}$ over the fluid particles in such a way that

$$\nabla\nabla A(\mathbf{r}_i) = \sum_j \frac{1}{d_j}[A_j - A_i]\frac{W'_{ij}}{r_{ij}} [1 - 5\mathbf{e}_{ij}\mathbf{e}_{ij}] + \mathcal{O}(\nabla^4 A h^2). \qquad (117)$$

In this way, we can approximate the second spatial derivatives of a function in terms of the value of the function in the neighbourhood. If we look at the second derivative terms that appear in (98) we can obtain the following discrete versions for them,

$$\eta \left[(\nabla^2 \mathbf{v})_i + \frac{1}{3}(\nabla\nabla\cdot\mathbf{v})_i \right] \approx \frac{5\eta}{3} d_i \sum_j \omega_{ij} [\mathbf{v}_{ij} + \mathbf{e}_{ij}\mathbf{e}_{ij}\cdot\mathbf{v}_{ij}],$$

$$\kappa(\nabla^2 T)_i \approx 2\kappa d_i \sum_j \omega_{ij} T_{ij}, \qquad (118)$$

where ω_{ij} is a geometrical object that depends only on the positions of the particles,

$$\omega_{ij} = \omega_{ji} = -\frac{W'_{ij}}{r_{ij}d_id_j} \geq 0.$$

By comparing (118) with the dissipative terms in (115) we may postulate

$$a_{ij} = \frac{5\eta}{3}\omega_{ij} \quad \text{and} \quad c_{ij} = 2\kappa\omega_{ij}. \tag{119}$$

In this way, we can understand the discrete model (115) with (119) as a discretization of the continuum equations (98). According to the approximation (117) the simulation of (115) will produce results that are compatible with a given viscosity η and thermal conductivity κ.

Equations (115) conserve mass, momentum, energy and have a positive production of entropy. Arbitrary equations of state can be introduced through the specific functional form of the entropy function and we have, therefore, full thermodynamic consistency. The physical meaning of every term in these equations is also very transparent: The particles move following the flow field and they exert forces to its neighbouring particles of two kinds. First, a reversible force due to the average pressure of the interacting particles, which is directed to the line joining the particles. Second, a dissipative force that tries to reduce the velocity differences between neighbouring particles, and with an overall amplitude given by the viscosity of the fluid. Because these forces try to stop the particles, the kinetic energy which is dissipated must be transformed into internal energy, a fact that is described by the term quadratic in the velocity in the energy equation. In the energy equation there is a term due to the reversible work done by the pressure forces, and another dissipative term that tries to reduce the temperature differences between neighbouring particles, describing the phenomenon of heat conductivity.

Of course, (115) are deterministic equations. In order to construct the stochastic differential equations, we have to compute the term $k_{\mathrm{B}}\nabla M$ appearing in (78) and add the stochastic forces $d\tilde{x}$. The resulting final SDE are

$$d\mathbf{r}_i = \mathbf{v}_i dt,$$

$$m d\mathbf{v}_i = -\sum_j W'(r_{ij})\left[\frac{P_i}{d_i^2}+\frac{P_j}{d_j^2}\right]\mathbf{e}_{ij}dt - \frac{5\eta}{3}\sum_j(1-d_{ij})\omega_{ij}(\mathbf{v}_{ij}+\mathbf{e}_{ij}\cdot\mathbf{v}_{ij}\mathbf{e}_{ij})\,dt + m d\tilde{\mathbf{v}}_i,$$

$$dE_i = P_i\sum_j W'(r_{ij})\mathbf{e}_{ij}\cdot\mathbf{v}_{ij}dt$$

$$+ \frac{5\eta}{6}\sum_j\left[1-d_{ij}-\frac{T_j}{T_i+T_j}\frac{k_{\mathrm{B}}}{C_i}\right]\omega_{ij}\left(\mathbf{v}_{ij}^2+(\mathbf{v}_{ij}\cdot\mathbf{e}_{ij})^2\right) - 2\kappa\sum_j\omega_{ij}(T_i-T_j)$$

$$- \frac{20k_{\mathrm{B}}}{3m}\sum_j\frac{T_iT_j}{T_i+T_j}\omega_{ij}dt - 2\kappa\frac{k_{\mathrm{B}}}{C_i}\sum_j\omega_{ij}T_jdt + d\tilde{E}_i, \tag{120}$$

where we have introduced the dimensionless quantity

$$d_{ij} = \frac{T_iT_j}{(T_i+T_j)^2}\left[\frac{k_{\mathrm{B}}}{C_i}+\frac{k_{\mathrm{B}}}{C_j}\right], \tag{121}$$

where C_i is the heat capacity at constant volume of particle i. which is an extensive quantity. Therefore, for large fluid particles the dimensionless ratio k_B/C_i is very small. Note that in the limit $k_B \to 0$, one recovers the deterministic equations (115). We have here an explicit example of what has been discussed in Sect. 7.3 about the size of thermal fluctuations. In this case, the importance of thermal fluctuations is dictated by the physical size of the fluid particles.

Equations (115) with (119) are just a version of Smoothed Particle Hydrodynamics (SPH). SPH is a well-known technique used to discretize continuum equations with the aid of suitable weight functions [48]. In that respect, (120) represent a generalization of SPH that include thermal fluctuations in a thermodynamically consistent way. This generalization will be necessary when the fluid particles are very small. As the size of the fluid particles is dictated essentially by the need of resolving with sufficient accuracy the length scales of the problem, the fluctuating SPH equations (120) will be required in microhydrodynamic problems as those appearing in microfluidic devices or in the simulation of the solvent in colloidal suspensions.

There are many different implementations of the SPH idea and every group seems to have its favourite [48]. The particular discretization (116) is new. Other formulations of the stochastic forces are possible and they lead to different structural forms for the irreversible part of the dynamics which, still, can be considered as discretizations of second derivative terms [44]. In particular, it is possible to introduce a stochastic stress tensor and heat flux associated to each fluid particle, and formulate discrete divergences of these quantities for the corresponding stochastic terms in the momentum and energy equations [49]. This procedure is closely related to the Landau and Lifshitz method of introducing thermal fluctuation in continuum hydrodynamics. The basic reason for the formulation of the discrete model for hydrodynamics in the form (120) is to obtain a set of discrete equations that involve dissipative forces that depend on the velocity differences between neighbouring fluid particles. This was also the motivation to write the stochastic force in terms of a matrix of Wiener processes in the form (110). In this way, one can derive a model that is closely related to the Dissipative Particle Dynamics (DPD) model.

The DPD model was introduced as a promising mesoscopic technique that would allow to simulate complex fluids efficiently [50]. For a couple of recent reviews of the method see [51,52]. The model is characterized by a set of point particles with positions $\mathbf{r}_i$ and velocities $\mathbf{v}_i$ variables (generalizations of the model with an internal energy variable were presented latter in [53,54]). The postulated DPD equations have the form

$$
\begin{aligned}
\mathrm{d}\mathbf{r}_i &= \mathbf{v}_i \mathrm{d}t, \\
m\mathrm{d}\mathbf{v}_i &= -\sum_j W_{ij}\mathbf{e}_{ij}\mathrm{d}t - \gamma \sum_j W_{ij}\mathbf{e}_{ij}\cdot\mathbf{v}_{ij}\mathbf{e}_{ij}\mathrm{d}t + m\mathrm{d}\tilde{\mathbf{v}}_i,
\end{aligned}
\tag{122}
$$

where $W_{ij} = W(r_{ij})$ is a bell-shaped function, γ is a friction coefficient, and $\mathrm{d}\tilde{\mathbf{v}}_i$ is a random term with a form similar to (110) where the matrix $\mathrm{d}\mathbf{W}_{ij}$ is replaced by a single scalar $\mathrm{d}W_{ij}$. The similitude of the DPD model and the SPH model is apparent.

We definitely prefer the SPH model over the DPD model as originally formulated. First of all, there are no restrictions in the DPD model about the form of the function $W(r)$ that determines the range of interaction between particles. The fluid particles have no physical size and the scale of the DPD particles is undetermined. There is no direct connection between the friction coefficient γ and the viscosity of the solvent and one has to resort to kinetic theory for such a connection [55]. The model is purely isothermal, as a consequence of the missing internal energy of the fluid particles, and energy is not conserved. The thermodynamic behaviour of the system is dictated by the form of $W(r)$, and cannot be varied (it produces a pressure that scales quadratically with the density [56])). Although attempts have been taken to introduce arbitrary equations of state [57], they are still isothermal. None of this problems appear in the stochastic SPH equations (78). The particles have definite sizes that fix the relevance of thermal fluctuations and the actual scale being simulated. The transport properties of the fluid being simulated are input of the model, arbitrary equations of state can be introduced, and non-isothermal situations can be studied.

8.2 Complex Fluids

In the previous sections we have discussed two discrete fluid particle models for the simulation of the Navier–Stokes equations. These models can be understood as suitable discretizations of the continuum equations, with the bonus of having thermodynamic consistency even at the discrete level. Of course, one is interested in the simulation of complex fluids, which may even not have a known constitutive equation. There are two possible strategies for the simulation of complex fluids. The first one has been used by the DPD community in order to study complex fluid systems like colloids, polymers, mixtures, amphiphilic systems, membranes, etc. The basic strategy has been to complexify the simple fluid particle model by introducing further conservative interactions between dissipative particles. By connecting fluid particles with springs one has a model for macromolecules, by moving the particles within a sphere like a rigid solid one can model solid spherical colloidal particles, by having two types of particles which repel in different ways one gets a binary mixture that can phase separate, etc. [51,52].

The second one is the introduction of additional internal variables characterizing the microstructural state of the fluid particles. These variables are coupled to the conventional hydrodynamic variables. The coupling renders the behaviour of the fluid non-Newtonian and complex. For example, polymer melts are characterized by additional conformation tensors, colloidal suspensions can be described by further concentration fields, mixtures are characterized by several density fields (one for each chemical specie), emulsions are described with the amount and orientation of interface, etc. All these continuum models rely on the hypothesis of local equilibrium and, therefore, the fluid particles are regarded as thermodynamic subsystems. The physical picture emerging from these fluid particles is that they represent large portions of the fluid and therefore, the scale of these fluid particles is supramolecular. This allows one to study time scales larger than those described with a DPD model with additional conservative interactions. The price, of course, is the need for a deep understanding of the physics at this more coarse-grained level, which appears in the form of entropy and energy functionals depending on internal variables, and kinematic and dissipative matrices $L(x), M(x)$ describing the complex coupling of the internal microstructure and flow.

In order to model polymer solutions, for example, we have developed a thermodynamically consistent model of fluid particles inspired by a previous model introduced

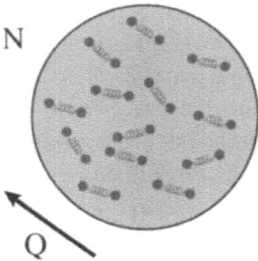

Fig. 6. A fluid particle containing N polymer molecules (dumbbells) with a typical orientation $\mathbf{Q}$. These two further variables, N and $\mathbf{Q}$ characterize, along with $\mathbf{r}_i, \mathbf{v}_i, \mathcal{V}_i, S_i$, the state of the fluid particle, and their dynamics is specified from simple physical grounds: fluid particles exchange dumbbells due to chemical potential differences and the vector $\mathbf{Q}$ changes due to the underlying Brownian motion of the dumbbells and the extensional state of the fluid particle

by Ten Bosch [58]. In this fluid particle model the dissipative particles include as additional variables the number N of polymer molecules within the fluid particle and an elongation vector $\mathbf{Q}$ representing the average elongation of polymer molecules within the fluid particle. A graphical view of the fluid particle for polymer solutions is given in Fig. 6.

The formulation of the equations of motion for the hydrodynamic and elastic variables in this model is very much simplified by the GENERIC structure of the dynamic equations and requires very simple assumptions about the underlying motion of the polymer molecules [59]. The model allows for the interchange of polymer molecules among fluid particles, leading to advection-diffusion behaviour. Also, the elastic state of the fluid particle produces forces on the fluid particles leading to a viscoelastic behaviour for the model.

Another example where the strategy of internal variables is successful is in the simulation of chemically reacting mixtures. Chemically reacting mixtures are not easily implemented with the usual approach taken by DPD in order to model mixtures. In DPD, mixtures are represented by red and blue particles. It is not trivial to specify a chemical reaction in which, for example, two red particles react with a blue particle to form a green particle. In this case, it is better to start from the well established continuum equations for chemical reactions [11,60]. The resulting discrete model consists on fluid particles that have as additional variable the fraction of component red and blue inside the fluid particle. There are diffusion terms that change this fraction depending on the difference between chemical potentials of neighbouring particles, and there are simple reaction terms, that can be expressed in measurable reaction rates [61].

9   Summary

In this chapter we have reviewed the theory of coarse graining, a well-established subject since the early 50's of last century under the name of non-equilibrium statistical mechanics. Under the basic assumption that the coarse-grained variables are slow, it is possible to obtain a Fokker–Planck equation that governs the full dynamics at the coarse-grained level. All the objects appearing in the FPE have a well-defined molecular interpretation and, in principle, it is straightforward to compute these objects from molecular dynamics simulations. In practice, however, this task is extremely difficult due to the high dimensionality of the coarse state space. One is condemned to model. However, the microscopic derivation provides a large number of restrictions on the possible forms of the objects in the FPE. This is particularly true for the case of GENERIC, a thermodynamically consistent structure that emerges as a particular case when the dynamical invariants can be expressed as functions of the coarse variables.

One of the focuses of this chapter is the formulation of fluid particle models for complex fluids. Rather than sticking to coarse-graining from the microscopic dynamics, it proves sufficient to construct models according to the GENERIC framework. In this way, we have shown how a fully thermodynamically consistent model of fluid particles can be devised in the spirit of the DPD model, but without any of its conceptual flaws. The resulting model belongs to the family of Smoothed Particle Hydrodynamics models with thermal fluctuations included consistently. Finally, we emphasize the usefulness of the GENERIC framework when devising new discrete models for complex fluids, which can be constructed by including additional structural variables describing the inner microstructure of the fluid and its coupling to hydrodynamic flow.

Acknowledgements

I am very grateful for the discussions with Hans Christian Öttinger during the years. These discussions have been always a source of understanding and inspiration. I would like to thank Marisol Ripoll and Mar Serrano for the very good times we have had discussing DPD.

References

1. J.G. Kirkwood: J. Chem. Phys. **14**, 180 (1946); J.H. Irwing, J.G. Kirkwood: J. Chem. Phys. **18**, 817 (1950)
2. M.S. Green: J. Chem. Phys. **20**, 1281 (1952); M.S. Green: J. Chem. Phys. **22**, 398 (1954)
3. R. Zwanzig: Phys. Rev. **124**, 983 (1961)
4. H. Mori: Prog. Theor. Phys. **33**, 423 (1965)
5. H. Grabert: *Projection Operator Techniques in Nonequilibrium Statistical Mechanics* (Springer-Verlag, Berlin 1982)
6. M. Grmela, H.C. Öttinger: Phys. Rev. E **56**, 6620 (1997); H.C. Öttinger, M. Grmela: Phys. Rev. E **56**, 6633 (1997)
7. P. Español, F. Vázquez: Phil. Trans. R. Soc. London Series A: A **360**, 1 (2002)
8. P. Resibois, M. De Leener: *Classical Kinetic Theory of Fluids* (Wiley 1977)
9. M.P. Allen, D.J. Tildesley: *Computer Simulations of Liquids* (Clarendon, Oxford 1987)
10. S.R. de Groot, P. Mazur: *Non-Equilibrium Thermodynamics* (North Holland, Amsterdam 1964)
11. L.D. Landau, E.M. Lifshitz: *Fluid Mechanics* (Pergamon Press 1959)
12. E.H. Hauge, A. Martin-Löf: J. Stat. Phys. **7**, 259 (1973)
13. D. Bedeaux, P. Mazur: Physica **76**, 235 (1974)
14. B.J. Alder, T.E. Wainwright: Phys. Rev. A **1**, 18 (1970)

15. Y.W. Kim, J.E. Matta: Phys. Rev. Lett. **31**, 208 (1973); G.L. Paul, P.N. Pusey: J. Phys. A: Math. Gen. **14**, 3301 (1981); J.X. Zhu, D.J. Durian, J. Müller, D.A. Weitz, D.J. Pine: Phys. Rev. Lett. **68**, 2559 (1992)
16. A.J.C. Ladd: Phys. Rev. Lett. **70**, 1339 (1993); A.J.C. Ladd: J. Fluid Mech. **271**, 285 (1994)
17. I. Pagonabarraga: Lattice Boltzmann Modeling of Complex Fluids: Colloidal Suspensions and Fluid Mixtures, Lect. Notes Phys. **640**, 275 (2004)
18. J.M.V.A. Koelman, P.J. Hoogerbrugge: Europhys. Lett. **21**, 363 (1993); E.S. Boek, P.V. Coveney, H.N.W. Lekkerkerker: J. Phys.: Condens. Matter **8**, 9509 (1996); E.S. Boek, P.V. Coveney, H.N.W. Lekkerkerker, P. van der Schoot: Phys. Rev. E **55**, 3124 (1997); E.S. Boek, P. van der Schoot: Int. J. Mod. Phys. C **9**, 1307 (1998)
19. A. Malevanets, R. Kapral: J. Chem. Phys. **110**, 8605 (1999); A. Malevanets, R. Kapral: J. Chem. Phys. **112**, 7260 (2000)
20. A. Malevanets, R. Kapral: Mesoscopic Multi-particle Collision Model for Fluid Flow and Molecular Dynamics, Lect. Notes Phys. **640**, 112 (2004)
21. H.C. Öttinger: *Stochastic Processes in Polymeric Fluids* (Springer, Berlin 1996)
22. T.J. Murphy, J.L. Aguirre: J. Chem. Phys. **57**, 2098 (1972)
23. J. García-Ojalvo, J.M. Sancho: *Noise in Spatially Extended Systems* (Springer, NY 1999)
24. D. Lhuillier: J. Non-Newt. Fluid Mech. **96**, 19 (2000)
25. P. Español, H.C. Öttinger: Zeitschrift für Physik B **90**, 377 (1993)
26. H. Goldstein: *Classical Mechanics* (Addison-Wesley 1950)
27. J. Español, F.J. de la Rubia: Physica A **187**, 589 (1992)
28. E.T. Jaynes: Phys. Rev. **106**, 620 (1957); *The Maximum Entropy Formalism*, ed. by R.D. Levine, M. Tribus (MIT Press, Cambridge, MIT 1979)
29. J. Español: Phys. Lett. A **146**, 21 (1990)
30. M. Spivak: *Calculus on Manifolds* (Benjamin, New York 1965)
31. J. Español, I. Zúñiga: J. Chem. Phys. **98**, 574 (1993)
32. V.G. Mavrantzas, H.C. Öttinger: Macromolecules **35**, 960 (2002)
33. N.J. Wagner, H.C. Öttinger, B.J. Edwards: AIChE J. **45**, 1169 (1999); H.C. Öttinger: Phys. Rev. D **60**, 103507 (1999); P. Ilg, H.C. Öttinger: Phys. Rev. D **61** 023510 (2000); B.J. Edwards, H.C. Öttinger: Phys. Rev. E **56**, 4097 (1997); H.C. Öttinger, A.N. Beris: J. Chem. Phys. **110**, 6593 (1999); H.C. Öttinger: J. Rheol. **43**, 1461 (1999)
34. H.C. Öttinger: Phys. Rev. E **57**, 1416 (1998)
35. H.C. Öttinger: private communication.
36. C.W. Gardiner: *Handbook of Stochastic Methods* (Springer Verlag, Berlin 1983)
37. H. Grabert, W. Weidlich: Phys. Rev. A **21**, 2147 (1980)
38. A.N. Kaufmann: Phys. Lett. **100**A, 419 (1984)
39. B.J. Edwards, H.C. Öttinger: Phys. Rev. E **56**, 4097 (1997)
40. E.G. Flekkøy, P.V. Coveney: Phys. Rev. Lett. **83**, 1775 (1999); E.G. Flekkøy, P.V. Coveney, G.D. Fabritiis: Phys. Rev. E **62**, 2140 (2000)
41. J.J. de Pablo, H.C. Öttinger: J. Non-Newt. Fluid Mech. **96**, 137 (2001)
42. P. Español, C.A.P. Thieulot: J. Chem. Phys. **118**, 9109 (2003)
43. P. Español: J. Chem. Phys. **115**, 5392 (2001)
44. M. Serrano, P. Español: Phys. Rev. E **65**, 46115 (2001)
45. S.V. Patankar: *Numerical Heat Transfer and Fluid Flow* (Hemisphere 1980)
46. M. Serrano, G. de Fabritiis, P. Español, E.G. Flekkøy, P.V. Coveney: J. Phys. A: Math. Gen. **35**, 1605 (2002)
47. Computational Geometry Algorithms Library at http://www.cgal.org
48. J.J. Monaghan: Annu. Rev. Astron. Astrophys. **30**, 543 (1992); H. Takeda, S.M. Miyama, M. Sekiya: Prog. Theor. Phys. **92**, 939 (1994); H.A. Posch, W.G. Hoover, O. Kum: Phys. Rev. E **52**, 1711 (1995); O. Kum, W.G. Hoover, H.A. Posch: Phys. Rev. E **52**, 4899 (1995); P.W. Cleary, J.J. Monaghan: J. Comp. Phys. **148**, 227 (1999)

49. P. Español, M. Serrano, H.C. Öttinger: Phys. Rev. Lett. **83**, 4552 (1999)
50. P.J. Hoogerbrugge, J.M.V.A. Koelman: Europhys. Lett. **19**, 155 (1992); P. Español, P. Warren: Europhys. Lett. **30**, 191 (1995)
51. P. Warren: Curr. Opin. Coll. Interface Sci. **3**, 620 (1998)
52. P. Español: Dissipative Particle Dynamics. In: *Trends in Nanoscale Mechanics: Analysis of Nanostructured Materials and Multi-Scale Modeling*, ed. by V. M. Harik, M. D. Salas (Kluwer 2003). An electronic version appeared in the 4th Simu Newsletter http://simu.ulb.ac.be/newsletters/newsletter.html
53. J. Bonet-Avalós, A.D. Mackie: Europhys. Lett. **40**, 141 (1997); J. Bonet-Avalós, A.D. Mackie: J. Chem. Phys. **11**, 5267 (1999)
54. P. Español: Europhys. Lett. **40**, 631 (1997); M. Ripoll, P. Español, M.H. Ernst: Int. J. Mod. Phys. C **9**, 1329 (1998)
55. C. Marsh, G. Backx, M.H. Ernst: Europhys. Lett. **38**, 411 (1997); C. Marsh, G. Backx, M.H. Ernst: Phys. Rev. E **56**, 1976 (1997); A.J. Masters, P.B. Warren: Europhys. Lett. **48**, 1 (1999); M. Ripoll, M.H. Ernst, P. Español: J. Chem. Phys. **115**, 7271 (2001)
56. R.D. Groot, P.B. Warren: J. Chem. Phys. **107**, 4423 (1997)
57. I. Pagonabarraga, D. Frenkel: J. Chem. Phys. **115**, 5015 (2001)
58. B.I.M. ten Bosch: J. Non-Newt. Fluid Mech. **83**, 231 (1999)
59. P. Español, E.G. Flekkøy, M. Ellero: submitted to Phys. Rev. E
60. L.E. Reichl: *A Modern Course in Statistical Physics* (Univ. of Texas Press, Austin 1980)
61. C. Thieulot, P. Español: preprint

Mesoscopic Multi-particle Collision Model for Fluid Flow and Molecular Dynamics

Anatoly Malevanets[1] and Raymond Kapral[2]

[1] Flow Software Technologies 3070 Jefferson Blvd., Windsor, ON N8T 3G9, Canada
[2] Chemical Physics Theory Group, Department of Chemistry, University of Toronto, Toronto, ON M5S 3H6, Canada

Abstract. Several aspects of modeling dynamics at the mesoscale level are discussed: (1) The construction of a mesoscopic description of fluid dynamics. The mesoscale dynamics consists of free streaming interrupted by multi-particle collisions. The multi-particle collisions are carried out by performing random rotations of particle velocities in predetermined cells in a manner that conserves mass, momentum and energy. The algorithmic implementation of the method is described and its theoretical basis is justified. Examples of simulation results on hydrodynamic flows are presented. (2) A hybrid molecular dynamics (MD)-Mesoscale Solvent model is described next. In this method full MD of solute particles is combined with the mesoscale dynamics of the surrounding fluid. The method is illustrated by considering the diffusive dynamics and hydrodynamic interactions among solute particles and clusters in the mesoscale solvent. (3) Finally, extensions of such schemes are outlined. In particular, generalizations to molecular solvents are presented and examples of solute dynamics in mesoscale water are given; also extensions to reactive flows are described.

1 Introduction

It is well known that the hydrodynamic equations of motion are a macroscopic manifestation of the microscopic conservation laws of mass, momentum and energy. This observation has prompted the construction of mesoscopic models whose dynamics may be simplistic in comparison to that of real fluids but which preserve these conservation laws and lead to the hydrodynamic equations on macroscopic distance and time scales. Perhaps the best known models of this class are lattice gas automata and lattice Boltzmann equations. Lattice gas automata for hydrodynamics often suffer from lattice artifacts that have limited their use, but they have proven their utility for the simulations of complex fluids or fluid flows in complex geometries [1,2]. Lattice Boltzmann methods [3] have been developed extensively and have been used to investigate a variety of problems ranging from hydrodynamic flows to fluid flows in complex geometries and complex systems [4–6].

In this chapter we discuss a particle-based mesoscopic model for the description of hydrodynamic fluid flow and solute molecular dynamics [7,8]. The fictitious particles of the model are not restricted to the sites of a lattice and thus particle positions and velocities can take on continuous values. The dynamics consists of free streaming interrupted by multi-particle collisions that change the particle velocities. Formally, the multi-particle collision dynamics is constructed as a superposition of propagators

A. Malevanets and R. Kapral, Mesoscopic Multi-particle Collision Model for Fluid Flow and Molecular Dynamics, Lect. Notes Phys. **640**, 116–149 (2004)
http://www.springerlink.com/

corresponding to these processes which act in position-momentum space and conserve momenta, energy and phase space volume. As a result, one may demonstrate that the exact full set of hydrodynamic equations is obtained in the macroscopic limit. We focus our attention on two specific types of propagator: Free-streaming propagator, arising from integration of the equations of motion for non-interacting molecules, and collision propagator, exchanging momenta among particles in a collision cell. The model differs from lattice Boltzmann models in that it is not a discrete simulation of a Boltzmann equation for a single particle distribution function, rather, it is a mesoscopic molecular dynamics which possesses the stability properties of particle-based methods. It is akin to Direct Simulation Monte Carlo (DSMC) methods [9] with a more efficient multi-particle collision dynamics.

The model should prove especially useful for applications to the dynamics of complex fluids and complex systems. Because of its particle nature, a hybrid version of the model that combines mesoscopic multi-particle collision dynamics with full molecular dynamics of embedded molecules or particles is easily constructed. Detailed features of the intermolecular forces between the mesoscopic solvent particles and the solute molecules are naturally taken into account. This permits potential applications of the model to large biomolecule or polymer dynamics in solution or to the dynamics of colloidal suspensions. For these systems the fluctuations intrinsic in the mesoscopic particle dynamics are an advantageous feature. Consequently, the model should see applications to a number of problems in biophysics and the rheology of complex systems.

The outline of the chapter is as follows: In Sect. 2 we describe the construction of the model and outline some of its main properties. Section 3 demonstrates how the hydrodynamic equations of motion can be deduced from the dynamics by the application of projection operator methods. In this section discrete Green–Kubo expressions for the transport properties are derived. Some illustrations of the utility of the scheme for simulations of fluid flow are described in Sect. 4. The hybrid scheme for molecular dynamics in the mesoscopic solvent is formulated in Sect. 5 while applications are discussed in Sect. 6. The conclusions are given in Sect. 7.

2   Multi-particle Collision Model for Fluid Flow

A simplified version of molecular collision dynamics in a fluid that yields the correct hydrodynamic equations on long distance and time scales can be constructed in the following way. We adopt a mesoscopic view of the fluid which involves discrete time updating of continuous particle positions and velocities through both free streaming and collisions. The dynamics is constructed so that the conservation laws of mass, momentum and energy are satisfied, an essential feature in any dynamical scheme.

Consider a system comprising N particles, each with mass m. [1] The particle positions and velocities are denoted by $\mathbf{X}^{(N)} = \{\mathbf{x}_1, \mathbf{x}_2, \ldots, \mathbf{x}_N\}$ and $\mathbf{V}^{(N)} = \{\mathbf{v}_1, \mathbf{v}_2, \ldots, \mathbf{v}_N\}$, respectively. The dynamics consists of free streaming,

$$\mathbf{x}_i(t + \tau) = \mathbf{x}_i(t) + \mathbf{v}_i(t)\tau \,, \tag{1}$$

---

[1] The mesoscopic collision dynamics is easily generalized to systems where the particles have different masses.

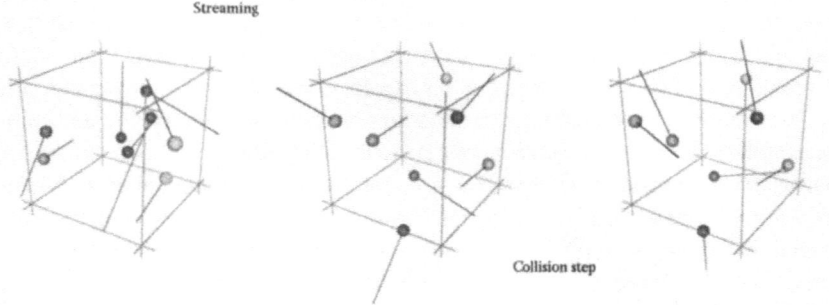

Fig. 1. Schematic representation of the division of a system into cells for the application of the multi-particle collision rule. The particle positions are continuous variables and are not confined to the cell centers. The velocities are also continuous and are denoted by lines in the figure

interspersed by multi-particle collisions at discrete time intervals τ. The post-collision velocities after multi-particle collisions [10] are determined by first dividing the system into cells as shown schematically in Fig. 1. At each time interval rotation operators $\hat{\omega}_\xi$ are chosen at random from a set Ω of rotation operators and assigned to the cells. Let V_ξ be the center of mass velocity of the particles in cell ξ,

$$V_\xi = \frac{1}{n_\xi} \sum_{i|\mathbf{x} \in \mathcal{V}} \mathbf{v}'_i \,,$$

where n_ξ is the number of particles in the cell with volume $\mathcal{V}$ and $\mathbf{v}'_i$ is the pre-collision value of the velocity. The post-collision velocity $\mathbf{v}_i$ of every particle i in cell ξ is given by

$$\mathbf{v}_i = V_\xi + \hat{\omega}_\xi(\mathbf{v}'_i - V_\xi) \,. \tag{2}$$

The same rotation operator is applied to every particle in the cell but it differs from cell to cell. This simple collision rule changes both the directions and magnitudes of the particle velocities in the cell as can be seen from the two-dimensional example for two particle velocities shown in Fig. 2. The collision rule conserves mass, momentum and energy within each cell. This can be seen by summing the post-collision momenta and energies in each cell. For the momentum and energy we have, respectively,

$$\sum_{i|\mathbf{x} \in \mathcal{V}} m\mathbf{v}_i = \sum_{i|\mathbf{x} \in \mathcal{V}} m\big(V + \hat{\omega}[\mathbf{v}'_i - V]\big) = \sum_{i|\mathbf{x} \in \mathcal{V}} m\mathbf{v}'_i \,, \tag{3}$$

$$\sum_{i|\mathbf{x} \in \mathcal{V}} \frac{m}{2}\|\mathbf{v}_i\|^2 = \sum_{i|\mathbf{x} \in \mathcal{V}} \frac{m}{2}\big\|V + \hat{\omega}[\mathbf{v}'_i - V]\big\|^2 = \sum_{i|\mathbf{x} \in \mathcal{V}} \frac{m}{2}\|\mathbf{v}'_i\|^2 \,. \tag{4}$$

In addition, one may show that the dynamics preserves phase space volumes [7].

As discussed above, in order to apply the multi-particle collision rule the system must be divided into cells where the collisions occur. The choice of cell size is dictated

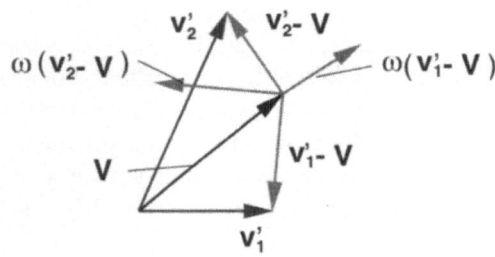

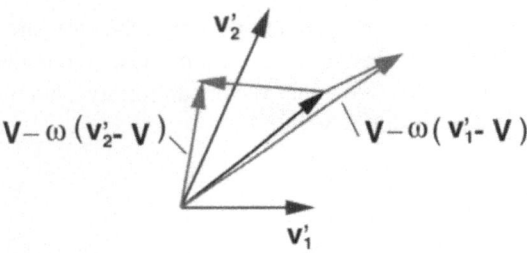

Fig. 2. Multi-particle collision rule applied to two particles in two dimensions for a rotation by $\pi/2$. The dark arrows labelled $\mathbf{v}_1'$ and $\mathbf{v}_2'$ denote the pre-collision velocities. The upper panel also shows the center of mass velocity and the pre-collision velocities relative to the center of mass before and after rotation by $\pi/2$. The lower panel shows the result of adding back the center of mass velocity to yield the post-collision velocities. Both the velocity directions and magnitudes are changed as a result of the collision

by the particle density and the mean velocity of the particles in the system. The particle density determines the mean number of particles per cell and thus affects the number of particles involved in a multi-particle collision event. The mean velocity determines, on average, how far particles travel between collision events. Multi-particle collisions will be efficient if large numbers of particles participate in each collision event. In order to avoid correlations between collisions, particles should travel on the order of a cell length between collisions. Consideration of these factors provides rough guides for the appropriate choice of the collision cell size.

Once the system has been partitioned into cells of suitable size, a specific form for the particle velocity rotation operators must be chosen. This choice will also influence the magnitudes of the transport properties of the fluid. Consider a three-dimensional system. It is convenient to take rotations about a randomly chosen direction, $\hat{\mathbf{n}}$, by an angle Φ chosen from a set of angles. For this choice, the contribution to the post-collision velocity $\mathbf{v}_i$ of particle i in cell $\boldsymbol{\xi}$, $\mathbf{v}_i = \mathbf{V}_\xi + \hat{\omega}_\xi(\mathbf{v}_i' - \mathbf{V}_\xi)$, arising from the rotation is given explicitly by

$$\hat{\omega}_\xi(\mathbf{v}_i' - \mathbf{V}_\xi) = \hat{\mathbf{n}}\hat{\mathbf{n}} \cdot (\mathbf{v}_i' - \mathbf{V}_\xi) + (\mathbf{I} - \hat{\mathbf{n}}\hat{\mathbf{n}}) \cdot (\mathbf{v}_i' - \mathbf{V}_\xi)\cos\Phi - \hat{\mathbf{n}} \times (\mathbf{v}_i' - \mathbf{V}_\xi)\sin\Phi.$$

$$(5)$$

The unit vector $\hat{n}$ may be sampled uniformly from the surface of a sphere, while the angles Φ may be chosen in any convenient way. For example, convenient choices are to select Φ randomly from the set $\{\pi/2, -\pi/2\}$, or from the set of all angles $0 \leq \Phi \leq \pi$.

A more elegant way to implement the collision rule is to represent a rotation matrix as a unit quaternion $\hat{\omega}_\xi$, and a velocity vector as a pure quaternion $\tilde{v} = (0, v_x, v_y, v_z)$. For this choice, the contribution to the post-collision velocity $\mathbf{v}_i$ of particle i in cell $\boldsymbol{\xi}$, with slight abuse of notation, is written as $\mathbf{v}_i = \mathsf{V}_\xi + Ad_{\hat{\omega}_\xi} \cdot (\mathbf{v}_i' - \mathsf{V}_\xi)$, where the conjugation operation is defined as

$$Ad_x \cdot y = xyx^* \,.$$

One can verify that the conjugation operation preserves the structure of the rotation group $SO(3)$ and transforms the vector space of pure quaternions into itself. A unit quaternion $\hat{\omega}_\xi$ can be generated by sampling vectors uniformly from unit ball $B \in R^4$ and subsequent scaling of the resulting vector.

2.1 Evolution Equation

In order to carry out a detailed analysis of the model, one may write the dynamics of the system in terms of an evolution equation for the phase space probability density, $\mathsf{P}(\mathbf{V}^{(N)}, \mathbf{X}^{(N)}, t)$, which takes the form,

$$\mathsf{P}(\mathbf{V}^{(N)}, \mathbf{X}^{(N)} + \mathbf{V}^{(N)}\tau, t + \tau) = \hat{\mathcal{C}}\mathsf{P}(\mathbf{V}^{(N)}, \mathbf{X}^{(N)}, t) \,. \tag{6}$$

The displaced position on the left hand side reflects the free streaming between collisions while the collision operator $\hat{\mathcal{C}}$ on the right hand side is defined by

$$\hat{\mathcal{C}}\mathsf{P}(\mathbf{V}^{(N)}, \mathbf{X}^{(N)}, t) = \frac{1}{\|\Omega\|^L} \sum_{\Omega^L} \int d\mathbf{V}'^{(N)} \mathsf{P}(\mathbf{V}'^{(N)}, \mathbf{X}^{(N)}, t)$$
$$\times \prod_{i=1}^N \delta(\mathbf{v}_i - \mathsf{V}_{\boldsymbol{\xi}} - \hat{\omega}_{\boldsymbol{\xi}}[\mathbf{v}_i' - \mathsf{V}_{\boldsymbol{\xi}}]) \,.$$

Here L is the number of cells and $\|\Omega\|^L$ is the number of rotation operators in the set. Introducing the free streaming Liouville operator,

$$i\mathcal{L}_0 = \mathbf{V}^{(N)} \cdot \nabla_{\mathbf{X}^{(N)}} \,, \tag{7}$$

we may write (6) in the alternative form,

$$e^{i\mathcal{L}_0 \tau}\mathsf{P}(\mathbf{V}^{(N)}, \mathbf{X}^{(N)}, t + \tau) = \hat{\mathcal{C}}\mathsf{P}(\mathbf{V}^{(N)}, \mathbf{X}^{(N)}, t) \,. \tag{8}$$

Using this expression it is instructive to write the evolution equation in continuous time with a delta function collision term,

$$\frac{\partial}{\partial t}P(\mathbf{X}^{(N)}, \mathbf{V}^{(N)}, t) = \left(-i\mathcal{L}_0 + \tilde{c}\right)P(\mathbf{X}^{(N)}, \mathbf{V}^{N)}, t) \,, \tag{9}$$

where the collision operator $\tilde{C}$ acts at discrete time intervals on the velocities of the particles and is defined as

$$\tilde{C}P(\mathbf{X}^{(N)}, \mathbf{V}^{(N)}, t) = \sum_{m=0}^{\infty} \delta(t - m\tau)(\hat{C} - 1)P(\mathbf{X}^{(N)}, \mathbf{V}^{(N)}, t) . \qquad (10)$$

The reduction of the evolution equation (9) to the discrete form (8) can be carried out by integrating (9) from $t = m\tau + \epsilon$ to $t + \tau$ where ϵ is an infinitesimal number.

The evolution equation (6) can be written in a more compact form by letting $\Gamma = (\mathbf{V}^{(N)}, \mathbf{X}^{(N)})$ denote a phase point and defining a transition operator $\mathcal{W}(\Gamma' \rightarrow \Gamma)$ that accounts for the streaming and collision steps. The discrete-time evolution equation may then be written as

$$\mathsf{P}(\Gamma, t + \tau) = \int d\Gamma' \mathcal{W}(\Gamma' \rightarrow \Gamma)\mathsf{P}(\Gamma', t) \equiv \hat{\mathcal{W}}\mathsf{P}(t), \qquad (11)$$

where the integral implies summation over any preimages of the state Γ. More explicitly, the transition operator has the definition,

$$\int d\Gamma' \mathcal{W}(\Gamma' \rightarrow \Gamma)\mathsf{P}(\Gamma', t) = \frac{1}{\|\Omega\|^L} \sum_{\Omega^L} \int d\mathbf{V}'^{(N)} d\mathbf{X}'^{(N)} \times$$

$$\times \prod_{i=1}^{N} \delta(\mathbf{v}_i - \mathbf{V}_{\boldsymbol{\xi}} - \hat{\omega}_{\boldsymbol{\xi}}[\mathbf{v}'_i - \mathbf{V}_{\boldsymbol{\xi}}])\delta(\mathbf{x}'_i - (\mathbf{x}_i + \mathbf{v}_i\tau))\mathsf{P}(\mathbf{V}'^{(N)}, \mathbf{X}'^{(N)}, t).$$

The distribution of this Markov chain is denoted by $P_0(\Gamma) = P_0(\mathbf{V}^{(N)}, \mathbf{X}^{(N)})$ in equilibrium. Assuming the system is ergodic, in view of the conservation laws obeyed by the dynamics, the stationary distribution is given by the microcanonical ensemble expression,

$$P_0(\Gamma) = \mathcal{N}\delta\left(\frac{1}{N}\sum_{i=1}^{N}\frac{m}{2}\|\mathbf{v}_i\|^2 - \frac{d}{2\beta}\right)\delta\left(\sum_{i=1}^{N}[\mathbf{v}_i - \mathbf{u}]\right), \qquad (12)$$

where $\mathbf{u}$ is the mean velocity of the system and $\mathcal{N}$ is a normalization constant. If (12) is integrated over the coordinates and velocities of particles with labels $i = 2, \ldots, N$, the Maxwell distribution,

$$P_{\mathrm{m}}(\mathbf{v}_1, \mathbf{x}_1) = \frac{1}{V}\left(\frac{m\beta}{2\pi}\right)^{d/2}\exp(-\beta m\|\mathbf{v}_1 - \mathbf{u}\|^2/2), \qquad (13)$$

is obtained in the limit of large N. Here $\beta = (k_{\mathrm{B}}T)^{-1}$, V is the system volume and d is the dimension.

2.2   H-Theorem

While the above arguments indicate that the equilibrium one-particle distribution function is Maxwellian, it is instructive to establish an H-theorem for relaxation to equilibrium for the multi-particle collision dynamics [7]. In the Boltzmann approximation where the full phase space probability distribution function P is a product of identical one-particle probability distributions,

$$P(\mathbf{V}^{(N)}, \mathbf{X}^{(N)}, t) = \prod_{i=1}^{N} P_1(\mathbf{v}_i, \mathbf{x}_i, t) , \qquad (14)$$

it is possible to derive such a relation as we now show.

Letting $f(\mathbf{v}, \mathbf{x}, t) = N P_1(\mathbf{v}, \mathbf{x}, t)$, the H-functional is defined in terms of the reduced one-particle distribution function as,

$$\begin{aligned} H(t) &= \int d\mathbf{v} d\mathbf{x}\, f(\mathbf{v}, \mathbf{x}, t) \ln f(\mathbf{v}, \mathbf{x}, t) \\ &= \sum_{\xi, n \in \mathbf{N}} \frac{e^{-\rho_\xi}}{n!} \int_{\mathcal{V}^n} d\mathbf{V}^{(n)} d\mathbf{X}^{(n)} \prod_{i=1}^{n} f(\mathbf{v}_i, \mathbf{x}_i, t) \ln \prod_{i=1}^{n} f(\mathbf{v}_i, \mathbf{x}_i, t) , \end{aligned} \qquad (15)$$

where the second equality follows from the representation of the system in terms of phase space cells and makes use of the resolution of identity $1 = \sum_{n \geq 0} \frac{e^{-x} x^n}{n!}$. That $H(t)$ decreases on each discrete evolution step may be proved using the convexity inequality,

$$\sum_s A(s) B(s) \ln B(s) \geq \left(\sum_s A(s) B(s) \right) \ln \left(\sum_s A(s) B(s) \right) , \qquad (16)$$

where A is normalized so that $\sum_s A(s) = 1$. We define $R^{(n)}$ by

$$R^{(n)}(\mathbf{V}^{(n)}, \mathbf{V}'^{(n)}) = \frac{1}{\|\Omega\|} \sum_{\hat{\omega} \in \Omega} \prod_{i=1}^{n} \delta\big(\mathbf{v}_i - \mathbf{V} + \hat{\omega}[\mathbf{V} - \mathbf{v}'_i]\big) , \qquad (17)$$

whose integral over $\mathbf{V}'^{(n)}$ is unity. Making use of $R^{(n)}$, we may write (15) in the form

$$\begin{aligned} H(t) &= \sum_{\xi, n \in \mathbf{N}} \frac{e^{-\rho_\xi}}{n!} \int_{\mathcal{V}^n} d\mathbf{V}^{(n)} d\mathbf{X}^{(n)} \prod_{i=1}^{n} f(\mathbf{v}_i, \mathbf{x}_i, t) \\ &\quad \times \ln \prod_{i=1}^{n} f(\mathbf{v}_i, \mathbf{x}_i, t) \int d\mathbf{V}'^{(n)} R^{(n)}(\mathbf{V}^n, \mathbf{V}'^{(n)}) . \end{aligned} \qquad (18)$$

Next, we exchange the order of the $\mathbf{V}'^{(n)}$ and $\mathbf{V}^{(n)}$ integrations in each term in the sum and use (16) to write

$$\begin{aligned} \int d\mathbf{V}^{(n)} R^{(n)}(\mathbf{V}^{(n)}, \mathbf{V}'^{(n)}) \prod_{i=1}^{n} f(\mathbf{v}_i, \mathbf{x}_i, t) \ln \prod_{i=1}^{n} f(\mathbf{v}_i, \mathbf{x}_i, t) \\ \geq \tilde{f}^{(n)}(\mathbf{V}'^{(n)}, \mathbf{X}^{(n)}, t) \ln \tilde{f}^{(n)}(\mathbf{V}'^{(n)}, \mathbf{X}^{(n)}, t) , \end{aligned} \qquad (19)$$

where

$$\tilde{f}^{(n)}(\mathbf{V}'^{(n)}, \mathbf{X}^{(n)}, t) = \int d\mathbf{V}^n R^{(n)}(\mathbf{V}^{(n)}, \mathbf{V}'^{(n)}) \prod_{i=1}^{n} f(\mathbf{v}_i, \mathbf{x}_i, t) .$$

As a result of these manipulations we obtain,

$$H(t) \geq \sum_{\xi, n \in \mathbf{N}} \frac{e^{-\rho_\xi}}{n!} \int_{\mathcal{V}^n} d\mathbf{V}'^{(n)} d\mathbf{X}^{(n)} \tilde{f}^{(n)}(\mathbf{V}'^{(n)}, \mathbf{X}^{(n)}, t) \ln \tilde{f}^{(n)}(\mathbf{V}'^{(n)}, \mathbf{X}^{(n)}, t). \quad (20)$$

To simplify this expression, we let

$$A = \prod_{i=1}^{n} \hat{f}_i^{(n)} , \qquad B = \frac{\tilde{f}^{(n)}(\mathbf{V}'^{(n)}, \mathbf{X}^n, t)}{Z \prod_{i=1}^{n} \hat{f}_i^{(n)}} ,$$

where Z and $\hat{f}^{(n)}$ are defined as

$$Z = \int_{\mathcal{V}^n} d\mathbf{V}'^{(n)} d\mathbf{X}^{(n)} \tilde{f}^{(n)}(\mathbf{V}'^{(n)}, \mathbf{X}^{(n)}, t) = \rho_\xi^n , \quad (21)$$

$$\hat{f}_i^{(n)}(\mathbf{v}'_i, \mathbf{x}_i, t) = \frac{1}{Z} \int_{\mathcal{V}^{[n-1]}} d\mathbf{v}'_1 d\mathbf{x}_1 \cdots d\hat{\mathbf{v}}'_i d\hat{\mathbf{x}}_i \cdots d\mathbf{v}'_n d\mathbf{x}_n \tilde{f}^{(n)}(\mathbf{V}'^{(n)}, \mathbf{X}^{(n)}, t), \quad (22)$$

where integrations over variables with a hat in (22) are omitted. Substituting A and B into (16) we find

$$\frac{1}{Z} \int_{\mathcal{V}^n} d\mathbf{V}'^{(n)} d\mathbf{X}^n \tilde{f}^{(n)}(\mathbf{V}'^{(n)}, \mathbf{X}^{(n)}, t) \ln \frac{\tilde{f}^{(n)}(\mathbf{V}'^{(n)}, \mathbf{X}^{(n)}, t)}{Z \prod_{i=1}^{n} \hat{f}_i^{(n)}} \geq$$

$$\int_{\mathcal{V}^n} d\mathbf{V}'^{(n)} d\mathbf{X}^{(n)} \frac{\tilde{f}^{(n)}(\mathbf{V}'^{(n)}, \mathbf{X}^{(n)}, t)}{Z} \ln \int_{\mathcal{V}^n} d\mathbf{V}'^{(n)} d\mathbf{X}^{(n)} \frac{\tilde{f}^{(n)}(\mathbf{V}'^{(n)}, \mathbf{X}^{(n)}, t)}{Z} = 0 . \quad (23)$$

Since the argument of the logarithm is unity given the definition of Z, the last term in (23) is equal to zero. From (23) it follows that

$$\int_{\mathcal{V}^n} d\mathbf{V}'^{(n)} d\mathbf{X}^{(n)} \tilde{f}^{(n)}(\mathbf{V}'^{(n)}, \mathbf{X}^{(n)}, t) \ln \tilde{f}^{(n)}(\mathbf{V}'^{(n)}, \mathbf{X}^{(n)}, t)$$

$$\geq \int_{\mathcal{V}^n} d\mathbf{V}'^{(n)} d\mathbf{X}^{(n)} \tilde{f}^{(n)}(\mathbf{V}'^{(n)}, \mathbf{X}^{(n)}, t) \ln Z \prod_{i=1}^{n} \hat{f}_i^{(n)} .$$

Finally, using the identity,

$$\sum_{i=1}^{n} \int_{\mathcal{V}^n} d\mathbf{v}'_i \mathbf{x}_i Z \hat{f}_i^{(n)}(\mathbf{v}'_i, \mathbf{x}_i, t) \ln Z^{1/n} \hat{f}_i^{(n)}(\mathbf{v}'_i, \mathbf{x}_i, t) =$$

$$= \int_{\mathcal{V}^n} d\mathbf{V}'^{(n)} d\mathbf{X}^{(n)} \tilde{f}^{(n)}(\mathbf{V}'^{(n)}, \mathbf{X}^{(n)}, t) \ln Z \prod_{i=1}^{n} \hat{f}_i^{(n)}(\mathbf{v}'_i, \mathbf{x}_i, t) ,$$

in (20) we have,

$$H(t) \geq \sum_{i,\xi,n \in \mathbf{N}1} \frac{e^{-\rho_\xi} \rho_\xi^{n-1}}{n!} \int_V dv' dx Z^{1/n} \hat{f}_i^{(n)}(\mathbf{v}', \mathbf{x}, t) \ln Z^{1/n} \hat{f}_i^{(n)}(\mathbf{v}', \mathbf{x}, t)$$

$$\geq \int dv' dx \mathcal{C}(f) \ln \mathcal{C}(f) = \int dv' dx f(\mathbf{v}', \mathbf{x}, t+1) \ln f(\mathbf{v}', \mathbf{x}, t+1) .$$

The second inequality follows from the application of (16) with A given by the Poisson distribution $A(n) = (e^{-\rho_\xi} \rho_\xi^{n-1})/[(n-1)!]$, and the last equality follows from the invariance of the integral with respect to translations by the streaming transformation. Therefore, the value of the H functional at time $t + \tau$ does not exceed its value at time t so that

$$H(t) \geq \int dv' dx f(\mathbf{v}', \mathbf{x}, t+\tau) \ln f(\mathbf{v}', \mathbf{x}, t+\tau) . \tag{24}$$

System Thermalization. In order to confirm these predictions, we study the thermalization of a system obeying multi-particle collision dynamics. We consider a three-dimensional system with density $\rho = 2$ in a box of length $L = 40$ with periodic boundary conditions which is partitioned into $40 \times 40 \times 40$ cells of unit length. The temperature in reduced units ($m = 1$, $\tau = 1$, and unit cell length) is taken to be $k_B T = 4/3$. We study the equilibration of a system initialized by $f(\mathbf{v}) = \frac{1}{2}\delta(\mathbf{v} + \mathbf{v}_0) + \frac{1}{2}\delta(\mathbf{v} - \mathbf{v}_0)$, where $\mathbf{v}_0 = (2, 0, 0)$. The results of such a simulation after relaxation to equilibrium are shown as a histogram of the x-component of the velocity in Fig. 3. One can see that a Boltzmann distribution of velocities is obtained.

It is instructive to examine the relaxation to equilibrium of the multi-particle collision dynamics in more detail. For this purpose, it is convenient to rewrite one-particle probability distribution as a sum of Hermite polynomials ,

$$f(v_x) = \zeta^{-1} e^{-(v_x/\zeta)^2} \sum \alpha_i H_i(v_x/\zeta), \tag{25}$$

where $\zeta = \sqrt{2k_B T/m}$. The expansion coefficients are easily obtained from the expectation values of Hermite polynomials,

$$(-1)^n n! 2^n \sqrt{\pi} \alpha_n = \langle H_n(v_x/\zeta) \rangle. \tag{26}$$

The expansion coefficients are less sensitive to fluctuations than histograms (Fig. 3) and demonstrate fast convergence to local equilibrium.

In Fig. 4 we plot the values of the expansion coefficients as a function of time. Using the above expansion and assuming the system is spatially homogeneous, we may write for velocity component of the one-particle probability distribution

$$-H_v = \int d\mathbf{v} f(\mathbf{v}) \ln f(\mathbf{v}) = \langle \ln f(\mathbf{v}) \rangle. \tag{27}$$

Approximating $f(\mathbf{v})$ by the Hermite polynomial expansion to obtain the results, we plot values of the H-function in a non-equilibrium run. As expected from the theoretical analysis, the functional increases monotonously with time and rapidly converges to the equilibrium value.

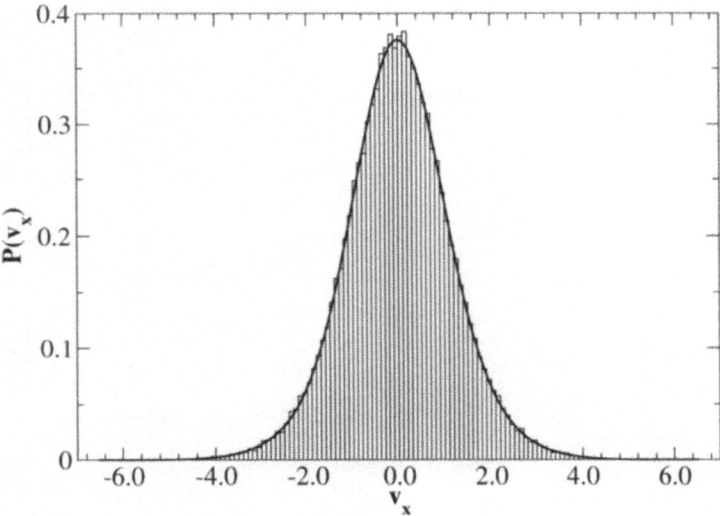

Fig. 3. Histogram of the velocity distribution in the system at equilibrium. The system has $N = 5.1 \times 10^6$ particles, $k_B T = 4/3$ and $\rho = 2.0$. The solid line is the calculated distribution based on an expansion of the distribution function using ten Hermite polynomials (see text)

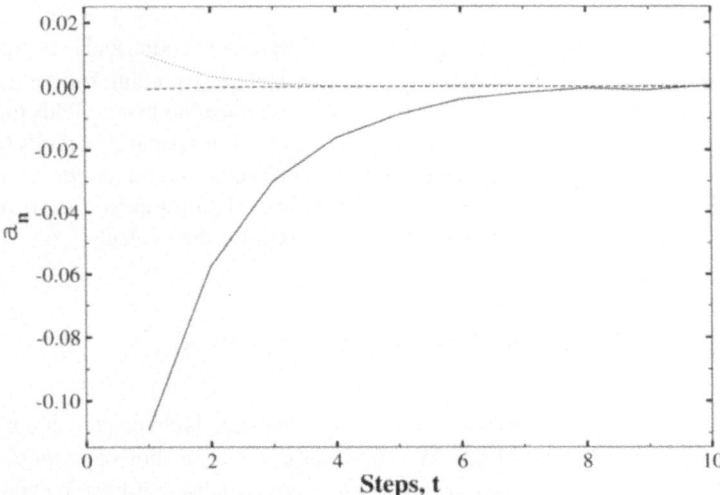

Fig. 4. Dynamics of the coefficients of the Hermite polynomials of a system relaxing to equilibrium. Coefficients of $H_2(v_x)$-(*solid line*), $H_4(v_x)$-(*dotted line*) and $H_6(v_x)$-(*dotted line*) are plotted. Odd coefficients vanish due to symmetry and the coefficient of H_0 is equal to one by definition

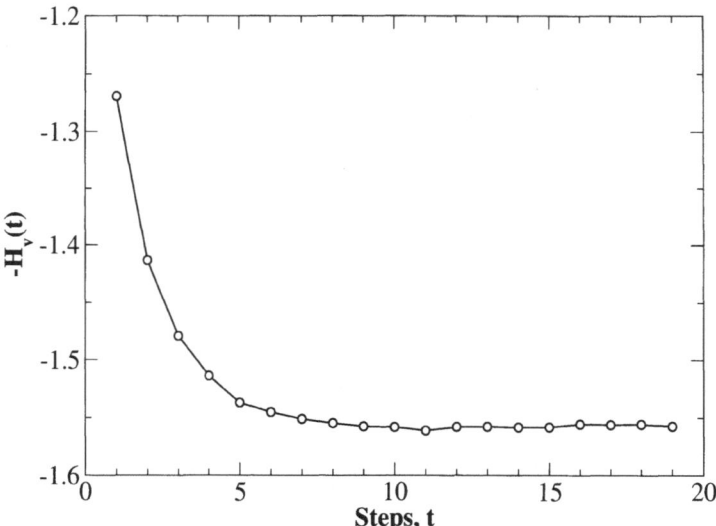

Fig. 5. Evolution of the velocity component of the H-function from non-equilibrium initial conditions. Details are given in the text

3   Hydrodynamic Equations and Transport Properties

The dynamics described above is very simple, both in its conception and in its implementation. However, it is essential to provide a theoretical underpinning for the mesoscale dynamics to be able to assess its properties and to demonstrate that it yields the correct hydrodynamic equations. In this section we use projection operator methods to reduce the multi-particle collision dynamics to hydrodynamic equations for the conserved fields. As a by-product of this derivation we also obtain correlation function expressions for the transport properties of the system that can be used for their calculation.

3.1   Evolution Equations for Mean Dynamical Variables

The projection operator methods we use are based on methods developed for continuous-time deterministic systems [11,12]. Modifications of the projection operator techniques must be made to account for the intrinsic stochasticity and discrete-time dynamics of the mesoscopic dynamics [13]. We first present the derivation of a general set of projected equations and then specialize the projection operator to one that depends on the conserved fields.

In order to derive a set of evolution equations for the mean values of a set of dynamical variables $\mathbf{a}$, we introduce a projection operator $\mathcal{P}$ by,

$$(\mathcal{P}\mathbf{h})(\Gamma) = \mathbf{a}^\dagger(\Gamma)P_0(\Gamma)\left\langle\mathbf{a}\mathbf{a}^\dagger\right\rangle^{-1}\int d\Gamma'\mathbf{a}(\Gamma')\mathbf{h}(\Gamma'),$$

where $\mathbf{h}(\Gamma)$ is any function of the phase space variables. The dagger denotes the adjoint and the angular brackets symbolize an average over the equilibrium distribution, i.e.,

$$\langle \cdots \rangle = \int d\Gamma \cdots P_0(\Gamma) \, .$$

The complementary operator $\mathcal{Q}$ is defined by $\mathcal{Q} = 1 - \mathcal{P}$. Applying the projection operators $\mathcal{P}$ and $\mathcal{Q}$ to (11) we obtain a system of two equations,

$$\mathsf{P}_\mathcal{P}(t+\tau) = \mathcal{P}\hat{\mathcal{W}}\mathsf{P}_\mathcal{P}(t) + \mathcal{P}(\hat{\mathcal{W}} - 1)\mathsf{P}_\mathcal{Q}(t), \tag{28}$$

$$\mathsf{P}_\mathcal{Q}(t+\tau) = \mathcal{Q}(\hat{\mathcal{W}} - 1)\mathsf{P}_\mathcal{P}(t) + \mathcal{Q}\hat{\mathcal{W}}\mathsf{P}_\mathcal{Q}(t), \tag{29}$$

where $\mathsf{P}_\mathcal{P}(t) = \mathcal{P}\mathsf{P}(t)$ and $\mathsf{P}_\mathcal{Q}(t) = \mathcal{Q}\mathsf{P}(t)$. To facilitate some of the calculations presented below, we have replaced $\hat{\mathcal{W}}$ by $\hat{\mathcal{W}} - 1$ in certain terms where this replacement has no effect because $\mathcal{P}\mathcal{Q} = 0$. Solving (29) by iteration leads to

$$\mathsf{P}_\mathcal{Q}(t) = \left[\mathcal{Q}\hat{\mathcal{W}}\right]^t \mathsf{P}_\mathcal{Q}(0) + \sum_{\ell=1}^{n} \left[\mathcal{Q}\hat{\mathcal{W}}\right]^{\ell-1} \mathcal{Q}(\hat{\mathcal{W}} - 1)\mathsf{P}_\mathcal{P}(t - \ell\tau) \, , \tag{30}$$

where $t = n\tau$. Substitution of this result into (28) gives

$$\mathsf{P}_\mathcal{P}(t+\tau) = \mathcal{P}\hat{\mathcal{W}}\mathsf{P}_\mathcal{P}(t) + \sum_{\ell=1}^{n} \mathcal{K}(\ell - 1)\mathsf{P}_\mathcal{P}(t - \ell\tau), \tag{31}$$

where the memory kernel is defined by

$$\mathcal{K}(\ell) = \mathcal{P}(\hat{\mathcal{W}} - 1)\left[\mathcal{Q}\hat{\mathcal{W}}\right]^{\ell} \mathcal{Q}(\hat{\mathcal{W}} - 1)\mathcal{P}. \tag{32}$$

The first term on right hand side of (30) was eliminated by the use of a specially prepared ensemble of initial conditions where deviations from equilibrium occur only in the dynamical variables in the set $\mathbf{a}$ so that $\mathsf{P}_\mathcal{Q}(0) = 0$.

From (31) we can easily derive a set of equations for the average values of the dynamical variables,

$$\bar{\mathbf{a}}(t) = \int d\Gamma \mathbf{a}(\Gamma)\mathsf{P}(\Gamma, t) \, , \tag{33}$$

by multiplying this equation from the left by $\mathbf{a}$ and integrating over the phase space variables to obtain,

$$\bar{\mathbf{a}}(t+\tau) - \bar{\mathbf{a}}(t) = \left\langle \mathbf{a}(\hat{\mathcal{W}} - 1)\mathbf{a}^\dagger \right\rangle \left\langle \mathbf{a}\mathbf{a}^\dagger \right\rangle^{-1} \bar{\mathbf{a}}(t) + \sum_{\ell=1}^{t} \mathbf{K}(\ell - 1)\bar{\mathbf{a}}(t - \ell\tau) \, , \tag{34}$$

where

$$\mathbf{K}(\ell) = \left\langle \mathbf{a}(\hat{\mathcal{W}} - 1)[\mathcal{Q}\hat{\mathcal{W}}]^\ell \mathcal{Q}(\hat{\mathcal{W}} - 1)\mathbf{a}^\dagger \right\rangle \left\langle \mathbf{a}\mathbf{a}^\dagger \right\rangle^{-1} \, .$$

For slowly decaying dynamical variables, which are our main concern, $\bar{\mathbf{a}}(t - \ell\tau)$ can be replaced by $\bar{\mathbf{a}}(t)$ and the upper limit on the sum can be replaced by infinity. In this approximation (34) becomes the kinetic equation

$$\bar{\mathbf{a}}(t + \tau) - \bar{\mathbf{a}}(t) = \left(\boldsymbol{\Omega} + \boldsymbol{\Phi}\right)\bar{\mathbf{a}}(t) , \tag{35}$$

where the matrices $\boldsymbol{\Omega}$ and $\boldsymbol{\Phi}$ are given by

$$\boldsymbol{\Omega} = \left\langle \mathbf{a}(\hat{W} - 1)\mathbf{a}^\dagger \right\rangle \left\langle \mathbf{a}\mathbf{a}^\dagger \right\rangle^{-1} , \tag{36}$$

$$\boldsymbol{\Phi} = \sum_{\ell=1}^{\infty} \mathbf{K}(\ell - 1) .$$

We now analyze the structure of (35) for the specific choice of conserved dynamical variables.

3.2   Kinetic Equations for Conserved Variables

The hydrodynamic equations describe the dynamics of the conserved mass, momentum and energy density fields. The microscopic fields corresponding to these variables, which are relevant for the multi-particle collision dynamics, are the particle density along with momentum (or, equivalently, velocity for equal mass particles) and energy densities in a cell,

$$\rho(\mathbf{x}) = \sum_{i=1}^{N} \delta(\mathbf{x}_i - \mathbf{x}) , \tag{37}$$

$$\boldsymbol{\mu}(\boldsymbol{\xi}) = \sum_{i=1}^{N} \mathbf{v}_i \theta(1/2 - |\mathbf{x}_i - \boldsymbol{\xi}|) , \tag{38}$$

$$\varepsilon(\boldsymbol{\xi}) = \sum_{i=1}^{N} \frac{m}{2} v_i^2 \theta(1/2 - |\mathbf{x}_i - \boldsymbol{\xi}|) .$$

Here θ is the Heaviside function. Since the hydrodynamic equations are valid on distance and time scales which are long compared to molecular scales, it is convenient to work with the Fourier transforms of these fields and then consider the small k limit. The Fourier transforms of these variables are given by,

$$\rho_{\mathbf{k}} = \int d\mathbf{x} \, e^{i\mathbf{k}\cdot\mathbf{x}} \rho(\mathbf{x}) = \sum_{i=1}^{N} e^{i\mathbf{k}\cdot\mathbf{x}_i} , \tag{39}$$

$$\boldsymbol{\mu}_{\mathbf{k}} = \sum_{\boldsymbol{\xi}} e^{i\mathbf{k}\cdot\boldsymbol{\xi}} \boldsymbol{\mu}(\boldsymbol{\xi}) = \sum_{i=1}^{N} \mathbf{v}_i \sum_{\boldsymbol{\xi}} e^{i\mathbf{k}\cdot\boldsymbol{\xi}} \theta(1/2 - |\mathbf{x}_i - \boldsymbol{\xi}|) , \tag{40}$$

$$\varepsilon_{\mathbf{k}} = \sum_{\boldsymbol{\xi}} e^{i\mathbf{k}\cdot\boldsymbol{\xi}} \varepsilon(\boldsymbol{\xi}) = \sum_{i=1}^{N} \frac{m}{2} v_i^2 \sum_{\boldsymbol{\xi}} e^{i\mathbf{k}\cdot\boldsymbol{\xi}} \theta(1/2 - |\mathbf{x}_i - \boldsymbol{\xi}|) .$$

Rather than dealing with this set of variables directly, it is useful to define an orthogonal set of dynamical variables, $\mathbf{a_k} = \{\rho_\mathbf{k}, \mu_\mathbf{k}, s_\mathbf{k}\}$, where the entropy density is defined as $s_\mathbf{k} = \varepsilon_\mathbf{k} - C_\mathrm{v} T \rho_\mathbf{k}$ with C_v the specific heat. In terms of this set of variables the matrix $\left\langle \mathbf{a_k a_k^\dagger} \right\rangle$ is diagonal and given by

$$\left\langle \mathbf{a_k a_k^\dagger} \right\rangle = N \begin{pmatrix} 1 & \mathbf{0} & 0 \\ \mathbf{0} & (k_\mathrm{B}T/m)\mathbf{1} & \mathbf{0} \\ 0 & \mathbf{0} & C_\mathrm{v} k_\mathrm{B} T^2 \end{pmatrix}, \tag{41}$$

where $C_\mathrm{v} = 3k_\mathrm{B}/2$ in three dimensions. Using this statistically independent set of variables simplifies the algebraic manipulations. When confusion is unlikely to arise, we shall simplify the notation and drop the subscripts $\mathbf{k}$ on the vectors of dynamical variables. In view of the fact that one is primarily interested in the evolution of the conserved variable fields, we specialize the general kinetic equation (35) to this case and, in particular, we determine the forms of the $\boldsymbol{\Omega}$ and $\boldsymbol{\Phi}$ matrices.

Structure of the $\boldsymbol{\Omega}$ Matrix. We first construct the form of the matrix $\boldsymbol{\Omega}$ to $\mathcal{O}(\mathbf{k}^2)$ for conserved fields. For this purpose it is convenient to introduce an operator $\mathcal{S}(\Gamma, t)$ which relates the state Γ at the initial time to the set of states at time t, weighted with the probability of transition to the corresponding state. In terms of this notation, equilibrium averages may be written as

$$\int d\Gamma \int d\Gamma' \mathbf{a}(\Gamma) W(\Gamma' \to \Gamma) \mathbf{h}^\dagger(\Gamma') P_0(\Gamma') = \left\langle \mathbf{a}(\mathcal{S}(\Gamma, \tau)) \mathbf{h}^\dagger(\Gamma) \right\rangle,$$

where summation over states is implied. Similarly, if we consider the equilibrium average of a set of dynamical variables, using the stationarity of the equilibrium distribution, $P_0(\Gamma) = \hat{W} P_0(\Gamma)$, we may write,

$$\int d\Gamma \mathbf{h}^\dagger(\Gamma) P_0(\Gamma) = \int d\Gamma \mathbf{h}^\dagger(\Gamma) \hat{W} P_0(\Gamma) = \int d\Gamma \mathbf{h}^\dagger(\Gamma) \int d\Gamma' W(\Gamma' \to \Gamma) P_0(\Gamma')$$

$$= \int d\Gamma' \left[\int d\Gamma \mathbf{h}^\dagger(\Gamma) W(\Gamma' \to \Gamma) \right] P_0(\Gamma'),$$

$$\equiv \int d\Gamma' \mathbf{h}^\dagger(\mathcal{S}(\Gamma', \tau)) P_0(\Gamma') .$$

From this result we may establish that

$$\left\langle \mathbf{a}(\mathcal{S}(\Gamma, \tau)) \mathbf{a}^\dagger(\mathcal{S}(\Gamma, \tau)) \right\rangle = \left\langle \mathbf{a}(\Gamma) \mathbf{a}^\dagger(\Gamma) \right\rangle.$$

Letting $\mathbf{b}(\Gamma) = \mathbf{a}(\mathcal{S}(\Gamma, \tau)) - \mathbf{a}(\Gamma)$ we may write $\boldsymbol{\Omega}$ as

$$\boldsymbol{\Omega} = \left\langle \mathbf{b} \mathbf{a}^\dagger \right\rangle \left\langle \mathbf{a} \mathbf{a}^\dagger \right\rangle^{-1} .$$

The elements of the **b** variable vector are given by

$$b_{\mathbf{k}}^{\rho} = i\mathbf{k} \cdot \sum_{i=1}^{N} \tau \mathbf{v}_i + o(\mathbf{k}), \tag{42}$$

$$b_{\mathbf{k}}^{\mu} = i\mathbf{k} \cdot \sum_{i=1}^{N} \Delta \boldsymbol{\xi}_i \mathbf{v}_i + o(\mathbf{k}), \tag{43}$$

$$b_{\mathbf{k}}^{\epsilon} = i\mathbf{k} \cdot \sum_{i=1}^{N} \Delta \boldsymbol{\xi}_i \frac{m}{2} v_i^2 + o(\mathbf{k}),$$

where we introduced notation $\Delta \boldsymbol{\xi}_i(t) = \boldsymbol{\xi}_i(t+\tau) - \boldsymbol{\xi}_i(t)$ with $\boldsymbol{\xi}_i(t)$ the coordinate of the cell in which particle i is located at time t. One can see that these elements are $\mathcal{O}(\mathbf{k})$.

It is useful to rewrite the numerator of Ω as a sum of symmetric and antisymmetric terms; thus, we have

$$\Omega = \mathcal{A} - \frac{1}{2} \langle \mathbf{bb}^{\dagger} \rangle \langle \mathbf{aa}^{\dagger} \rangle^{-1}, \tag{44}$$

where we defined $\mathcal{A}$ as

$$\mathcal{A} = \frac{1}{2} \langle \mathbf{a}(\mathcal{S}(\Gamma,\tau))\mathbf{a}^{\dagger}(\Gamma) - \mathbf{a}(\Gamma)\mathbf{a}^{\dagger}(\mathcal{S}(\Gamma,\tau)) \rangle \langle \mathbf{aa}^{\dagger} \rangle^{-1}.$$

Using the explicit forms of the conserved mass, momentum and energy density fields, $\mathcal{A}$ may be computed to give,

$$\mathcal{A} = N\tau \frac{k_{\mathrm{B}}T}{m} \begin{pmatrix} 0 & i\mathbf{k} & 0 \\ i\mathbf{k}^T & 0 & i\mathbf{k}^T k_{\mathrm{B}}T \\ 0 & i\mathbf{k}k_{\mathrm{B}}T & 0 \end{pmatrix} \langle \mathbf{aa}^{\dagger} \rangle^{-1}. \tag{45}$$

In order to write $\langle \mathbf{bb}^{\dagger} \rangle$ in the second term of (44) in an alternative form, we introduce a projection operator onto a dynamical variable as

$$P\mathbf{h} = \langle \mathbf{ha}^{\dagger} \rangle \langle \mathbf{aa}^{\dagger} \rangle^{-1} \mathbf{a}. \tag{46}$$

In terms of this new projection operator we have

$$\frac{1}{2} \langle \mathbf{bb}^{\dagger} \rangle = \frac{1}{2} \langle \mathbf{b}Q\mathbf{b}^{\dagger} \rangle + \frac{1}{2} \langle \mathbf{b}P\mathbf{b}^{\dagger} \rangle = \frac{1}{2} \langle \mathbf{f}(0)\mathbf{f}^{\dagger}(0) \rangle + \frac{1}{2} \langle \mathbf{b}P\mathbf{b}^{\dagger} \rangle, \tag{47}$$

where we have defined the random forces,

$$\mathbf{f}(0) = \mathbf{b} - \langle \mathbf{ba}^{\dagger} \rangle \langle \mathbf{aa}^{\dagger} \rangle^{-1} \mathbf{a} = \mathbf{a}(\mathcal{S}(\Gamma,\tau)) - \langle \mathbf{a}(\mathcal{S}(\Gamma,\tau))\mathbf{a}^{\dagger} \rangle \langle \mathbf{aa}^{\dagger} \rangle^{-1} \mathbf{a}. \tag{48}$$

The explicit expressions for the random forces corresponding to the conserved variable fields are, to lowest order in $\mathbf{k}$,

$$f_{\mathbf{k}}^{\rho}(t) = 0 + o(\mathbf{k}), \tag{49}$$

$$f_{\mathbf{k}}^{\mu}(t) = \tau \sum_{i} \left(\mathbf{v}_i(t) \left[i\mathbf{k} \cdot \frac{\Delta \boldsymbol{\xi}_i(t)}{\tau} \right] - \frac{1}{d} i\mathbf{k} v_i(t)^2 \right) + o(\mathbf{k}), \tag{50}$$

$$f_{\mathbf{k}}^{\epsilon}(t) = \tau i\mathbf{k} \cdot \sum_{i} \left[\frac{\Delta \boldsymbol{\xi}_i(t)}{\tau} \left(\frac{m}{2} v_i(t)^2 - C_{\mathrm{v}}T \right) - \mathbf{v}_i(t)(C_{\mathrm{p}} - C_{\mathrm{v}})T \right] + o(\mathbf{k}).$$

The term $\langle \mathbf{b}P\mathbf{b}^\dagger \rangle \langle \mathbf{aa}^\dagger \rangle^{-1}$ may be expressed as

$$\langle \mathbf{b}P\mathbf{b}^\dagger \rangle \langle \mathbf{aa}^\dagger \rangle^{-1} = \langle \mathbf{ba}^\dagger \rangle \langle \mathbf{aa}^\dagger \rangle^{-1} \langle \mathbf{ab}^\dagger \rangle \langle \mathbf{aa}^\dagger \rangle^{-1} .$$

From (44) we have,

$$\langle \mathbf{ba}^\dagger \rangle \langle \mathbf{aa}^\dagger \rangle^{-1} = \mathcal{A} + o(\mathbf{k}), \quad \langle \mathbf{ab}^\dagger \rangle \langle \mathbf{aa}^\dagger \rangle^{-1} = -\mathcal{A} + o(\mathbf{k}).$$

Consequently, it follows that

$$\langle \mathbf{b}P\mathbf{b}^\dagger \rangle \langle \mathbf{aa}^\dagger \rangle^{-1} = -\mathcal{A}^2 + o(\mathbf{k}^2).$$

Assembling all of these results, the net effect of these manipulations is that the Ω matrix may be written as

$$\Omega = \mathcal{A} + \frac{1}{2}\mathcal{A}^2 - \frac{1}{2} \langle \mathbf{f}(0)\mathbf{f}^\dagger(0) \rangle \langle \mathbf{aa}^\dagger \rangle^{-1} + o(\mathbf{k}^2).$$

Structure of the Φ Matrix. In order to write the Φ matrix in a convenient form, we first show that the projected dynamics can be replaced by ordinary dynamics in the small $\mathbf{k}$ limit for conserved variables. If the dynamics is given by a composition of streaming and collision operators in that order, we may write,

$$a_{\mathbf{k}} = \sum_i a_i(t)e^{i\mathbf{k}\cdot\mathbf{x}_i(t)} = \sum_i a_i(t-\tau)e^{i\mathbf{k}\cdot\mathbf{x}_i(t)}, \tag{51}$$

where a_i is a conserved variable. This result follows from the conservation of the quantities $\mathbf{a}$ under collisions at time $t - \tau$. Using the identity (51) and expanding $\mathbf{a}$ in powers of $\mathbf{k}$, we may write $\mathcal{P}[\hat{W} - 1]\mathbf{h}^\dagger$, where $\mathbf{h}^\dagger$ is an arbitrary function, as

$$(\mathcal{P}[\hat{W} - 1]\mathbf{h}^\dagger)(\Gamma) = \mathbf{a}^\dagger P_0(\Gamma) \langle \mathbf{aa}^\dagger \rangle^{-1} \int d\Gamma' \mathbf{b}(\Gamma')\mathbf{h}^\dagger(\Gamma')$$

$$= \mathbf{a}^\dagger P_0(\Gamma) \langle \mathbf{aa}^\dagger \rangle^{-1} \int \int \Gamma' \left[\mathbf{h}^\dagger(\Gamma') \right.$$

$$\left. \sum_i (i\mathbf{k} \cdot [\mathbf{x}'_i(t+\tau) - \mathbf{x}'_i(t)]a'_i(t) + o(\mathbf{k})) \right]. \tag{52}$$

Similarly, one may show that $[\hat{W} - 1]\mathcal{P} = \mathcal{O}(\mathbf{k})$.

Using these results we may prove by induction that

$$[Q\hat{W}]^\ell Q = QW^\ell Q + \mathcal{O}(\mathbf{k}). \tag{53}$$

For $\ell = 0$ relation (53) holds; we assume that it holds for ℓ and prove the relation for $\ell + 1$. We write $[Q\hat{W}]^{\ell+1}Q = [Q\hat{W}]^\ell Q\hat{W}Q$. Then

$$[Q\hat{W}]^{\ell+1}Q = [QW]^\ell Q\hat{W}Q = QW^\ell Q\hat{W}Q + \mathcal{O}(\mathbf{k})$$

$$= QW^\ell [Q + (\hat{W} - 1) + \mathcal{O}(\mathbf{k})]Q + \mathcal{O}(\mathbf{k})$$

$$= Q\hat{W}^{\ell+1}Q + \mathcal{O}(\mathbf{k}),$$

where we expressed $\mathcal{Q}\hat{\mathcal{W}}$ in the equivalent form

$$\mathcal{Q}\hat{\mathcal{W}} = \mathcal{Q} + (\hat{\mathcal{W}} - 1) - \mathcal{P}(\hat{\mathcal{W}} - 1) = \mathcal{Q} + (\hat{\mathcal{W}} - 1) + \mathcal{O}(\mathbf{k}). \qquad (54)$$

This proves the assertion of the recursion relation and, thus, the validity of (53).

The correlation function in the definition of the $\mathbf{\Phi}$ matrix can now be written as

$$\mathbf{K}(\ell - 1)\langle \mathbf{a}\mathbf{a}^\dagger \rangle = \langle \mathbf{a}(\hat{\mathcal{W}} - 1)\mathcal{Q}\hat{\mathcal{W}}^{\ell-1}\mathcal{Q}(\hat{\mathcal{W}} - 1)\mathbf{a}^\dagger \rangle = \langle \mathbf{f}(\ell)\tilde{\mathbf{f}}(0) \rangle , \qquad (55)$$

with

$$\mathbf{f}(\ell) = \mathbf{a}(\mathcal{S}(\Gamma, (\ell+1)\tau)) - \langle \mathbf{a}(\mathcal{S}(\Gamma, \tau))\mathbf{a}^\dagger \rangle \langle \mathbf{a}\mathbf{a}^\dagger \rangle^{-1} \mathbf{a}(\mathcal{S}(\Gamma, \ell\tau)) , \qquad (56)$$

and

$$\tilde{\mathbf{f}}(0) = \mathbf{a}^\dagger - \mathbf{a}^\dagger(\mathcal{S}(\Gamma, \tau)) \langle \mathbf{a}\mathbf{a}^\dagger \rangle^{-1} \langle \mathbf{a}(\mathcal{S}(\Gamma, \tau))\mathbf{a}^\dagger \rangle .$$

The last step in the analysis of $\mathbf{\Phi}$ is to rewrite the expression for $\tilde{\mathbf{f}}(0)$ for conserved variables fields for small $\mathbf{k}$. We may write,

$$\mathbf{f}^\dagger(0) + \tilde{\mathbf{f}}(0) = -\mathbf{b}^\dagger \langle \mathbf{a}\mathbf{a}^\dagger \rangle^{-1} \mathcal{A} \langle \mathbf{a}\mathbf{a}^\dagger \rangle + \frac{1}{2}(\mathbf{a}^\dagger + \mathbf{a}^\dagger(\mathcal{S}(\Gamma, \tau))) \langle \mathbf{a}\mathbf{a}^\dagger \rangle^{-1} \langle \mathbf{b}\mathbf{b}^\dagger \rangle$$

$$= \mathcal{O}(\mathbf{k}^2) .$$

Making use of this result, the $\mathbf{\Phi}$ matrix to order $\mathcal{O}(\mathbf{k}^2)$ takes the form,

$$\mathbf{\Phi} = -\sum_{\ell=1}^{\infty} \langle \mathbf{f}(\ell)\mathbf{f}^\dagger(0) \rangle \langle \mathbf{a}\mathbf{a}^\dagger \rangle^{-1} .$$

3.3 General Form of the Kinetic Equation

In order to use these results to write the kinetic equations in the form of the hydrodynamic equations, we consider the passage from discrete-time to continuous-time dynamics. From the analysis presented above, the sum of the $\mathbf{\Omega}$ and $\mathbf{\Phi}$ matrices may be written in terms of contributions of $\mathcal{O}(\mathbf{k})$ and $\mathcal{O}(\mathbf{k}^2)$, respectively, as

$$\mathbf{\Omega} + \mathbf{\Phi} = \mathcal{A} - \mathcal{B} ,$$

where

$$\mathcal{B} = -\frac{1}{2}\mathcal{A}^2 + \frac{1}{2}\langle \mathbf{f}(0)\mathbf{f}^\dagger(0) \rangle \langle \mathbf{a}\mathbf{a}^\dagger \rangle^{-1} - \mathbf{\Phi} .$$

The matrices $\mathcal{A}$ and $\mathcal{B}$ are of the first and second orders in $\mathbf{k}$, respectively. The evolution equation (35) can be written as

$$\bar{\mathbf{a}}(t + \tau) = e^{\tau \frac{\partial}{\partial t}} \bar{\mathbf{a}}(t) = (1 + \mathcal{A} - \mathcal{B})\bar{\mathbf{a}}(t) ,$$

so that we have the operator identity

$$e^{\tau \frac{\partial}{\partial t}} = 1 + \mathcal{A} - \mathcal{B} .$$

Taking the logarithm of this operator identity and expanding the logarithm in a Taylor series up to second order in k we obtain

$$\tau \frac{\partial}{\partial t} = \mathcal{A} - \frac{1}{2}\mathcal{A}^2 - \mathcal{B},$$

$$= \mathcal{A} - \frac{1}{2}\langle \mathbf{f}(0)\mathbf{f}^\dagger(0)\rangle \langle \mathbf{a}\mathbf{a}^\dagger\rangle^{-1} - \sum_{\ell=1}^{\infty} \langle \mathbf{f}(\ell)\mathbf{f}^\dagger(0)\rangle \langle \mathbf{a}\mathbf{a}^\dagger\rangle^{-1} . \qquad (57)$$

Using this relation, the continuous-time limit of (35) is given by

$$\tau \partial_t \bar{\mathbf{a}} = \mathcal{A}\bar{\mathbf{a}} - \left[\lim_{\mathsf{T}\to\infty} \frac{1}{2\mathsf{T}} \sum_{\ell,\ell''<\mathsf{T}} \langle \mathbf{f}(\ell)\mathbf{f}^\dagger(\ell')\rangle \langle \mathbf{a}\mathbf{a}^\dagger\rangle^{-1}\right] \bar{\mathbf{a}} .$$

$$(58)$$

The second term on the right hand side provides expressions for the transport coefficients in terms of discrete-time sums of autocorrelation functions. In writing this equation we used the notation,

$$\lim_{\mathsf{T}\to\infty} \frac{1}{2\mathsf{T}} \sum_{\ell,\ell'<\mathsf{T}} \langle \mathbf{f}(\ell)\mathbf{f}^\dagger(\ell')\rangle = \frac{1}{2}\langle \mathbf{f}(0)\mathbf{f}^\dagger(0)\rangle + \sum_{\ell=1}^{\infty} \langle \mathbf{f}(\ell)\mathbf{f}^\dagger(0)\rangle . \qquad (59)$$

3.4 Hydrodynamic Equations

All that remains in the derivation is to substitute the explicit forms of the conserved fields to obtain the hydrodynamic equations. To separate the shear and bulk viscosity contributions one may write the momentum random force as a sum terms which are parallel and perpendicular to k,

$$f_{\mathbf{k}}^{\mu}(t) = \tau \sum_i \left(\mathbf{v}_i^\perp(t)\mathrm{i}\mathbf{k}\cdot\frac{\Delta\boldsymbol{\xi}^\perp_i(t)}{\tau} + \mathrm{i}\mathbf{k}\left[\mathbf{v}_i^\parallel(t)\frac{\Delta\boldsymbol{\xi}^\parallel_i(t)}{\tau} - \frac{1}{d}v_i(t)^2\right]\right) + o(\mathbf{k}),$$

where $\mathbf{v}_i^\parallel$ and $\mathbf{v}_i^\perp$ are the parallel and perpendicular components of the velocity, respectively. Then, using this decomposition of the random force and the explicit expression for the $\mathcal{A}$ matrix, we find that the linearized hydrodynamic equations take the form,

$$\partial_t \rho_{\mathbf{k}} = \mathrm{i}\mathbf{k}\cdot\boldsymbol{\mu}_{\mathbf{k}}, \qquad (60)$$

$$\partial_t \boldsymbol{\mu}_{\mathbf{k}} = \mathrm{i}\mathbf{k}\cdot\left[k_{\mathrm{B}}T\rho_{\mathbf{k}} + \frac{s_{\mathbf{k}}}{c_v}\right] - \frac{\eta}{m\rho}\left[\mathbf{k}\mathbf{k} - \frac{1}{d}k^2\mathbf{1}\right] : \boldsymbol{\mu}_{\mathbf{k}} - \frac{\eta_b}{m\rho}\mathbf{k}\mathbf{k} : \boldsymbol{\mu}_{\mathbf{k}} \qquad (61)$$

$$\partial_t s_{\mathbf{k}} = k_{\mathrm{B}}T\mathrm{i}\mathbf{k}\cdot\boldsymbol{\mu}_{\mathbf{k}} - \frac{\lambda}{m\rho}k^2 s_{\mathbf{k}} .$$

The viscosity coefficient is obtained from the auto-correlation of the transverse component of $f_{\mathbf{k}}^{\mu}(t)$:

$$\eta = \lim_{\mathsf{T}\to\infty} \frac{m^2\rho}{2k_{\mathrm{B}}TNT\tau} \sum_{\ell,\ell'<\mathsf{T}} \sum_{i,j} v_{xi}(\ell\tau)\Delta\xi_{yi}(\ell\tau)v_{xj}(\ell'\tau)\Delta\xi_{yj}(\ell'\tau) . \qquad (62)$$

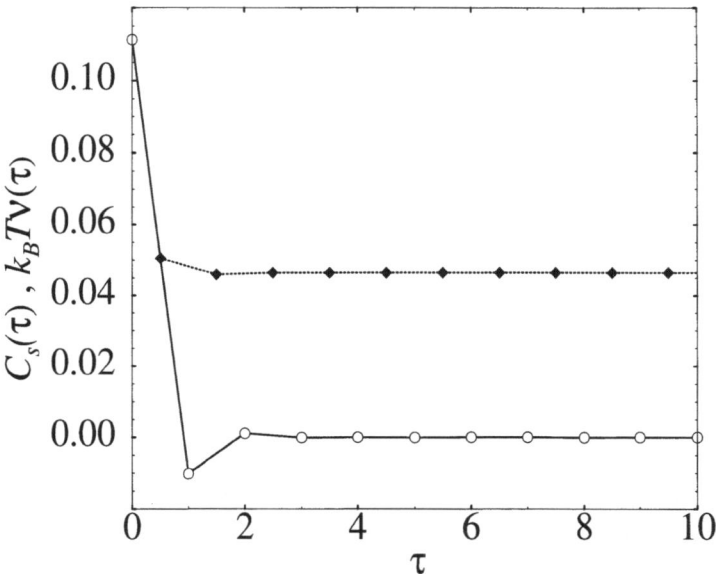

Fig. 6. Shear viscosity and stress autocorrelation function as a function of time. The circles on the solid line show computed values of the stress autocorrelation function. The filled diamonds on the dotted line are the results of the partial summation of the autocorrelation function. The parameter values are $\rho = 10.0$ and $k_\mathrm{B}T = 1/3$

Similarly, the bulk viscosity η_b and heat conductivity λ transport coefficients are given by the correlation function expressions,

$$\eta_b = \lim_{\mathsf{T} \to \infty} \frac{\tau m \rho}{2 k_\mathrm{B} T N \mathsf{T}} \sum_{\ell, \ell' < \mathsf{T}} \sum_{i,j} \left[\mathbf{v}_i^{\|}(\ell \tau) \frac{\Delta \boldsymbol{\xi}^{\|}{}_i(\ell \tau)}{\tau} - \frac{1}{d} v_i(\ell \tau)^2 \right]$$

$$\times \left[\mathbf{v}_j^{\|}(\ell' \tau) \frac{\Delta \boldsymbol{\xi}^{\|}{}_j(\ell' \tau)}{\tau} - \frac{1}{d} v_j(\ell' \tau)^2 \right] , \tag{63}$$

$$\lambda = \lim_{\mathsf{T} \to \infty} \frac{m \rho}{4 C_\mathrm{v} k_\mathrm{B} T^2 N \mathsf{T} \tau} \sum_{\ell, \ell' < \mathsf{T}} f_\mathbf{k}^\varepsilon(\ell \tau) f_\mathbf{k}^\varepsilon(\ell' \tau) . \tag{64}$$

This rather detailed calculation has served to establish that the mesoscopic multi-particle collision model does lead to the correct full set of hydrodynamic equations on long distance and time scales. Furthermore, we have derived expressions for the transport properties of the fluid in terms of discrete-time autocorrelation function expressions that can serve as the starting points for computing these properties from simulations of the dynamics.

As an illustration of the calculation of a transport property using the discrete Green-Kubo expression, in Fig. 6 we present the results of numerical simulations of the stress-stress autocorrelation function and shear viscosity. A scattering rule was used where the velocity was rotated by $\pi/2$ in random directions. The simulations were carried out on a three-dimensional system of size $32 \times 32 \times 32$ lattice cells. The figure shows both the stress-stress autocorrelation function and its time integral, whose asymptotic value is the solvent viscosity. We note that the stress autocorrelation decays to zero in about two discrete time units, setting the time scale for solvent relaxation.

4   Simulations of Fluid Flow

While the demonstration of the properties of the mesoscopic multi-particle collision model and the derivation of the hydrodynamic equations are rather involved, the dynamics may be simulated in a simple and efficient manner. Below we give some examples of the simulations of fluid flow in order to demonstrate the utility of the model for investigations of hydrodynamic flows.

We first show the results of simulations of three-dimensional flow past a cylinder. A solid cylinder of radius $2R = 100$ parallel to the z axis was created by imposing bounce-back conditions on particles colliding with the cylinder. Periodic boundary conditions were imposed on the system. The system was initialized by assigning velocities from Maxwell distribution

$$\sqrt{\frac{m}{2\pi k_B T}} \; \exp\left(-mu^2/(2k_B T)\right).$$

The flow along the x-direction was established by assigning velocities from a Maxwell distribution with the velocity shifted u_x in the domain $x < 5$. One may demonstrate that the system relaxes to a uniform steady flow in the absence of dissipation.

After approximately 2000 time steps the flow past the cylinder stabilizes and a periodically oscillating pattern is established. In Fig. 7 a snapshot of the flow pattern is plotted. The length of an arrow at each position is proportional to the flow velocity at this point and we have color coded the direction of the flow with varying hues. We note that spatiotemporal averaging obscures fine details of flow. In Fig. 7 the fine structure of flow pattern supporting the two large vortexes is not visible.

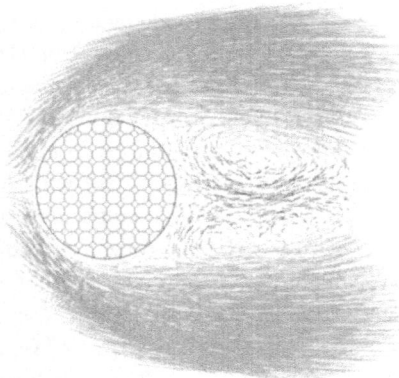

Fig. 7. Simulation of fluid flow past a cylinder showing the development of vortices. The cylinder diameter, system size and density are $D = 2R = 100$, $500 \times 600 \times 5$ and $\rho = 4.0$, respectively. The temperature and velocity at the origin are $T = 1.0$ and $u_x = 0.3$. The corresponding kinematic viscosity and Reynolds number are $\nu = 0.58$ and $Re = 52.0$

Fig. 8. Development of the boundary layer at the rear of a disc suddenly set in motion

As another illustration of the method, in Fig. 8 we show the results of simulations of stages in the development of the boundary layer at the rear of a disk suddenly set in motion. The system size is 400×400 cells of unit length and the disc diameter $2R = 100$. The flow velocity at $x = 0$, the density of the system and the Reynolds number are $u_x = 0.4$, $\rho = 20.0$ and $\mathrm{Re} = 1520$, respectively. Periodic boundary conditions are imposed on the system and bounce-back conditions on the disc. The initially symmetric flow separates from the disk and a backflow at the far end of the disk develops. At the contact line between the normal flow and the oppositely-directed backflow, a system of vortices appears which later expands into the full-scale boundary layer [14].

A series of more detailed studies of the properties of hydrodynamic flows and fluid flow around obstacles of different shapes using the mesoscopic multi-particle collision model were carried out by Ihle and Kroll [15] and Lamura et al. [16,17]. These studies provide further evidence of the utility of the method for the investigation of problems in hydrodynamics.

5   Mesoscopic Model for Solute Molecular Dynamics

In this section we show how one may study the dynamics of molecules and other types of molecular aggregates or particles embedded in a mesoscopic solvent. There are many circumstances when such an approximate treatment of solvent dynamics is appropriate or necessary. This is the case if one's primary interest is in the properties of the solute molecules; then the details of the solvent molecule motions need not be analyzed except in so far as they influence solute molecule dynamics. This view of solvent dynamics underlies all Langevin and generalized Langevin approximations to condensed phase systems where the solvent is not explicitly included in the description of the system. Rather than adopting such a extreme reduction of description, we show how one may construct a hybrid molecular dynamics-mesoscopic solvent scheme that eliminates the need for some of the restrictive assumptions in more phenomenological models.

The system we now consider comprises N solvent or bath molecules (labelled b) with phase space coordinates$(\mathbf{V}^{(N)}, \mathbf{X}^{(N)})$ introduced earlier, which will be described at the mesoscopic level, and M solute molecules (labelled s) with phase space coordinates $\mathbf{X}^{(M)} = (\mathbf{x}_{N+1}, \mathbf{x}_{N+2}, \ldots, \mathbf{x}_{N+M})$ and $\mathbf{V}^{(M)} = (\mathbf{v}_{N+1}, \mathbf{v}_{N+2}, \ldots, \mathbf{v}_{N+M})$, which will be described microscopically. The system is demonstrated schematically in Fig. 9. We assume the solute molecules interact through the intermolecular potential $V_{ss}(\mathbf{X}^{(M)})$, while the solute-solvent interactions are governed by the potential energy $V_{sb}(\mathbf{X}^{(M)}, \mathbf{X}^{(N)})$. Since the solvent will be treated at the mesoscopic level using multi-particle collision dynamics, solvent-solvent interactions are set to zero.

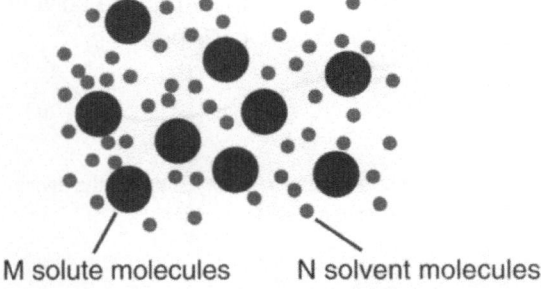

M solute molecules   N solvent molecules

Fig. 9. Schematic representation of a system following hybrid mesoscopic-molecular dynamics. The small particles are treated at the mesoscopic level while the large particles are treated by full molecular dynamics that includes solute-solute and solute-solvent interactions

If we let $(\mathbf{X}^{(K)}, \mathbf{V}^{(K)}) = (\mathbf{X}^{(N)}, \mathbf{X}^{(M)}, \mathbf{V}^{(N)}\mathbf{V}^{(M)})$ for the $K = N+M$ particles, we can define the classical evolution operator for streaming in the potential energy function, $V(\mathbf{X}) = V_{ss}(\mathbf{X}^{(M)}) + V_{sb}(\mathbf{X})$ as

$$i\mathcal{L} = \mathbf{V}^{(K)} \cdot \nabla_{\mathbf{X}^{(K)}} + \mathbf{F} \cdot \mathbf{M}^{-1} \cdot \nabla_{\mathbf{V}^{(K)}} \ ,$$

where $\mathbf{F} = -\nabla_{\mathbf{X}^{(K)}} V(\mathbf{X}^{(K)})$ is the force and $\mathbf{M}$ is a diagonal matrix of masses. The equation for the evolution of the $N + M$ particle phase space density in the hybrid molecular-mesoscopic dynamics is easily written using this classical evolution operator and the collision operator $\mathcal{C}$ introduced earlier. We have

$$\frac{\partial}{\partial t}\mathsf{P}(\mathbf{X}^{(K)}, \mathbf{V}^{(K)}, t) = \left(-i\mathcal{L} + \tilde{\mathcal{C}} \right)\mathsf{P}(\mathbf{X}^{(K)}, \mathbf{V}^{K}, t) \ ,$$

where the collision operator $\tilde{\mathcal{C}}$ was defined in (10). If we integrate this equation from $t_0 = m\tau + \epsilon = t$ to $t + \tau$ we obtain

$$e^{i\mathcal{L}\tau}\mathsf{P}(\mathbf{X}^{(K)}, \mathbf{V}^{(K)}, t + \tau) = \hat{\mathcal{C}}\mathsf{P}(\mathbf{X}^{(K)}, \mathbf{V}^{(K)}, t) \ . \tag{65}$$

If the solute molecules are not present, the potential contribution to the evolution operator $i\mathcal{L}$ is zero, and the evolution operator describes free streaming. In this limit (65) reduces to (6).

To implement the dynamics described by (65), we imagine that the phase space trajectory of the entire system is partitioned into time segments of length τ within which one evolves the system by Newton's equations of motion,

$$\dot{\mathbf{x}}_i = \mathbf{v}_i$$
$$m_i\dot{\mathbf{v}}_i = -\frac{\partial V}{\partial \mathbf{x}_i} = \mathbf{F}_i \ ,$$

where m_i is the mass of particle i. Such a partitioned phase space trajectory is depicted schematically in Fig. 10. During this Newtonian portion of the evolution the solvent-solvent intermolecular potential is zero; thus, solvent molecules undergo free streaming in the solute-solvent intermolecular potential (assuming they are within range of this

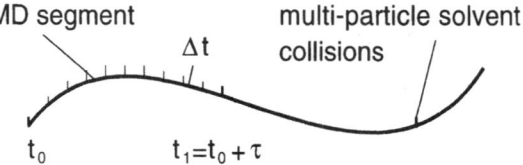

Fig. 10. Schematic representation of a system trajectory showing the division into MD and segments separated by multi-particle solvent molecule collisions

potential) but do not interact with each other. At discrete time intervals τ the solvent molecules are assumed to undergo multi-particle collisions as discussed in Sect. 2 which change the velocities of the solvent molecules. These multi-particle collisions replace the full solvent-solvent interactions. Since there are no solvent-solvent interactions this molecular dynamics evolution may be carried out efficiently since it scales with $M \times (N+M)$ rather than $(M+N)^2$. Since typically $N \gg M$ this will lead to short simulation times compared to full molecular dynamics.

6   Simulations of Hybrid Dynamics

In this section we give some simple illustrations of the implementation of the hybrid mesoscopic-molecular dynamics algorithm. In order to carry out such simulations, one must select a values for the molecular dynamics and multi-particle collision time steps, Δt and τ, respectively. During the molecular dynamics segments Δt for the integration of Newton's equations of motion must be chosen to resolve motion in all the forces, including solvent-solute forces. In addition, since the multi-particle collision rule produces momentum jumps in the solvent momenta, an integration scheme must be selected that involves the velocities of the molecules, such as the velocity Verlet or leap-frog algorithms [18]. The time τ for multi-particle collisions is dictated by several factors. From the perspective of the solvent molecules, τ must be sufficiently large that solvent molecules stream on the order of a cell length in order to ensure sufficient dynamical changes in multi-particle collisions. Furthermore, τ should be sufficiently small to incorporate correctly the effect of solvent dynamics on the solute molecules. In such applications, typically we are interested in both microscopic (mesoscopic) and collective effects on the solute dynamics. We now show how this hybrid model can be used to study some familiar problems in condensed matter physics.

6.1   Brownian Motion

The classical theory of Brownian motion is perhaps one of the best known models of solute motion in a solvent whose dynamics is treated approximately [19]. In this theory the Langevin equation for a Brownian particle takes the form [20,21]

$$M\frac{\mathrm{d}\mathbf{u}(t)}{\mathrm{d}t} = -\zeta\mathbf{u}(t) + \mathbf{f}(t) , \qquad (66)$$

where **r** and **u** are the position and velocity of the Brownian particle with mass M, ζ is the friction coefficient and **f** is a random force which is usually taken to be a Gaussian random process with white noise spectrum,

$$\langle \mathbf{f}(t) \rangle = 0 \quad \text{and} \quad \langle \mathbf{f}(t)\mathbf{f}(t') \rangle = 2k_B T \zeta \delta(t - t') \ .$$

In this mesoscopic description all information about solute-solvent interactions is contained in the friction coefficient which must be evaluated by other means; it is a parameter in phenomenological Brownian motion theory. In general, the friction coefficient contains both microscopic and macroscopic (hydrodynamic) contributions whose relative magnitudes depend on the relative sizes of the Brownian and solvent molecules. Thus, the Langevin approach is not complete unless the friction coefficient and stochastic properties of the random force are determined from the molecular dynamics of the solute and solvent molecules.

In order to study Brownian motion in the mesoscopic solvent, we consider the diffusion of solute particles that interact with the mesoscopic solvent molecules through continuous intermolecular forces. The Hamiltonian that governs the molecular dynamics segments is

$$\mathcal{H} = \frac{1}{2}M\mathbf{u}^2 + \sum_{i=1}^{N} \frac{1}{2}m\mathbf{v}_i^2 + \sum_{i=1}^{N} V_{sb}(|\mathbf{x}_i - \mathbf{r}|) \ . \tag{67}$$

For the simulation results presented below, the Brownian particle-solvent interactions, V_{sb}, are given by truncated LJ potentials,

$$V_{sb}(r) = \begin{cases} 4\epsilon \left[\dfrac{\sigma^{12}}{r^{12}} - \dfrac{\sigma^6}{r^6} + \dfrac{1}{4} \right], & r < 2^{1/6}\sigma \\ 0, & r > 2^{1/6}\sigma \end{cases} \ , \tag{68}$$

with $\sigma = 3.0$ and $\epsilon = 1.0$. In the Newtonian trajectory segments the equations of motion were integrated using the velocity Verlet algorithm [18] with a time step of $\Delta t = 0.02\tau$, which is sufficient to resolve the intermolecular forces. The mass of the Brownian particle was taken to be $M = 250$ while the solvent particle mass was $m = 1$. The solvent density was $\rho = 10$ and reduced temperature was $k_B T = 1/3$. For these conditions, both the mass density ratio, $m/M = 0.004$ and the mass density ratio, $m\rho/(M/V_B) = 0.45$, where the volume of the Brownian particle is $V_B = 4\pi\sigma^3/3$, are small so that one expects simple dynamics [22]. The multi-particle collision dynamics was carried out using random rotations by $\pm\pi/2$ about randomly chosen axes as discussed earlier.

The velocity autocorrelation function of the Brownian particle is defined as

$$C_u(t) = \frac{1}{3} \langle \mathbf{u}(t) \cdot \mathbf{u} \rangle \ .$$

The phenomenological theory of Brownian motion based on the Langevin equation (66) predicts exponential decay, $C_u(t) = (k_B T/M) \exp -\zeta t/M$, for all times. For systems with continuous forces, the $C_u(t)$ must have an initial slope of zero and behave as

$$C_u(t) \sim \frac{k_B T}{M} - \frac{\langle F^2 \rangle}{3M^2} \frac{t^2}{2} \ ,$$

for short times. Simulation results using the hybrid mesoscopic model confirm this short time behavior. It is more interesting to study the long time behavior of the velocity correlation function. While a simple Langevin model predicts exponential decay for all times, it is known from both full molecular dynamics simulations [23], kinetic theory [24] and mode coupling theories [25] that the $C_u(t) \sim t^{-3/2}$ for long times in three dimensions. The long time tail arises from the coupling between the Brownian particle density field and the viscous modes of the solvent. Such an effect can also be captured in a generalized Langevin model,

$$M\frac{d\mathbf{u}(t)}{dt} = -\int_0^t dt' \zeta(t - t')\mathbf{u}(t') + \mathbf{f}(t) , \qquad (69)$$

where the time dependent friction is evaluated using hydrodynamics [26]. In particular, using the expression for the friction coefficient for a macroscopic particle oscillating with frequency ω in an incompressible continuum fluid, [27]

$$\zeta(\omega) = \zeta_h \left[\frac{(1+\alpha\sigma)}{(1+\alpha\sigma/3)} + \frac{1}{6}(\alpha\sigma)^2 \right] , \qquad (70)$$

where the hydrodynamic friction coefficient is $\zeta_h = 4\pi\eta\sigma$ for slip boundary conditions appropriate for our central our system with central forces. The Brownian particle radius has been set equal to the Lennard-Jones σ parameter for Brownian particle-solvent particle interactions. Here $\alpha^2 = -i\omega\rho M/\eta$. The time-dependent diffusion coefficient $D(t)$ is defined by the finite time integral of the velocity correlation function,

$$D(t) = \int_0^t dt' C_u(t') .$$

Using (69) with (70), the long time behavior of $D(t)$ is given by

$$D(t) \approx D - \frac{\alpha_1}{\sqrt{t}}, \text{ where } \alpha_1 = \frac{2}{3}(4\pi\eta)^{-3/2}\sqrt{M\rho} .$$

This result is valid for either slip or stick boundary conditions. The coefficient of $t^{-1/2}$ depends only on the Brownian particle mass and the solvent density and viscosity. Figure 11 plots the hybrid mesoscopic model simulation results for $D(t)$ versus $t^{-1/2}$. One can see from this figure that $D(t)$ does indeed possess a $t^{-1/2}$ long time tail. Furthermore, the coefficient of this long time decay is in agreement with the predictions of hydrodynamics: The predicted hydrodynamic value using the mesoscopic solvent viscosity is $\alpha_1 = 0.0114$ while the simulation value is approximately 0.0104. These results show that the mesoscopic multi-particle collision dynamics correctly captures the collective hydrodynamic component of the Brownian particle diffusion coefficient.

It is also of interest to compare the magnitude of the diffusion coefficient obtained from the simulation with that predicted by simple theories that account for both microscopic and hydrodynamic contributions to this transport coefficient. One can view the environment of the Brownian particle as being composed of two parts: A boundary layer of microscopic dimensions where details of the intermolecular forces and collision dynamics are essential and an outer region where a continuum hydrodynamic description

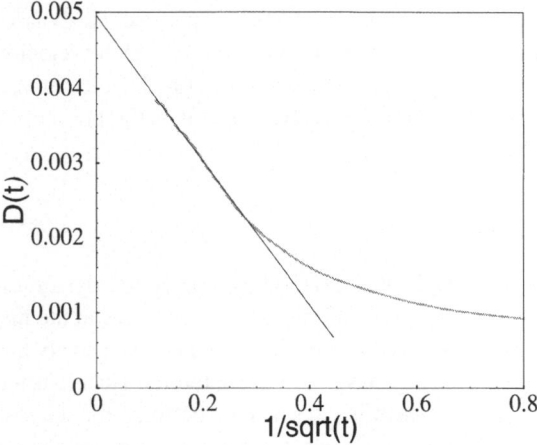

Fig. 11. Long time behavior of the time dependent diffusion coefficient

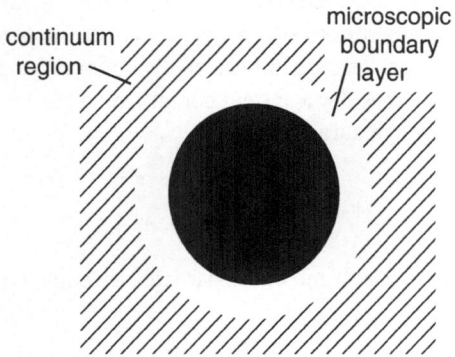

Fig. 12. Boundary layer around a microscopic particle

of the solvent is appropriate. (See Fig. 12 for a schematic picture of this decomposition of the environment.) Using a generalized boundary condition in conjunction with a kinetic theory description of the boundary layer one may derive an approximate expression for the diffusion coefficient of the form, [28]

$$D = D_0 + D_h \ , \tag{71}$$

where D_0 is a microscopic contribution arising from binary collision events whose approximate from can be estimated from kinetic theory,

$$D_0 = \frac{3}{8\rho\sigma^2} \left(\frac{k_{\mathrm{B}}T}{2\pi M} \right)^{1/2} \ ,$$

and a hydrodynamic contribution $D_h = k_{\mathrm{B}}T/\zeta_h$. The estimates of these quantities are: $D_0 = 1.0 \times 10^{-3}$ and $D_h = 4.5 \times 10^{-3}$ giving $D = 5.5 \times 10^{-3}$. The hydrodynamic component dominates the microscopic component for a large Brownian particle. The value of the diffusion coefficient determined from an extrapolation of the simulation data is $D = 4.9 \times 10^{-3}$ which is close to that from the approximate theory.

From these results we conclude that the hybrid mesoscopic dynamics is able capture important microscopic and hydrodynamic contributions to the velocity correlation function and diffusion coefficient. It provides a description of the dynamics that goes beyond that of simple or even generalized Langevin models with approximate expressions for the time dependent friction.

6.2 Cluster Dynamics

As a somewhat more complicated example we study the dynamics and equilibrium structure of an aggregate or cluster of microscopic particles in the mesoscopic solvent. Clusters are interesting systems that have been studied extensively both experimentally and theoretically [29,30]. They have attracted attention since clusters with nanoscale dimensions lie in a regime which is intermediate between the microscopic and macroscopic domains and exhibit unusual properties as a result of the strong competition between bulk and surface forces. Clusters of Lennard–Jones (LJ) particles with parameters that model argon atoms have been studied in vacuum using full molecular dynamics [31–33]. Studies of the properties of such LJ clusters in the mesoscopic solvent provide information on how such clusters are influenced by a thermalizing environment [34].

More specifically, we consider a system comprising a cluster whose particles have mass m_s and interact through attractive LJ forces,

$$V_{ss} = 4\epsilon_{ss}\left[\frac{\sigma_{ss}^{12}}{r^{12}} - \frac{\sigma_{ss}^{6}}{r^{6}}\right] ,$$

embedded in the mesoscopic solvent whose particles have mass m. The system Hamiltonian is

$$\mathcal{H} = \sum_{i=N+1}^{K}\frac{1}{2}m_s\mathbf{u}_i^2 + \sum_{i<i'=N+1}^{K}V_{ss}(|\mathbf{r}_i - \mathbf{r}_{i'}|) + \sum_{i=N+1}^{K}\sum_{j=1}^{N}V_{sb}(|\mathbf{r}_i - \mathbf{x}_j|) + \sum_{j=1}^{N}\frac{1}{2}m\mathbf{v}_j^2$$
$$\equiv \mathcal{H}_s + V_{sb} + \mathcal{H}_b , \tag{72}$$

where $\mathcal{H}_s$ is the cluster Hamiltonian, $\mathcal{H}_b$ is the bath Hamiltonian which only has a kinetic energy contribution since solvent-solvent forces are zero, and V_{sb} is again the cluster particle-solvent interaction potential which is taken to be a truncated LJ potential (68). The cluster particle interaction parameters are are taken to mimic argon: $\sigma_{ss} = 0.34$ nm and $\epsilon_{ss} = 1.00604$ kJ/mol and the values for the cluster particle-solvent molecule interactions are $\epsilon_{sb} = 1.00604$ with σ_{sb} taking either of two values, $\sigma_{sb} = 0.17$ nm or $\sigma_{sb} = 0.221$ nm. The masses of the cluster particles are $m_s = 39.948$ g/mol and the solvent molecules have masses $m = 3.9948$ g/mol.

The cubic simulation box had length $L = 5.44$ nm with periodic boundary conditions and contained $N = 327680$ solvent molecules with number density $\rho_s = 2035.42$ nm^{-3} and a cluster with either $M = 25$ or 123 particles. The molecular dynamics time step for the velocity Verlet integrator was $\Delta t = 0.002$ ps. The system was divided into $32 \times 32 \times 32$ cells for the multi-particle collision dynamics and $\tau = 0.1$ ps. The same random rotation rule as that described for Brownian motion was employed. In Fig. 13 we show a picture of a cluster with 123 particles at temperature $T = 48.4$.

Fig. 13. Picture of a $M = 123$ atom cluster in the mesoscopic solvent. The cluster particles are depicted as large atoms while the solvent atoms are small. Only solvent molecules close to the cluster surface are shown in the figure for clarity

The cluster structure can be described by the radial distribution function for cluster and solvent molecules relative to the center of mass of the cluster,

$$g_{\mathrm{CM}-\alpha}(r) = \frac{1}{4\pi r^2 \rho_\alpha} \left\langle \sum_i^{N_\alpha} \delta(|\mathbf{x}_i - \mathbf{R}_{\mathrm{CM}}| - r) \right\rangle ,$$

where $\alpha = s$ or b designates a cluster or solvent molecule, $\mathbf{R}_{\mathrm{CM}}$ is the center of mass of the cluster and ρ_α is the number density of cluster or solvent molecules. Here $\rho_s = \sigma_{\mathrm{ss}}^{-3} = 25.44 \, \mathrm{nm}^{-3}$.

For clusters in vacuum $g_{\mathrm{CM}-\mathrm{c}}(r)$ shows liquid-like distributions of cluster particles and structural ordering within the clusters [32,33]. The cluster structure is modified when it is embedded in the mesoscopic solvent and the modifications depend on the cluster-solvent interactions. We fix $\epsilon_{\mathrm{sb}} = \epsilon_{\mathrm{ss}}$ and vary σ_{sb}. The $M = 25$ cluster radial distribution functions are shown in Fig. 14 (left) and (right) for $\sigma_{\mathrm{sb}} = 0.17$ and $\sigma_{\mathrm{sb}} = 0.221$, respectively. The radial distribution functions $g_{\mathrm{CM}-\mathrm{s}}(r)$ for the solvent molecules relative to the cluster center of mass are also shown in the figure. The structures of the radial distribution functions are similar in vacuum and in the mesoscopic solvent for $\sigma_{\mathrm{sb}} = 0.17 \, \mathrm{nm}$ but are quite different for $\sigma_{\mathrm{sb}} = 0.221 \, \mathrm{nm}$ where the cluster is compressed and adopts a solid-like configuration.

This difference is signalled in the structure of the solvent molecule-cluster particle distributions. For $\sigma_{\mathrm{sb}} = 0.17$ the solvent molecules are able to penetrate into the cluster. For $\sigma_{\mathrm{sb}} = 0.221$ they are not able to do so and the solvent provides a larger external force on the cluster which compresses it and induces a solid-like structural configuration.

The center of mass diffusion coefficients of the clusters can be analyzed using the approximate formula (71) presented in the previous subsection. Since the cluster is a composite object we can determine its effective radius R from the radial distribution function. For the $M = 25$ cluster with $\sigma_{\mathrm{sb}} = 0.17$ we find $R \approx 2.25$. Furthermore, the central LJ forces now act on the individual cluster atoms so for macroscopic sizes the cluster will appear to have stick boundary conditions. Given these facts, we may estimate D for the cluster and compare the value with that obtained from the simulation of the

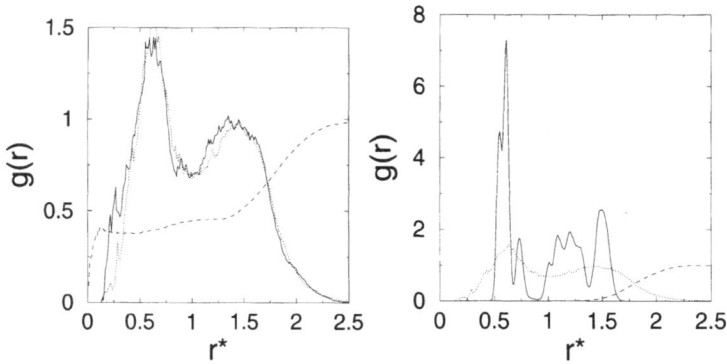

Fig. 14. Radial distribution function $g_{CM-c}(r)$ versus r^* for $\sigma_{sb} = 0.221$ nm (left) and $\sigma_{sb} = 0.17$ nm (right) for $M = 25$ ($T = 40.33$) clusters in the mesoscale solvent (*solid lines*); vacuum cluster (*dotted lines*). Also shown is the solvent radial distribution function $g_{CM-s}(r)$ (*dashed lines*)

mean square displacement of the cluster. We find, in units of cm^2/s, $D_0 = 2.8 \times 10^{-7}$, $D_h = 7.8 \times 10^{-7}$, so that $D(\text{theory}) \approx 1.1 \times 10^{-6}$. The simulation result is $D(\text{sim}) = 9.6 \times 10^{-7}$.

6.3 Polymer Dynamics

The dynamics of a long polymer chain in solution is determined by macroscopic properties of the solvent [35] rather than the microscopic details of the chain-solvent interactions. Indeed, time scale separation between the microscopic collision processes and hydrodynamic flows that define the long time evolution of a polymer chain makes direct simulations of polymer dynamics a challenging task. Therefore, in simulations, it is suitable to replace molecular solvents with mesoscale models [36–38].

Multi-particle collision dynamics can be adapted to model polymer flows [38]. In one version of the model the system is extended to include a polymer chain and "inert" solvent. The polymer chain itself is modelled using standard molecular dynamics. The architecture of the chain and the interactions within it can be varied. The free-streaming propagation step is modified to include evolution of the chain by integrating the chain equations of motion on the time interval τ,

$$(\mathbf{X}^{(M)}(t+\tau), \mathbf{V}^{(M)}(t+\tau)) = e^{i\mathcal{L}\tau}(\mathbf{X}^{(M)}(t), \mathbf{V}^{(M)}(t)) . \qquad (73)$$

Here $(\mathbf{X}^{(M)}(t), \mathbf{V}^{(M)}(t))$ denotes a set velocities and coordinates of a polymer chain with M monomers and $i\mathcal{L}$ is defined by the relation $i\mathcal{L}u = \{\mathcal{H}, u\}$ for any dynamical variable u. The Hamiltonian $\mathcal{H}$ of the system is given by

$$\mathcal{H} = \sum_{i \in \text{solvent}} \frac{1}{2}m_i v_i^2 + \sum_{i \in \text{chain}} \frac{1}{2}M_i V_i^2 + \sum_{i \leq j \in \text{chain}} U_{ij}(\mathbf{R}_i, \mathbf{R}_j) . \qquad (74)$$

During the propagation step the system evolves according to Newton's equation of motion. For chain atoms (73) is solved by integration of the equation of motions of a

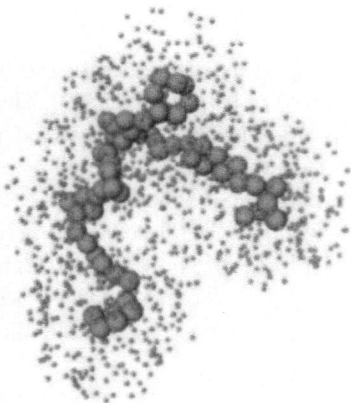

Fig. 15. Snapshot of a chain configuration showing a few of the surrounding mesoscopic solvent molecules

free chain using the Verlet algorithm

$$
\begin{aligned}
\mathbf{R}_i(t + \Delta t) &= \mathbf{R}_i(t) + \Delta t \mathbf{V}_i(t) + \frac{(\Delta t)^2}{2m_i} \mathbf{F}_i\big(\mathbf{R}_i(t)\big) , \\
\mathbf{V}_i(t + \Delta t) &= \mathbf{V}_i(t) + \frac{\Delta t}{2M_i}\left[\mathbf{F}_i\big(\mathbf{R}_i(t)\big) + \mathbf{F}_i\big(\mathbf{R}_i(t + \Delta t)\big) \right] ,
\end{aligned}
\tag{75}
$$

with a time step of Δt to evolve the positions and velocities in the MD step. As no force is exerted on the solvent between collision steps we propagate solvent particles with a time step $\Delta t = 1$.

To capture hydrodynamic interactions between molecules comprising the chain it is necessary to add chain-solvent interactions. This can done within the multi-particle collision scheme approach without destroying the conservation laws. To this end, during the collision step an additional collision step (2) exchanging momenta and energies of the polymer beads and solvent particles in the same cell is performed. By choosing a suitable class of rotation matrices $\hat{\omega}$ one may tune the bead-solvent friction coefficient. A snapshot of a polymer configuration in the mesoscopic solvent is shown in Fig. 15.

The numerical results of simulations were shown to agree well with an equation of motion for a chain which assumed a linear coupling to the velocity of the surrounding fluid via viscous friction [38]. Simulations demonstrated that the velocity-velocity correlation function comprises two contributions; an initial exponential decay resulting from Brownian collisions of monomers with uncorrelated solvent molecules and, at longer times, a hydrodynamic contribution as monomers interact via hydrodynamic modes propagated through the solvent. Analysis of chain diffusion coefficient given by the integral of the velocity-velocity correlation function demonstrated Zimm scaling $D \sim R_G^{-1}$ for the hydrodynamic contribution to the diffusion constant.

6.4 Complex Fluids

The multi-particle collision model has been extended to the treatment of binary immiscible fluids with a conserved order parameter [39]. The extension is based on the
Rothman–Keller model [40] for immiscible lattice gases. The original Rothman–Keller
model considers two kinds of particle, called *red* and *blue*, and introduces two fields
related to the colors of the particles: A color flux and a color field. The immiscible
lattice-gas model is constructed to account for cohesion in real fluids arising from short
range attractive intermolecular forces by allowing particles in neighboring sites to influence the configuration of particles at a chosen site. This is accomplished by constructing
collision rules where the "work" performed by the color flux against the color field is a
minimum.

This idea can be transcribed to the multi-particle collision rule by altering the nature
of the rotation operator that effects the collisions in a cell [39]. Again, we let $\boldsymbol{\xi}_\nu$ denote
the coordinate of the center of cell ν, and introduce a characteristic function $H_i(\boldsymbol{\xi}_\nu)$ for
particle i in cell ν which takes the value +1 if the particle is red and -1 if the particle is
blue. Using this notation the color flux is defined as

$$\mathbf{q}_\nu = \sum_{i=1}^{n(\xi_\nu)} H_i(\boldsymbol{\xi}_\nu)(\mathbf{v}_i' - \mathbf{V}_{\xi_\nu}) \,, \tag{76}$$

where $n(\xi_\nu)$ is the number of particles in cell ν. This is the color-weighted relative
velocity of the particles in cell ν. The color field is a color gradient arising from the
color differences in neighboring cells,

$$\mathbf{f}_\nu = \sum_{\nu' \in \mathcal{N}(\nu)} \mathbf{w}_{\nu\nu'} \hat{\boldsymbol{\xi}}_{\nu\nu'} \sum_{i=1}^{n(\xi_{\nu'})} H_i(\boldsymbol{\xi}_{\nu'}) \,, \tag{77}$$

where $\hat{\boldsymbol{\xi}}_{\nu\nu'}$ is a unit vector along the relative separation between cells ν and ν', that
is $\hat{\boldsymbol{\xi}}_{\nu\nu'} = (\boldsymbol{\xi}_\nu - \boldsymbol{\xi}_{\nu'})/|\boldsymbol{\xi}_\nu - \boldsymbol{\xi}_{\nu'}|$, and $\mathcal{N}(\nu)$ denotes the cells neighboring cell ν. The
weight function $\mathbf{w}_{\nu\nu'}$ is taken to be $\mathbf{w}_{\nu\nu'} = |\boldsymbol{\xi}_\nu - \boldsymbol{\xi}_{\nu'}|^{-1}$.

Given these definitions, the rotation matrix, $\hat{\omega}_\nu^c$, for multi-particle collisions in a cell
with coordinates $\boldsymbol{\xi}_\nu$ is constructed so that the rotated color flux lies along the color field,
$\hat{\mathbf{f}}_\nu = \hat{\omega}_\nu^c \hat{\mathbf{q}}_\nu$. If we let $\hat{\mathbf{h}}_\nu = \hat{\mathbf{f}}_\nu \times \hat{\mathbf{q}}_\nu$, which is normal to both the color flux and color
field vectors, the rotation operator effects a rotation of $\hat{\mathbf{q}}_\nu$ by an angle θ about $\hat{\mathbf{h}}_\nu$, where
θ is the angle between $\hat{\mathbf{q}}_\nu$ and $\hat{\mathbf{f}}_\nu$, $\cos\theta = \hat{\mathbf{f}}_\nu \cdot \hat{\mathbf{q}}_\nu$. The multi-particle collision rule again
takes the form, $\mathbf{v}_i = \mathbf{V}_\xi + \hat{\omega}^c(\mathbf{v}_i' - \mathbf{V}_\xi)$, where now, from (5),

$$\hat{\omega}^c(\mathbf{v}_i' - \mathbf{V}_\xi) = \hat{\mathbf{h}}\hat{\mathbf{h}} \cdot (\mathbf{v}_i' - \mathbf{V}_\xi) + (\mathbf{I} - \hat{\mathbf{h}}\hat{\mathbf{h}}) \cdot (\mathbf{v}_i' - \mathbf{V}_\xi) \cos\theta - \hat{\mathbf{h}} \times (\mathbf{v}_i' - \mathbf{V}_\xi) \sin\theta \,. \tag{78}$$

This rule has been used to simulate phase segregation in binary fluids in two and three dimensions [39]. Simulations on the model have shown that Laplace's law for the pressure
difference inside and outside a droplet of radius R, $\Delta p = 2\sigma/R$, where σ is the surface
tension, is satisfied and that this generalization of the collision rule reproduces the main

features of phase segregation dynamics. However, we note that this generalization no longer preserves phase space volumes and is not time reversible. As a result the equilibrium distribution is not Maxwellian. The model has also been extended to include surfactant molecules so that microemulsions may be simulated using such dynamics [41,42].

7   Conclusion and Perspectives

The multi-particle collision model provides a simple way to simulate the dynamics of fluids at the mesoscopic level. We have demonstrated that it possesses some basic ingredients which are desirable in such models: Mass, momentum and energy are conserved, the equilibrium distribution is Maxwellian, an H-theorem exists and the full set of hydrodynamic equations are obtained on long distance and time scales. Consequently, the model should find applications to problems in fluid flow and turbulence, especially in complex geometries where solutions of the Navier–Stokes equation are difficult. In this case it may provide an approach that complements DSMC, lattice-gas or lattice Boltzmann methods.

The hybrid mesoscopic multi-particle collision model marries full molecular dynamics of embedded particles (solutes) with a mesoscopic treatment of the solvent. We have shown that this hybrid scheme is able to capture essential features of both microscopic and hydrodynamic contributions arising from coupling between the solute and solvent degrees of freedom. The model is flexible enough to allow one to model large colloidal particles by using bounce-back [8] and other suitable generalizations of the collision rule to account for solid objects [39,43] or molecular degrees of freedom using explicit solute-solvent intermolecular potentials. The simple applications presented above have demonstrated the utility of this approach for the Brownian motion of molecules, molecular aggregates and polymer molecules in solution. This hybrid approach is likely to provide a promising route to investigate the dynamics of large biopolymers in solution where both specific details of solute-solvent forces and hydrodynamic solvent effects play important roles.

The development of the multi-particle collision model is still at an early stage. Further generalizations of the model to more complex situations; for example, molecular fluids, chemically reacting flows and more complex systems are possible. The model should not only provide a means to simulate efficiently complex systems and help bridge microscopic and macroscopic time scales but should also permit one to understand some of the important features of solute-solvent dynamics without recourse to full molecular dynamics of very large systems.

Acknowledgements

This work was supported in part by a grant from the Natural Sciences and Engineering Research Council of Canada.

References

1. D.H. Rothman, S. Zaleski: *Lattice-Gas Cellular Automata: Simple Models of Complex Hydrodynamics* (Cambridge University Press, Cambridge 1997)
2. J.-P. Rivet, J.-P. Boon: *Lattice Gas Hydrodynamics* (Cambridge University Press, Cambridge 2001)
3. I. Pagonabarraga: Lattice Boltzmann Modeling of Complex Fluids: Colloidal Suspensions and Fluid Mixtures, Lect. Notes Phys. **640**, 275 (2004)
4. For a review, see, S. Chen, G. Doolen: Ann. Rev. Fluid Mech. **30**, 329 (1998)
5. L. S. Luo: Phys. Rev. E **62**, 4982 (2000)
6. S. Succi: *The Lattice Boltzmann Equation – For Fluid Dynamics and Beyond* (Clarendon Press, Oxford 2001)
7. A. Malevanets, R. Kapral: J. Chem. Phys. **110**, 8605 (1999)
8. A. Malevanets, R. Kapral: J. Chem. Phys. **112**, 7260 (2000)
9. G. A. Bird: *Molecular Gas Dynamics* (Clarendon Press, Oxford 1976); G. A. Bird: Comp. & Math. with Appl. **35**, 1 (1998)
10. The multi-particle collision rule was first introduced in the context of a lattice model with a stochastic streaming rule in: A. Malevanets, R. Kapral: Europhys. Lett. **44**, 552 (1998)
11. R. Zwanzig: J. Chem. Phys. **33**, 1338 (1960)
12. H. Mori: Prog. Theor. Phys. **33**, 423 (1965)
13. Related derivations for lattice-gas automata have been carried out in D. d'Humières, B. Hasslacher, P. Lallemand, Y. Pomeau, J.-P. Rivet: Complex systems **1**, 839, (1987); M. H. Ernst. In: *Microscopic Simulations of Complex Hydrodynamics Phenomena*, ed. by M. Mareschal, B. Holian (Plenum Press, New York 1992) p. 153
14. G. K. Batchelor: J. Fluid Mech. **74**, 1 (1976)
15. T. Ihle, D. M. Kroll: Phys. Rev. E **63**, 020201 (2001)
16. A. Lamura, G. Gompper, T. Ihle, D. M. Kroll: Europhys. Lett. **56**, 768 (2001)
17. A. Lamura, G. Gompper, T. Ihle, D. M. Kroll: Europhys. Lett. **56**, 319 (2001)
18. W. C. Swope, H. C. Andersen, P. H. Berens, K. R. Wilson: J. Chem. Phys. **76**, 673 (1982); M. Tuckerman, B. J. Berne, G. J. Martyna: J. Chem. Phys. **97**, 1990 (1992)
19. A. Einstein: *Investigations on the Theory of Brownian Movement*, ed. by R. Fürth (Dover, New York 1956)
20. S. Chandrasekhar: Rev. Mod. Phys. **15**, 1 (1943)
21. R. Kubo: The Fluctuation-Dissipation Theorem. In: *Many-Body Problems*, ed. by W. E. Parry *et al.* (W. A. Benjamin, New York 1969), p. 235
22. M. Tokuyama, I. Oppenheim: Physica A **94**, 501 (1978)
23. B. J. Alder, T. E. Wainwright: Phys. Rev. Lett. **18**, 988 (1967)
24. J. R. Dorfman, E. G. D. Cohen: Phys. Rev. Lett. **25**, 1257 (1970)
25. K. Kawasaki: Prog. Theor. Phys. **45**, 1691 (1971)
26. R. Zwanzig, M. Bixon: Phys. Rev. A **2**, 2005 (1970)
27. L. D. Landau, E. M. Lifshitz: *Fluid Mechanics* (Pergamon Press, New York 1959)
28. J. T. Hynes, R. Kapral, M. Weinberg: J. Chem. Phys. **70** 1456 (1979)
29. A. W. Castleman, Jr., R. G. Keese: Chem. Rev. **86**, 589 (1986); Annu. Rev. Phys. Chem. **37**, 525 (1986); Science **241**, 36 (1988); A. W. Castleman, Jr., S. Wei: Ann. Rev. Phys. Chem. **45**, 685 (1994)
30. R. S. Berry, T. L. Beck, H. I. Davis, J. Jellinek: Adv. Chem. Phys. **70**, 75 (1988); M. Y. Hahn, R. L. Whetten: Phys. Rev. Lett. **61**, 1190 (1988); H.-P. Cheng, X. Li, R. L. Whetten, R. S. Berry: Phys. Rev. A **46**, 791 (1992)
31. J. D. Honeycutt, H. C. Andersen: J. Phys. Chem. **91**, 4950 (1987)
32. B. G. Moore, A. A. Al-Quraishi: Chem. Phys. **252**, 337 (2000)

33. A. S. Clarke, R. Kapral, B. Moore, G. Patey, X.-G. Wu: Phys. Rev. Lett. **70**, 3283 (1993); A. S. Clarke, R. Kapral, G. Patey: J. Chem. Phys. **101**, 2432 (1994)
34. S.-H. Lee, R. Kapral: Physica A **298**, 56 (2001)
35. M. Doi: *Introduction to Polymer Physics* (Clarendon Press, Oxford 1996)
36. P. Ahlrichs, B. Dünweg: J. Chem. Phys. **111**, 8225 (1999)
37. Y. Kong, C. W. Manke, W. G. Madden, A. G. Schlijper: J. Chem. Phys. **107**, 592 (1997)
38. A. Malevanets, J. M. Yeomans: Europhys. Lett. **52**, 231 (2000)
39. Y. Hashimoto, Y. Chen, H. Ohashi: Comput. Phys. Commun. **129**, 56 (2000)
40. D. H. Rothman, J. M. Keller: J. Stat. Phys.**52**, 1119 (1988)
41. Y. Inoue, Y. Chen, H. Ohashi: Colloids and Surfaces A **201**, 297 (2002)
42. T. Sakai, Y. Chen, H. Ohashi: Phys. Rev. E **65**, 031503 (2002)
43. T. Sakai, Y. Chen, H. Ohashi: J. Stat. Phys. **107**, 85 (2002)

Molecular Dynamics of Complex Systems: Non-Hamiltonian, Constrained, Quantum-Classical

Giovanni Ciccotti and Galina Kalibaeva

INFM and Dipartimento di Fisica, Universita degli Studi di Roma La Sapienza Piazzale Aldo Moro 5, Roma, Italy

Abstract. A theoretically sound and computationally tractable treatment for non-Hamiltonian molecular dynamics is needed for simulations of complex systems. Here, statistical mechanics of non-Hamiltonian systems is derived and it is applied to Nosé-Hoover and related isostats. Then, the most rigorous family of algorithms for integration of equations of motion, Trotter-derived unitary integrators (e.g. SHAKE), are discussed in detail. In the last part we address the generalization of molecular dynamics to treat interacting systems composed of quantum and classical subsystems.

1 Introduction

In the last years it has become apparent that an efficient and sound treatment of molecular dynamics simulation of complex systems can not be obtained without a suitable generalization of statistical mechanics to dynamical systems which are not derivable from a Hamiltonian. This generalization is needed in order to understand and to correctly apply new dynamics in generalized ensembles (e.g., constant temperature, constant pressure, etc.). Another consequence of this new approach is a better understanding of the nature of mechanical systems subjected to holonomic constraints. The latter have been of wide use in simulations of macromolecular systems and their statistical mechanical behavior deserves a correct treatment.

This article is organized in the following way. In Sects. 2 and 3 we will derive the statistical mechanics of non-Hamiltonian systems, apply it to Nosé–Hoover and related 'iso'-stats (thermostat, barostat, etc.), and interpret the statistical behavior of constrained dynamics in this new language. As an important aside we will also discuss the most rigorous family of algorithms for integration of equations of motion: Trotter-derived unitary integrators, e.g., SHAKE and related implementations. Section 4 is devoted to a discussion of the most-needed generalization of molecular dynamics to treat interacting systems composed of quantum and classical subsystems. This problem is far from trivial but recent progress has raised the hope that there will soon be at least tractable if not rigorous algorithms to compute the statistical properties of such systems.

2 Non-Hamiltonian Molecular Dynamics

2.1 Nosé–Hoover "Demonstration"

NVE (constant particle number, volume and energy) molecular dynamics describes the simplest case of a dynamical system. The equations of motion are the well-known Newton equations. If r_i is the position and p_i the momentum of particle i of mass m_i, these

G. Ciccotti and G. Kalibaeva, Molecular Dynamics of Complex Systems: Non-Hamiltonian, Constrained, Quantum-Classical,
Lect. Notes Phys. **640**, 150–189 (2004)

equations read

$$\dot{r}_i = \frac{p_i}{m_i} \quad \text{and} \quad \dot{p}_i = F_i, \tag{1}$$

where F_i is the total force on particle i.

To simulate a system at constant temperature and volume (NVT ensemble) which is more realistic and describes typical experimental conditions better than NVE, Nosé has proposed a system with additional variables characterizing a thermostat associated to the particles. The equations of motion for this extended space $\{r, p, \eta, P_\eta\}$ are

$$\dot{r}_i = \frac{p_i}{m_i}$$

$$\dot{p}_i = F_i - \frac{P_\eta}{Q} p_i$$

$$\dot{\eta} = \frac{P_\eta}{Q}$$

$$\dot{P}_\eta = \sum_i \frac{p_i^2}{2m_i} - Lk_B T_{\text{ext}}, \tag{2}$$

where k_B is the Boltzmann constant, L the number of degrees of freedom, and T_{ext} the desired external temperature of the system. The instantaneous internal temperature is

$$T_{\text{int}} = \frac{1}{Lk_B} \sum_{i=1}^{N} \frac{p_i^2}{m_i}.$$

The thermostat is characterized by a 'mass' Q and dynamical variables η and P_η. The only conserved quantity for this system in the case of $\sum_i F_i \neq 0$ is

$$H' = H(r,p) + \frac{P_\eta^2}{2Q} + Lk_B T\eta = C_1, \tag{3}$$

where $H(r,p)$ is the usual Hamiltonian of an N-particle system. As we know, if $\sum_i F_i = 0$, the most common physical case, we have an additional vectorial conserved quantity

$$K = Pe^\eta = C_2, \tag{4}$$

where P is the total momentum of the system, i.e., $P = \sum_i p_i$.

Next we present the "Nosé–Hoover demonstration". This was first carried out by Hoover in 1985 [1] and then by Nosé in 1986 [2]. As our starting point, we use the Liouville equation for the normalized distribution function, f. In Eulerian form it reads

$$\frac{\partial f}{\partial t} + \nabla \dot{x} f = 0,$$

and in Lagrangian form it can be written as

$$\frac{df}{dt} = -f\nabla \dot{x} = fdN\frac{P_\eta}{Q}, \tag{5}$$

where x denotes the phase space, d is the dimension of the system, and we have used the equations of motion (2). Calling

$$H''(r, p, P_\eta) = H(r, p) + \frac{P_\eta^2}{2Q} = H' - Lk_B T\eta,$$

we obtain

$$\frac{dH''}{dt} = -Lk_B T \frac{P_\eta}{Q}.$$

Then, by choosing $L = dN$ in (5) we obtain $df = -f\beta dH''$, and therefore $f(r, p, P_\eta) \propto e^{-\beta H''}$. We can now write the reduced distribution function as

$$f^{\text{reduced}}(r, p) = \int dP_\eta f(r, p, P_\eta) \propto e^{-\beta H}.$$

Apparently, the use of the standard Liouville equation applied to Nosé–Hoover dynamics has given us the correct canonical probability distribution for the NVT ensemble. As we will see in the following (Sect. 2.5), however, the statistical distribution obtained from a practical implementation of the Nosé–Hoover equations appears to be very different from what could be expected from the present (pseudo)-derivation.

2.2 Invariant Measure for Non-Hamiltonian Dynamical Systems

The use of non-Hamiltonian dynamics introduces the problem of describing the statistical distribution sampled by the dynamical variables in the phase space. In 1999 Tuckerman et al. [3] presented a classical statistical mechanics approach to non-Hamiltonian systems. They provided a definite procedure for derivation of the ensemble generated by time averages and generalized the usual Hamiltonian based statistical mechanical phase space principles to non-Hamiltonian systems. In this section we summarize the non-Hamiltonian formalism of Tuckerman et al. [3,4].

We start by considering a non-Hamiltonian dynamical system of form

$$\dot{x} = \xi(x, t), \tag{6}$$

where $x = \{x_i\}|_{i=1,\nu}$ are the ν dynamical variables. In order to define correctly the relevant phase space, the first step is to identify and eliminate linearly dependent and driven variables. Linearly dependent variables are those which can be written in the form $x_3(t) = A_1 x_1(t) + A_2 x_2(t)$, where A_1 and A_2 are constants. Such variables must be removed from the phase space since they do not add any information about the statistical behavior of the system. Driven variables are those such as x_b when $\dot{x}_a = \xi_a(x_a)$ and $\dot{x}_b = \xi_b(x_a, x_b)$, with conservation laws only of the form $\Lambda_k(x_a) = 0$ and $\Lambda_{k'}(x_b) = 0$. Driven variables, such as x_b, must be eliminated, since the phase space distribution of x_a contains all statistically relevant information and is not affected by x_b.

Working now with the remaining variables, x', we have to solve our equations of motion, for given initial conditions $x_i'(0)$. The solution, $x_i'(t) = x_i'(t, x_1'(0), \dots, x_\nu'(0))$, can be seen as a coordinate transformation from the coordinates at time zero to those at time t. In general, we can write

$$x'_t = x'_t(x'_0), \tag{7}$$

where $t \in (-\infty, +\infty)$. We now determine how the initial phase space volume element $\mathrm{d}^N x'_0$ transforms under (7). In general,

$$\mathrm{d}^N x'_t \neq \mathrm{d}^N x'_0 \quad \text{and} \quad \mathrm{d}^N x'_t = J_{t,0} \mathrm{d}^N x'_0,$$

where $J_{t,0}$ is the Jacobian of the coordinate transformation,

$$J_{t,0} = J(x'(t); x'(0)) = \frac{\partial(x'_t)}{\partial(x'_0)}$$

and it satisfies the following equation [1]

$$\frac{\mathrm{d}J_t}{\mathrm{d}t} = J_t \sum \frac{\partial \dot{x}'}{\partial x'} = J_t \sum \frac{\partial \xi(x')}{\partial x'} = J_t \kappa(x') \tag{8}$$

subject to the initial condition $J_0 \equiv J(x'(0); x'(0)) = 1$.

The quantity $\kappa(x')$ in (8) is called phase space compressibility of a non-Hamiltonian system. For a Hamiltonian system $\kappa(x')$ vanishes, such that $J_{t,0} = 1$ at all times. This leads to the familiar conservation of the Euclidean phase space volume and indicates that the phase space is a flat manifold. For a non-Hamiltonian system, however, $\kappa(x')$ needs not to vanish. Then it follows that the usual phase space measure is no longer an invariant measure under dynamical evolution. In fact, the general solution to (8) is

$$J_{t,0} = \mathrm{e}^{\int_0^t \mathrm{d}\tau \kappa(x'_\tau)} = \mathrm{e}^{w(x'_t) - w(x'_0)}. \tag{9}$$

This can be written as

$$\mathrm{e}^{-w(x'_t)} \mathrm{d}^N x'_t = \mathrm{e}^{-w(x'_0)} \mathrm{d}^N x'_0 = \sqrt{g} \mathrm{d}^N x'_t, \tag{10}$$

where $w(x'_t)$ is the time integral of κ and $\sqrt{g}$ the metric factor. Equation (10) defines an invariant measure associated with non-Hamiltonian dynamics given by (6).

What we have gained in the above analysis is that (8)–(10) permit us to carry out the statistical analysis of any non-Hamiltonian ergodic system in an arbitrary set of coordinates. Please note that in the next section we will drop the primes from the relevant phase space variables for simplicity.

---

[1] We can easily prove (8). We start by considering a matrix M with elements μ_{ij}. The corresponding diagonalized matrix is $D = U^{-1}MU$ with elements $\mu_i \delta_{ij}$. Property $\mathrm{Tr}(\dot{M}M^{-1}) = \mathrm{Tr}(\dot{D}D^{-1})$ follows from $D^{-1} = U^{-1}M^{-1}U$, $\dot{D} = \dot{U}^{-1}MU + U^{-1}\dot{M}U + U^{-1}M\dot{U}$ and $U^{-1}U = 1$. A straightforward calculation gives $U\dot{U}^{-1} + U^{-1}\dot{U} = 0$. Writing $\det(\dot{M}) = (\prod_i \dot{\mu}_i) = \sum_j \frac{\dot{\mu}_j}{\mu_j}(\prod_i \mu_i) = (\det M)\sum_j \frac{\dot{\mu}_j}{\mu_j} = (\det M)\mathrm{Tr}(\dot{D}D^{-1}) = (\det M)(\mathrm{Tr}(\dot{M}M^{-1})$, and remembering that $J = \det M$, $\mu_{ij} = \partial x_i(t)/\partial x_j(0)$, $\dot{\mu}_{ij} = \partial \dot{x}_i(t)/\partial x_j(0)$ we get $\mathrm{Tr}(\dot{M}M^{-1}) = \mathrm{Tr}(\sum_j [\partial \dot{x}_i(t)/\partial x_j(0) \cdot \partial x_j(0)/\partial x_k(t)]) = \mathrm{Tr}(\partial \dot{x}_i(t)/\partial x_k(t)) = \sum_i \partial \dot{x}_i/\partial x_i$, which proves (8).

2.3   Liouville Equation and Its Stationary Solutions

For a Hamiltonian system, the phase space distribution function $f(x,t)$ satisfies the standard Liouville equation

$$\frac{\partial f}{\partial t} + \nabla(f\dot{x}) = 0,$$

which expresses the fact that $f(x,t)$ is a conserved probability distribution function. This equation is not valid for a non-Hamiltonian system. In the light of the previous section, the elementary probability for the system to stay around x is $\wp(x,\mathrm{d}x) = f(x,t)\sqrt{g_t}\mathrm{d}x_t$ and the Liouville equation takes the form

$$\frac{\partial f\sqrt{g}}{\partial t} + \nabla(f\sqrt{g}\dot{x}) = 0. \tag{11}$$

This equation was derived from the balance between the rate of decrease in the number of ensemble members in phase space volume and the flux through the boundary surface taking into account the geometry of the space. It is, therefore, valid for ensembles generated by non-Hamiltonian systems and it reduces to the standard Liouville equation for Hamiltonian systems since then the metric $\sqrt{g_t}$, is equal to unity.

In general, the ensemble average of any property $A(x)$ is determined from the invariant measure and the ensemble distribution function as

$$\langle A \rangle_t = \frac{\int \mathrm{d}x \sqrt{g} A(x) f(x,t)}{\int \mathrm{d}x \sqrt{g} f}.$$

Since, from (8) and (10), $\frac{\mathrm{d}}{\mathrm{d}t}\sqrt{g} = \sqrt{g}\kappa$, we have

$$\frac{\mathrm{d}f}{\mathrm{d}t} = 0$$

even for non-Hamiltonian systems, i.e., the ensemble distribution function $f(x,t)$ is conserved along the trajectory.

Now let us assume that $\xi(x) \equiv \xi(x,t)$ is an autonomous dynamical system. Then a stationary solution for f exists and the system can go to equilibrium. Let us further assume that our dynamical system possesses n_c conserved quantities:

$$\Lambda_k(x_t) = \Lambda_k(x_0) = C_k,$$

where $k = 1, ..., n_c$. Then, for an ergodic system, the distribution function is the product of many δ-functions, one for each conservation law, i.e.,

$$f_{eq}(x) = \prod_{k=1}^{n_c} \delta(\Lambda_k(x_t) - C_k).$$

The function constructed considering a reduced set of conservation laws

$$f^{rd}(x) = \prod_{k=1}^{n_c' < n_c} \delta(\Lambda_k(x_t) - C_k)$$

still satisfies the generalized Liouville equation, (11), but does not represent the equilibrium ensemble. Actually, to satisfy the Liouville equation is a necessary, but not sufficient condition for an equilibrium ensemble. The partition function for such dynamical system is written in terms of a metric factor and conservation laws as

$$\Omega(C_1, ..., C_k) = \int \mathrm{d}x \sqrt{g} \prod_{k=1}^{n_c} \delta(\Lambda_k(x) - C_k).$$

Illustration: Equilibrium Ensemble for a System Subject to Holonomic Constraints.
Molecular systems consist typically of N atoms of one or more species, which can be grouped into n molecules. Each atom interacts with all the other atoms in its own molecule as well as those in other molecules via some kind of potentials. Sometimes it is convenient to replace some strong intramolecular interactions by holonomic constraints. Constraints can be defined by $\sigma_l(\{r\}) = 0$, where $l = 1, ..., g$, and g is the total number of constraints in the system. The total force acting on atom i can now be more conveniently written as $F_i + G_i$, where F_i is the contribution to the total force coming from the potential energy and G_i is the constraint force on atom i. The statistical behavior of a system subject to holonomic constraints can be analysed using the approach of Tuckerman et al. [4]. This case was first discussed by Melchionna [5].

The equations of motion for a simple dynamical system subject to constraints are

$$\dot{r}_i = \frac{p_i}{m_i} \quad \text{and} \quad \dot{p}_i = F_i + G_i, \tag{12}$$

where G_i is the constraint force on atom i. Using the constraint condition G_i takes the form

$$G_i = \sum_{\alpha} \lambda_{\alpha} \nabla_{r_i} \sigma_{\alpha},$$

where λ_{α} are Lagrange multipliers. In Hamiltonian language the constraint conditions $\sigma_{\alpha}(r, p) = 0$ lead to another set of constraints (independent, since r and p are independent variables in the Hamiltonian description)

$$\dot{\sigma}_{\alpha}(r, p) = \sum_{i} \frac{p_i}{m_i} \nabla_{r_i} \sigma_{\alpha} = 0.$$

To obtain the explicit expressions for the Lagrange multipliers λ we have to differentiate the constraint equations twice with respect to time and substitute $\dot{p}_i$ from the equations of motion. This yields

$$\ddot{\sigma}_{\alpha} = \sum_{i,j} \frac{p_i p_j}{m_i m_j} \nabla_{r_i} \nabla_{r_j} \sigma_{\alpha} + \sum_{i} \left(F_i + \sum_{\beta} \lambda_{\beta} \nabla_{r_i} \sigma_{\beta} \right) \frac{\nabla_{r_i} \sigma_{\alpha}}{m_i} = 0. \tag{13}$$

From the above equation, we obtain an expression for λ as

$$\lambda_{\alpha}(r, p) = -\sum_{\beta} Z_{\alpha\beta}^{-1} \left(\sum_{i,j} \frac{p_i p_j}{m_i m_j} \nabla_{r_i} \nabla_{r_j} \sigma_{\beta} + \sum_{i} \frac{F_i}{m_i} \nabla_{r_i} \sigma_{\beta} \right),$$

where

$$Z_{\alpha\beta} = \sum_{i,j} \frac{\nabla_{r_i} \sigma_\alpha \nabla_{r_i} \sigma_\beta}{m_i}.$$

Substituting the above expression for λ in the equations of motion we obtain a non-Hamiltonian system whose evolution satisfies the $2g$ conditions $\sigma = \dot{\sigma} = 0$. Therefore, the constraints appear now as additional constants of motion. Once we have an explicit expression for the Lagrange multipliers, we can calculate the metric factor of the system as follows,

$$\frac{\mathrm{d}}{\mathrm{d}t} \ln \sqrt{g} = -\sum_i \frac{\partial \dot{p}_i}{\partial p_i} = \sum_{\alpha,\beta} Z_{\alpha\beta}^{-1} \dot{Z}_{\beta\alpha} = \mathrm{Tr}(Z^{-1}\dot{Z}) = \frac{\mathrm{d}}{\mathrm{d}t} \ln |Z|.$$

The probability density $f(r, p)$ is given as

$$f(r, p) = \delta(H - E) \prod_\alpha \delta(\sigma_\alpha) \delta(\dot{\sigma}_\alpha)$$

and the elementary probability is

$$f \sqrt{g} \mathrm{d}\Gamma = \delta(H - E) \prod_\alpha \delta(\sigma_\alpha) \delta(\dot{\sigma}_\alpha) \times |Z| \mathrm{d}\Gamma,$$

which is a well-known formula first obtained by Ryckaert and Ciccotti in 1983 [6].

2.4 The Correct Rules to Construct the Equilibrium Ensemble of Extended Variables Dynamical Systems

Based on the previous results we can now describe a general procedure for constructing partition and distribution functions for the equilibrium ensemble generated by generic dynamics:

1. Determine ALL conservation laws satisfied by the equations of motion (ergodic problem).
2. Using these conservation laws and equations of motion, identify and eliminate linearly dependent and driven variables from the phase space, i.e., identify variables of no statistical importance. Example: The center of mass in complex systems.
3. For the remaining variables calculate the phase space compressibility factor $\nabla_{x'}\dot{x}'$ and the invariant measure $\sqrt{g(x')}$. The invariant volume element is $\sqrt{g}\mathrm{d}x'$.
4. Construct the microcanonical partition function

$$\Omega(C_1, ...C_{n_c}) = \int \mathrm{d}x' \sqrt{g(x')} \prod_{k=1}^{n_c} \delta(\Lambda_k(x') - C_k).$$

   Number n_c includes all the conservation laws $\Lambda_k(x') = C_k$ and vector x' describes the physically meaningful extended phase space.
5. Reduce, by integration, the distribution function of the extended to the physical phase space variables (e.g., by integrating over η and P_η in Nosé–Hoover case).

Step 5 assumes ergodicity of the x' phase space, subject to the conservation laws. Steps 1-4 are merely necessary but not sufficient conditions to ensure ergodicity.

2.5   The True Statistics of the Nosé–Hoover Ensemble

To correctly analyse the statistical distribution corresponding to the Nosé–Hoover ensemble, let us consider the physically meaningful case in which there are no external forces on the system ($\sum_i F_i = 0$). We have two conserved quantities given by (3) and (4). Following the rules described in the previous section, we first have to reduce the set of variables. The new phase space variables will be $r_i' = r_i - r_N$, $p_i' = p_i - p_N$, the center of mass of the system $R = \frac{1}{M}\sum_i m_i r_i$ (where $M = \sum_i m_i$ is the total mass of the system), and the total momentum $P = \sum_i p_i$. Therefore, we can write $6(N-1)$ equations of motion for r' and p', three equations for the center of mass position and three for the total momentum, and the same equations we had before for the thermostat and barostat variables. The center of mass of the system is a driven variable and two components of the total momentum are linearly dependent, since the three components of the total momentum are subjected, thanks to (4), to two linear conditions. So we have to take only one variable for the total momentum, its magnitude $|P|$. This leaves us with $6(N-1)$ equations for the atomic coordinates and momenta, one equation for the modulus of the total momentum P, and equations for η and P_η, i.e.,

$$\dot{r}_i' = \frac{p_i'}{m_i}$$

$$\dot{p}_i' = F_i' - \frac{P_\eta}{Q}p_i'$$

$$\dot{P} = -\frac{P_\eta}{Q}P$$

$$\dot{\eta} = \frac{P_\eta}{Q}$$

$$\dot{P}_\eta = \sum \frac{p_i^2}{2m_i} + \frac{P^2}{2M} - Lk_BT.$$

The calculation of the phase space compressibility gives

$$\nabla_x \dot{x} = -[3(N-1)+1]\dot{\eta},$$

where x denotes vector $\{r', p', \eta, P_\eta\}$ from which we obtain the metric factor

$$\sqrt{g} = e^{[3(N-1)+1]\eta}.$$

The partition function $\Omega \equiv \Omega_{NH}(N, V, C_1, C_2)$ is given by

$$\Omega_{NH} = \int d^{3(N-1)}p' dP \int d^{N-1}r' \int dP_\eta d\eta \sqrt{g}\left(H + \frac{P_\eta^2}{2Q} + Lk_BT\eta - C_1\right)\delta(e^\eta P - C_2)$$

which, after integrating over η and P_η, gives

$$\Omega_{NH} = \frac{1}{C_2}\int d^{N-1}p' dP \int d^{N-1}r' \int dP_\eta \left(\frac{C_2}{P}\right)^{d(N-1)+1}\delta\left(H + \frac{P_\eta^2}{2Q} + Lk_BT\ln\frac{C_2}{P} - C_1\right)$$

$$= \frac{\sqrt{2Q}}{C_2}\int d^{N-1}p' dP \int d^{N-1}r' \left(\frac{C_2}{P}\right)^{d(N-1)+1}\frac{1}{\sqrt{C_1 - H - Lk_BT\ln\frac{C_2}{P}}}.$$

As we see, this expression is completely wrong for a canonical ensemble.

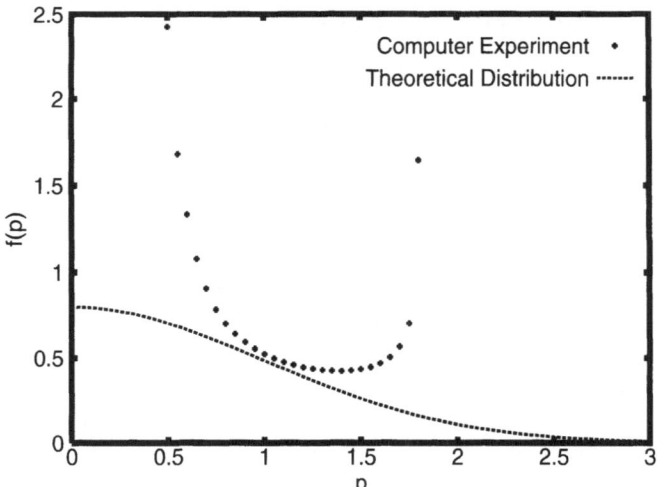

Fig. 1. The momentum distribution function (*dots*) for a 1D free particle coupled to a Nosé–Hoover thermostat obtained by computer simulation with initial conditions $m = 1$, $k_B T = 1$, $p(0) = 1$, $x(0) = 0$, $Q = 1$, and $P_\eta(0) = 1$, and using the time step 0.05, compared to the canonical distribution (*line*)

This distribution can be verified numerically. As an illustration of the Nosé–Hoover pathology just discussed, let us take the case of a one-dimensional free particle coupled to a Nosé–Hoover thermostat. The equations of motion are

$$\dot{r} = \frac{p}{m}$$

$$\dot{p} = -\frac{P_\eta}{Q}p$$

$$\dot{\eta} = \frac{P_\eta}{Q}$$

$$\dot{P}_\eta = \frac{p^2}{2m} - k_B T$$

with conserved quantities

$$H' = \frac{p^2}{2m} + \frac{P_\eta^2}{2Q} + k_B T\eta = C_1 \quad \text{and} \quad K = pe^\eta = C_2.$$

By performing steps 1 to 5 shown in Sect. 2.4, we obtain the following distribution

$$f(p) = \frac{\sqrt{2Q}}{\sqrt{p^2 \left[C_1 - \frac{p^2}{2m} + k_B T \ln \frac{p}{C_2} \right]}}.$$

Figure 1 shows the total momentum distribution obtained from a computer simulation of this model and a comparison to the canonical distribution. The results obtained from the simulation agree perfectly with the above analytical prediction which is far from canonical!

2.6   New Atomic "iso"-stats (NVT, NPT)

A "good" NVT. The difficulties resulting from the presence of multiple conservations in the Nosé–Hoover system can be eliminated via the use of the Nosé–Hoover chain algorithm [7,8]. In this approach the problem is solved by a smart but simple generalization of the original Nosé dynamics, where the kinetic energy fluctuations of the thermostat variable are coupled to another thermostat variable. The kinetic energy fluctuations of the second thermostat are, in turn, controlled by coupling it to a third thermostat and so on, to form a chain of M thermostats of 'masses' Q_i, where $i = 1, ..., M$.

The Nosé–Hoover chain algorithm gives the following non-Hamiltonian dynamical system:

$$\dot{r}_i = \frac{p_i}{m_i}$$

$$\dot{p}_i = F_i - \frac{P_{\eta_1}}{Q_1} p_i$$

$$\dot{\eta}_k = \frac{P_{\eta_k}}{Q_k}$$

$$\dot{P}_{\eta_1} = \sum_1^N \frac{p_i^2}{2m_i} - Lk_\mathrm{B}T - \frac{P_{\eta_2}}{Q_2} P_{\eta_1}$$

$$\dot{P}_{\eta_k} = \frac{P_{\eta_{k-1}}^2}{Q_{k-1}} - k_\mathrm{B}T - \frac{P_{\eta_{k+1}}}{Q_{k+1}} P_{\eta_k}$$

$$\dot{P}_{\eta_M} = \frac{P_{\eta_{M-1}}^2}{Q_{M-1}} - k_\mathrm{B}T,$$

where $k = 1, ..., M - 1$. In the case $\sum_i F_i = 0$ the only conserved quantity is

$$H' = H(r, p) + \sum_k \frac{P_{\eta_k}^2}{2Q_k} + Lk_\mathrm{B}T\eta_1 + k_\mathrm{B}T \sum_{k=2}^M \eta_k = C_1$$

and there are no linearly dependent variables. Only the first thermostat, η_1, and the thermostat center $\eta_c = \sum_{k=2}^M \eta_k$ are independently coupled to the dynamics. The remaining thermostat variables are driven. The compressibility and the metric factor are

$$\kappa = -3N\dot{\eta}_1 - \dot{\eta}_c \quad \text{and} \quad \sqrt{g} = e^{3N\eta_1 + \eta_c},$$

respectively. Substituting the metric factor into the expression for the partition function and integrating over the extended space variables, we can easily obtain the canonical distribution (taking $L = 3N$).

Instead, if $\sum_i F_i = 0$, there will be an additional vectorial conserved quantity

$$K = Pe^{\eta_1} = C_2,$$

where P is the total momentum, with its three components. As in the previous case, only two η's appear, i.e., η_1 and $\eta_c = \sum_{k=2}^M \eta_k$. Moreover, the center of mass of the system

(R), is driven, while two components of total momentum are linearly dependent. This leaves us only one additional equation describing the magnitude of the total momentum $|P|$. This gives for the metric factor

$$\sqrt{g} = e^{(3(N-1)+1)\eta_1 + \eta_c},$$

such that the partition function $\Omega_T \equiv \Omega_T(N, V, C_1, C_2)$ is

$$\Omega_T = \int d^{N-1}p' dP \int d^{N-1}r' \int d^M P_\eta d\eta_1 d\eta_c \sqrt{g} \delta(H' - C_1)\delta(e^{\eta_1}P - C_2).$$

By choosing $L = 3N$ and integrating over η_1 and η_c the above equation reduces to

$$\Omega_T = \frac{e^{C_1/k_B T}}{C_2 k_B T} \int d^M P_\eta e^{-\beta \sum_{k=1}^M \frac{P_{\eta_k}^2}{2Q_k}} \int d^{N-1}p' dP \int d^{N-1}r' e^{-\beta H(p',r',P)}.$$

The correct canonical (NVT) distribution function is obtained, since Ω_T can be rewritten as

$$\Omega_T \propto \int d^{N-1}p' dP P^2 \int d^{N-1}r' \int dR e^{-\beta H(p',r',P)} = \int d^N p \int d^N r e^{-\beta H(p,r)}.$$

The presence of two extra parameters permits us to integrate the distribution function containing two delta-functions and to obtain the correct canonical distribution. This distribution has been obtained numerically for the same one-dimensional free particle system as described in Sect. 2.5. A comparison of the computer simulation and the canonical distribution is shown in Fig. 2. The results show that this time we are getting the correct distribution.

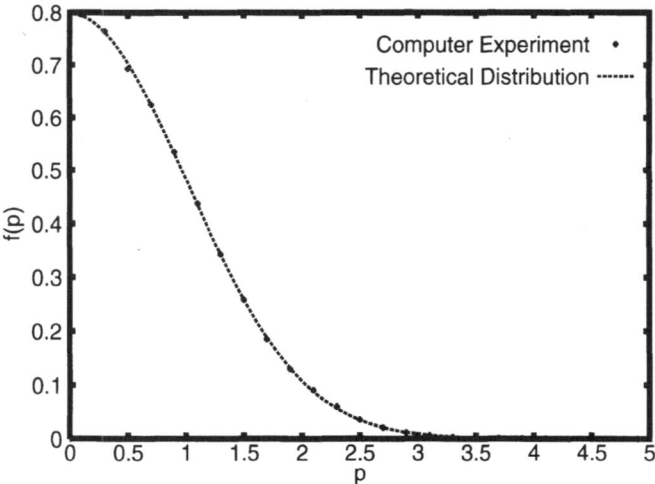

Fig. 2. The momentum distribution function (*dots*) for a 1D free particle coupled to a Nosé-Hoover chain of length $M = 3$ obtained by computer simulation with initial conditions $m = 1$, $kT = 1$, $p(0) = 1$, $x(0) = 0$, $Q_i = 1$, $P_{\eta_1}(0) = P_{\eta_3}(0) = 1$, $P_{\eta_2}(0) = -1$ using the time step 0.05, compared to the canonical distribution (*line*)

A "good" NPT. In an effort to correct some of the deficiencies of the original Hoover formulation of the NPT ensemble, Martyna, Tobias and Klein [9] proposed a new set of equations of motion to carry out NPT simulations. These equations can be added to a Nosé–Hoover chain. In comparison with NVT, the equations have additional barostat variables to guarantee constant pressure. The barostat is characterized by a 'mass' W and dynamical variables ϵ and P_ϵ,

$$\dot{r}_i = \frac{p_i}{m_i} + \frac{P_\epsilon}{W} r_i$$

$$\dot{p}_i = F_i - \left(1 + \frac{1}{N}\right) \frac{P_\epsilon}{W} p_i - \frac{P_\eta}{Q} p_i$$

$$\dot{V} = \frac{3V P_\epsilon}{W}$$

$$\dot{P}_\epsilon = 3V(\Pi_{\text{int}} - \Pi_{\text{ext}}) + \frac{1}{N} \sum_{i=1}^{N} \frac{p_i^2}{m_i} - \frac{P_\eta}{Q} P_\epsilon$$

$$\dot{\eta} = \frac{P_\eta}{Q}$$

$$\dot{P}_\eta = \sum_{1}^{N} \frac{p_i^2}{m_i} + \frac{P_\epsilon^2}{W} - (3N + 1)k_B T. \tag{14}$$

If $\sum_i F_i \neq 0$ the only conserved quantity is

$$H' = H(r, p) + \frac{P_\eta^2}{2Q} + (3N + 1)k_B T\eta + \Pi_{\text{ext}} V + \frac{P_\epsilon^2}{2W} = C_1. \tag{15}$$

In (14) and (15) Π_{ext} is the constant external pressure. Π_{int} indicates the internal pressure of the system which can vary with time, but must be kept close to Π_{ext}. It can be written as

$$\Pi_{\text{int}} = \frac{1}{3V} \sum_{i=1}^{N} \left[\frac{p_i^2}{m_i} + F_i \cdot r_i \right].$$

The calculation of the metric factor gives

$$\sqrt{g} = e^{(3N+1)\eta}$$

which, after the substitution in the partition function and integration over the extended space variables, conduces to the correct isothermal-isobaric distribution. For $\sum_i F_i = 0$, there is an additional vectorial conserved quantity

$$K = Pe^{(1+\frac{1}{N})\epsilon+\eta} = C_2,$$

where P is the total momentum. In this case we have to introduce the Nosé–Hoover chain, since without it we would not be able to integrate the partition function over the extended phase space variables. Our equations of motion, after elimination of all driven and linearly dependent variables, are

$$\dot{r}_i = \frac{p_i}{m_i} + \frac{P_\epsilon}{W} r_i$$

$$\dot{p}_i = F_i - \left(1 + \frac{1}{N}\right) \frac{P_\epsilon}{W} p_i - \frac{P_{\eta_1}}{Q_1} p_i$$

$$\dot{P} = -\left(1 + \frac{1}{N}\right) \frac{P_\epsilon}{W} P - \frac{P_{\eta_1}}{Q_1} P$$

$$\dot{V} = \frac{3V P_\epsilon}{W}$$

$$\dot{P}_\epsilon = 3V(\Pi_{\text{int}} - \Pi_{\text{ext}}) + \frac{1}{N} \sum_{i=1}^{N} \frac{p_i^2}{m_i} - \frac{P_{\eta_1}}{Q_1} P_\epsilon$$

$$\dot{\eta}_k = \frac{P_{\eta_k}}{Q_k}$$

$$\dot{P}_{\eta_1} = \sum_{1}^{N} \frac{p_i^2}{m_i} + \frac{P_\epsilon^2}{W} - (3N+1)k_{\mathrm{B}}T - \frac{P_{\eta_2}}{Q_2} P_{\eta_1}$$

$$\dot{P}_{\eta_k} = \frac{P_{\eta_{k-1}}^2}{Q_{k-1}} - k_{\mathrm{B}}T - \frac{P_{\eta_{k+1}}}{Q_{k+1}} P_{\eta_k}$$

$$\dot{P}_{\eta_M} = \frac{P_{\eta_{M-1}}^2}{Q_{M-1}} - k_{\mathrm{B}}T.$$

The first conserved quantity is now

$$H' = H(r', p', P) + \sum_{k=1}^{M} \frac{P_{\eta_k}^2}{2Q_k} + (3N+1)k_{\mathrm{B}}T\eta_1 + \Pi_{\text{ext}}V + \frac{P_\epsilon^2}{2W} + k_{\mathrm{B}}T\eta_c = C_1,$$

where $\eta_c = \sum_{k=2}^{M} \eta_k$ is same as in the NVT case. The calculation of the metric factor for the case $\sum_i F_i = 0$ is straightforward, although long, and gives

$$\sqrt{g} = \frac{1}{V^{[1/N - 1/3 - 1/3N]}} e^{(3(N-1)+2)\eta_1 + \eta_c},$$

which, after the substitution in the partition function and integration over the extended space variables, provides the correct isothermal-isobaric distribution.

In this section we have seen that there are good and simple methods to model NVT and NPT ensemble for atomic systems. In the following section we generalize the NPT dynamics to molecular systems with constraints.

2.7 Molecular NPT with Constraints

In simulations of molecular systems, we can write the equations of motion in two different forms, i.e., using the atomic or molecular expression for the virial. As it is known, the atomic virial for such systems is

$$\Pi^A = \frac{1}{3V} \sum_{i=1}^{N} \left[\frac{p_i^2}{m_i} + F_i \cdot r_i \right],$$

while the molecular virial is

$$\Pi^M = \frac{1}{3V} \sum_{\mu=1}^{n} \left[\frac{P_\mu^2}{M_\mu} + F_\mu \cdot R_\mu \right].$$

Here R_μ, P_μ, M_μ and F_μ are the center of mass position, total momentum, mass and force of the molecule μ, respectively.

While the atomic virial is based on momenta and forces associated with individual atoms, the molecular virial depends only on momenta and forces associated with the molecules. The time averages of molecular and atomic pressures have been shown to be the same in the thermodynamic limit [10]. Time correlation functions, however, are different and only their integrals are equal [11]. For the NVT case there are no special effects or difficulties caused by constraints. For NPT the solution of the atomic coupling is rather complex [12], while the molecular coupling does not present any special effect or difficulty due to constraints [13].

Atomic Coupling. The constrained NPT molecular dynamics with fully atomic virial has been essentially solved by Kneller and Mulders [14] and then perfected by Ciccotti et al. [12]. In this part we present the equations of reference [12]. For simplicity we denote the partial derivatives of constraints with respect to atomic positions as follows: $\frac{\partial \sigma^\alpha}{\partial r_i}$ as $\partial_i \sigma^\alpha$, and $\frac{\partial}{\partial r_i} \frac{\partial \sigma^\alpha}{\partial r_j}$ as $\partial_{ij}^2 \sigma^\alpha$.

The equations of motion proposed in [12] are the Kneller–Mulders equations slightly modified for our three-dimensional N atom system:

$$\dot{r}_i = \frac{p_i}{m_i} + \frac{P_\epsilon}{W} \left[r_i - \sum_{\alpha,\beta} \frac{\partial_i \sigma^\alpha}{m_i} (Z^{-1})^{\alpha\beta} \sum_j \partial_j \sigma^\beta r_j \right]$$

$$\dot{p}_i = F_i + \sum_\alpha \lambda^\alpha \partial_i \sigma^\alpha - \left(1 + \frac{1}{N} \right) \frac{P_\epsilon}{W} p_i - \frac{P_\eta}{Q} p_i - \frac{P_\epsilon}{W} \sum_j H_{ij} \frac{p_j}{m_j}$$

$$\dot{V} = \frac{N_f}{N} V \frac{P_\epsilon}{W}$$

$$\dot{P}_\epsilon = 3V \left(\Pi^A - \frac{N_f}{3N} \Pi_{\text{ext}} \right) + \sum_\alpha \sum_i \lambda^\alpha \partial_i \sigma^\alpha r_i - \frac{P_\eta}{Q} P_\epsilon - \frac{P_\epsilon}{W} \sum_{i,j} r_i H_{ij} \frac{p_j}{m_j} + \sum_{i,j} \frac{1}{N} \frac{p_i p_j}{m_i}$$

$$\dot{\eta} = \frac{P_\eta}{Q}$$

$$\dot{P}_\eta = \sum_i \frac{p_i^2}{m_i} + \frac{P_\epsilon^2}{W} - (N_f + 1) k_B T, \tag{16}$$

where Π^A is the internal atomic pressure in the system, $Z^{\alpha\beta}$ is the same matrix as in Sect. 2.3, H_{ij} is a suitable symmetric matrix and N_f is the number of degrees of freedom with g total number of constraints,

$$R^{\alpha\beta} = (Z^{-1})^{\alpha\beta}$$

$$H_{ij} = -\sum_{\alpha,\beta} R^{\alpha\beta} \sum_k \partial_k \sigma^\beta r_k \partial^2_{ij} \sigma^\alpha$$

$$N_f = 3N - g.$$

Here, Q and W are the 'masses' associated to the thermostat and barostat of the system, respectively. The time derivative of the constraints is

$$\dot{\sigma}^\gamma = \sum_i \partial_i \sigma^\gamma \dot{r}_i = \sum_i \partial_i \sigma^\gamma \frac{p_i}{m_i} = 0,$$

where we have used (16) and the fact that $\sum_i \{\partial_i \sigma^\gamma [r_i - \sum_{\alpha,\beta} \frac{\partial_i \sigma^\alpha}{m_i}]\} = 0$. To compute the Lagrange multipliers, we have to differentiate the equations of constraints twice with respect to time. That yields

$$\ddot{\sigma}^\alpha = \sum_i F_i \frac{\partial_i \sigma^\alpha}{m_i} + \sum_\beta \lambda^\beta Z^{\alpha\beta} - \sum_i \left(1 + \frac{1}{N}\right) \frac{P_\epsilon}{W} \frac{\partial_i \sigma^\alpha}{m_i} p_i - \sum_{i,j} \frac{P_\epsilon}{W} \frac{\partial_i \sigma^\alpha}{m_i} H_{ij} \frac{p_j}{m_j}$$

$$- \sum_i \frac{P_\eta}{Q} p_i \frac{\partial_i \sigma^\alpha}{m_i} + \sum_{i,j} \frac{p_i}{m_i} \frac{p_j}{m_j} \partial^2_{ij} \sigma^\alpha + \sum_{i,j} \frac{P_\epsilon}{W} \frac{p_i}{m_i} \partial^2_{ij} \sigma^\alpha r_j$$

$$- \sum_{i,j} \frac{P_\epsilon}{W} \frac{p_i}{m_i} \partial^2_{ij} \sigma^\alpha \sum_{\beta,\gamma} \frac{\partial_j \sigma^\beta}{m_i} R^{\beta\gamma} \sum_k \partial_k \sigma^\gamma r_k = 0.$$

Solving in λ we find

$$\lambda^\delta(r, p, P_\eta, P_\epsilon) = -\sum_{i,j} \sum_\alpha F_i \frac{\partial_j \sigma^\alpha}{m_i} R^{\alpha\delta}$$

$$+ \left[\left(1 + \frac{1}{N}\right) \frac{P_\epsilon}{W} + \frac{P_\eta}{Q}\right] \sum_i \sum_\alpha \frac{p_i}{m_i} \partial_i \sigma^\alpha R^{\alpha\delta}$$

$$- \sum_{i,j} \frac{p_i}{m_i} \sum_\alpha \partial^2_{ij} \sigma^\alpha R^{\alpha\delta} \left(\frac{p_j}{m_j} + \frac{P_\epsilon}{W} r_j\right)$$

$$+ \frac{P_\epsilon}{W} \sum_{i,j} \sum_{\alpha,\beta} \frac{\partial_j \sigma^\alpha}{m_i} \partial^2_{ij} \sigma^\beta \sum_\gamma \frac{p_j}{m_j} \sum_k \partial_k \sigma^\gamma r_k (R^{\alpha\gamma} R^{\beta\delta} - R^{\alpha\delta} R^{\beta\gamma}).$$

By substituting the expressions for the Lagrange multipliers in the equations of motion, we can find the metric factor using the standard procedure as described before. Calculations are long but straightforward and the result is

$$\sqrt{g} = e^{(N_f+1)\eta} |Z|.$$

Using the metric factor, we can compute the partition function once all conservation laws are defined. The first conserved quantity for these equations, i.e.,

$$H = \frac{1}{2} \sum_i \frac{p_i^2}{m_i} + U(r) + \frac{P_\eta^2}{2Q} + \frac{P_\epsilon^2}{2W} + (N_f + 1)k_B T\eta + P_{ext}V = C_1$$

is the only conserved quantity in the case $\sum_i F_i \neq 0$. Instead, if $\sum_i F_i = 0$, we have to take into account another conservation law

$$PV^{(N+1)/N_f}e^\eta = C_2.$$

For $\sum_i F_i \neq 0$ substituting the metric factor and performing the integration over the extended space variables yields

$$\sqrt{g}f\mathrm{d}x \propto |Z|e^{-\frac{H(p,r)+P_{\mathrm{ext}}V}{k_BT}}\prod_\alpha \delta(\sigma_\alpha)\dot{\delta}(\sigma_\alpha)\mathrm{d}x,$$

which is the probability corresponding to the correct isothermal-isobaric distribution.

For the case $\sum_i F_i = 0$ we have to add a Nosé–Hoover chain of length M, using the same procedure as in Sect. 2.6 in order to obtain the correct NPT ensemble. Despite the fact that these equations generate the correct NPT ensemble, their complexity is such that the implementation by a reversible integrator [15] is quite complex, so that they are of little practical use.

Molecular Coupling. An alternative to atomic scaling is molecular scaling based on the molecular virial [13]. The equations of motion we will use here are written such that the volume (and not its logarithm) is taken as the extra dynamical variable needed to keep the pressure constant. Moreover, the use of the Nosé–Hoover chain is replaced by just two thermostats: One for the momenta and another for the momentum associated with the volume.

The system we consider consists of N atoms grouped into n molecules, each containing n_μ atoms, such that $N = \sum_{\mu=1}^n n_\mu$. The number of constraints inside molecule μ is f_μ. Here we will denote the atomic positions, momenta and masses with $r_{\mu i}$, $p_{\mu i}$ and $m_{\mu i}$, respectively. The double subscript indicates the molecule to which the atom belongs to. The equations of motion proposed for the system are

$$\dot{r}_{\mu i} = \frac{p_{\mu i}}{m_{\mu i}} + R_\mu \frac{P_V}{3VQ_V}$$

$$\dot{p}_{\mu i} = F_{\mu i} + G_{\mu i} - p_{\mu i}\frac{P_\eta}{Q_\eta} - \frac{m_{\mu i}}{M_\mu}P_\mu\frac{P_V}{3VQ_V}$$

$$\dot{V} = \frac{P_V}{Q_V}$$

$$\dot{P}_V = (\Pi^M - \Pi_{\mathrm{ext}}) - \frac{P_\xi}{Q_\xi}P_V$$

$$\dot{\eta} = \frac{P_\eta}{Q_\eta}$$

$$\dot{\xi} = \frac{P_\xi}{Q_\xi}$$

$$\dot{P}_\eta = Lk_B(T - T_{\mathrm{ext}})$$

$$\dot{P}_\xi = \frac{P_V^2}{Q_V} - k_BT,$$

where Π^M is the internal molecular pressure, and R_μ and P_μ are, respectively, the position and momentum of the center of mass of molecule μ. The thermostat for the atomic momenta is characterized by 'mass' Q_η and dynamical variables η and P_η, the barostat by 'mass' Q_V associated with the variables V and P_V, and the additional thermostat for the volume by 'mass' Q_ξ and variables ξ and P_ξ. $G_{\mu i}$ is the constraint force acting on atom i in molecule μ,

$$G_{\mu i} = \sum_{\alpha=1}^{f_\mu} \lambda_\mu^\alpha \nabla_{r_{\mu i}} \sigma_\mu^\alpha. \tag{17}$$

The λ's in (17) are Lagrange multipliers. As before, in order to obtain their explicit expressions we have to differentiate the equations for the constraints twice with respect to time

$$\sum_{i=1}^{n_\mu} \sum_{j=1}^{n_\mu} \left(\frac{p_{\mu i}}{m_{\mu i}} + \frac{R_\mu P_V}{3V Q_V} \right) \left(\frac{p_{\mu j}}{m_{\mu j}} + \frac{R_\mu P_V}{3V Q_V} \right) \cdot \nabla_{r_{\mu j}} \nabla_{r_{\mu i}} \sigma_\mu^\beta$$

$$+ \sum_{i=1}^{n_\mu} \left(\frac{\dot{p}_{\mu i}}{m_{\mu i}} + \frac{\dot{R}_\mu P_V}{3V Q_V} + \frac{R_\mu \dot{P}_V}{3V Q_V} - \frac{R_\mu P_V \dot{V}}{3V^2 Q_V} \right) \cdot \nabla_{r_{\mu i}} \sigma_\mu^\beta =$$

$$= \sum_{i=1}^{n_\mu} \sum_{j=1}^{n_\mu} \frac{p_{\mu i}}{m_{\mu i}} \frac{p_{\mu j}}{m_{\mu j}} \cdot \nabla_{r_{\mu j}} \nabla_{r_{\mu i}} \sigma_\mu^\beta + \sum_{i=1}^{n_\mu} \frac{\dot{p}_{\mu i}}{m_{\mu i}} \cdot \nabla_{r_{\mu i}} \sigma_\mu^\beta = 0.$$

Inserting $\dot{p}_{\mu i}$ from the equations of motion and solving for the λ's, we get

$$\lambda_\mu^l = \sum_{m=1}^{f_\mu} (Z^{-1})_\mu^{lm} \sum_{i=1}^{n_\mu} \left[\frac{-1}{m_{\mu i}} \nabla_{r_{\mu i}} \sigma_\mu^m \cdot \left(F_{\mu i} + p_{\mu i} \frac{P_\eta}{Q_\eta} \right) + \sum_{j=1}^{n_\mu} \frac{p_{\mu i} p_{\mu j}}{m_{\mu i} m_{\mu j}} \nabla_{r_{\mu i}} \nabla_{r_{\mu j}} \sigma_\mu^m \right],$$

where

$$Z_\mu^{lm} = \sum_{i=1}^{n_\mu} \frac{\nabla_{r_{\mu i}} \sigma_\mu^l \nabla_{r_{\mu i}} \sigma_\mu^m}{m_{\mu i}}.$$

Inserting the expression for λ into the equations of motion, we obtain the explicit dynamics in the presence of constraints. For the case $\sum F \neq 0$ there is, in addition to the $2g$ quantities coming from the constraint conditions, only one other conserved quantity for the dynamical system, i.e.,

$$H' = H(p,r) + L k_B T \eta + k_B T \xi + \Pi_{ext} V + \frac{P_V^2}{2Q_V} + \frac{P_\eta^2}{2Q_\eta} + \frac{P_\xi^2}{2Q_\xi} = C_1.$$

The calculation of the phase space compressibility gives

$$\kappa = \sum \frac{\partial \dot{x}}{\partial x} = -\frac{3N P_\eta}{Q_\eta} - \frac{P_\xi}{Q_\xi} + \sum_{mu=1}^{n} \sum_{i=1}^{n_\mu} \sum_{l=1}^{f} \nabla_{p_i} \lambda_\mu^l \cdot \nabla_{r_i} \sigma_\mu^l \tag{18}$$

with

$$
\nabla_{p_i}\lambda_\mu^l = \sum_{m=1}^{f} (Z^{-1})_\mu^{lm} \left[\frac{1}{m_i} \nabla_{r_i}\sigma_\mu^l \frac{P_\eta}{Q_\eta} - \sum_{j=1}^{n_\mu} \frac{2}{m_i m_j} p_j \cdot \nabla_{r_j}\nabla_{r_i}\sigma_\mu^l \right]. \tag{19}
$$

By substituting (19) into (18) and simplifying, we obtain

$$
\begin{aligned}
\kappa &= -\frac{3NP_\eta}{Q_\eta} - \frac{P_\xi}{Q_\xi} + g\frac{P_\eta}{Q_\eta} - \sum_{mu=1}^{n}\sum_{i=1}^{n_\mu}\sum_{l=1}^{f}(Z^{-1})_\mu^{lm}\dot{Z}_m^{lm}u \\
&= -(3N-g)\dot{\eta} - \dot{\xi} - Tr(Z^{-1}\dot{Z}) \\
&= -(3N-g)\dot{\eta} - \dot{\xi} - \frac{\mathrm{d}\ln|Z|}{\mathrm{d}t}
\end{aligned}
$$

from which

$$
\sqrt{g} = |Z|\mathrm{e}^{(3N-g)\eta+\xi}.
$$

Using $L = 3N - g$ we can write the partition function as

$$
\Omega = \int\!dV \int\!dr^N dp^N \int\!dP_\eta dP_\xi dP_V \int\!d\xi \int\!d\eta |Z| \mathrm{e}^{L\eta+\xi}\delta(H'-C_1)\prod_{l=1}^{G}\delta(\sigma^l)\delta(\dot{\sigma}^l).
$$

By integrating over η we obtain

$$
\Omega = \int dV \int dr^N dp^N \int dP_\eta dP_\xi dP_V \int d\xi \int d\eta |Z| \mathrm{e}^{-\frac{H(p,r)}{k_B T}}
$$
$$
\times \mathrm{e}^{-\frac{\Pi_{ext}V}{k_B T}}\mathrm{e}^{-\frac{P_V^2}{2k_B T Q_V}}\mathrm{e}^{-\frac{P_\eta^2}{2k_B T Q_\eta}}\mathrm{e}^{-\frac{P_\xi^2}{2k_B T Q_\xi}}\prod_{l=1}^{G}\delta(\sigma^l)\delta(\dot{\sigma}^l),
$$

which, after integration over P_V, P_η, P_ξ and ξ corresponds to the correct isothermal-isobaric distribution in the presence of constraints. For the case $\sum_i F_i = 0$ we have the additional vectorial conserved quantity

$$
K = PV^{1/3}\mathrm{e}^\eta = C_2.
$$

However, we get the correct isothermal-isobaric distribution for volume and momentum also in this case [13].

3   Implementations

3.1   Reversible Integrators

Explicit reversible multiple time step integrators have been developed to treat efficiently problems with separation in time scales (e.g., problems involving stiff vibrations) and forces that can be separated into short and long-range components [16]. These methods are originally based on the Liouville operator formulation of Hamiltonian dynamics. The non-Hamiltonian equations of motion, presented in Sect. 2, also possess time reversibility. Moreover, Martyna et al. [15] have shown in 1996 how to generalize reversible integrators to extended systems dynamics, such as canonical and isothermal-isobaric ensembles. These integrators are time reversible and conserve the phase space measure of dynamical systems. In this section we give a brief review of this method with special emphasis on its application to non-Hamiltonian systems.

One of the best (fast and stable) ways to solve the classical Hamiltonian equations of motion is the Verlet algorithm. In its standard version it is obtained by combining backward and forward Taylor expansions

$$r_i(t \pm h) = r_i(t) \pm \dot{r}_i(t)h + \frac{1}{2}h^2\ddot{r}_i(t) \pm \frac{h^3}{6}\dddot{r} + \mathcal{O}(h^3)$$

which gives

$$r_i(t+h) = -r_i(t-h) + 2r_i(t) + h^2\ddot{r}_i(t)$$
$$\dot{r}_i(t) \quad = \frac{r_i(t+h) - r_i(t-h)}{2h}.$$

Its equivalent velocity version is

$$r_i(t+h) = r_i(t) + \dot{r}_i(t)h + \frac{1}{2}\ddot{r}_i(t)h^2$$
$$\dot{r}_i(t+h) = \dot{r}_i(t) + \frac{h}{2}[\ddot{r}_i(t) + \ddot{r}_i(t+h)]. \tag{20}$$

This algorithm is known to be very stable and accurate in comparison with other algorithms (for example, Gear predictor-corrector algorithm [17]). Let us see why.

The Trotter Expansion of the Classical Liouville Propagator. The Liouville operator L for a system which possesses f degrees of freedom is defined, using the Hamiltonian in the Cartesian coordinates, as

$$iL = \{..., H\} = \sum_{j=1}^{f}\left(\dot{x}_j\frac{\partial}{\partial x_j} + F_j\frac{\partial}{\partial p_j}\right) = \dot{\Gamma}\frac{\partial}{\partial \Gamma}, \tag{21}$$

where $\Gamma = (x_j, p_j)$ are the positions and the conjugate momenta of the system, F_j is the force on the jth degree of freedom, and $\{..., ...\}$ is the Poisson bracket of the system. The system at time t can be defined, knowing its initial state at time zero, as

$$\Gamma(t) = U(t)\Gamma(0)$$

where $U(t)$ is the classical propagator obtained from the Liouville operator

$$U(t) = e^{iLt}. \tag{22}$$

The propagator $U(t)$ has the property $U(-t) = U^{-1}(t)$. Let us now decompose the Liouville operator L into two parts such that

$$iL = iL_1 + iL_2.$$

For this decomposition the Trotter theorem gives (for large N)

$$e^{i(L_1+L_2)t} = \left(e^{i(L_1+L_2)t/N}\right)^N = \left(e^{iL_2(\Delta t/2)}e^{iL_1(\Delta t)}e^{iL_2(\Delta t/2)}\right)^N + \mathcal{O}\left(\left(\frac{t}{N}\right)^3\right),$$

where $\Delta t = t/N$. Using this expression, we can define the discrete time propagator (for each time step) as

$$G(\Delta t) = U_2\left(\frac{\Delta t}{2}\right)U_1(\Delta t)U_2\left(\frac{\Delta t}{2}\right) = e^{iL_2(\Delta t/2)}e^{iL_1(\Delta t)}e^{iL_2(\Delta t/2)}.$$

Applying this decomposition to the Liouville operator (21), we obtain

$$iL_1 = \dot{x}\frac{\partial}{\partial x} \quad \text{and} \quad iL_2 = F(x)\frac{\partial}{\partial p}$$

and our discrete time propagator is

$$G(\Delta t) = U_2\left(\frac{\Delta t}{2}\right)U_1(\Delta t)U_2\left(\frac{\Delta t}{2}\right) = e^{(\Delta t/2)F(x)\partial/\partial p}e^{(\Delta t)\dot{x}\partial/\partial x}e^{(\Delta t/2)F(x)\partial/\partial p}.$$

Since $G(-t) = G^{-1}(t)$, this integrator is exactly time reversible, a fact which guarantees its stability.

The Implementation in a Computer Code. The solution of the differential equation $\dot{x} = f(x)$ can be written as

$$x(t) = e^{t\dot{x}\frac{\partial}{\partial x}}x = e^{tf(x)\frac{\partial}{\partial x}}x = \sum_N \frac{t^N}{N!}\left(\frac{d}{dt}\right)^N x.$$

Moreover, if $\dot{x} = a = $ constant, we have

$$x(t) = e^{at\frac{\partial}{\partial x}}x = x + at.$$

Instead, if $\dot{x} = ax$, we get

$$x(t) = e^{axt\frac{\partial}{\partial x}}x = e^{at\frac{\partial}{\partial \ln x}}e^{\ln x} = e^{\ln x + at} = e^{at}x.$$

We will use these two rules to obtain our algorithm.

Let us consider two variables, x and y, such that

$$\dot{x} = F(x,y) = f(y) + ax$$
$$\dot{y} = G(x,y) = g(x) + by,$$

where a and/or b can possibly be zero. We can write the evolution operator for this system as:

$$\begin{pmatrix} x(t) \\ y(t) \end{pmatrix} = e^{t(F(x,y)\frac{\partial}{\partial x}+G(x,y)\frac{\partial}{\partial y})}\begin{pmatrix} x \\ y \end{pmatrix}. \tag{23}$$

But (23) is only a formal expression without explicit algorithmic consequences. However, if the time step $\Delta t = h \to 0$, then by using the Trotter formula we can write

$$e^{\Delta t (F(x,y)\frac{\partial}{\partial x} + G(x,y)\frac{\partial}{\partial y})} = e^{\frac{h}{2} F(x,y)\frac{\partial}{\partial x}} e^{h G(x,y)\frac{\partial}{\partial y}} e^{\frac{h}{2} F(x,y)\frac{\partial}{\partial x}} + \mathcal{O}(h^3). \qquad (24)$$

There are two different ways to compute the effect of applying the previous three operators (more generally n operators) to a phase space point. The first one is

$$A(x)B(x)C(x)f(x) = A(x)B(x)f(x_c(x)) = f(x_c(x_b(x_a(x)))). \qquad (25)$$

Equation (25) can be implemented only analytically, since at each successive application of the operators we have to write the explicit equations for the evolution of the phase space variables. That yields very complex analytical expressions for the equation $f(x_c(x_b(x_a(x))))$. Instead, by applying the operators in reversed order, first A then B, etc., we obtain the same result

$$A(x)B(x)C(x)f(x) = B(x_a(x))C(x_a(x))f(x_a(x)) = f(x_c(x_b(x_a(x)))), \qquad (26)$$

but in a much more useful form. In fact, (26) represents a recursive procedure in which the successive transformations are automatically taken into account. In particular, this feature is easy to program in a computer code, where, instead of writing long analytical expressions at every application of each operator, we just update the variables successively. This simplification is possible since, as we have written in (26), the application of the first operator from the left updates the variables of both the other operators and the function to which they are applied.

Let us now see how this rule works in the case described at the beginning of Sect. 2 by (1). The discretized time propagator (24) applied to a phase space point $\{r, p\}$ gives

$$e^{ihL}\begin{pmatrix} r \\ p \end{pmatrix} = e^{\frac{h}{2}\dot{p}\frac{\partial}{\partial p}} e^{h\dot{r}\frac{\partial}{\partial r}} e^{\frac{h}{2}\dot{p}\frac{\partial}{\partial p}} \begin{pmatrix} r \\ p \end{pmatrix} = e^{h\dot{r}'\frac{\partial}{\partial r'}} e^{\frac{h}{2}\dot{p}'\frac{\partial}{\partial p'}} \begin{pmatrix} r' = r \\ p' = p + \frac{h}{2}F(r) \end{pmatrix}$$

$$= e^{\frac{h}{2}\dot{p}''\frac{\partial}{\partial p''}} \begin{pmatrix} r'' = r' + h\frac{p}{m} \\ p' = p + \frac{h}{2}F(r) \end{pmatrix} = \begin{pmatrix} r''' = r'' \\ p''' = p'' + \frac{h}{2}F(r'') \end{pmatrix},$$

from which we obtain

$$r(t + h) = r''' = r(t) + h\frac{p(t)}{m} + \frac{1}{2}h^2\frac{F(r(t))}{m}$$

$$p(t + h) = p''' = p(t) + \frac{h}{2}\left[F(r(t)) + F(r(t + h))\right],$$

i.e., the velocity Verlet algorithm (20).

Illustration: Application to the NPT Ensemble. For the NPT equations, (14), of an N-particle system, the Liouville operator iL is split into nine parts

$$iL = iL_1 + iL_2 + iL_3 + iL_4 + iL_5 + iL_6 + iL_7 + iL_8 + iL_9 \qquad (27)$$

where the partial Liouville operators are:

$$iL_1 = \frac{p_i}{m_i}\frac{\partial}{\partial r_i}$$

$$iL_2 = \frac{P_\epsilon}{W}r_i\frac{\partial}{\partial r_i}$$

$$iL_3 = F_i\frac{\partial}{\partial p_i}$$

$$iL_4 = -\left[\left(1+\frac{1}{N}\right)\frac{P_\epsilon}{W}+\frac{P_\eta}{Q}\right]p_i\frac{\partial}{\partial p_i}$$

$$iL_5 = 3\frac{P_\epsilon}{W}V\frac{\partial}{\partial V}$$

$$iL_6 = -\frac{P_\eta}{Q}P_\epsilon\frac{\partial}{\partial P_\epsilon}$$

$$iL_7 = \left[3V(\Pi_{\text{int}}-\Pi_{\text{ext}})+\frac{1}{N}\sum_{i=1}^{N}\frac{p_i^2}{m_i}\right]\frac{\partial}{\partial P_\epsilon}$$

$$iL_8 = \left[\sum_{i=1}^{N}\frac{p_i^2}{m_i}+\frac{P_\epsilon^2}{W}-(3N+1)k_BT_{\text{ext}}\right]\frac{\partial}{\partial P_\eta}$$

$$iL_9 = \frac{P_\eta}{Q}\frac{\partial}{\partial \eta}$$

By successive application of Trotter formula,

$$e^{ihL} = e^{iL_9h/2}...e^{iL_2h/2}e^{iL_1h}e^{iL_2h/2}...e^{iL_9h/2}$$

the application of this Liouvillian reduces to 17 elementary translation/multiplication operations updating phase space variables. This can be represented as a simple table

$$\eta^I = \eta + \frac{P_\eta}{Q}\frac{h}{2}$$

$$P_\eta^I = P_\eta + \left[\sum_{i=1}^{N}\frac{p_i^2}{m_i}+\frac{P_\epsilon^2}{W}-(3N+1)k_BT_{\text{ext}}\right]\frac{h}{2}$$

$$P_\epsilon^I = P_\epsilon + \left[3V(\Pi_{\text{int}}-\Pi_{\text{ext}})+\frac{1}{N}\sum_{i=1}^{N}\frac{p_i^2}{m_i}\right]\frac{h}{2}$$

$$P_\epsilon^{II} = P_\epsilon^I \exp\left[-\frac{h}{2}\frac{P_\eta^I}{Q}\right]$$

$$V^I = V\exp\left[\frac{h}{2}\frac{3P_\epsilon^{II}}{W}\right]$$

$$p_i^{\text{I}} = p_i' \exp\left[-\frac{h}{2}\left(\left(1+\frac{1}{N}\right)\frac{P_\epsilon^{\text{II}}}{W} + \frac{P_\eta^{\text{I}}}{Q}\right)\right]$$

$$p_i^{\text{II}} = p_i^{\text{I}} + F_i\frac{h}{2}$$

$$r_i^{\text{I}} = r_i \exp\left[\frac{h}{2}\frac{P_\epsilon^{\text{II}}}{W}\right]$$

$$r_i^{\text{II}} = r_i^{\text{I}} + \frac{p_i^{\text{II}}}{m_i}h$$

$$r_i^{\text{III}} = r_i^{\text{II}} \exp\left[\frac{h}{2}\frac{P_\epsilon^{\text{II}}}{W}\right]$$

$$p_i^{\text{III}} = p_i^{\text{II}} + F_i'\frac{h}{2}$$

$$p_i^{\text{IV}} = p_i^{\text{III}} \exp\left[-\frac{h}{2}\left(\left(1+\frac{1}{N}\right)\frac{P_\epsilon^{\text{II}}}{W} + \frac{P_\eta^{\text{I}}}{Q}\right)\right]$$

$$V^{\text{II}} = V^{\text{I}} \exp\left(\frac{h}{2}\frac{3P_\epsilon^{\text{II}}}{W}\right)$$

$$P_\epsilon^{\text{III}} = P_\epsilon^{\text{II}} \exp\left[-\frac{h}{2}\frac{P_\eta^{\text{I}}}{Q}\right]$$

$$P_\epsilon^{\text{IV}} = P_\epsilon^{\text{III}} + \left[3V^{\text{II}}(\Pi_{\text{int}}' - \Pi_{\text{ext}}) + \frac{1}{N}\sum_{i=1}^{N}\frac{(p_i^{\text{IV}})^2}{m_i}\right]\frac{h}{2}$$

$$P_\eta^{\text{II}} = P_\eta^{\text{I}} + \left[\sum_{i=1}^{N}\frac{(p_i^{\text{IV}})^2}{m_i} + \frac{(P_\epsilon^{\text{IV}})^2}{W} - (3N+1)k_{\text{B}}T_{\text{ext}}\right]\frac{h}{2}$$

$$\eta^{\text{II}} = \eta^{\text{I}} + \frac{P_\eta^{\text{II}}}{Q}\frac{h}{2}.$$

The calculation of forces at time $t+h$ (indicated as $F' = F(r(t+h))$) is carried out after the calculation of the new atomic positions $r_i^{\text{III}} = r_i(t+h)$, and the calculation of the new pressure, Π_{int}', and temperature after the calculation of the new atomic momenta, $p_i^{\text{IV}} = p_i(t+h)$. The result, as one can see, is a 'simple' (for the computer!) algorithm.

3.2 Constraints (SHAKE)

We have already derived the statistical equilibrium properties of a dynamical system subjected to constraints in Sect. 2. Here, we will explain how to implement constraints in molecular dynamics.

Understanding SHAKE. To describe a constrained system we have to take into account the constraint forces $G_i = \sum_\alpha \lambda^\alpha \frac{\partial \sigma^\alpha}{\partial r_i}$, where λ^α are Lagrange multipliers, i.e., unknown parameters to be determined afterward, and σ^α are the constraint conditions. The constraint conditions can take a variety of forms, depending on the nature of the

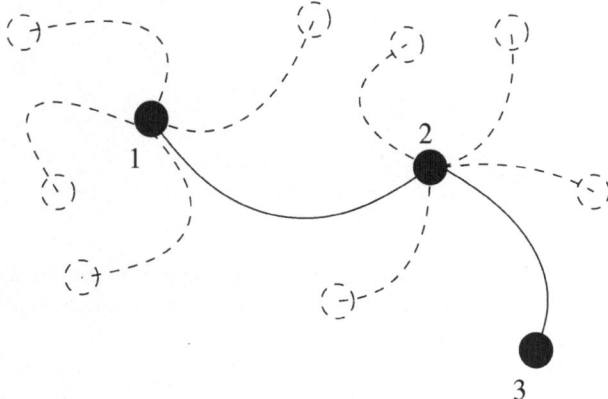

Fig. 3. Schematic representation of the swarm of trajectories generated by SHAKE for different values of the λ's. The continuous line indicates the chosen prediction

constraint. As it is well known, explicit expressions for Lagrange multipliers, λ^α, can be obtained using the condition $\ddot{\sigma}^\alpha = 0$, which determines the λ's as functions of the coordinates and velocities of the atoms. If the initial conditions satisfy $\sigma^\alpha = 0$ and $\dot{\sigma}^\alpha = 0$, the constraints themselves should be satisfied at any later time. This is true for the exact solution, but is only approximately so for any numerical integration algorithm. The consequences are fatal since after a finite time no constraint condition will be satisfied any more.

This difficulty can be solved considering the Lagrange multipliers as free parameters and obtaining a prediction for the atomic coordinates at each time step as an explicit function of the λ's. Plugging the predicted coordinates back into the constraint condition, we get the λ's such that the atomic coordinates satisfy exactly the constraints at each time step. The simple and efficient method which treats the constraints in this way is known as SHAKE [10].

Let us, first of all, try to visualize the method. All integration algorithms generate a prediction of the phase space point at time $t + h$, given the initial point at time t. However, if the system possesses constraints, the prediction depends on the values of the parameters λ^α, one for each constraint. Therefore, the prediction generated by the algorithm will be a function of the λ's, which means that our algorithm does not generate, at each time step, one point along a trajectory, but instead a set of points characterized by the different values of λ's. In addition, from these points we have to choose the point which simultaneously satisfies all the constraint conditions.

SHAKE is a procedure which examines this set of points at time $t + h$ and fixes the point which satisfies the constraints. Figure 3 illustrates how SHAKE works. For each time step we consider all the possible predictions and find the appropriate λ's after which we update our configuration in the phase space and repeat the process. Therefore, at every step we have new values for the λ's. The dotted lines in Fig. 3 correspond to the possible predictions which were not chosen and the solid lines correspond to the predictions which completely satisfy the constraints at time t and $t + h$.

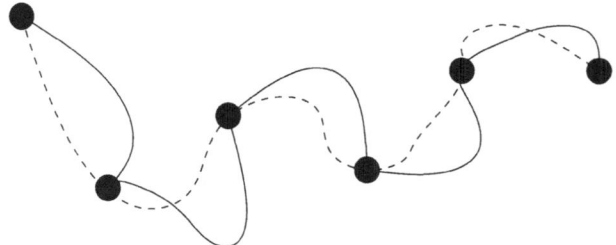

Fig. 4. Schematic representation of SHAKE trajectory compared to the true trajectory of the system. The dotted line indicates the real trajectory and the continuous line the trajectory generated by SHAKE

Figure 4 shows a schematic representation of a discretized SHAKE trajectory compared with the true trajectory of the system. Using SHAKE, the trajectory between the times t and $t + h$ does not coincide with a true trajectory of the system. The important result is that, although our trajectory is not the true one, it coincides with the true one at every time step.

Illustration: A Simple Diatomic System. Let us illustrate the solution of the constraints with a simple diatomic system in the NVE ensemble (see (12)). In this system we have only one bond constraint between each two atoms, $\sigma_{12} = r_{12}^2 - d^2 = 0$, where d is the bond length and $r_{12} = r_1 - r_2$. The constraint forces are

$$G_1 = -G_2 = -2\lambda r_{12}. \tag{28}$$

The Liouvillian of this system can be split into three parts (separating the constraints from the intermolecular forces) as

$$iL_1 = \frac{p_i}{m_i}\frac{\partial}{\partial r_i} \quad \text{and} \quad iL_2 = (F_i + G_i)\frac{\partial}{\partial p_i}.$$

Using the Trotter formula $e^{ihL} = e^{iL_2 h/2}e^{iL_1 h}e^{iL_2 h/2}$ we can start writing down the integration algorithm

$$p_i^I = p_i + (F_i + G_i)\frac{h}{2}$$

$$r_i^I = r_i + \frac{p_i^I}{m_i}h,$$

where $r_i^I = r_i(t + h)$ are the predictions for the final positions of the atoms as functions of the unknown parameter G_i. Actually, we can write down the explicit dependence of the final positions on the constraint forces as

$$r_i(t + h) = r_i + \frac{p_i + (F_i + G_i)h/2}{m_i}h = r_i^{nc} + \frac{h^2 G_i}{2m_i}, \tag{29}$$

where r_i^{nc} denotes the atomic positions which we would get if there were no constraints. Substituting (28) for G_i in (29) for each atom of our diatomic molecule we obtain

$$r_1(t + h) = r_1^{nc} - 2\frac{\lambda}{m_1}r_{12}(t)$$

$$r_2(t + h) = r_2^{nc} + 2\frac{\lambda}{m_2}r_{12}(t). \tag{30}$$

Now, by substituting the positions at time $t + h$ into the constraint condition $\sigma_{12} = r_{12}^2 - d^2 = 0$ we can derive an analytical expression for the Lagrange multiplier for this simple case,

$$\lambda = \frac{-(r_{12} \cdot r_{12}^{nc}) + \sqrt{(r_{12} \cdot r_{12}^{nc})^2 - d^2(d^2 - (r_{12}^{nc})^2}}{2h^2\nu d^2},$$

where $\nu = \frac{1}{m_1} + \frac{1}{m_2}$. Substitution of this Lagrange multiplier into (30) gives us the final atomic positions.

To complete the integration step, we also have to find the correct final momenta of the particles. Once we obtain the value of λ, we also know the values of the constraint forces G_i, and, therefore, the values of the atomic momenta p_i^I. After that, we write down the last line of our integration algorithm as

$$p_i^{II} = p_i^I + (F_i' + G_i')\frac{h}{2} \equiv p_i^{nc} + G_i'\frac{h}{2}, \tag{31}$$

where $F' = F(r(t + h))$ and G_i' are the unknown constraint forces at time $t + h$. In this equation $p_i^{II} = p_i(t + h)$ are the predictions for the final momenta of the atoms.

Plugging the prediction given by (31) into the constraint conditions given by the time derivatives of the constraints, $\dot{\sigma}_{12} = 0$, we get, in our case,

$$\dot{\sigma}_{12}(t + h) = 2r_{12} \cdot (\dot{r}_1 - \dot{r}_2) = 2r_{12} \cdot \left(\frac{p_1}{m_1} - \frac{p_2}{m_2}\right) = 0. \tag{32}$$

As for the constraint force, we can write it as $G_i' = (-1)^i 2\lambda' r_{12}$ with $i = 1, 2$. Substituting this expression in the equation for the predicted momenta (31) and then in (32), we get

$$2r_{12} \cdot \left(\frac{p_1^{nc} - 2\lambda' r_{12}h/2}{m_1} - \frac{p_2^{nc} + 2\lambda' r_{12}h/2}{m_2}\right) = 0$$

from which we obtain the new value for the Lagrange multiplier λ', G_i' and the final constrained momenta.

SHAKE Implementation. We have just shown how to implement the constraints in the NVE algorithm for diatomic molecules. The procedure described in the previous subsection can be applied to any molecule, μ, containing n_μ atoms and k_μ constraints per molecule, $\{\sigma^l = 0\}_{l=1,k_\mu}$. However, the calculations can be time consuming.

Let us describe some details of such implementation. For any atom i belonging to molecule μ we write

$$r_{\mu i}(t+h) = r_{\mu i}^{nc} + \frac{h^2}{2m_{\mu i}} \sum_{l=1}^{k_\mu} \lambda^l \frac{\partial \sigma^l}{\partial r_{\mu i}}$$

$$p_{\mu i}(t+h) = p_{\mu i}^{nc} + \frac{h}{2} \sum_{l=1}^{k_\mu} \left(\lambda^l \left. \frac{\partial \sigma^l}{\partial r_{\mu i}} \right|_{r_{\mu i}(t)} + \lambda'^l \left. \frac{\partial \sigma^l}{\partial r_{\mu i}} \right|_{r_{\mu i}(t+h)} \right). \qquad (33)$$

The substitution of the above expression for $r_{\mu i}(t+h)$ into the constraint conditions gives us a set of k_μ algebraic equations for each atom belonging to molecule μ

$$\sigma^l(\{r_{\mu i}(t+h)\})_{i=1,n_\mu} = T_l(\{\lambda^j\}_{j=1,k_\mu}) = 0.$$

SHAKE [18] is an iterative algorithm developed to solve this set of equations. It was first developed to treat bond constraints and has been later generalized to arbitrary constraints [10]. During the iterative process each loop treats all connected constraints individually in succession.

To explain the procedure, let us fix our attention on a concrete constraint l' at the nth cycle of iteration. Let us denote by $r_{\mu i}^{\text{old}}$ the atomic coordinates of each atom participating in the constraint $\sigma^{l'}$ at the $[(n-1)k_\mu + l']$-th step of the iteration. From (33), the new atomic coordinates for atom i can be obtained from the old coordinates by adding the constraint force contribution due to this constraint to $r_{\mu i}^{\text{old}}$

$$r_{\mu i}^{\text{new}} = r_{\mu i}^{\text{old}} + \frac{h^2}{2m_{\mu i}} \beta_{(n)}^{l'} \left. \frac{\partial \sigma^{l'}}{\partial r_{\mu i}} \right|_{r_{\mu i}(t)}. \qquad (34)$$

The unknown parameter $\beta_{(n)}^{l'}$, a contribution to $\lambda^{l'}$, is now evaluated as a first order solution of the scalar equation $\sigma^{l'}(\{r_{\mu i}^{\text{new}}\}) = 0$. To evaluate it, we can expand the last equation to the first order in $r_{\mu i}^{\text{new}} - r_{\mu i}^{\text{old}}$,

$$\sigma^{l'}(\{r_{\mu i}^{\text{new}}\}) = \sigma^{l'}(\{r_{\mu i}^{\text{old}}\}) + \sum_i (r_{\mu i}^{\text{new}} - r_{\mu i}^{\text{old}}) \left. \frac{\partial \sigma^{l'}(\{r_{\mu i}\})}{\partial r_{\mu i}} \right|_{r_{\mu i}^{\text{old}}} = 0. \qquad (35)$$

The sum in (35) runs over all the atoms participating in the constraint. Solving for $\beta_{(n)}^{l'}$, we get

$$\beta_{(n)}^{l'} = -h^{-2} \sigma^{l'}(\{r_{\mu i}^{\text{old}}\}) / \left[\sum_i \frac{1}{2m_{\mu i}} \left(\frac{\partial \sigma^{l'}}{\partial r_{\mu i}} \right) \Bigg|_{r_{\mu i}^{\text{old}}} \cdot \left(\frac{\partial \sigma^{l'}}{\partial r_{\mu i}} \right) \Bigg|_{r_{\mu i}(t)} \right]. \qquad (36)$$

If the iterative process starts from the atomic positions obtained from the integration algorithm without taking into account the constraints, $\{r_{\mu i}^{nc}\}$, the atomic coordinates for the atom i at the lth constraint of the n:th loop of the iteration are

$$r_{\mu i}^{\text{new}} = r_{\mu i}^{nc} + \frac{h^2}{2m_{\mu i}} \sum_{l=1}^{k_\mu} \lambda_{(n,l')}^l \left. \frac{\partial \sigma^l}{\partial r_{\mu i}} \right|_{r_{\mu i}(t)}$$

with

$$\lambda^l_{(n,l')} \equiv \left\{ \begin{array}{l} \sum_{n'=1}^{n} \beta^l_{(n')}, l \le l' \\ \sum_{n'=1}^{n-1} \beta^l_{(n')}, l > l' \end{array} \right\}.$$

The convergence of the iterative process is controlled by the maximum deviation of $\sigma^l(\{r^{\text{new}}_{\mu i}\})$ from zero for all constraints ($l = 1, \dots, k_\mu$) within the nth loop. When this deviation becomes smaller than a tolerance value, the iteration is stopped and the final atomic coordinates are accepted as the new configuration at time $t + h$, $\{r_{\mu i}(t + h)\}$.

To implement SHAKE, we have to program (34) and (36), and, considering all constraints in succession in a cyclic way, evaluate the parameters $\beta^{l'}_{(n)}$ at the nth iterative loop for the constraint l'. Then we compute $r^{\text{new}}_{\mu i}$ and either find convergence or update $r^{\text{old}}_{\mu i}$ and proceed to the next iteration. As for the momenta, we have to use (33) for $\{p_{\mu i}(t+h)\}$, and follow the same procedure as in the SHAKE process for the coordinates, but using as constraint the time derivatives of the σ's which represent a constraint on the momenta:

$$\dot{\sigma}^l(\{r_{\mu i}\}) = \sum_{l=1}^{k_\mu} \frac{\partial \sigma^l}{\partial r_{\mu i}} \cdot \dot{r}_{\mu i}(r, p) = 0.$$

At convergence, we get the values of the λ'^l's and the final momenta of the particles, $\{p_{\mu i}(t + h)\}$.

If we do not have an NVE ensemble, there may be certain changes in the implementation of the constraints. The details of constraint implementation for a molecular NPT dynamics are presented in [13].

3.3 Concluding Remarks for Non-Hamiltonian Molecular Dynamics

To conclude, we summarize the results obtained thus far. In Sect. 2 we have introduced the procedure developed by Tuckerman et al. [3,4] to derive the statistical behavior of non-Hamiltonian systems. In Sect. 3 we have seen how to integrate the dynamics of non-Hamiltonian systems by means of reversible integrators. If these systems are not too far from Hamiltonian, e.g. forces depending on momenta only linearly, we have exact and stable algorithms to integrate their evolution numerically. The algorithm, based on the Trotter expansion of the Liouville propagator, gives us a simple integration procedure which is straightforward to implement. As for constrained systems, we have discussed SHAKE [10], a working algorithm which permits the treatment of different types of constraints and can be combined with reversible integrators for general equations of motion.

4 Dynamics and (Some) Statistical Mechanics of Quantum-Classical Systems

4.1 Quantum-Classical Non-adiabatic Dynamics

At present we possess a statistical mechanical theory and algorithms in order to implement classical molecular dynamics. In addition, we also have sufficient resources to describe quantum statistical mechanics of equilibrium systems (except fermions) by path integrals, which may sometimes be a long and complex calculation, though. As for quantum statistical dynamics, a full calculation can only be done for very simple systems comprising very few degrees of freedom. To extend the dynamical calculation to other quantum systems, we can try to reduce the dimension of the quantum problem to a minimum and to combine it with classical dynamics. This can be applied, for example, to the case of a light quantum particle or a small set of quantum degrees of freedom interacting with more massive particles [19–21]. In this case we can try to solve the problem by treating the massive degrees of freedom classically and using a quantum description for the light ones. This means that we have to consider the dynamics of a quantum subsystem coupled to a classical bath.

In this section we first present a scheme for carrying out quantum-classical evolution of many-body systems based on a partial Wigner transform. Then, we formulate their statistical mechanical description, and, last, we indicate how to compute equilibrium and time-dependent expectation values. The evolution equations for the density matrix, or the dynamical variables, are expressed in an adiabatic basis and the evolution is determined by an ensemble of surface-hopping trajectories. The quantum-classical form of the canonical equilibrium density is discussed (considering possible ways to circumvent the difficulties associated with lacking an explicit expression for it). The formulation of non-equilibrium statistical mechanics and time-dependent equilibrium properties is given together with a discussion of the properties of time correlation functions.

Quantum Dynamics and Statistical Mechanics. Let us start by summarizing the formal nature of full quantum statistical dynamics. For a quantum system characterized by the Hamiltonian $\hat{H}(\hat{R}, \hat{P})$ (carets are used for operators), we can write the von Neumann evolution equation for the density matrix as

$$\frac{\partial \hat{\rho}(t)}{\partial t} = -\frac{i}{\hbar}[\hat{H}, \hat{\rho}]. \tag{37}$$

Knowing the initial conditions at time $t = 0$, the formal solution is

$$\hat{\rho}(t) = e^{-i\hat{L}t}\hat{\rho}(0) \equiv e^{-i\hat{H}t/\hbar}\hat{\rho}(0)e^{i\hat{H}t/\hbar},$$

where $i\hat{L} = (i/\hbar)[\hat{H}, ...]$ is the quantum Liouville operator. Then, for any observable $\hat{A}(t)$ having no explicit time-dependence the evolution equation can be written as

$$\frac{d\hat{A}(t)}{dt} = i[\hat{H}, \hat{A}].$$

Solving this equation, we get $\hat{A}(t) = e^{i\hat{L}t}\hat{A}(0) = e^{i\hat{H}t/\hbar}\hat{A}(0)e^{-i\hat{H}t/\hbar}$. The expectation value of an observable A at time t can be written in terms of density matrix as

$$\overline{A(t)} = \text{Tr}\,\hat{A}\hat{\rho}(t) = \text{Tr}\,\hat{A}(t)\hat{\rho}(0),$$

where the trace is taken over the quantum system. The equilibrium density matrix is the solution of $[\hat{H}, \hat{\rho}_{\text{eq}}] = 0$ and, therefore, $\hat{\rho}_{\text{eq}} \propto e^{-\beta\hat{H}}$.

Now, let us use the linear response theory. It assumes that a system in thermal equilibrium has been subjected to a time dependent external force $F(t)$. According to the linear response theory, the response of a system to an external perturbation can be determined by calculating the time-dependent density matrix up to the first order. To do that, we assume that the Hamiltonian of the system in the presence of an external force can be written as

$$\hat{H}_T(t) = \hat{H} - \hat{B}^+ F(t),$$

where $\hat{B}^+$ ($^+$ is the adjoint) is the observable coupled to the external field. The von Neumann equation (37) for this case reads as

$$\frac{\partial \hat{\rho}(t)}{\partial t} = -\frac{i}{\hbar}[\hat{H}_T, \hat{\rho}] = -(i\hat{L} - i\hat{L}_B F(t))\hat{\rho}(t),$$

where $i\hat{L}_B = i[\hat{B}^+, ...]$. Now, the solution for the density matrix can be written in terms of the initial conditions, i.e., the equilibrium density matrix

$$\hat{\rho}(t) = \hat{\rho}_{eq}(t) + \int_{-\infty}^{t} dt' e^{-i\hat{L}(t-t')} i\hat{L}_B \hat{\rho} F(t').$$

Then, the expectation value of an observable A at time t reads

$$\overline{A(t)} = \frac{i}{\hbar}\int_{-\infty}^{t} dt' \mathrm{Tr}(\hat{A}(t-t')[\hat{B}^+, \hat{\rho}_{eq}])F(t') \equiv \int_{-\infty}^{t} dt' \Phi_{AB} F(t'), \qquad (38)$$

where Φ_{AB} is the response function. Inserting the Kubo identity [22]

$$\frac{i}{\hbar}[\hat{B}^+, \hat{\rho}_{eq}] = \int_0^\beta d\lambda \hat{\rho}_{eq} \dot{B}^+(-i\hbar\lambda) \qquad (39)$$

in (38), we can rewrite $\overline{A(t)}$ as

$$\overline{A(t)} = \int_{-\infty}^{t} dt' \mathrm{Tr}(\hat{A}(t-t')\left\{\int_0^\beta d\lambda \hat{\rho}_{eq} \dot{B}^+(-i\hbar\lambda)\right\} F(t'),$$

which is computationally a much simpler formula.

Wigner Representation. Let us consider an alternative description of quantum mechanics which uses the Wigner [23] transform of the density matrix instead of a wavefunction. This approach is not useful for computing the dynamics of a system but it can be used to derive the classical limit.

For an N-particle quantum system, the Wigner description defines a distribution function $f(R, P)$ as the Fourier transform of the elements of the density matrix in the configuration representation,

$$\rho_W(R, P) = (2\pi\hbar)^{-3N} \int dz e^{iPz/\hbar} \langle R - \frac{z}{2}|\hat{\rho}|R + \frac{z}{2}\rangle.$$

The Wigner transform of a quantum operator $\hat{A}(R, P)$, instead, is defined as

$$A_W(R, P) = \int dz e^{-iPz/\hbar} \langle R + \frac{z}{2} | \hat{A} | R - \frac{z}{2} \rangle.$$

The Wigner-transformed density matrix is a normalized and real, but not a positive, definite function. For this reason it cannot be identified with a probability density. The properties of the Wigner-transformed density matrix are the following:

$$\int dP \rho_W(R, P) = \langle R | \hat{\rho} | R \rangle \quad \text{and} \quad \int dR \, dP \rho_W(R, P) = \text{Tr}\hat{\rho} = 1.$$

The expectation value of an observable A at time t is given as

$$\overline{A(t)} = \text{Tr}\hat{\rho}\hat{A} = \int dR \, dP \rho_W(R, P, t) A_W(R, P).$$

To derive the dynamics of a system, we use the fundamental property of the Wigner representation of operators $\hat{A}$ and $\hat{B}$ [23]

$$(AB)_W(R, P) = \hat{A}_W(R, P) e^{\frac{\hbar \Lambda}{2i}} \hat{B}_W(R, P) \tag{40}$$

with

$$\Lambda = \frac{\overleftarrow{\partial}}{\partial P} \frac{\overrightarrow{\partial}}{\partial R} - \frac{\overleftarrow{\partial}}{\partial R} \frac{\overrightarrow{\partial}}{\partial P}$$

the negative Poisson bracket operator. [2] Here the arrow over the partial derivative indicates in which direction the derivative is acting. Using this property, the von Neumann equation for the Wigner-transformed density matrix becomes

$$\frac{\partial \rho_W(R, P, t)}{\partial t} = -\frac{i}{\hbar} \left(H_W e^{\frac{\hbar}{2i}\Lambda} \rho_W(t) - \rho_W(t) e^{\frac{\hbar}{2i}\Lambda} H_W \right) =$$

$$= -\frac{i}{\hbar} \left(\overrightarrow{H}_\Lambda \rho_W - \rho_W(t) \overleftarrow{H}_\Lambda \right) \equiv -iL_W \rho_W, \tag{41}$$

where $\overrightarrow{H}_\Lambda = H_W e^{\frac{\hbar}{2i}\Lambda}$ and $\overleftarrow{H}_\Lambda = e^{\frac{\hbar}{2i}\Lambda} H_W$. The solution for the density matrix at time t as a function of its value at time $t = 0$ reads as

$$\rho_W(t) = e^{-iL_W t} \rho_W(0).$$

---

[2] To prove (40), first Fourier transform $A_W(R, P)$: $\alpha(\sigma, \tau) = \int dR dP e^{-i(\sigma R + \tau P)} A_W(R, P)$, and let $A_W(R, P) = (2\pi)^{-6N} \int d\sigma d\tau e^{i(\sigma R + \tau P)} \alpha(\sigma, \tau)$. First, we write $(AB)_W = \int dz e^{-iPz} \langle R + z/2 | AB | R - z/2 \rangle$. This gives, $(AB)_W = \int dz \int dR' e^{-iPz} \langle R + z/2 | A | R' \rangle \langle R' | B | R - z/2 \rangle = (2\pi)^{-12N} \int dz \int dR' e^{-iPz} \int d\sigma d\tau e^{i\sigma(R + R' + z/2)/2} \times \alpha(\sigma, R' - R - z/2) e^{i\sigma'(R + R' - z/2)/2} \beta(\sigma', R - R' - z/2)$. By defining $\tau = R' - R - z/2$ and $\tau' = R - R' - z/2$, we obtain $(AB)_W = \int d\sigma d\sigma' d\tau d\tau' e^{i(\sigma R + \tau P)} \alpha(\sigma, \tau) e^{i(\sigma' \tau - \tau' \sigma)/2} \beta(\sigma', \tau') e^{i(\sigma' R + \tau' P)} = A_W e^{\hbar \Lambda/(2i)} B_W$, since in the integral $e^{i(\sigma' \tau - \tau' \sigma)/2}$ can be substituted by $e^{\hbar \Lambda/(2i)}$.

Now, let us obtain the classical limit ($\hbar \to 0$) for this expression. Expanding the exponential to the first order in $\hbar$, we find

$$e^{\frac{\hbar A}{2i}} \cong 1 + \frac{\hbar}{2i} A,$$

such that $iL_W \to iL_C = (H_W A \rho_W - \rho_W A H_W)/2 = -\{H_W, \rho_W\}_{PB}$, where

$$-\{H_W, \rho_W\}_{PB} = -\sum \left[\frac{\partial H_W}{\partial R} \frac{\partial \rho_W}{\partial P} - \frac{\partial H_W}{\partial P} \frac{\partial \rho_W}{\partial R} \right] = -\sum \left[\dot{P} \frac{\partial \rho_W}{\partial P} - \dot{R} \frac{\partial \rho_W}{\partial R} \right]$$

and the solution for the density matrix is

$$\rho_C(R, P, t) = e^{-iL_C t} \rho_C(R, P, 0) = \rho_C(R(-t), P(-t), 0),$$

i.e., the classical solution of the Liouville equation.

Quantum-Classical Dynamics. Let us consider a quantum system composed of a light particle interacting with a set of more massive particles. The light particle is characterized by mass m and position $\hat{r}$, and the massive particles by mass M and positions $\hat{R} = (\hat{R}_1, ..., \hat{R}_N)$, where N is the number of massive particles. Moreover, we assume that $\frac{m}{M} \to 0$. The Hamiltonian of this system is

$$\hat{H} = \frac{\hat{P}^2}{2M} + \frac{\hat{p}^2}{2m} + \hat{V}(\hat{r}, \hat{R}),$$

where vector $\hat{P} = \{\hat{P}_1, ..., \hat{P}_N\}$ is the momentum operator of the system of massive particles, $\hat{p}$ the momentum of the light particle, and $\hat{V}$ the potential energy depending on the positions of all particles. As we already mentioned, standard quantum dynamics of such system can not be calculated. Let us separate this system into two interacting subsystems, a quantum subsystem containing the light particle obeying quantum laws and a bath containing the massive particles obeying classical laws. To describe the dynamics of such a system, we use a partial Wigner transform, retaining the Hilbert space description of the quantum subsystem, and carry out the Wigner transform only over the coordinates of the N-particle bath [24]. We define the partial Wigner transform of any operator $\hat{A}$ as

$$\hat{A}_W(R, P) = \int dz e^{-iPz/\hbar} \langle R + \frac{z}{2} | \hat{A} | R - \frac{z}{2} \rangle$$

and the partial Wigner transform of the density matrix as

$$\hat{\rho}_W(R, P) = (2\pi\hbar)^{-3N} \int dz e^{iPz/\hbar} \langle R - z/2 | \hat{\rho} | R + z/2 \rangle.$$

Applying the partial Wigner transform to our Hamiltonian, we obtain

$$\hat{H}_W(R, P) = \frac{P^2}{2M} + \frac{\hat{p}^2}{2m} + \hat{V}(\hat{r}, R) = \frac{P^2}{2M} + \hat{h}_W,$$

a fully quantum Hamiltonian represented partly by standard operators, partly by Wigner representation. In this representation the Hamiltonian $\hat{h}_W(R)$ characterizes the quantum subsystem inside the bath. The bath is characterized by the phase space coordinates (R, P) defined by the real parameters appearing in the Wigner transform. Now, let us introduce the quantum-classical approximation. Considering the fact that $\frac{m}{M} \to 0$, we can expand the exponential factor of (40) to the first order in Taylor series as

$$e^{\frac{\hbar \Lambda}{2i}} = 1 + \frac{\hbar}{2i}\Lambda + O(\frac{m}{M})^{\frac{1}{2}}. \tag{42}$$

As for the von Neumann equation, using (40) and (42), and taking the partial Wigner transform of (37), we obtain

$$\begin{aligned}
\frac{\partial \hat{\rho}_W}{\partial t} &= -\frac{i}{\hbar}([\hat{H}, \hat{\rho}])_W \\
&= -\frac{i}{\hbar}[\hat{H}_W, \hat{\rho}_W] + \frac{1}{2}\left(\{\hat{H}_W, \hat{\rho}_W\} - \{\hat{\rho}_W, \hat{H}_W\}\right) \\
&\equiv -i\hat{L}\hat{\rho}_W.
\end{aligned}$$

The eigenvalue problem of a Hamiltonian $\hat{h}_W$ is

$$\hat{h}_W(R)|\alpha; R\rangle = E_\alpha(R)|\alpha; R\rangle. \tag{43}$$

For each bath configuration R we can construct an adiabatic set of eigenfunctions $|\alpha; R\rangle$ with corresponding eigenvalues $E_\alpha(R)$, such that

$$\sum_\alpha |\alpha; R\rangle\langle\alpha; R| = 1.$$

In this adiabatic basis, the density matrix elements are given by

$$\rho_W^{\alpha,\alpha'}(R, P) = \langle\alpha; R|\hat{\rho}_W(R, P)|\alpha'; R\rangle, \tag{44}$$

while for any operator $\hat{O}$ we write

$$\hat{O}_W^{\alpha\alpha'} = \langle\alpha; R|\hat{O}_W|\alpha'; R'\rangle \equiv \hat{O}_W^s,$$

where, for the system which has a finite number of eigenstates, we have simplified the notation by defining $s = \alpha n + \alpha'$ for a quantum subsystem with n states.

Using the representation (44) for the density matrix elements, we can write

$$\frac{\partial \rho_W^{s_j}}{\partial t} = \sum_{s_k} -i\mathcal{L}_{s_j s_k} \rho_W^{s_k}(R, P, t),$$

where

$$-i\mathcal{L}_{s_j s_k} = -(i\omega_{s_j} + iL_{s_j})\delta_{s_j s_k} + J_{s_j s_k}. \tag{45}$$

In these equations $s_i = (\alpha_i, \alpha_i')$ denotes the different values of α and α' that occur in the course of dynamics. The first term of (45) is diagonal, $\omega_{s_j}(R)$ being the adiabatic frequency difference defined by

$$\omega_{s_j}(R) = \frac{E_{\alpha_j}(R) - E_{\alpha_j'}(R)}{\hbar},$$

while the Liouville operator iL_{s_j} involves the mean of the Hellmann–Feynman forces of the two quantum states labeling the density matrix elements and is given by

$$iL_{s_j} = \frac{P}{M}\frac{\partial}{\partial R} + \frac{1}{2}(F_W^{\alpha_j} + F_W^{\alpha_j'})\frac{\partial}{\partial P}$$

(here $F_W^{\alpha_j}$ is the Hellmann–Feynman force). The second term in (45) is off-diagonal (in a real basis $|\alpha; R\rangle$). It couples different adiabatic states and is written as

$$J_{s_j s_k} = -\frac{P}{M}d_{\alpha_j \alpha_k}\left(1 + \frac{1}{2}S_{\alpha_j \alpha_k}\frac{\partial}{\partial P}\right)\delta_{\alpha_j' \alpha_k'}$$
$$- \frac{P}{M}d_{\alpha_j' \alpha_k'}^*\left(1 + \frac{1}{2}S_{\alpha_j' \alpha_k'}^*\frac{\partial}{\partial P}\right)\delta_{\alpha_j \alpha_k}, \qquad (46)$$

where

$$S_{\alpha_j \alpha_k} = \left(E_{\alpha_j} - E_{\alpha_k}\right)d_{\alpha_j \alpha_k}\left(\frac{P}{M}d_{\alpha_j \alpha_k}\right)^{-1} \quad \text{and} \quad d_{\alpha_j \alpha_k} = \langle \alpha_{ji}R|\frac{\partial}{\partial R}|\alpha_{kj}R\rangle,$$

the latter is a non-adiabatic coupling matrix (a small parameter in the cases of interest). Assuming a small non-adiabatic coupling, we can write

$$d(1 + \frac{1}{2}S_{\alpha_j \alpha_k}\frac{\partial}{\partial P}) \cong e^{\frac{1}{2}S_{\alpha_j \alpha_k}\frac{\partial}{\partial P}}$$

and, therefore, simplify J given by (46). This approximation originates from what we can call momentum jumps approximation [24].

Due to the presence of $J_{s_j s_k}$, a description of a single Newtonian trajectory is no longer possible. The solution for such systems may be represented by an ensemble of surface-hopping trajectories, where classical evolution segments are interrupted by quantum transitions [25]. To see that, let us develop the density matrix expression using the Dyson identity with the diagonal part as unperturbed and the off-diagonal as perturbations in the evolution operator. It takes the form

$$\rho_W^{s_0}(R, P, t) = e^{-(i\omega_{s_0} + iL_{s_0})t}\rho_W^{s_0}(R, P, 0) +$$
$$+ \sum_{s_1}\int_0^t dt' e^{-i(\omega_{s_0} + iL_{s_0})(t-t')}J_{s_0 s_1}\rho_W^{s_1}(R, P, t').$$

In this expression the exponential factor can be re-written as [24]

$$e^{-(i\omega_{s_k} + iL_{s_k})t} \equiv U_{s_k}^+(t) = e^{-is\int_0^t d\tau \omega_{s_k}(R_{s_k, \tau}^+)}e^{-iL_{s_k}t}, \equiv W_{s_k}(0, t)e^{-iL_{s_k}t},$$

i.e., a phase factor $W_{s_k}(0, t)$ times the "almost" classical Liouvillian. W_{s_k} is a function of the time-reversed phase point evolved under the "classical" evolution operator L_{s_k}:

$$(R^t_{s_k,0}, P^t_{s_k,0}) = e^{-iL_{s_k}t}(R, P).$$

The full iterative solution for the density matrix is

$$\rho^{s_0}_W(R, P, t) = U^+_{s_0}(t)\rho^{s_0}_W(R, P, 0) +$$

$$+ \sum_{n=1}^{\infty} \sum_{s_1,\dots,s_n} \int_0^{t_0} dt_1 \dots \int_0^{t_{n-1}} dt_n \prod_{k=1}^{n} [U^+_{s_{k-1}}(t_{k-1} - t_k)J_{s_{k-1},s_k}]U^+_{s_n}(t)\rho^{s_n}(R, P, 0).$$

Transporting the evolution from the density matrix to the observables, we can write the expectation value of an observable $\hat{O}$ as

$$\overline{O(t)} = \text{Tr}' \int dR\, dP \hat{O}_W(R, P)\hat{\rho}_W(R, P, t)$$

$$= \text{Tr}' \int dR\, dP(e^{i\hat{L}t}\hat{O}_W(R, P))\hat{\rho}_W(R, P, 0)$$

$$= \text{Tr}' \int dR\, dP \hat{O}_W(R, P, t)\hat{\rho}_W(R, P, 0),$$

where $\hat{O}_W(R, P, t) = e^{i\hat{L}t}\hat{O}_W(R, P, 0)$ is the forward (adjoint) time evolution of the observable, Tr' is the trace over the quantum subsystem, and the integral $\int dRdP...$ is computed by importance sampling based on $\hat{\rho}_W(R, P, 0)$.

In equilibrium the condition $\overline{O_W(t + \tau)} = \overline{O_W(t)}$ is satisfied. The expression for the density matrix for mixed quantum-classical systems can be evaluated using numerical methods, e.g., a hybrid Molecular Dynamics-Monte Carlo Scheme [25].

As an application of this approach, we have recently solved [26,27] a spin-boson system, which consists of a two-level system bilinearly coupled to a bath of harmonic oscillators. By that we have demonstrated that the quantum-classical equations of motion can be solved accurately using the surface-hopping algorithm for a quantum subsystem interacting with a many-body bath, although not for long times. The solution method provides a dynamical description of the bath in terms of an ensemble of coherently evolving surface-hopping trajectories, and, therefore, both quantum subsystem and bath dynamical correlation functions can be computed directly using this ensemble. For the spin-boson model the implementation of the quantum-classical dynamics is exact and the results of surface-hopping simulations are in accordance with previous numerically exact results for the model [28,29].

4.2   Quantum-Classical Statistical Mechanics

Quantum vs Quantum-Classical Dynamics. Next, we will (i) give a compact formulation of mixed quantum-classical dynamics using its similarity with full quantum mechanics and (ii) compare the most important properties of the two approaches.

The first important property to explore is associativity. For quantum dynamics, we can deduce it using the expression for the Wigner transform of the product of two operators

(40). In fact, it is easy to check (and to guarantee for $n > 3$) that the Wigner transform of the product of three operators is

$$(\hat{A}\hat{B}\hat{C})_W \equiv ((\hat{A}_W e^{\frac{\hbar}{2i}\Lambda}\hat{B}_W)e^{\frac{\hbar}{2i}\Lambda}\hat{C}_W) = (\hat{A}_W e^{\frac{\hbar}{2i}\Lambda}(\hat{B}_W e^{\frac{\hbar}{2i}\Lambda}\hat{C}_W)).$$

As for the quantum-classical case, we use the approximation (42) valid for $m/M \to 0$. We obtain

$$\left[\left(\hat{A}_W\left(1 + \frac{\hbar}{2i}\Lambda\right)\hat{B}_W\right)\left(1 + \frac{\hbar}{2i}\Lambda\right)\hat{C}_W\right]$$

$$= \left[\hat{A}_W\left(1 + \frac{\hbar}{2i}\Lambda\right)\left(\hat{B}_W\left(1 + \frac{\hbar}{2i}\Lambda\right)\hat{C}_W\right)\right]$$

$$- \frac{\hbar^2}{4}\left[\left(\left(\hat{A}_W\Lambda\hat{B}_W\right)\Lambda\hat{C}_W\right) - \left(\hat{A}_W\Lambda\left(\hat{B}_W\Lambda\hat{C}_W\right)\right)\right]$$

$$\equiv \left[\hat{A}_W\left(1 + \frac{\hbar}{2i}\Lambda\right)\left(\hat{B}_W\left(1 + \frac{\hbar}{2i}\Lambda\right)\hat{C}_W\right)\right] + \mathcal{O}(\hbar^2)$$

which means that associativity is valid only up to the order $\hbar^2$ in the quantum-classical limit.

The second important difference between the full quantum and quantum-classical dynamics is the time evolution of a product of two operators. If we use the quantum approach, $\hat{C}(t) = \hat{A}(t)\hat{B}(t)$ gives

$$\hat{C}_W(t) = \hat{A}_W(t)e^{\frac{\hbar}{2i}\Lambda}\hat{B}_W(t),$$

whereas the quantum-classical approximation yields

$$\hat{C}_W(t) = \hat{A}_W(t)(1 + \frac{\hbar}{2i}\Lambda)\hat{B}_W(t) + \mathcal{O}(\hbar)$$

instead. This means that the evolution of a composite operator in quantum-classical dynamics cannot be determined exactly in terms of the quantum-classical evolution of its constituent operators. The result is only valid up to terms of $\mathcal{O}(\hbar)$, in contrast to both quantum and classical dynamics. This has serious implications when computing time correlation functions.

Third, an important property of any known exact dynamics is the Jacoby identity: The quantum mechanical Lie bracket, either in its original form or in the partially Wigner transformed form, satisfies it. Indeed,

$$(\hat{A}_W, (\hat{B}_W, \hat{C}_W)_Q)_Q + (\hat{C}_W, (\hat{A}_W, \hat{B}_W)_Q)_Q + (\hat{B}_W, (\hat{C}_W, \hat{A}_W)_Q)_Q = 0.$$

However, the Jacoby identity involving the quantum-classical bracket is valid only up to terms $\mathcal{O}(\hbar)$,

$$(\hat{A}_W, (\hat{B}_W, \hat{C}_W)) + (\hat{C}_W, (\hat{A}_W, \hat{B}_W)) + (\hat{B}_W, (\hat{C}_W, \hat{A}_W)) = \mathcal{O}(\hbar).$$

Thus, the quantum-classical bracket is not a Lie bracket. As a consequence, the Poisson theorem which says that the Poisson, or the quantum Lie, bracket of any two constants of motion is also a constant of motion, fails. As a consequence, products or powers of constants of motion are not constants of motion, since the quantum-classical evolution is not generated by the Lie bracket.

Equilibrium Density Matrix. The quantum mechanical canonical equilibrium density matrix can be written in the form

$$\hat{\rho}_{eq}^Q = Z^{-1}e^{-\beta\hat{H}},$$

where Z is the partition function, i.e., $Z = \mathrm{Tr}(e^{\beta\hat{H}})$. Its partial Wigner transform becomes

$$\hat{\rho}_{Weq}^Q(R, P) = \int dz e^{iPz/\hbar}\langle R - \frac{z}{2}|\hat{\rho}_{eq}^Q|R + \frac{z}{2}\rangle.$$

$\hat{\rho}_{Weq}^Q(R, P)$ is invariant under quantum dynamics, but not under quantum-classical dynamics. The quantum-classical equilibrium density matrix, $\hat{\rho}_{Weq}$, is an approximation to $\hat{\rho}_{Weq}^Q(R, P)$, that must be stationary under quantum-classical dynamics

$$i\hat{L}\hat{\rho}_{Weq} = \frac{i}{\hbar}(\hat{H}_W(1 + \frac{\hbar}{2i}\Lambda)\hat{\rho}_{Weq} - \hat{\rho}_{Weq}(1 + \frac{\hbar}{2i}\Lambda)\hat{H}_W) = 0. \quad (47)$$

There is no explicit general solution in sight, but we can find it recursively in the following way: Write $\hat{\rho}_{Weq}$ as a power series in $\hbar$ (or in the mass ratio $(m/M)^{1/2}$ if scaled variables are used)

$$\hat{\rho}_{Weq} = \sum_{n=0}^{\infty} \hbar^n \hat{\rho}_{Weq}^{(n)}. \quad (48)$$

Substituting (48) in (47) and grouping in powers of $\hbar$, we obtain a set of recursive relations. For $n = 0$ we have

$$i[\hat{H}_W, \hat{\rho}_{Weq}^{(0)}] = 0, \quad (49)$$

and for $n > 0$

$$i[\hat{H}_W, \hat{\rho}_{Weq}^{(n)}] = \frac{1}{2}[\hat{H}_W, \hat{\rho}_{Weq}^{(n-1)}] - \frac{1}{2}[\hat{\rho}_{Weq}^{(n-1)}, \hat{H}_W]. \quad (50)$$

To obtain the solutions of the recursive equations, we use the adiabatic basis consisting of the adiabatic eigenfunctions $|\alpha; R\rangle$ of (43), with the corresponding eigenvalues E_α. The elements of the density matrix in this adiabatic basis are given by (44). Therefore, taking the matrix elements of (49) and (50), respectively, for $n = 0$ we find

$$iE_{\alpha\alpha'}\rho_{Weq}^{(0)\alpha\alpha'} = 0$$

and for $n > 0$

$$iE_{\alpha\alpha'}\rho_{Weq}^{(n)\alpha\alpha'} = -iL_{\alpha\alpha'}\rho_{Weq}^{(n-1)\alpha\alpha'} + \sum_{\nu\nu'} J_{\alpha\alpha',\nu\nu'}\rho_{Weq}^{(n-1)\nu\nu'}.$$

The formal solution of these equations can be worked out explicitly up to the first order [30], while higher order terms are much more difficult to evaluate. The unpleasant conclusion of the entire argument is that a closed simple expression for $\hat{\rho}_{Weq}$ is not available.

Linear Response Theory. As for the linear response theory, the partially Wigner transformed Hamiltonian of the system in the presence of an external, time-dependent, force can be written as:

$$\hat{\mathbf{H}}_W(\mathbf{t}) = \hat{H}_W - \hat{A}_W^+ F(t)$$

where $\hat{A}^+$ is the observable coupled to the external field. The evolution equation for the density matrix can be obtained substituting $\hat{\mathbf{H}}_W(\mathbf{t})$ instead of $\hat{H}_W$ in (41). We get

$$\frac{\partial \hat{\rho}_W(t)}{\partial t} = -\frac{i}{\hbar}(\hat{\mathbf{H}}_W(t)e^{\frac{\hbar}{2i}\Lambda}\hat{\rho}_W(t) - \hat{\rho}_W(t)e^{\frac{\hbar}{2i}\Lambda}\hat{\mathbf{H}}_W(t)), = -i(\hat{L} - \hat{L}_A F(t))\hat{\rho}_W(t),$$

where $i\hat{L}_A$ has a form analogous to $i\hat{L}$ (see (41)) with $\hat{A}^+$ replacing $\hat{H}_W$. The formal solution of this equation is

$$\hat{\rho}_W(t) = e^{-i\hat{L}(t-t_0)}\hat{\rho}_W(t_0) + \int_{t_0}^t dt' e^{-i\hat{L}(t-t')}i\hat{L}_A\hat{\rho}_W(t')F(t'). \tag{51}$$

If we choose for $\hat{\rho}_W(t_0)$ as an initial condition at time t_0 for the equilibrium density matrix $\hat{\rho}_W(t_0) = \hat{\rho}_{We}$, which is invariant under quantum-classical dynamics, the first term on the right side of (51) coincides with the equilibrium density matrix. The response of the system to the external force $F(t)$ is estimated by computing the average value of an operator $\hat{B}_W$ over the density matrix $\hat{\rho}_W(t)$,

$$\overline{B_W}(t) = \mathrm{Tr}' \int dRdP \hat{B}_W \hat{\rho}_W(t). \tag{52}$$

In this equation the prime on the trace denotes that the trace is over the quantum subsystem. To compute $\overline{B_W}(t)$, we have to substitute the solution for the density matrix (51) in (52). After the substitution we see that we can express the value $\overline{B_W}(t)$ as a sum of two terms, the first being equal to $\langle \hat{B}_W \rangle_{eq}$ and the second, denoted by $\overline{\Delta B_W}(t)$, can be written as

$$\overline{\Delta B_W}(t) = -\int_{-\infty}^t dt' \langle (\hat{A}_W^+, \hat{B}_W(t-t')) \rangle_{eq} F(t') = \int_{-\infty}^t dt' \Phi_{BA}(t-t')F(t').$$

The last line defines the response function

$$\Phi_{BA}(t) = -\langle (\hat{A}_W^+, \hat{B}_W(t)) \rangle_{eq},$$

where we have an equilibrium average of the quantum-classical bracket.

For both quantum and classical equilibrium systems the time evolution of the observables generates a stationary random process. In both cases the ensemble averages of observables are independent of time, and time correlation functions depend only on time differences. For quantum-classical dynamics the ensemble averages are also time independent:

$$\mathrm{Tr}' \int dRdP \hat{B}_W(t)\hat{\rho}_{Weq} = \mathrm{Tr}' \int dRdP (e^{i\hat{L}t}\hat{B}_W)\hat{\rho}_{Weq} = \mathrm{Tr}' \int dRdP \hat{B}_W \hat{\rho}_{Weq}.$$

Instead, time correlation functions do not obey this law rigorously, since

$$\langle(\hat{A}_{\mathrm{W}}, \hat{B}_{\mathrm{W}}(t))\rangle = \langle(\hat{A}_{\mathrm{W}}(\tau), \hat{B}_{\mathrm{W}}(t+\tau))\rangle + \mathcal{O}(\hbar).$$

Thus, the standard time translation invariance is valid only approximately in quantum-classical dynamics. However,

$$\langle e^{i\tau\hat{L}}(\hat{A}_{\mathrm{W}}, \hat{B}_{\mathrm{W}}(t))\rangle = \langle(\hat{A}_{\mathrm{W}}, \hat{B}_{\mathrm{W}}(t))\rangle,$$

which is a weak (but sufficient) form of stationarity.

At the beginning of this section we saw that the expectation value of an observable for a quantum system subject to a small perturbation is given by (38) and the response function $\Phi_{BA}(t)$ can be re-written using Kubo identity, (39). However, the quantum-classical version of Kubo identity is valid only to $\mathcal{O}(\hbar)$ [30]:

$$[\hat{B}^+, \hat{\rho}_{\mathrm{Weq}}] = \int_0^\beta \mathrm{d}\lambda \hat{\rho}_{\mathrm{Weq}}(1 + \frac{\hbar\Lambda}{2i})\dot{B}^+(-i\hbar\lambda) + \mathcal{O}(\hbar).$$

Due to this, the correlation functions cannot be calculated exactly making use of this expression, as we can do in quantum case. Thus, the numerical calculation of quantum-classical time correlation functions is still difficult to achieve.

4.3 Concluding Remarks for Quantum-Classical Systems

We have seen that the time evolution of quantum-classical systems can be rationalized and computed. Recently, the spin-boson model has been solved [26,27] for physically interesting, although not too long, times. This indicates that we are in a good position to solve spin-real solvent models, which provides good physical insight. However, realistic quantum subsystems-real solvents are still a dream. We have also seen that the statistical (static) equilibrium properties of quantum-classical systems can be properly defined and computed (although indirectly). In Sect. 4.1 we have formulated the quantum-classical linear response theory and introduced time-dependent equilibrium properties. We still have difficulties in computing them exactly. The development of the topics should focus on discovering ways to circumvent the theoretical problems and on finding reliable and efficient numerical algorithms to compute the 'long' time dynamics of quantum-classical systems.

Acknowledgements

We are grateful to T. O. White, M. E. Tuckerman and Y. Liu for providing useful material for this work. This work has been partially supported by a MIUR Cofin 2000 grant.

References

1. W.G. Hoover: Phys. Rev. A **31**, 1695 (1985)
2. S. Nosé: Mol. Phys. **57**, 187 (1986)
3. M.E. Tuckerman, G.J. Mundy, G.J. Martyna: Europhys. Lett. **45**, 149 (1999)
4. M.E. Tuckerman, Y. Liu, G. Ciccotti, G.J. Martyna: J. Chem. Phys. **115**, 1678 (2001)
5. S. Melchionna: Phys. Rev. E **61**, 6165 (2000)
6. J.P. Ryckaert, G. Ciccotti: J. Chem. Phys. **78**, 7368 (1983)
7. G.J. Martyna, M.L. Klein, M.E. Tuckerman: J. Chem. Phys. **97**, 2635 (1992)
8. M.E. Tuckerman, B.J. Berne, G.J. Martyna, M.L. Klein: J. Chem. Phys. **99**, 2796 (1993)
9. G.J. Martyna, D.J. Tobias, M.L. Klein: J. Chem. Phys. **101**, 4177 (1994)
10. G. Ciccotti, J.P. Ryckaert: Comp. Phys. Rep. **4**, 345 (1986)
11. G. Marechal, J.P. Ryckaert: Chem. Phys. Lett. **101**, 548 (1983)
12. G. Ciccotti, G.J. Martyna, S. Melchionna, M.E. Tuckerman: J. Phys. Chem. B **105**, 6710 (2001)
13. G. Kalibaeva, M. Ferrario, G. Ciccotti: Mol. Phys. **101**, 765 (2003)
14. G.R. Kneller, T. Mulders: Phys. Rev. E **54**, 6825 (1996)
15. G.J. Martyna, M.E. Tuckerman, D.J. Tobias, M.L. Klein: Mol. Phys. **87**, 1117 (1996)
16. M.E. Tuckerman, B.J. Berne, G.J. Martyna: J. Chem. Phys. **97**, 1990 (1992)
17. S.W. Gear: *Numerical Initial Value Problems in Ordinary Differential Equations* (Prentice Hall, Englewood Cliffs 1971)
18. J.P. Ryckaert, G. Ciccotti, J.C. Berendsen: J. Comp. Phys. **23**, 327 (1977)
19. J.C. Tully. In: *Modern Methods for Multidimensional Dynamics Computation in Chemistry*, ed. by D.L. Thompson (World Scientific, New York 1998)
20. M.F. Herman: Annu. Rev. Phys. Chem. **45**, 83 (1994); J. Chem. Phys. **87**, 4779 (1987)
21. *Classical and Quantum Dynamics in Condensed Phase Simulations*, ed. by B.J. Berne, G. Ciccotti, D.F. Coker (World Scientific, Singapore 1998)
22. R. Kubo: Rep. Prog. Phys. **29**, 255 (1966)
23. E. Wigner: Phys. Rev. **40**, 749 (1932); K. Imre, E. Ozizmir, M. Rosenbaum, P.F. Zweifel: J. Math. Phys. **5**, 1097 (1967)
24. R. Kapral, G. Ciccotti: J. Chem. Phys. **110**, 8919 (1999)
25. S. Nielsen, R. Kapral, G. Ciccotti: J. Stat. Phys. **101**, 225 (2000)
26. D. Mac Kernan, G. Ciccotti, R. Kapral: J. Chem. Phys. **116**, 2346 (2002)
27. D. Mac Kernan, R. Kapral, G. Ciccotti: J. Phys.: Condens. Matter **14**, 9069 (2002)
28. N. Makri, K. Thompson: J. Phys. Chem. **291**, 101 (1998); K. Thompson, N. Makri: J. Chem. Phys. **110**, 1343 (1999)
29. D.E. Makarov, N. Makri: Chem. Phys. Lett. **221**, 482 (1994); E. Sim, N. Makri: Comput. Phys. Commun. **99**, 335 (1997); N. Makri: J. Phys. Chem. **102**, 4414 (1998)
30. S. Nielsen, R. Kapral, G. Ciccotti: J. Chem. Phys. **115**, 5805 (2001)

Hybrid Models: Bridging Particle and Continuum Scales in Hydrodynamic Flow Simulations

Eirik G. Flekkøy[1], Sean McNamara[2], Knut Jørgen Måløy[1], Jens Feder[1], and Geri Wagner[3]

[1] Department of Physics, University of Oslo, PB 1048 Blindern, 0316 Oslo, Norway
[2] ICA1, University of Stuttgart Pfaffenwaldring 27, 70569 Stuttgart, Germany
[3] School of Astronomy and Physics, Raymond and Beverly Sackler Faculty of Exact Sciences, Tel Aviv University, Ramat Aviv, 69978 Tel Aviv, Israel

Abstract. Different models for the coupling of field and particle descriptions are introduced and examined. For the purpose of establishing how a molecular description may be coupled to a continuum description of the same physical system, we study a molecular dynamics system coupled to a Navier–Stokes description within the same physical space. A simple toy model version of this system is studied as well, i.e., a system of random walkers coupled to the diffusion equation. These coupling schemes are shown to work in the sense that they provide a seamless coupling between the different representations. In order to establish a sufficiently computationally efficient method for the simulation of gas–grain flow, we introduce a model where the grains are described explicitly but where the gas is described only through its continuum pressure field. It is shown that this model easily produces macroscopic structures, such as the bubbles in fluidized beds. The model is also used to study a novel bubble instability observed experimentally in the flow of gas–grain systems in simple tubes.

1 Introduction

In a simple Newtonian fluid the macroscopic description in terms of the Navier–Stokes equations contains no direct information of the microscopic dynamics that goes on at the molecular scale. In complex fluids, on the other hand, processes at the microscopic length-scales will typically influence the macroscopic behavior. The simulation of such fluids must therefore contain the proper elements of this microscopic behavior. In a molecular fluid the need for a full microscopic description may arise in the case of complex boundary conditions, or when polymers are present at interfaces or as micelles. Famous examples of such processes include the moving contact line [1], the breakup and merging of fluid droplets [2], strong shear localization, dynamic melting processes [3] and the evolution of a fracture tip [4–6]. In all of these examples dense systems self-organize to produce strong gradients on the atomic scale, thus coupling the macroscopic behavior to microscopic processes in ways that are not easily captured by constitutive relations or other averaged descriptions.

In a granular fluid the need for a microscopic, or particle based, description arises as soon as the fluid is not strongly excited: Granular flows at high packing densities exhibit flow characteristics that directly reflect the nature of the grains, i.e., their shapes and friction properties, size distribution, and the extent to which they dissipate energy in collisions. Although very promising progress on continuum descriptions of granular

E.G. Flekkøy et al., Hybrid Models: Bridging Particle and Continuum Scales in Hydrodynamic Flow Simulations, Lect. Notes Phys. **640**, 190–218 (2004)
http://www.springerlink.com/   © Springer-Verlag Berlin Heidelberg 2004

shock fronts has recently been made [7], the question whether such descriptions can be obtained for weakly excited flows remains open.

Since full molecular, or particle based simulations will usually be computationally too expensive for most interesting applications, the challenge is to choose the right level of description for the problem at hand. This may amount to using adaptive grids, resolving molecular detail only where it is needed, say, at some interface, or to choose the correct overall coarse graining level.

In the present paper the last two of these strategies are implemented: In order to model hydrodynamic flow on a molecular basis only where this level of detail is needed, we introduce a scheme where a particle description is directly coupled to a continuum description [8–12]. For the purpose of studying the principal aspects of this scheme in a simpler context, we couple a lattice gas of random walkers and the numerical solution of the diffusion equation [13].

In order to model the combined flow of a grains and a gas we introduce a model where particles are described explicitly but where the gas is described by a continuum pressure field only [14,15]. In this approach the hydrodynamic detail in between the grains is sacrificed for greater numerical efficiency.

All the three models that we discuss are based on the coupling between continuum fields and particles. However, while the particle-field coupling between two distinct spatial regions involves symmetric exchange of fluxes of all relevant conserved quantities (mass and momentum in this case), the gas–grain coupling is only partial: The discrete particles feel a force field from the gas but the force acting back onto the gas is not in the description. Moreover, there is no mass exchange between the gas and the grains.

2 A Hybrid Model for Diffusion

To describe a diffusive process on the microscopic scale, we consider a set of random walkers moving on the sites of a one-dimensional lattice. The random walkers move with equal probability to the left or to the right, hopping from one site to a neighboring site during each time step. This discrete, microscopic description is coupled to a continuous, macroscopic description, obtained by solving a discretized version of the diffusion equation

$$\frac{\partial \rho_c}{\partial t} = D\nabla^2 \rho_c \qquad (1)$$

on a one-dimensional array of nodes. Here, ρ_c is the density of random walkers or particles on a continuum node, and D is the macroscopic diffusion constant. The two descriptions overlap to some extent, in the sense that some of the continuum nodes and some sites of the particle lattice cover the same region of space. This is shown in Fig. 1. One continuum node corresponds to $W > 1$ lattice sites of unit length on which the particles move. As units we shall take the lattice constant of the particle lattice and the time step of the particles, i.e., every particle moves a unit length in a unit time. The lattice constant of the continuum lattice (on which we discretize the diffusion equation) is therefore W, which is the number of particle step lengths per continuum node. The density ρ_c is defined on the continuum nodes, and the particle density ρ_p is the number

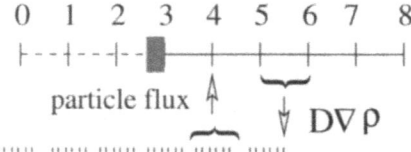

Fig. 1. Sketch of the coupling scheme of a hybrid model in the special case of 9 sites on the continuum lattice and $W = 6$ particle sites per continuum site. The particle system (P) is coupled to the continuum (C) at the location of the right arrow, and the continuum to the particles at the left arrow

of particles per site on the more fine-grained lattice. In equilibrium the averages of ρ_p and ρ_c will be equal.

The general idea of particle–continuum hybrid models is to resolve finer space and time scales in the particle system than in the continuum, which thus is taken to represent a coarse grained description of the particle system. One time step in the iteration of (1) thus corresponds to τ microscopic time steps. Since particle time steps are taken to have unit length, the time step of (1) is simply τ. In fact a main virtue of this coupling scheme is that it allows the separation of both the space and time scales of the two domains.

Since the particle system has intrinsic fluctuations, it would appear that one would need to add a fluctuating term to the above diffusion to get a fully consistent picture. It is clearly possible to do this along the lines of fluctuating hydrodynamics [16]. However, the main effect of the continuum on the particle number fluctuations is linked to averaging of the particles that is needed in the coupling region. Moreover, the general idea is to link a fine scale description (the particles) to a region of coarser scale where fluctuations, in general, play a less important role. This has been the view taken in previous studies.

The basic idea of the coupling scheme is to impose the flux of the particle system as a boundary condition on the continuum at one location, and vice versa at another location. This is illustrated in Fig. 1, where an overlap zone is defined between sites 4 and 6. At site 4 the particle flux is imposed on the continuum, and at the rightmost site on the particle system the flux of the continuum system is imposed on the particle system. The sites 0 to 3 are represented by the particle system only and the continuum equation is not updated there.

Figure 1 shows where these fluxes are imposed. The mass flux $D\nabla\rho$ in the continuum, where $\nabla\rho$ is evaluated from sites 5 and 6 is imposed as a source term on the second outermost site of the particle lattice (at the outermost site which is not shown the particles just bounce back to the left).

Once the continuum flux is computed, it is imposed as a particle-source which is constant over the τ particle updates. Then at the last of these updates the particle flux j_p is averaged over W sites and used to define the source on the continuum.More precisely, $D\nabla\rho_c$ particles are added to the particle system at every update. In doing this $D\nabla\rho_c$ is rounded to the closest integer. Correspondingly, the continuum receives a mass $j_p\tau$ by adding this value at ρ_c at the site of the leftmost arrow of Fig. 1, i.e., site 4. The particle flux j_p is measured simply as the number of right moving minus the number of left moving particles at the given time.

Of course, the flux boundary condition on the continuum could also have been imposed as a condition on the concentration gradient. However, since this condition represents a noise source that imposes variations on all wavelengths down to the lattice scale it does not conserve mass (the integrated concentration) to a high accuracy. This was checked in independent simulations using the continuum equation solver and a random boundary condition. In these simulations the imposition of a source term conserves mass to a higher accuracy than the imposition of a concentration gradient.

As usual in the numerical treatment of differential equations, the gradient of the density must be expressed as a difference across nodes. In particular, to express the continuum flux density the difference gradient

$$\nabla \rho_c \approx \frac{\rho_c(x_6) - \rho_c(x_5)}{W}, \tag{2}$$

where x_5 and x_6 are shown in the figure, should be used. Note that since the site of the particle source is located right between x_5 and x_6 the difference expression above is really a *centered* difference. The discrete value of $\nabla \rho_c$ then defines the flux density at the boundary of the continuous domain.

However, (2) does not prevent discontinuities at the discrete-continuous interface. Indeed, a configuration with $\rho_p = $ const. everywhere in the discrete domain, and $\rho_c = $ const. everywhere in the continuous domain, would lead to vanishing averaged fluxes, even for $\rho_p \neq \rho_c$. Equilibration is enforced if a hybrid gradient ∇' of the form

$$\nabla' \rho_c \approx \frac{\rho_c(x_6) - \rho_p(x_5)}{W} \tag{3}$$

is employed to define the flux $D\nabla \rho_c$ into the particle system. In (3), the macroscopic continuum density is *replaced* by the corresponding microscopic density characterizing the node x_5. The imposed flux of random walkers, based on (3), enforces that, on average, $\rho_p = \rho_c$ at the discrete-continuous interface. When the coupling scheme is applied to a physical system involving both mass and momentum transport [8], one may do without this device and use (2), as density mismatch at the interface will lead to momentum and mass flux exchange and equilibration by default.

As an illustration of the fact that continuum descriptions generally capture only the large scale behavior of the particle system, note that the average particle evolution is not exactly described by (1). Equation (1) only contains the lowest order in the gradient terms. In fact, it is generally possible – for instance by applying a simple version of the standard Chapman–Enskog expansion technique [17,18] – to show that the diffusion equation contains correction terms of higher order in ∇^2, that become important when density gradients are large. When gradients are small on the lattice scale, or the scale of the mean free path, Fick's law is generally valid to an excellent approximation.

2.1 Simulations and Results

The simulations focus on two main questions: (i) Is the time-dependent mass transport across the discrete-continuous interface continuous and smooth? (ii) In what sense does the continuum domain represent a continued thermodynamic bath for the discrete system of random walkers?

If not stated otherwise, $W = 20$ lattice sites were taken to correspond to one continuum node, and between five and 20 random walkers were employed per lattice site. The
continuum description was evolved by means of a Cranck–Nicholson finite difference
scheme, using the diffusion constant $D = 0.5$ lattice constant^2/time step. D is then
equal to the diffusivity of a single random walker.

2.2 Transport Properties and Continuity
of the Discrete-Continuous Interface

Is the coupling mechanism able to propagate the diffusive current continuously across
the discrete-continuous interface? Figure 2a shows results from simulations intended
to test the transport properties of the hybrid model. A discrete domain containing 210
sites was patched together with a continuum extending on 33 nodes, counting also the
leftmost sites that are not updated. The continuum time step was set to $\tau = 50$ (in units
of microscopic time steps), and each run was over 30 000 particle time steps. A Gaussian
density profile was imposed initially and left to relax. Averaging over an ensemble of 100
independent runs, the density profile runs continuously across the discrete-continuous
interface, and the microscopic fluctuations in the discrete domain are barely visible. Note
that the microscopic densities of every W sites were averaged to yield one data point,
corresponding to one node position.

Figure 2b shows the same data as Fig. 2a but only the difference $\Delta\rho(x/W) =
\rho(x/W) - \rho(32 - x/W)$. Each run conserved the total mass to within 0.5% . The
main particle–continuum discontinuity is present in the initial configuration. The particle
system was initialized with steps of width W and constant density.

For comparison with the case where mass flows out of, and not into, the particle
system Fig. 3 shows density profiles measured during a single run, using the same

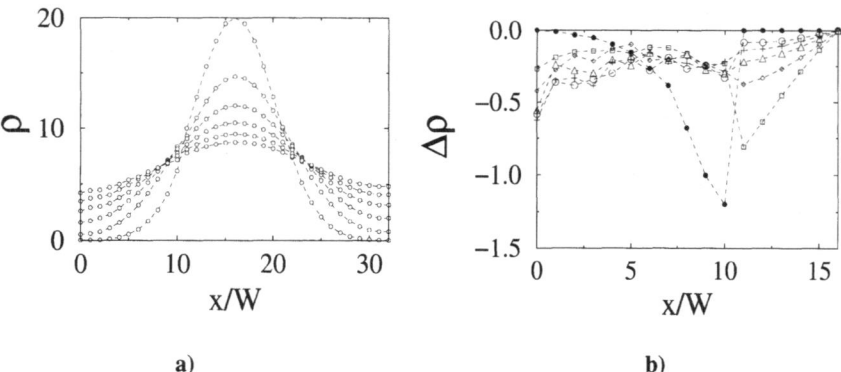

a)   b)

Fig. 2. **a)** A relaxing Gaussian density profile in a one-dimensional box of size $32W$, shown at
six different times $t = 0, 6, 12, 18, 24$, and 30×10^3 particle time steps. The particle domain
covers the leftmost $x/W \leq 10$ positions. The data were averaged over 100 independent runs
and the particle density data were averaged over a space of W sites at each node position. **b)**
The difference $\Delta\rho(x/W) = \rho(x/W) - \rho(32 - x/W)$ as a function of x, where $\rho(x)$ is shown
in Fig. 2a. The sequence of symbols $\bullet$, $\square$, $\diamond$, $\triangle$, $+$ and $\circ$ corresponds to the time sequence
$t = 0, 6, 12, 18, 24,$ and 30×10^3 particle time steps

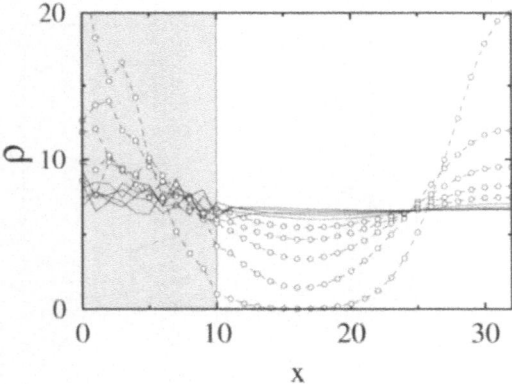

Fig. 3. Two relaxing Gaussian density profiles in a one-dimensional box, shown at various stages. The same system and plot scale as in Fig. 2a was used, and the discrete domain is indicated by the shaded area. The microscopic density data were averaged over a space of W sites at each node position

combined discrete-continuous system. Two Gaussian profiles, centered at the extreme ends of the domains, were imposed initially and left to relax. Here the microscopic fluctuations are clearly visible, in sharp contrast to the smooth profile obtained in the continuous domain.

In the simulations illustrated in Figs. 2a and 3, the initial state was symmetric around the center of the hybrid system, and the extreme end boundary conditions were reflective. Hence density profiles, which initially were symmetric in space, should preserve symmetry. The coupling scheme may be tested by directly comparing the left- and right-hand portions of the graph. These were indeed found to evolve symmetrically, up to the effect of microscopic fluctuations. As an independent check, it was verified that the microscopic density profile of random walkers alone indeed evolved with a diffusivity of $D = 0.5$.

In another independent test of the coupling scheme in the more realistic case of source terms present, random walkers were injected at the left hand side of the combined discrete-continuous system, starting from $\rho_p = \rho_c = 0$ everywhere. Figure 4 shows resulting density profiles, obtained by injecting one walker every 20 microscopic time steps and averaging the profiles from ten independent runs.

In this, as in the former cases, analytic results are readily available. Let us denote the source by s, which is the number of injected particles per time step at the left-most node position (s may be less than 1). Equation (1) then takes the form

$$\frac{\partial \rho}{\partial t} = D\nabla^2\rho + 2s(t)\delta(x),\tag{4}$$

where δ is the Dirac δ-function and the factor 2 is due to the reflecting boundary condition at $x = 0$ (in the open-space scenario described by the diffusion equation, half the particles escape to the left, in contrast to the simulated scenario). Using the Green's function of

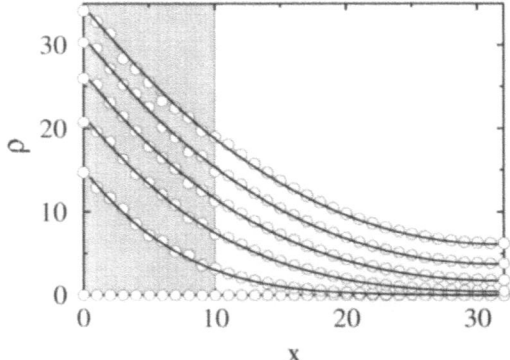

Fig. 4. Increasing density profiles in a one-dimensional box. The same system and plot scale as in Fig. 2a was used, and the discrete domain is indicated by the shaded area. Random walkers were slowly fed from the left, starting with zero density everywhere. Circles indicate the results of the hybrid simulation, averaged over ten independent runs, and solid lines show the theoretical prediction given in (5). The microscopic density data were averaged over a space of W sites at each node position

the diffusion equation, the solution to (4) is easily written down as

$$\rho(x,t) = \int_0^t dt' \; \frac{1}{\sqrt{4\pi D t'}} \exp\left(\frac{-x^2}{4Dt'}\right) 2s(t') \; . \tag{5}$$

This expression was evaluated numerically to give the solid lines in Fig. 4. We note that in this figure the diffusion equation is simultaneously solved in three different ways, and the agreement between the results is seen to be good.

Switching off the source, it was found that the density profile relaxed to a constant value everywhere in the hybrid system. In the final state the fluctuations in the discrete domain were seen to propagate into the continuous domain where they damped out as they progressed right-wards.

2.3 Equilibrium Fluctuations

In what sense does the continuum define a thermodynamic reservoir for the discrete system of random walkers? Ideally, the continuum should behave as an extended particle system. But, in contrast to an extended particle system, the continuum density does not fluctuate on the microscopic time and length scales. While it would be possible to add fluctuations to the continuum along the lines of *fluctuating hydrodynamics* [16], in the following we will study the coupling to the fluctuationless continuum.

The coupling scheme is characterized by two parameters, the ratio W macroscopic to microscopic length scales, and the ratio τ of macroscopic to microscopic time scales. We shall examine how the particle number fluctuations depend on W and τ. The particle flux density j_p that is imposed on the continuum in the $P \to C$ region is given by the sum of the particle currents J_i that characterize each of the W sites in the underlying microscopic lattice,

$$j_{\mathrm{p}} = \frac{1}{W} \sum_{i=1}^{W} \overline{J_i} = \frac{\overline{J}}{W} \tag{6}$$

where the line denotes time-averaging over τ microscopic time steps. The instantaneous net current across the $P \to C$ region is $J = R - L$ where R and L are, respectively, the total number of right- and left moving random walkers in the region, and J_i is the corresponding quantity on particle site i.

Since the particle number fluctuations are a result of the fluctuations in j_{p}, we need to compute $\overline{\langle j_{\mathrm{p}}^2 \rangle}$, where the brackets $\langle \ldots \rangle$ denote an ensemble average. First we evaluate $\langle J^2 \rangle$ in the equilibrium state when $\langle J \rangle = 0$. Denoting the particle number in the $P \to C$ region by $N_{\mathrm{W}} = R + L$ and making use of $\langle R \rangle = \langle N_{\mathrm{W}} \rangle / 2$, we can compute $\langle J^2 \rangle$ as

$$\langle J^2 \rangle = \langle R^2 - 2RL + L^2 \rangle = \langle 4R^2 - N_{\mathrm{W}}^2 \rangle = \sum_{N_{\mathrm{W}}=0}^{\infty} Q(N_{\mathrm{W}}) \sum_{R=0}^{N_{\mathrm{W}}} P_N(R)(4R^2 - N^2)$$

where the probability of finding N_{W} random walkers in the $P \to C$ region is given by the Poisson distribution [19]

$$Q(N_{\mathrm{W}}) = \frac{\mathrm{e}^{-\langle N_{\mathrm{W}} \rangle} \langle N_{\mathrm{W}} \rangle^{N_{\mathrm{W}}}}{N_{\mathrm{W}}!} \tag{7}$$

and the probability that R of these N_{W} random walkers are moving right is given by the distribution

$$P_{N_{\mathrm{W}}}(R) = \frac{1}{2^{N_{\mathrm{W}}}} \binom{N_{\mathrm{W}}}{R}$$

where

$$\binom{N_{\mathrm{W}}}{R} = \frac{R!}{N_{\mathrm{W}}!(N_{\mathrm{W}} - R)!}$$

is the binomial coefficient. We may then compute the current fluctuation

$$\langle J^2 \rangle = \sum_{N_{\mathrm{W}}=0}^{\infty} Q(N_{\mathrm{W}}) \sum_{R=0}^{N_{\mathrm{W}}} P_{N_{\mathrm{W}}}(R)(4R^2 - N_{\mathrm{W}}^2) = \sum_{N_{\mathrm{W}}=0}^{\infty} Q(N_{\mathrm{W}})N_{\mathrm{W}} = \langle N_{\mathrm{W}} \rangle . \tag{8}$$

It may be noted from the last equality that the same result for $\langle J^2 \rangle$ would have been obtained if we had assumed a fixed particle number $\langle N_{\mathrm{W}} \rangle$ and computed the fluctuations due to the flipping between right and left moving particles only. In other words, the microcanonical and grand canonical ensembles produce the same current fluctuations. Now, in order to obtain the fluctuations in the averaged current according to (6), we need to carry out the time average as well. But this is an easy task since there is no difference between the independent events that occur over time and those that occur over space. If in (8) the average is also taken over a time span τ, we only need to replace the number

N_W of particles by $N_W \tau$. Since a time average of the current J implies the division by a factor τ, the total result of the averaging is with (6)

$$\langle j_p^2 \rangle = \frac{\langle N_W \rangle}{\tau W^2} = \frac{\rho_p}{\tau W} . \tag{9}$$

We now assume that the continuum is transmitting the fluctuations imposed on its boundary region without damping or distortion to the $C \to P$ region, where the boundary flux of microscopic random walkers is imposed. Indeed, the diffusive current that enters the continuous domain at $P \to C$ must create a corresponding flux further to the right. However, the coarse nature of the continuum description is likely to cause incorrect estimates of this current since the imposed fluctuations cause relatively large density variations among adjacent continuum nodes; this in turn leads to poor approximations for the flux derivatives. This discretization effect is viewed as the main source of discrepancies in what follows. The ensemble-averaged fluctuations of the number of random walkers, in response to the imposed boundary conditions, are given by the particle densities as

$$\langle \delta N_W{}^2 \rangle = \int_V dx dx' \langle \delta\rho_p(x,t)\delta\rho_p(x',t)\rangle. \tag{10}$$

The integrals run over the discrete domain of volume V, and $\delta\rho_p$ denotes the continuum-induced deviation of the particle density (ρ_p) from its mean value. To study the effect of fluctuations in the $C \to P$ region, the current density j_p imposed by the fluctuating continuum is included in the coarse-grained description of the discrete domain,

$$\partial_t \rho_p(x,t) = \nabla \cdot (D\nabla\rho_p(x,t) + j_c(x,t)) . \tag{11}$$

Here ρ_p is the number of particles per particle site, and j_c is the imposed current which is assumed to obey the same statistics as the averaged microscopic current j_p. We need, however, its full correlations $\langle j_c(x,t)j_c(x',0)\rangle$ now. Clearly there are no equilibrium space correlations. Time correlations are caused by the numerical scheme, since j_c does not change for τ microscopic time steps. As a result of this invariance and of (9), we obtain

$$\langle j(x,t)j(x',0)\rangle = \frac{\rho}{\tau}\sum_{i=1}^{\tau} \delta(t-t_i)\delta(x-x'), \tag{12}$$

where t_i runs through integers from 1 to τ. Equations (11) and (12) are fully analogous to corresponding equations of the theory of fluctuating hydrodynamics. From a theoretical viewpoint it is interesting to note that it is possible to arrive at (12) by considering the entropy production associated with the entropy $S = -\int dx\rho(x)\log\rho(x)$ to identify the thermodynamic fluxes and forces. From that result a fluctuation-dissipation theorem that coincides with (12) may be derived.

Equation (11) may be solved by the introduction of the Fourier transforms

$$j_k(t) = \frac{1}{V}\int dx\, j(x,t)e^{-ikx} \quad \text{and} \quad j(x,t) = \sum_k j_k(t)e^{ikx}, \tag{13}$$

where $\int dx\, e^{i(k-k')x} = V\delta_{kk'}$ and V denotes the volume of the discrete domain.

Upon the application of this transform (11) takes the form

$$\partial_t \rho_k = -k^2 D\rho_k + ikj_k,$$

which is easily solved to give

$$\rho_k(t) = ik \int_{-\infty}^{t} dt'\, j_k(t') e^{-k^2 D(t-t')} . \tag{14}$$

The Fourier transform of (12) gives

$$\langle j_k(t) j_{k'}^*(0) \rangle = \frac{\rho}{WV} \delta_{kk'} \frac{1}{\tau} \sum_{i=1}^{\tau} \delta(t - t_i) . \tag{15}$$

Equation (10) may now be solved by the combination of (13), (14), and (15). This gives the somewhat involved expression

$$\langle \delta N^2 \rangle = \int dx\, dx' \sum_{kk'} \int_{-\infty}^{t} dt' \int_{-\infty}^{t} dt''\, k^2 \langle j_k(t) j_{k'}^*(0) \rangle \exp(-k^2 D(2t - t' - t'') + i(kx - k'x')).$$

Using (15) we obtain

$$\langle \delta N^2 \rangle = \frac{\rho}{WV} \int dx\, dx' \sum_k \frac{1}{\tau} \sum_{i=1}^{\tau} \int_{-\infty}^{t} dt'\, k^2 \exp\left(-2k^2 D(t - t' + t_i) + ik(x - x')\right)$$

which upon time-integration becomes

$$\langle \delta N^2 \rangle = \frac{\rho}{2DWV} \int dx\, dx' \sum_k \frac{1}{\tau} \sum_{i=1}^{\tau} \exp\left(-2k^2 D t_i + ik(x - x')\right) .$$

The integrals over x and x' are easily carried out. They give

$$\langle \delta N^2 \rangle = \frac{\rho_p V}{2DW} \sum_k \delta_{k0} \frac{1}{\tau} \sum_{i=1}^{\tau} e^{-2k^2 D t_i} = \frac{N}{W}, \tag{16}$$

where in the last step we used $D = 1/2$, $N = \rho_p V$ and $(1/\tau)\sum_{i=1}^{\tau} 1 = 1$. Note that since in the end only the $k = 0$ contribution was projected out by the x-integration, $\langle \delta N^2 \rangle$ does not depend on the averaging time τ. Physically this is because the reduction in the current fluctuations due to time-averaging is exactly balanced by the increased correlation time of the fluctuating current.

The above formalism could easily be extended to deal with space-time correlations of the density. However, these correlations will depend strongly on the underlying conservation laws in the system – in the present case, mass conservation only. For that reason these will be of less interest for the comparison with hydrodynamic systems [8], in which also momentum is conserved. The result of (16) is expected to be robust under changes of the dynamical rules. It is of interest to note that in the $W \rightarrow 1$ limit (16) reduces to $\langle \delta N^2 \rangle = N$ which is the fluctuation of a system in touch with a real particle

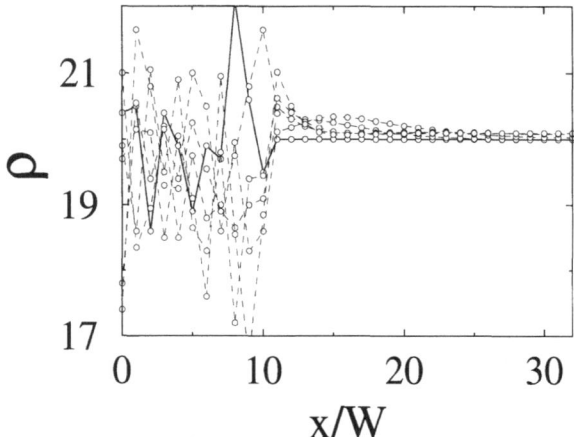

Fig. 5. The density profiles in the left half of the system measured at the different times $t = 0, 2, 4, 6, 8,$ and 10×10^4 particle time steps. Here $W = 20 \; \tau = 10$ and the solid line shows the initial state

reservoir: It is the statistical mechanical result that follows from the Poisson distribution of (7).

In order to compare predictions and measurements, a sequence of equilibrium simulations with a flat initial density profile was carried out for different ratios W of macroscopic to microscopic length scales. The hybrid system parameters were the same as those used in the previous section, except that the particle system only occupied $1/8$ of an entire system of width $65 \, W$. Conservation of total mass M_{tot} only holds to within $1 \, \%$ in these simulations. The measurements of the fluctuations were corrected for this drift by measuring the deviations in particle number from the instantaneous, rather than initial, value of $M_{\text{tot}}/8$. However, as is noted below, the drift still seems to have an effect.

In Fig. 5 the density is shown at 5 consecutive stages. As before, ρ is obtained from the particle data at $x/W \leq 10$. Note how the fluctuations are damped in the continuum part of the system. Care was taken so that the fluctuations did not significantly affect the right edge of the continuum $x/W = 64$, thus creating unwanted finite-size effects.

Figure 6 shows the fluctuations for different W-values on a log-log scale. The fluctuations were ensemble averaged over 100 independent runs using $\tau = 10$. The prediction (16) gives $\log(\langle \delta N^2 \rangle / N) = - \log W$. The gray dashed line shows (16) with a correction term added to it, i.e.,

$$\langle \delta N^2 \rangle = \frac{N}{W} + \alpha N^2 \, , \qquad (17)$$

with $\alpha = 10^{-5}$. While the main trend of the data in Fig. 6 is to confirm the theoretical prediction, the discrepancies between the measurements of Fig. 6 and the theory of (16) occur at small and large W, and to a lesser extent at intermediate values.

The theory assumes that the particle and continuum fluxes coincide for all wavelengths and frequencies. This assumption is expected to work better when W is large

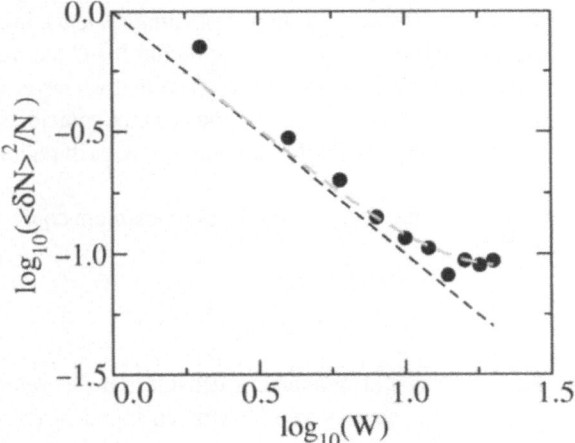

Fig. 6. The particle number fluctuation as a function of the number W of particle sites per continuum node. The black dashed line shows the theoretical result of (16) and the gray dashed line the result of (17)

and some of the rapid, short wavelength behavior of the particle system is averaged away. This may explain the small W departure between theory and measurement. At $W = 1$ the noise level of the continuum boundary condition made the Cranck–Nicholson solver unstable. Hence $W = 2$ is the smallest value shown.

Only the integer part of the flux $D\nabla\rho$ is imposed on the particle system. This error source is likely to be visible at all W, and most so when W is large. For the largest W it also appears that the small drift in particle number has an effect. Assuming that the drift in N is proportional to N we get an N^2 contribution to δN^2 which is given in (17). This equation seems to give a good fit to the large W data, indicating that the crossover behavior at $W = 14$ is due to the imperfect mass conservation.

3   A Hybrid Model for the Navier–Stokes Equation

The present model couples the Navier–Stokes equation to a Lennard–Jones molecular dynamics description. Both the continuum and particle descriptions represent the same underlying physical system, just as in the diffusive hybrid model. For this reason we need the transport coefficients to be the same in the two descriptions. Li et al. [20] studied the general problem of obtaining boundary conditions from a particle ensemble. Garcia et al. [21] have coupled multigrid continuum equations to a Direct Simulation Monte Carlo particle simulator. This sophisticated approach includes the exchange of mass, momentum and energy but is limited to dilute systems.

The hybrid model presented here is based fully on flux boundary conditions and is conservative in general. This is demonstrated to good a accuracy in two-dimensional flow simulations in which the particle system cannot easily be described in terms of a single Newtonian viscosity. Figure 1 illustrates how the discrete and continuous phases couple. While both phases are thought to represent the same underlying physical system, they may simultaneously be considered as two interacting systems. In region P←C the continuum fluxes are imposed on the particles; in region P→C the particle fluxes are measured and imposed on the continuum. This ensures that whatever flows out of one representation, flows into the other. The two regions cannot coincide since this would impose a hierarchy given by the order of flux exchanges; in each region the interaction is one-way.

The continuum equations that describe mass and momentum conservation are given in the general form [16]

$$\frac{\partial \rho}{\partial t} + \nabla \cdot (\rho \mathbf{u}) = 0 \text{ and } \frac{\partial \rho \mathbf{u}}{\partial t} + \nabla \cdot \Pi = 0, \tag{18}$$

where ρ and $\rho \mathbf{u}$ are the mass and momentum densities, $\mathbf{u}$ the velocity, and Π the momentum flux tensor. For a compressible Newtonian liquid in two dimensions we have

$$\Pi = \rho \mathbf{u}\mathbf{u} + P - \mu \left(\nabla \mathbf{u} + (\nabla \mathbf{u})^T - \nabla \cdot \mathbf{u}\right) - \lambda \nabla \cdot \mathbf{u}, \tag{19}$$

where μ and λ are the dynamic and bulk viscosities respectively, P is the pressure, and $(\nabla \mathbf{u})^T$ denotes the transpose of the $\nabla \mathbf{u}$ tensor. For simplicity we take both the continuum and particles systems to be at a constant temperature T. In order for the above description to hold for the particle system, the value of μ as well as the equation of state $P = P(\rho, T)$ must be measured in separate particle simulations for the relevant range of ρ-values.

Imposing boundary conditions on the continuum is straightforward. In the continuum region the derivatives of the flux densities, $\nabla \cdot \rho \mathbf{u}$ and $\nabla \cdot \Pi$, are computed as finite differences across nodes; these differences are used to propagate the continuity equations (18) in time. Flux boundary conditions are imposed by substituting the fluxes at the boundary by the averaged particle fluxes. We measure and coarse grain the mass and momentum flux density [22] of the particles in region P→C. This results in average flux densities that replace the original continuum flux densities on the nodes in region P→C:

$$\frac{1}{V} \sum_i m\langle \mathbf{v}_i \rangle \cdot \mathbf{n}_\perp \rightarrow \rho \mathbf{u} \cdot \mathbf{n}_\perp \text{ and } \frac{1}{V} \sum_i \left(m\langle \mathbf{v}_i \mathbf{v}_i \rangle + \frac{1}{2} \sum_{j \neq i} \langle \mathbf{F}_{ij} \mathbf{r}_{ij} \rangle\right) \cdot \mathbf{n}_\perp \rightarrow \Pi \cdot \mathbf{n}_\perp$$

where i labels individual MD particles of mass m in a subregion of volume V pertaining to one node, $\mathbf{v}_i$ is their velocity, $\mathbf{F}_{ij}$ is the force acting from particle j on particle i and $\mathbf{r}_{ij}$ is their separation vector. The unit vector $\mathbf{n}_\perp$ is defined in Fig. 1 and the average $\langle .. \rangle$ is taken over the time step Δt with which the continuum equations are integrated in time. In addition, the velocity $\mathbf{u}$ on the boundary nodes is replaced by the coarse grained particle velocity, in order to compute the velocity gradients in Π on the nodes next to region P→C in a consistent fashion.

Imposing boundary conditions on the particles implies that the few degrees of freedom described by the continuous fields must be used to set the many degrees of the particles in region P←C. For the mass flux into the particle system to equal the mass flux out of the continuum, we introduce $s(\mathbf{x}, t)$ particles per unit time in region P←C according to $ms(\mathbf{x}, t) = A\rho\mathbf{u} \cdot \mathbf{n}_\perp$, where A is the surface area (or length) corresponding to one node spacing L. Note that s may be negative, corresponding to the removal of particles from region P←C. Likewise, to obtain momentum flux continuity, we impose the averaged condition $ms(x, t)\langle\mathbf{v}'\rangle + \sum_i \mathbf{F}_i = A\,\Pi \cdot \mathbf{n}_\perp$ where $\mathbf{v}'$ is the velocity of the introduced particles, and $\mathbf{F}_i$ an external force acting on particle i in region P←C. The left hand side is thus the momentum per unit time introduced in the particle system averaged over Δt. Now, inserting the result for $s(\mathbf{x}, t)$ and comparing with (19) we observe from the last equation that momentum continuity is satisfied if the particles are introduced with the average velocity $\langle\mathbf{v}'\rangle = \mathbf{u}$ and that the overall force $\sum_i \mathbf{F}_i$ on the particles equals the stress force $A\sigma \cdot \mathbf{n}_\perp$ where $\sigma = \Pi - \rho\mathbf{u}\mathbf{u}$.

Since only the sum of the forces on the particles from the continuum is determined, it may be shown that the particles may spread arbitrarily far to the right of region P←C if $\mathbf{F}_i$ is position-independent. To fix the volume available to the particles we introduce an arbitrary weight function $g(x)$ that obeys $g(x) = g'(x) = 0$ for $x \leq 0$, and diverges as $g(x \to L/2) \sim (/2L - x)^{-1}$ on the edge of region[1] P←C (see Fig. 1). The coordinate x runs parallel to $\mathbf{n}_\perp$, and $x = 0$ in the middle of the $P \to C$ region. The precise form of the weight function is not important. The stress force acting on the i-th particle is $\mathbf{F}_i = (g(\mathbf{x}_i)/\sum_i g(\mathbf{x}_i))A\sigma \cdot \mathbf{n}_\perp$, where the sum includes all N_A particles in a given section A of region P←C. By construction, the fraction $\mathbf{F}_i$ of the total stress force imposed by the continuum increases as the ith particle approaches the edge of region P←C, driving the particle back into the bulk. This solves the potential stacking problem posed by the insertion of particles as these may always be placed sufficiently deep into region P←C, and yet avoid the vicinity of other particles.

In this work we disregard energy flux exchange and limit ourselves to an isothermal continuum at temperature T. Future applications will include energy exchange on the same footing as mass and momentum exchange. Each particle must then have an average kinetic energy $k_B T$, where k_B is Boltzmann's constant. New particles are inserted with a velocity vector $\mathbf{v}'$ picked randomly from the Maxwellian distribution $P(\mathbf{v}') \propto \exp[m(\mathbf{v}' - \mathbf{u})^2/(2k_B T)]$.

The imposed continuum stress forces will perform work on the particles and thus change their thermal energy. To compensate the stress work, it is necessary to thermalize the particles in region P←C. This is done by adding a Langevin force $\mathbf{F}_{Li} = -\alpha(\mathbf{v}_i - \mathbf{u}) + \tilde{\mathbf{F}}$ to $\mathbf{F}_i$. Here $\mathbf{v}$ is the particle velocity, α is a friction coefficient and $\tilde{\mathbf{F}}$ is a fluctuating force of zero mean and a Gaussian distribution with the correlation function $\langle\tilde{\mathbf{F}}(t)\tilde{\mathbf{F}}(t')\rangle = 2k_B T\alpha\delta(t - t')$. Further, to correct for fluctuations the term $-\left(\sum_{i\in(P\leftarrow C)} \mathbf{F}_{Li}/\sum_{i\in(P\leftarrow C)} 1\right)$ is added to all $\mathbf{F}_{Li}$. In this way an independent net momentum input due to the Langevin forces is prevented.

There is an inherent asymmetry in the coupling scheme as only the particle system supplies fluctuations, while the continuum is intrinsically non-fluctuating. As a consequence, the continuum does not fully play the role of a thermodynamic bath. For instance, fluctuations in the total particle number will be smaller than those predicted by statistical mechanics, and they will in general decrease as L is increased.

---

[1] In the simulations reported here, we used $g(x) = 2[(L - 2x)^{-1} - L^{-1} - 2xL^{-2}]$ for $0 \leq x \leq L/2$.

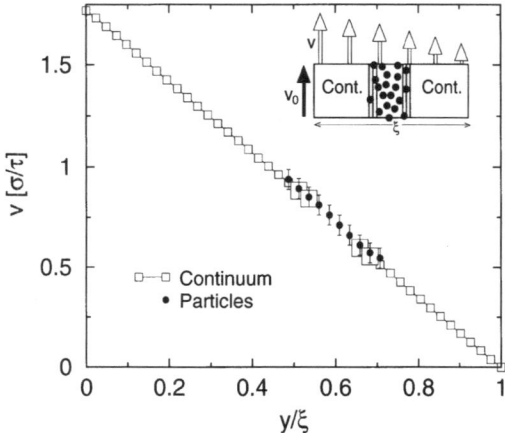

Fig. 7. Plot of the steady-state velocity profile $v(y)$ as a function of distance y/ξ, where ξ is the width of the channel in the Couette flow setting. Squares and and circles indicate continuum and particle averages, respectively. Error bars and square sizes indicate the standard deviation of the mean

The present coupling scheme was implemented and tested for elementary flow scenarios. The interaction forces governing the molecular dynamics were derived from the shifted Lennard-Jones potential [22] $V_{\mathrm{LJ}} = 4\varepsilon \left[(\sigma/r)^{12} - (\sigma/r)^6 \right] - V_{\mathrm{c}} - B(r - r_{\mathrm{c}})$, where ε is the characteristic interaction energy and σ the characteristic interaction distance, and the constants V_{c} and B are chosen so as to ensure vanishing forces at the cut-off distance $r_{\mathrm{c}} = 2.5\sigma$. The Newtonian equations of particle motion were integrated using the velocity Verlet algorithm [22] with a time step $\Delta t_{\mathrm{MD}} = 0.0017\tau$, where $\tau = (m\sigma^2/\varepsilon)^{1/2}$ is the characteristic time of the potential. At $T = 0.7\varepsilon/k_{\mathrm{B}}$ and a density of $\rho = 0.4\sigma^{-2}$, the MD particles formed a fluid with $\mu = 0.58\epsilon\tau\sigma^{-2}$.

The continuum equations were integrated using the MacCormack predictor-corrector algorithm [23] on a grid of spacing $L = 7.6\sigma$, a time step $\Delta t = 100\Delta t_{\mathrm{MD}}$, and the measured shear viscosity. The bulk viscosity was set to $\lambda = \mu/3$. The magnitude of the time steps reflect the stability requirements of the MacCormack scheme and the steepness of the Lennard–Jones potential. The size of Δt_{MD} could be increased if a softer, non-divergent potential were used. The length L must be sufficiently large that the fluctuations in the measured particle fluxes do not cause stability problems in the continuum solver (here the MacCormack scheme). A central region in the grid was used only by the particles, and boundary sections $3L$ wide each formed two continuum-particle interfaces according to Fig. 1.

Couette shear flow parallel to the continuum-particle interfaces was imposed by fixing the velocities at the impermeable walls of the channel. Periodic boundary conditions were used in flow direction. Initially both walls were at rest. In steady state one moved with velocity $v = 1.77\sigma/\tau$. Figure 7 shows that the fluctuations in the continuum velocities decreased away from the particle region. This effect is due to the viscous damping of continuum velocity fluctuations induced by fluctuations in the particle velocities. A

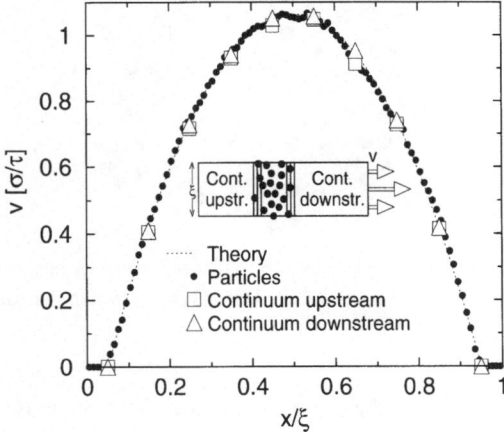

Fig. 8. Plot of velocity profile v across the flow channel in the Poiseuille flow setting, measured in the upstream and downstream continuum region and in the particle region. Also shown is the parabolic prediction

small difference in the slope in the particle and continuum regions is attributed to the discrepancy between the initially measured particle viscosity, which is used in the continuum simulations, and the actual value that results under the present flow conditions. This is the typical, rather than exceptional, situation and does not bear on the validity of the scheme. The good equality of the slopes on both sides of the particle region demonstrates that the model works as this reflects the correct transfer of shear forces. Correct treatment of mass and pressure transfer was tested in the Poiseuille flow simulations, illustrated in Fig. 8. Here the flow direction was perpendicular to the continuum-particle interfaces. In the continuum, no-slip boundary conditions were used at the channel walls, and periodic boundaries in flow direction. In the particle region, boundaries of fixed particles were used parallel to the flow. Good agreement is found between the velocity profiles obtained by the averaging of particle motion and the up- and downstream continuum solutions. Also the parabolic prediction obtained from the continuum viscosity and forcing agrees well with the measurements.

4   A Hybrid Model for the Coupling of Gas and Grains

Just as in Sects. 2 and 3 the present model is based on the exchange between the particle and continuum fields. In this case however, the exchange happens everywhere in space and it does not involve all the fluxes. The particle field only defines a density and velocity field for the continuum system. These fields are used to define a permeability field for the pressure equation. On the other hand, the pressure field defines a force field for the particles. This scheme is made possible by the fact that the inertia of the gas may be neglected when compared to the inertia of the particles. Hence, one may assume an instantaneous balance between the pressure gradient of the gas and the force acting from the gas on the particle system. Figure 9 illustrates conceptually the basis for the

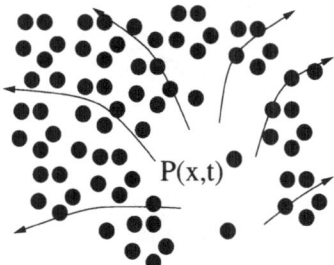

Fig. 9. A conceptual picture of the dynamics. The MD particles move according to Newton's laws while the pressure evolves according to a local Darcy law for which the particles form a porous medium

model. The gas flow, shown by the arrows, takes place between grains that define a local permeability $\kappa(\rho_s)$.

4.1 Gas Dynamics

Although the equation of motion is derived in Ref. [24] we re-derive it briefly here for completeness. We shall take the permeability to depend on the local volume fraction $\rho_s = 1 - \phi$ (ϕ is the porosity) of the solid phase according to the Carman–Kozeny relation [25]

$$\kappa(\rho_s) = \frac{a^2}{9K} \frac{(1 - \rho_s)^3}{\rho_s^2}, \tag{20}$$

where a is the (spherical) particle radius and the constant $K \simeq 5$ is obtained experimentally for a packing of spheres.

The evolution equation for the gas pressure is based on the conservation laws for the gas and grain masses. The conservation of the granular volume fraction ρ_s may be written as

$$\frac{\partial \rho_s}{\partial t} + \nabla \cdot (\rho_s \mathbf{u}) = 0, \tag{21}$$

where $\mathbf{u}$ is the granular velocity. The conservation of the mass density of the air, ρ_a, may be written as

$$\frac{\partial \rho_a}{\partial t} + \nabla \cdot \left(\rho_a \left[\mathbf{u} - \frac{\kappa(\rho_s)}{\mu} \nabla P \right] \right) = 0, \tag{22}$$

where P and μ is the gas pressure and viscosity respectively. Here the gas current has both an advective term, caused by the motion of the grains, and a diffusive term describing the Darcy flow in the local rest frame of reference for the sand. Substituting $\rho_s = 1 - \phi$ in (21) we get

$$-\frac{\partial \phi}{\partial t} + \nabla \cdot ((1 - \phi)\mathbf{u}) = 0 . \tag{23}$$

Using the isothermal equation of state for an ideal gas, $\rho_a \propto \phi P$, we can write (22) as

$$\frac{\partial(\phi P)}{\partial t} + \nabla \cdot \left(\phi P \left[\mathbf{u} - \frac{\kappa}{\mu} \nabla P \right] \right) = 0 \,. \tag{24}$$

By eliminating $\partial \phi / \partial t$ between (24) and (23) a small manipulation gives

$$\phi \left(\frac{\partial P}{\partial t} + \mathbf{u} \cdot \nabla P \right) = \nabla \cdot \left(\phi P \frac{\kappa}{\mu} \nabla P \right) - P \nabla \cdot \mathbf{u} \,. \tag{25}$$

Here, the left hand side is just the substantial derivative of the pressure. The first term on the right hand side describes the Darcy flow in the local rest frame of reference of the grains. The last term describes pressure changes due to changes in the grain density. See Ref. [26] for a more elaborate discussion of continuum equations like (25).

Finally, the numerical solution of (25) can be simplified by dividing the pressure into an average and fluctuating part: $P = P_0 + P'$. In the experiments we wish to study, the changes in pressure are only a small fraction of atmospheric pressure, so $P' \ll P_0$. Neglecting terms of order $O(P'/P_0)$ in (25) leads to

$$\phi \frac{\partial P'}{\partial t} = P_0 \nabla \cdot \left(\phi \frac{\kappa}{\mu} \nabla P' \right) - P_0 \nabla \cdot \mathbf{u} \,. \tag{26}$$

The simplifications leading to this equation are by no means crucial. In applications where it is needed, the neglected terms may well be re-inserted.

For the numerical implementation it is convenient to non-dimensionalize (26). Writing the characteristic magnitude of the permeability $\kappa_0 = a^2/45$, we may introduce the characteristic grain velocity $U_0 = (\kappa_0/\mu)\rho_g g$, where ρ_g is the mass density of the material making up the grains. Introducing a characteristic length scale l, a characteristic time scale $\tau \equiv l/U_0$ follows. The dimensional quantities may then be written in terms of non-dimensional (primed) quantities as $P = P_0 P'$, $\mathbf{u} = U_0 \mathbf{u}'$, $x = l x'$ and $t = \tau t'$. Substituting these relations in (26) we obtain

$$\phi \frac{\partial P'}{\partial t'} = \text{Pe}^{-1} \nabla' \cdot \left(\frac{\phi^4}{(1-\phi)^2} \nabla' P' \right) - \nabla' \cdot \mathbf{u}' \,,$$

where the Peclet number is defined as

$$\text{Pe} = \frac{U_0 l \mu}{P_0 \kappa_0} = \frac{\rho_g l g}{P_0} \,.$$

The Peclet number derives its name from the fact that it may be interpreted as the ratio between a diffusive and an advective (l/U_0) time scale. Note that it reduces to the ratio between the hydrostatic pressure $\rho_g l g$ caused by the grains and the background pressure P_0. The simplifications leading to (26) imply a small Peclet number. In the simulations that follow we use $\text{Pe} = 2 \times 10^{-4}$.

Equations (25) and (20) describe fluid flow where the fluid inertia may be neglected. This is generally possible when the Reynolds number [16] is small or the problem at hand is such that inertia effects are not important. The Reynolds number is small when the particles are small. With air as the fluid and gkass spheres as particles, a freely falling particle will acquire a Reynolds number of the order 1 when the particle diameter is around 0.1 mm. This implies that particles must be small. However, when particles become larger the description does not break down in a dramatic way as the first corrections in the Reynolds number amounts to small corrections in the pressure forces on the particles. As fluid inertia becomes increasingly important, however, it affects not only the fluid–particle coupling but also the fluid dynamics itself. In these cases an equation describing the flow of fluid momentum, like the Euler equation [27] is needed.

4.2   Particle Dynamics

The particles evolve according to Newton's second law

$$m\frac{d\mathbf{v}}{dt} = m\mathbf{g} + \mathbf{F}_I - \frac{\nabla P}{\rho},\tag{27}$$

where ρ is the number density of particles, $\mathbf{g}$ is the gravity, m the particle mass, $\mathbf{F}_I$ the inter-particle force, and $\rho = \rho_s\rho_g/m$ is the number density. What distinguishes the present model from conventional models of granular materials is the pressure force per particle $\nabla P/\rho$. It is the pressure gradient obtained from the continuum equation (25), distributed over the particles present in that volume.

Non-dimensionalizing (27) gives

$$\mathrm{Fr}\frac{d\mathbf{v}'}{dt'} = -\hat{\mathbf{z}} + \frac{\mathbf{F}_I}{mg} - \mathrm{Pe}^{-1}\frac{\nabla'P'}{\rho_s},$$

where Fr has the form of a Froude number

$$\mathrm{Fr} \equiv \frac{U_0^2}{gl} = \frac{U_0/g}{\tau}.\tag{28}$$

The Froude number can be considered as the ratio between two time scales: U_0/g, the time it takes for a falling particle to accelerate from rest to U_0 and $\tau = l/U_0$, the time for a particle falling at speed U_0 to travel a distance l.

In this paper, we use a version of the "TC model" [28], which is an event-driven algorithm to solve (27). Soft sphere molecular dynamics [29] and contact dynamics [30] could be used instead. In the event driven method, the output velocities after a collision are computed directly in terms of the input velocities. If $\mathbf{v}$ is the relative velocity between two particles, and $\hat{\mathbf{n}}$ is a normal vector pointing along the line of centers, the normal component of the output velocity $\mathbf{v}'$ is $\mathbf{v}' \cdot \hat{\mathbf{n}} = -r\mathbf{v} \cdot \hat{\mathbf{n}}$ where $r \leq 1$ is the restitution coefficient. When $r < 1$ the collisions are dissipative, and setting $r = 1$ conserves energy. The velocities perpendicular to $\hat{\mathbf{n}}$ are left unchanged: The particles are perfectly smooth, so we can ignore their rotation. Collisions between a particle and the walls can be considered in the same way, except that the wall has infinite mass.

The algorithm for computing the grain trajectories is outlined below:

1. Advance all particles by the time step Δt, assuming the particles do not interact. Since Δt must be chosen so that the particles move only a small fraction of their

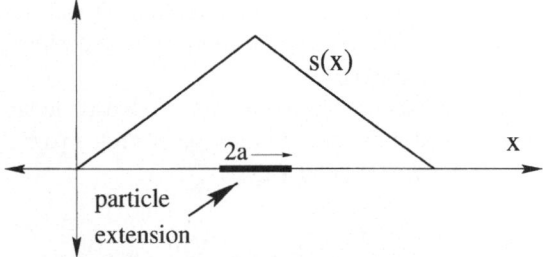

Fig. 10. The particle halo function

diameter each time step, the pressure force can be taken as a constant during the short time Δt. Therefore, the particles are advanced along parabolic line segments, corresponding to free flight in a constant force field.

2. Compile a list of all overlapping particles.
3. Scan through the list of overlapping particles. If any pair has relative velocities such that the two particles are approaching each other, implement a collision between these two particles. There are two types of collisions: Energy conserving and dissipative. If a particle has already suffered a collision in the current or preceding time step, then all collisions involving that particle are energy conserving. Otherwise, they are dissipative. This rule is necessary to avoid inelastic collapse (an infinite number of collisions in finite time) [28]. Repeat this step until all pairs of particles are separating. Go to step 1.

We now need to define ρ and $\mathbf{u}$ in terms of the particle positions and velocities $\mathbf{v}_i$ (i labels individual particles). In order to obtain a continuous density field we will distribute the particle mass in a halo which goes continuously to zero around the particle. This is a standard procedure and is done by introducing the halo function

$$s(\mathbf{x} - \mathbf{x}_0) = \begin{cases} \left(1 - \frac{|x-x_0|}{l}\right)\left(1 - \frac{|y-y_0|}{l}\right), & |x - x_0|, |y - y_0| < l \\ 0 & \text{otherwise,} \end{cases} \tag{29}$$

where $\mathbf{x}$ is the particle position and $\mathbf{x}_0$ is the position of a site in the lattice on which P is computed. This function is shown in Fig. 10. The lattice constant is l, and s has the property that it distributes a particle mass over the four nearest lattice sites, i.e., over a region bigger than the particle interaction radius. This follows from the observation that

$$\sum_k s(\mathbf{x} - \mathbf{x}_k) = 1,$$

where k labels the nearest lattice sites ($k = 1, .., 4$ on a 2d square lattice). The on-site mass density and velocity $\mathbf{u}$ are now easily defined as

$$\rho(\mathbf{x}_0) \equiv \sum_{i=1}^{N} s(\mathbf{x}_i - \mathbf{x}_0) \tag{30}$$

$$\mathbf{j} \equiv \rho\mathbf{u}(\mathbf{x}_0) \equiv \sum_{i=1}^{N} s(\mathbf{x}_i - \mathbf{x}_0)\mathbf{v}_i, \tag{31}$$

where N is the particle number and $\mathbf{v}_i$ the particle velocity. The advantage of these definitions is that ρ and $\mathbf{u}$ now vary smoothly as functions particle positions. The divergence $\nabla \cdot \mathbf{u}$ is evaluated as a finite difference.

Just as the halo function may be used to obtain smooth particle input to (25), it may be used the other way, to distribute the pressure forces on the particles. The pressure gradient term in (27) is evaluated at point $\mathbf{x}$ as

$$\frac{\nabla P}{\rho} = \sum_k s(\mathbf{x} - \mathbf{x}_k)(\nabla p)_k / \rho_k, \tag{32}$$

where the sum is taken over all grid points, $(\nabla p)_k$ is the pressure gradient at grid point k, and ρ_k is the density. Since the halo function is nonzero only at the four nearest grid points, there will be only four terms in the sum. Recall that ∇p is the rate at which momentum is being transferred from the fluid to the particles. Equation (32) simply says that a particle's share of the momentum deposited at a certain grid point is proportional to that particle's contribution to the density at that same grid point.

4.3 Implementation

For the model to work in practice it is necessary to introduce a cutoff $\rho_{\min}$ on the density. This has both physical and numerical reasons. Physically there is no sense in defining a permeability field if the particle density is too low. The Carman–Kozeny permeability gives a reasonable prediction only when $\rho_s > 0.25$ [31]. Numerically the pressure computations will encounter problems in the form of instabilities both when the permeability becomes too high and when the source term becomes too erratic, as will happen when $\rho_s \to 0$. Therefore we shall take $\rho_s = \rho_{\min}$ wherever the measured density in (14) is less than $\rho_{\min}$. This introduces a cutoff on the permeability, $\kappa < \kappa(\rho_{\min})$. Likewise, when the pressure force on the dilute particles are computed we shall use $\rho_{\min}$ in place of the actual density when it is too small. This implies that the pressure feels a permeability corresponding to a higher than actual particle density. Correspondingly, the particles are subjected to the force $\nabla P/\rho = \nabla P/\rho_{\min}$, when $\rho_s < \rho_{\min}$. This means that the particles in a cell of volume ΔV corresponding to a lattice site will not absorb the entire force $\nabla P \Delta V$ when $\rho_s < \rho_{\min}$. However, due to the overestimate made by $\kappa(\rho_s)$ in dilute regions, the force per particle will still be larger than the single particle Stokes drag [16]. This means that the error made by introducing the cut-off is mainly that dilute particles fall somewhat more slowly than they should.

Although the practical implementation of the present model in three dimensions is not significantly harder than in two dimensions we wish to simulate a two-dimensional system because it is numerically less expensive. However, the Carman–Kozeny equation (20) is a three dimensional relation as it gives the permeability in terms of the volume fraction of spheres ρ_s, and at the end we wish to compare our results to real three dimensional experiments. Consequently we need to transform the area fraction of grains in the simulations $\rho_s^{(2D)}$ to the volume fraction ρ_s in such a way that the close packed value of $\rho_s^{(2D)}$ corresponds to the close packed value of ρ_s. This is approximately achieved by the transformation $\rho_s = (2/3)\rho_s^{(2D)}$, which we use in the following.

In the simulations we use a distribution of particle sizes to avoid the two-dimensional hexagonal ordering. To improve the relation between the two- and three-dimensional packing densities the close packed value of $\rho_s^{(2D)}$ will eventually be measured and compared to the three dimensional random close packed values. For the present validation process however, this is not needed and we use the $2/3$ factor.

4.4   Sedimentation

The simplest application of our model, and one that corresponds to a common experimental measurement, is the determination of sedimentation velocities. A volume is initially filled with a uniform mixture of gas and particles. Then the particles are allowed to settle. Experimentally, one observes a sharp boundary between the clear fluid at the top of the container and a particle phase below. This front moves with a well defined velocity, called the sedimentation velocity, which depends on the initial density of the particles. This experiment is easy to simulate. In Fig. 11, we show a typical snapshot

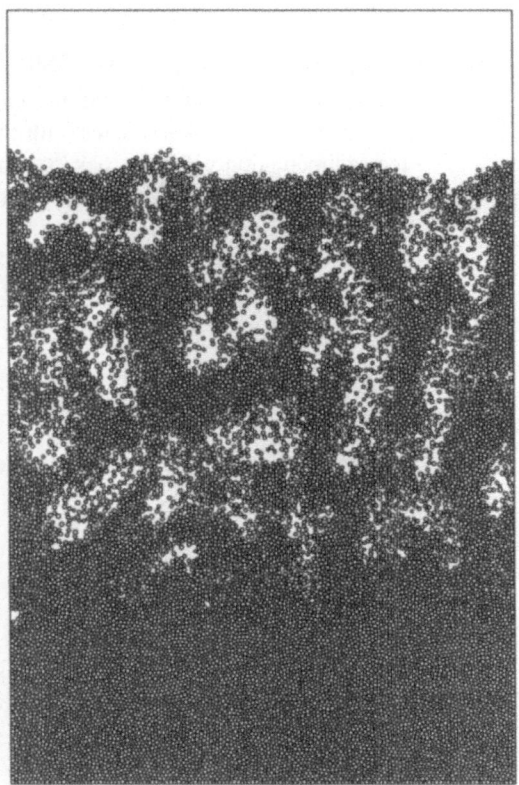

Fig. 11. Simulation of a sedimentation experiment. There are 18104 particles, and the size of the domain is $62l \times 92l$. The values of the dimensionless numbers are $Pe = 0.0002$ and $Fr = 1$. This picture shows a snapshot at $t = 30\tau$. The walls are perfectly smooth

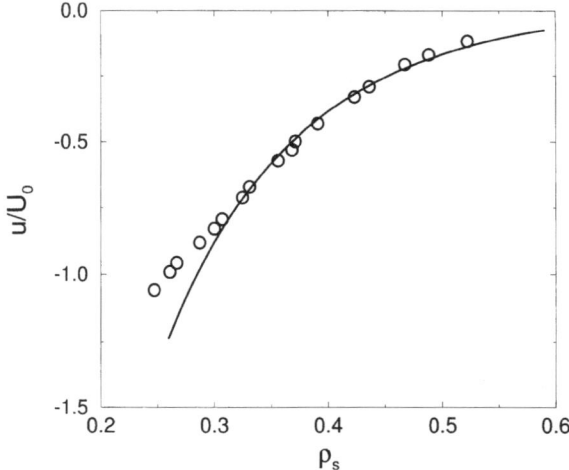

Fig. 12. The sedimentation velocity u, scaled by the typical velocity U_0, as a function of granular volume fraction ρ_s. Each data point corresponds to the results of one simulation at a different initial density. The solid line shows the theoretical prediction (33)

from a simulation: The sharp front at the top of the vessel is visible. Another feature is the variations of particle density which form below the front, not unlike the structure formation observed in Batchelors sedimentation experiments with two particles sizes [32]. In Fig. 12, we compare the sedimentation velocities observed in the simulations with theoretical values predicted by assuming that the density below the sedimentation front is uniform and the particles fall without accelerating.

Since the particle volume that is transported downward must be compensated by an equal volume upflow of fluid, it is a simple matter to derive a theoretical value for the sedimentation velocity. The conservation of volume means that

$$(1 - \phi)\mathbf{u} = -\mathbf{u} + \frac{\kappa(\phi)}{\mu}\nabla P,$$

where $\mathbf{u}$ is the local velocity of the grains. The other quantities are as defined previously. Solving this equation for $\mathbf{u}$ we obtain

$$\mathbf{u} = \frac{\kappa(\phi)}{\mu(2 - \phi)}\nabla P .$$

Now, the neglect of inertial effects means that the pressure forces must balance the grain weight, i.e., $\nabla P = \rho_s\rho_g\mathbf{g} = \rho_g(1 - \phi)\mathbf{g}$ where $\rho_s\rho_g$ is the bulk density of the grains, ρ_g is the density of the material that makes up the grains and $\mathbf{g}$ is the acceleration of gravity. Using (20) we get the sedimentation velocity

$$\mathbf{u} = U_0\frac{(1 - \rho_s)^3}{\rho_s(1 + \rho_s)} , \tag{33}$$

where as before $U_0 = a^2\rho_g\mathbf{g}/(45\mu)$.

Figure 11 shows a snapshot of a sedimentation simulation, and Fig. 12 the sedimentation velocity $v_y = u/U_0$ which is measured by averaging over a series of such simulations. Each simulation runs for 50τ, and data is recorded every $\Delta t = 10\tau$. First, average density and velocity profiles are obtained by averaging horizontally. Then, to exclude the settled material at the bottom of the simulation, points with a velocity of less than half the maximum are excluded. To exclude the data points contaminated by the clear region above the settling material, the maximum density of the remaining points is found, and points with less than $2/3$ of this density are also excluded. The remaining points are averaged to give a single (ρ, v_y) pair for each time. Then, the five points for each simulation are combined to give a single point on the graph. Each simulation has 18104 particles, a container width of $60l$, with the height adjusted to give the desired density (it ranges from $60l$ to $124l$). The particles diameters are uniformly distributed in $[0.7d_{max}, d_{max}]$, where $d_{max} = 0.5l$. The coefficient of restitution is $r = 0.8$.

The full line in Fig. 12 shows the theoretical result for homogeneous sedimentation (33). In spite of the density inhomogeneities shown in Fig. 11, the observed sedimentation values agree well with the theoretical prediction when $\rho_s > 0.3$. We attribute the discrepancy between measurement and theory for $\rho_s < 0.3$ to the presence of regions where the local density ρ falls below $\rho_{min} = 0.25$. These regions are caused by local density fluctuations and here the permeability $\kappa(\rho_{min})$ is smaller than $\kappa(\rho)$. Consequently, the measured settling velocity becomes smaller than the velocity predicted by the theory which does not include the effect of the density cut-off.

4.5 Fluidized Beds

As the next test of the model we inject a constant flux of air at the bottom while removing an equivalent flux from the top in order to produce the bubbling behavior as observed experimentally in gas fluidized beds [33–35]. The air is injected by adjusting the pressure in the bottom row of grid points according to the assumption that the air is isothermal. We specify the amount of air by the volume it would occupy if $P = P_0$. Thus a volume injection of ΔV is implemented by a pressure increase $\Delta P = -P\Delta V/V$ where V is the volume of the region where the pressure is increased. Thus, in Fig. 13 a total of $240l^2$ is injected at the bottom and the equivalent amount is removed from the top.

The formation of bubbles is a salient phenomenon in gas–grain flow, of which the model captures the main features. Figure 13 shows a time series of a fluidization simulation with an initial alternating layers of black and white particles. The particles differ by their color only. The computation was carried out with 18 104 particles and the size of the domain is $60l \times 84l$. A (two-dimensional) volume of air of $240l^2$ was injected uniformly across the bottom during the first 20τ of the simulation. The entire simulation took about 3 hours on a workstation. Figure 14 shows the last stage of the bubble formation process.

Three main features of this simulation should be noted. First the bubbles form spontaneously when the pressure difference is turned on. Second, small bubbles coalesce as they move upwards forming larger bubbles. And, finally the shape of the bubbles have the same qualitative features that are observed experimentally [35]. The two first features are also seen experimentally [33,35]. The first and last of the above features were also observed in the simulations of Kawaguchi et al. [27]. We note that due to the lack of

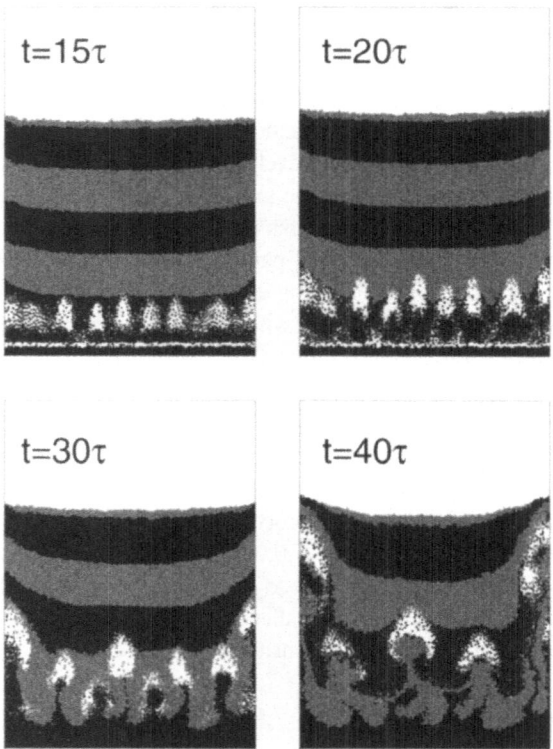

Fig. 13. A time sequence showing the simulation of fluidization. The times are given in dimensionless units. The dimensionless numbers were Pe = 0.0002 and Fr = 1

sliding friction between the particles in the present simulations, the system may start forming bubbles at a denser state than in real systems. Furthermore, the bubbles next to the walls rise abnormally quickly because there is no friction at the walls. The stationary layer of particles at the very bottom is caused by the discretization of the pressure field, and is a numerical rather than physical effect.

4.6 Experimental Verification

Finally we apply the gas–grain model to simulate an experiment that exhibits an instability. When a bubble of air is released at the bottom of a straight glass tube a train of precursor bubbles form spontaneously in front of the main bubble [15]. The tube is inclined at different angles Θ to the horizontal. In Fig. 15 the angle is set to $\Theta = 70°$.

The experimental setup consists of a glass tube of internal diameter 0.5 cm and length 100 cm filled with spherical glass particles of diameter $d = (180 \pm 15)\,\mu$m. The relative air humidity was kept within $27 \pm 3\,\%$ during the filling procedure. The initially horizontal tube has a 15 cm void space at one end. When the tube is tilted quickly to a preset angle the bubble rises and starts to form the precursor train of bubbles. Video

Fig. 14. Fully developed bubbles. The dimensionless numbers are $Pe = 0.0002$ and $Fr = 1$

recording and direct observation of the tube allow the determination of the propagation speed. As the main bubble starts to rise, a set of bubbles form sequentially away from the main bubble. The train of bubbles rapidly selects a characteristic spacing between them and while there appears to be a certain effective repulsion between bubbles, merging events do occur [36]. From Fig. 15 it is seen that the main qualitative effects are captured by the simulations. It has been demonstrated, however, that also the quantitative aspects of the experiment are captured [15]. The measured bubble velocity as a function of Θ is reproduced by the simulations, and the fact that the precursor bubbles disappear when the surrounding air pressure is lowered or the particle size is reduced, is also explained by the simulations.

5   Conclusions

We have studied three different practical applications of one simple idea, the coupling of descriptions of different intrinsic scales. Working our way from a simple toy model for

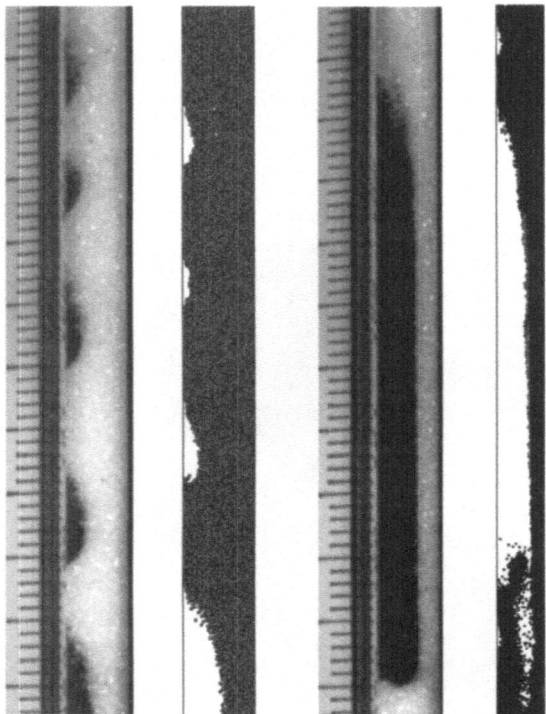

Fig. 15. The two first images show the ripple instability as it emerges from experiments and the friction-less 2d simulations, respectively. The two last images show the structure of the main bubble. The main ticks on the experimental scale are separated by 1 cm

1d diffusion to the more elaborate model of the Lennard–Jones / Navier–Stokes equation, we have tested the performance of the models in simulations. While there are clearly several ways to implement the particle–continuum coupling, we have demonstrated that flux exchange emerges as a natural requirement for the simulation of equations that express the conservation of these fluxes.

Diffusion: Here we have studied both the equilibrium and non-equilibrium behavior of a one-dimensional diffusive hybrid system. The coupling worked well after the introduction of a modified computation of the continuum density gradient in the overlap region. The non-equilibrium behavior, which is globally described by the diffusion equation, agreed well both with consistency checks and analytic predictions. In the equilibrium case we studied the particle fluctuations in order to establish the extent to which the continuous domain played the role of an established thermodynamic reservoir. It was found theoretically, and partly confirmed by simulations, that the role of the continuum approximates the role of a grand canonical reservoir in the small W limit, while it completely suppresses fluctuations in particle number in the large W limit.

Navier–Stokes–Lennard-Jones: In conclusion we have established and tested a general strategy for the consistent coupling of different physical representations of hydrodynamic systems by means of the exchange of *fluxes* of the conserved quantities. The scheme is in principle applicable to the coupling of conservative descriptions in general. It may be used to couple partial differential equations and molecular dynamics in arbitrary dimensions, as in the present case, or it could be applied to couple different particle descriptions, such as dissipative particle dynamics and molecular dynamics [37,38]. A scheme where energy fluxes are successfully exchanged in addition to the mass and momentum fluxes is under development.

Gas–Grains: In both fluidization phenomena and granular flows there is often a multitude of different scales that need to be resolved. For that reason numerical efficiency is often crucial, and the problem may only be modeled if less relevant information and details of the process is discarded. We have introduced a model where grain and gas flow couple. This has been done with a guiding principle of maximizing the conceptual and numerical simplicity. For that reason the particles move without rotation and friction, though there is a coefficient of restitution, and no detail of the fluid flow field at the sub-particle scale is kept. The fluid inertia is neglected. This allows for a very simple description as the fluid is described in terms of Darcy's law, and the particles in terms of event driven molecular dynamics. The result of the simplification made in the present model is a relatively efficient and robust computational scheme.

The model has been implemented for several test problems – sedimentation, fluidization and the ripple instability. The latter applications also showed good quantitative agreement between simulations and experiment. The simulations reproduce the experimentally observed ripple formation, qualitatively as well as quantitatively for the speed of the bubbles as a function of the tilt angle. The absence of friction in these simulations and the agreement with the experiments have stimulated theoretical investigations that shed light on the existence of the crossover effect that ripples disappear at a certain value of the surrounding pressure.

As for the other applications, the sedimentation velocity measured in the simulations agreed with the anticipated theoretical velocity, save for corrections due to the existence of a cut-off density in the permeability function. In the case of fluidized beds the simulations reproduced the key qualitative features known from experiments, i.e., spontaneous bubble formation, coalescence of bubbles and the characteristic bubble shapes.

References

1. J. Koplik, J.R. Banavar: Ann. Rev. Fluid Mech. **27**, 257 (1995)
2. M.P. Brenner, X.D. Shi, S.R. Nagel: Phys. Rev. Lett. **73**, 3391 (1994)
3. P.A. Thompson, M.O. Robbins: Science **250**, 792 (1990)
4. F.F. Abraham, J.Q. Broughton, N. Bernstein, E. Kaxiras: Comput. Phys. **12**, 538 (1998)
5. H. Rafii, L. Hua, M. Cross: J. Phys.: Condens. Matter **10**, 2375 (1998)
6. L.B. Freund: *Dynamical Fracture Mechanics* (Cambridge University Press, NY 1990)
7. J. Bougie, S.J. Moon, J.B. Swift, H.L. Swinney: Phys. Rev. E **66**, 051301 (2002)
8. E.G. Flekkøy, G. Wagner, J.G. Feder: Europhys. Lett. **52**, 271 (2000)
9. S.T. O'Connel, P.A. Thompson: Phys. Rev E **52**, 5792 (1995)
10. N.G. Hadjiconstantinou, A.T. Patera: Int. J. Mod. Phys. C **8**, 967 (1997)

11. N.G. Hadjiconstantinou: Phys. Rev. E **59**, 2475 (1999)
12. N.G. Hadjiconstantinou: J. Comp. Phys. **154**, 245 (1999)
13. E.G. Flekkøy, G. Wagner, J.G. Feder: Phys. Rev. E. **64**, 066302 (2002)
14. S. McNamara, E.G. Flekkøy, K.J. Måløy: Phys. Rev. E **61**, 4054 (2000)
15. E.G. Flekkøy, K.J. Måløy, S. McNamara: Phys. Rev. Lett. **87**, 134302 (2001)
16. L.D. Landau, E.M. Lifshitz: *Fluid Mechanics* (Pergamon Press, NY 1959)
17. D.A. McQuarrie: *Statistical Mechanics* (Harper and Row, NY 1976)
18. J.-P. Rivet: Complex Systems **1**, 838 (1987)
19. L.D. Landau, E.M. Lifshitz: *Statistical Physics* (Pergamon Press, NY 1959)
20. J. Li, D. Liao, S. Yip: Phys. Rev. E **57**, 7259 (1999)
21. A.L. Garcia, J.B. Bell, W.Y. Crutchfield, B.J. Alder: J. Comp. Phys. **154**, 134 (1999)
22. D.J. Tildesley, M.P. Allen: *Computer Simulations of Liquids* (Clarendon, Oxford 1987)
23. D.A. Anderson: *Computational Fluid Dynamics* (McGraw Hill, Singapore 1995)
24. T. Le Pennec, K.J. Måløy, E.G. Flekkøy, J.C. Messager, M. Ammi: Phys. Fluids **10**, 3072 (1998)
25. P. Carman: Trans. Inst. Chem. Eng. Lond. **15**, 150 (1937)
26. D. Gidaspau: *Multiphase Flow and Fluidization* (Academic Press, San Diego 1994)
27. T. Kawaguchi, T. Tanaka, Y. Tsuji: Powder technology **96**, 129 (1998)
28. S. Luding, S. McNamara: Granular Matter, **1**, 111 (1998)
29. *Physics of Dry Granular Media*, ed. by H.J. Herrmann, J.-P. Hovi, S. Luding (Kluwer Academic Publishers, Dordrecht 1998) p. 313
30. F. Radjai, D. Wolf: Granular Matter **1**, 3 (1998)
31. A. Zick, G. Homsy: J. Fluid Mech. **115**, 13 (1982)
32. G.K. Batchelor, R.W. Janse van Rensburg: J. Fluid Mech. **166**, 379 (1986)
33. J.F. Davidson: Bubbles in Fluidized Beds. In: *Mobile Particulate Systems*, ed. by E. Guazzelli, L. Oger (Kluwer Academic Publisher, New York 1995) p. 197
34. K.S. Lim, J.X. Zhu, J.R. Grace: Int. J. Multiphase Flow **21**, 141 (1995)
35. P.N. Rowe: *Fluidization* (Academic Press, London and New York 1971) Ch. 4, p. 121
36. D. Gendron, H. Troadec, K.J. Måløy, E.G. Flekkøy: preprint, 2000
37. E.G. Flekkøy, P.V. Coveney: Phys. Rev. Lett. **83**, 1775 (1999)
38. E.G. Flekkøy, P.V. Coveney, G. de Fabritiis: Phys. Rev. E **62**, 2140 (2000)

On the Reduction of Molecular Degrees of Freedom in Computer Simulations

Alexander P. Lyubartsev and Aatto Laaksonen

Division of Physical Chemistry, Arrhenius Laboratory, Stockholm University,
106 91, Stockholm, Sweden

Abstract. Molecular simulations, based on atomistic force fields are a standard theoretical tool in materials, polymers and biosciences. While various methods, with quantum chemistry incorporated, have been developed for condensed phase simulations during the last decade, there is another line of development with the purpose to bridge the time and length scales based on coarse-graining. This is expected to lead to some very interesting breakthroughs in the near future. In this lecture we will first give some background to common atomistic force fields. After that, we review a few common simple techniques for reducing the number of motional degrees of freedom to speed up the simulations. Finally, we present a powerful method for reducing uninteresting degrees of freedom. This is done by solving the "Inverse Problem" to obtain the interaction potentials. More precisely, we make use of the radial distribution functions, and by using the method of Inverse Monte Carlo [Lyubartsev & Laaksonen, Phys. Rev. E. **52**, 3730 (1995)], we can construct effective potentials which are consistent with the original RDFs. This makes it possible to simulate much larger system than would have been possible by using atomistic force fields. We present many examples: How to simulate aqueous electrolyte solutions without any water molecules but still having the hydration structure around the ions – at the speed of a primitive electrolyte model calculation. We demonstrate how a coarse-grained model can be constructed for a double-helix DNA and how it can be used. It is accurate enough to reproduce the experimental results for ion condensation around DNA for several different counterions. We also show how we can construct site-site potentials for large-scale atomistic classical simulations of arbitrary liquids from smaller scale *ab initio* simulations. This methodology allows us to start from a simulation with the electrons and atomic nuclei, to construct a set of atomistic effective interaction potentials, and to use them in classical simulations. As a next step we can construct a new set of potentials beyond the atomistic description and carry out mesoscopic simulations, for example by using Dissipative Particle Dynamics. In this way we can tie together three different levels of description. The Dissipative Particle Dynamics method appears as a very promising tool to use with our coarse-grained potentials.

1 Introduction

Atoms and molecules, the building blocks of matter from a chemist's point of view, interact with each other. They are attracted at long distances, whereas at short distances the interactions become strongly repulsive. In fact, there is no need to seek for a proof from the underlying physics to confirm that this is really the case as we can observe it indirectly on a macroscopic level with our own eyes every day. For example, due to the attraction water molecules nucleate to form drops, which in turn rain down to fill rivers. Thanks to the strong repulsion, on the other hand, one cubic meter of that very same water weighs basically always about 1000 kg, regardless of how much pressure we use in trying to compress it.

A.P. Lyubartsev and A. Laaksonen, On the Reduction of Molecular Degrees of Freedom in Computer Simulations, Lect. Notes Phys. **640**, 219–244 (2004)
http://www.springerlink.com/ © Springer-Verlag Berlin Heidelberg 2004

Today, molecular modelling is carried out by following and analysing interacting structural models in computers and visualizing the results using computer graphics. Historically, modelling of molecules started long before computers were invented. Already in the mid 1800's structural models were suggested for molecules as an aid to explain their chemical and physical properties. Similarly, the first attempts to model interacting molecules go back to the pioneering work of van der Waals [1]. With his simple equation of state where the molecules are approximated as spheres attracting each other, he was able to predict a first-order gas-liquid phase transition. Many of the conceptual ideas from over a century ago are still used in modern methods of molecular modelling and simulations.

The interactions involving molecules can be explained based on quantum mechanics. Close to a contact distance between molecules, there is a complex mixture of forces of quantum mechanical origin [2]. Although the physical background is known, the mathematical complexity effectively prohibits any analytical attempts to describe condensed matter systems on the basis of equations for interacting particles. Fortunately, thanks to computers, there are now approximate but powerful methods to treat many-body systems and extract macroscopic properties from the data. In their simplicity these methods and models work amazingly well, based on man-made force fields, and applied on atomistic molecular models together with a marriage of classical physics and statistical thermodynamics.

1.1 Atomistic Force Fields

The methods of atomistic force-field based computer simulations of condensed matter have become a standard tool to investigate practically everything from metallic solids to complex biological systems in their physiologically natural environments, i.e. aqueous solutions of ions and organic molecules. Quite surprisingly though, these tools are still basically the same as they were when they first were introduced. It may have been the phenomenally rapid development of computer technology during the last three decades hampering the development of molecular simulation methods, or that these methods were very robust already from the very beginning when presented by Alder and Wainwright [3] and Rahman [4] for simple liquids, and developed further by Rahman and Stillinger [5] for molecular liquid water and McCammon and coworkers for proteins [6]. Most likely it is the combination of both. Both the time scales covered by routine molecular dynamics simulations and the number of particles in the simulation cell have increased roughly by one order of magnitude each decade from the days of the pioneering paper by Rahman and Stillinger [5].

In molecular computer simulations, the internal energy of the whole system, at any moment during the simulation, can be calculated by summing up the contributions from a number of separate potential energy functions with the actual spatial and geometrical coordinates as parameters. This collection of potential functions is called the *force field* because it provides a mathematical tool to guide each atom in a system towards a new direction under the influence of all the other atoms in its surrounding. Conceptually these terms are divided to so-called bonded and non-bonded, or alternatively, intra- and intermolecular interactions. For small rigid molecules the intermolecular potentials are often enough.

Bonded Interactions. In the most commonly used force fields [7–10] the intra-molecular interactions are divided to terms involving two, three and four connected atoms, respectively. For example, there are the harmonic terms describing the distortions from equilibrium positions in bond-stretching and angle-bending. The "energies" calculated from these terms, however, are simply just penalties due to deviations from natural bond lengths and angles rather than any real binding energies. Therefore these are only relative quantities. Their use in the context of a force field was introduced by Hill [11] to describe molecular deformations due to steric interactions. Additionally, there is a term (constructed from four connected atoms) to describe the periodic torsional motion of dihedral angles. This torsional model goes back to the investigations of Kemp and Pitzer [12] on the ethane molecule to describe the rotation around the sp^3–sp^3 bond. The periodic nature is implemented naturally in trigonometric cosine expansions with the amplitudes related to corresponding rotational barrier heights to hinder the rotation around a covalent bond due to steric interactions and/or the internal structure of the molecular orbitals forming the bond.

The above three terms can be considered as a minimum description of the intramolecular interactions. Although it could be in principle possible to omit the torsional terms and treat them with non-bonded interactions between the outermost atoms describing the dihedral angle (see below), an additional term to describe the out-of-plane motion and inversion should be mentioned. In some cases it is important to keep part of the molecule planar (for example involving an sp^2 carbon). In other cases an umbrella type of motion may take place (e.g. an amine group). As this term is also constructed from four atoms (not in a sequence, but three of the atoms connected to the fourth central atom) it is normally known as the improper torsion term.

Non-Bonded Interactions. While considering what would make the non-bonded interactions we could simply imagine for a moment that there are no bonds between the atoms of the molecules, only a collection of moving atoms. The atoms have specific volumes and can exert interactions on each other. While the non-ideal origin of the interactions was recognized by van der Waals already, it can also be rigorously derived using quantum mechanics. These van der Waals interactions are basically a mixture of repulsion due to overlapping valence electrons and attraction due to induction and dispersion forces [2]. These interactions are commonly approximated all together in the 80 years old 12–6 potential [13], where the distance dependence of the repulsion term is proportional to r^{-12} (mimicing the exponential soft-wall behaviour) while it is proportional to r^{-6} for the attraction, r being the inter-atomic distance.

Certain atoms can be considered as more electronegative [14] than others. They gain electrons, while their neighbouring, less electronegative atoms will consequently suffer from a corresponding lack of electrons. Although the whole molecule is always electroneutral (unless it is an ion), the individual atoms in molecules mostly are not. Common force fields normally represent this redistribution of the total electron density by collecting the local net charges to single sites. Normally the charge is attached to the site of an atom. Sometimes it may also be placed along the bonds, or outside an atom to mimic a lone pair. Atomic charges are practical as they allow the use of Coulomb's law to describe their mutual interactions at the same time as they provide multipoles for molecules or individual functional groups (charge groups). The electrostatic interactions, however, decay as r^{-1} with respect to the charge-charge distance, making them very long-ranged compared to the van der Waals interactions. Correct treatment of Coulombic interactions is, however, crucial in almost all categories of molecular simulations because they play an important role at both short and long distances. Hill [11] used the Lennard–Jones 12–6 potential [13] for non-bonded interactions, but did not consider the electrostatic interactions. The use of partial charges in force fields was later introduced by Kitaigorodskii [15].

Normally the short-range van der Waals and long-range Coulombic interactions are computed together because they both require calculations of the same site-site distances. Distance calculation is the most computer-intensive part of every simulation as it requires computation of a square root and an inverse ($1/\sqrt{r^2}$) for $N(N-1)/2$ pairs, where N is the number of sites in the system to be simulated. In practice it is not quite an $\mathcal{O}(N^2)$ problem as it is possible to restrict the interactions within a cut-off sphere measured from each atom. This is obvious for the fast decaying Lennard–Jones terms but is applied even for the Coulombic interactions when methods based on Ewald summation techniques are used. Also, modern computers often have the inverse and square root computations implemented in the hardware.

At this point we can put all the covalent bonds back again. Based on the underlying philosophy behind the atomistic force fields, a number of closest "non-bonded" pairs will be excluded because they have already been treated as "bonded" interactions, although this is only a very small fraction of all the $N(N-1)/2$ pairs. Now, there is one specific term in most force fields that should be given some more attention as it turns out to be an intermediate case: a pair of atoms in the outermost positions of a dihedral angle consisting of four atoms and three bonds. In many cases a compromise is made to treat this particular pair partially as a bonded and partially as a non-bonded interaction, and be called the "1–4" interaction. The reason is that steric interactions affect rotations around a bond. However, they are rather rough and in most cases lack further details of local internal conformational degrees of freedom. However, due to the short distance between the 1–4 atoms, both the Lennard–Jones and the Coulombic interactions are normally scaled down substantially. The 1–2 or 1–3 non-bonded interactions are, although somewhat arbitrarily, omitted completely. It is assumed that they are properly described with bond and angle potentials, most often within the harmonic approximation.

It should be mentioned that hydrogen bonds are sometimes enhanced by terms containing a narrow well to adjust its spatial length and directional terms to adjust the angle of the hydrogen bond. Both are parameterized to fit into the geometric definition of a hydrogen bond. In most cases hydrogen bonds are naturally formed due to the Coulombic interactions.

There are many suggestions to improve the terms discussed above; simply too many to review them all here. To mention just a few, the harmonic bond-stretching term is sometimes replaced by the anharmonic Morse potential [16]. This is a realistic model containing the anharmonicity and dissociation behaviour. It is required in calculations of spectroscopic properties as harmonic potentials do not give, for example, overtones. Morse potential is shown to describe the important O–H stretching fairly well. The angle in the angle-bending term can sometimes be replaced by the distance and described as a bond thereby giving a somewhat more realistic behaviour. A nice example of these both features is the flexible water model by Toukan and Rahman [17] where the water molecule is made to act like a triangle by an additional H–H "bond", while O–H bonds are described by a Morse potential. The Lennard–Jones 12–6 term has occasionally been replaced by the physically more correct exponential term (e^{-6}) [18]. The above improvements become important if the molecules are simulated in more extreme conditions away from equilibrium geometries or configurations.

In general the very simple functional forms of the potential functions discussed above are adopted as they approximately describe the specific isolated physical interactions in situations close to equilibrium conformations and configurations. It is also assumed that the various terms do not interfere each other too much under normal conditions. If they do, cross terms can be introduced. Higher order terms are normally not used in force fields as they can easily lead to unexpected behaviour outside equilibrium conditions.

An additional problem is the treatment of non-bonded interactions between unlike atoms. Most often so-called simple (Lorentz–Berthelot) combination rules are applied for the Lennard–Jones potentials characterized by ε and σ for energy and length scales, respectively. Then, given ε_i to describe the potential energy scale between particles of type "i" (and ε_j accordingly), the parameter for unlike atoms is obtained by geometric mean: $\varepsilon_{ij} = \sqrt{\varepsilon_i \varepsilon_j}$. For σ, one applies arithmetic mean, i.e. $\sigma_{ij} = (\sigma_i + \sigma_j)/2$. Strictly speaking, the Lennard–Jones parameters apply only for pure substances. Any type of non-ideal behaviour between molecular mixtures is difficult to incorporate into force fields. Simple combination rules normally give too attractive cross-interaction. Other empirical combination rules than the above mentioned Lorentz–Berthelot description exist in the literature.

In summary, the whole force-field can be visualized as a collection of soft balls and springs to build the molecules, with the size of the balls determined by the Lennard–Jones 12-term, an elastic collision parameter. The strength of the springs (for covalent bonds and angles) is set by force constants, while rotations around bonds are not completely free but hindered by periodic frictions (barriers). The balls attract or repel each other pair-wisely depending on their mutual distances and/or the charges they carry. Force fields, consisting of continuous functions, are simple to use in modelling software to search for the energy minimum conformations, or in simulation programs to move molecules to sample the phase space, either randomly or deterministically by solving the equations of motions for each mass point (atom). In the latter case, derivatives of force fields are needed. A typical force field as discussed above with a minimum number of terms, used in computer simulations of flexible molecules, looks then like the following:

$$
V = \sum_{\text{covalent bonds}} K_{\text{b}}(r - r_{\text{eq}})^2 + \sum_{\text{covalent angles}} K_{\text{a}}(\theta - \theta_{\text{eq}})^2
$$

$$
+ \sum_{\text{torsional angles}} \frac{1}{2} K_{\text{t}} \left(1 + \cos(m_{\text{t}}\phi - \gamma_{\text{t}})\right)
$$

$$
+ \sum_{i<j} \left\{ 4\varepsilon_{ij} \left[\left(\frac{\sigma_{ij}}{r_{ij}}\right)^{12} - \left(\frac{\sigma_{ij}}{r_{ij}}\right)^{6} \right] \frac{q_i q_j}{r_{ij}} \right\}.
$$

1.2 Parameterization of Force Fields

Although the functional forms of atomistic force fields have been the same for several decades, force fields do differ from each other. The real value of a force field lies on its parameterization. What makes the parameterization a non-trivial matter is the demand for a force field to be transferable from one molecular system to another. Otherwise it would not have much practical use. To be more realistic, this requirement has been softened slightly so that the parameterization should be transferable from one system to another similar system. Historically, such general force fields were first constructed for organic molecules, thereafter for proteins, nucleic acids, membranes and carbohydrates. All of them require some special care to be taken to work well, and to give results in good or reasonable agreement with experiments. New revisions appear frequently in the literature.

There is no simple way to derive force field parameters from the underlying physics (quantum mechanics). They are therefore deduced empirically using a variety of sources by combining data from suitable experiments with *ab initio* electronic calculations. Functionality normally requires large sets of atomic types with their parameters. This means that instead of one carbon atom with a mass of 12 atomic mass units, most force fields have several of them depending on their hybridization. Same is true for oxygen depending on its position (–O–, =O or –O–H). There are no reliable potentials for multivalent metal ions due to a substantial three-body interactions and strong polarizing effects. It should be mentioned that there are several main philosophies to assign the atomic types and carry out the parameterization [19].

Among the sources to parameterize force fields are the experimental information about the molecular structure and geometry coming from X-ray, electron and neutron diffraction, NMR spectroscopy as well as optical spectroscopy for small molecules. Other types of experimental data are vibrational frequencies, internal rotational barriers, conformational differences and thermodynamic data such as heat of formation and solvation free energies. Although it seems that plenty of experimental information can be used, the problem is to find it. It is also possible to obtain structures, conformations, and vibrational frequencies from electronic structure calculations. Particularly the modern density functional theory (DFT) methods have proven to be reliable. Besides the two main lines of methodologies, there are some empirical rules one can apply as well. Also, computer simulations can be used to optimise the potential parameters. For example, thermodynamic properties such as density or pressure can be corrected by adjusting van der Waals interactions. Clearly the information is gathered from all possible sources. Only with the final results from a simulation at hand, it is possible to judge the quality of the force field.

Besides the important question of transferability from one system to another there is a question of applicability at different physical conditions like temperature or pressure. Also it is important to remember that there is no single universal force field applicable at all the three states of matter, not even for the smallest molecules. An illustrating example of this point is a work where an interaction potential was constructed for $Na^+ Cl^-$ at a very high level based on electronic structure calculations (multi-reference CI) [20]. It was fitted accurately to a very flexible function and applied to calculate gas, liquid and solid phase properties of sodium chloride. Because the potential function was constructed from $Na^+ - Na^+$, $Cl^- - Cl^-$ and $Na^+ - Cl^-$ interaction energies from gas state (vacuum) conditions it reproduced all spectroscopic (cross- and anharmonic) constants very accurately in a Dunham analysis. It worked fairly well for phonon dispersion curves and reproduced the polymorphic FCC $\rightarrow$ BCC phase transition. However, it failed to produce the interaction energies for molten salts of NaCl by 30 %. Clearly, because it was a pure pair-potential, it was lacking the many-body character needed in liquid state simulations. The empirically parameterized force fields, however, are effective pair potentials as they get some of the many-body interactions into the parameters when fitted on experimental data obtained from macroscopic systems. Another critical test for a potential function is to use it in simulations of ordered systems or close to interfaces. For example, water models doing well in simulations of liquid water sometimes fail when used in clusters (droplets) of water.

Last but certainly not least there is the problem how to assign the atomic fractional charges. As there is no such a thing as the atomic charge (it is neither an experimental nor a theoretical observable) there is no unique way to determine it. Normally when a force field is constructed there remains the problem where to put the charges. For small molecules they can be obtained by fitting the charges to reproduce a set of molecular multipole moments (dipole, quadrupole, octopole, hexadecapole, etc.). For example, in the case of HCl [21] the charges were fitted to reproduce the first four multipole moments of HCl. However, even in this case the procedure was nothing but trivial: accurate enough fitting required one to add a new site behind the chlorine atom, and to vary the distance from the chlorine atom to the virtual charged site. This illustrates that by assigning the charges entirely on atomic sites is not always sufficient. For large molecules charges are a major headache. Various schemes are discussed for example in [19].

2   Reduction of Molecular Degrees of Freedom

Dynamical molecular systems made up of interacting particles are characterized by numerous degrees of freedom. All of them are important to include in a simulation model, but not always at the same time. It is neither practical nor even possible today to simulate, for example, a protein in water solution over nanosecond time scales using *ab initio* MD methods. Therefore the time scale and/or the length scale associated with phenomena one aims to study using computer simulations are decisive in the consideration of a suitable level of approximations. In other words, which degrees of freedom should be kept while still hoping to get reliable results.

The thermal motion of molecules can be divided between translational, rotational and vibrational degrees of freedom based on the classical equipartition principle. Quantum mechanically we also have electronic and nuclear degrees of freedom. While internal nuclear degrees of freedom fall outside the regime of condensed matter simulations, the much slower motion of the nuclei themselves compared to that of the surrounding electrons is the essence of the Born–Oppenheimer approximation. It is also the basis of molecular dynamics simulations with quantum mechanics incorporated. Examples of other degrees of freedom are the description of the environment, such as solvent molecules. In all, this issue gives rise to a number of general questions. Do the solvent molecules interact with the solute in some specific way like water molecules or ions with biomolecules? Alternatively, is it sufficient to describe the solvent as a viscous or dielectric medium? Are there external fields applied on the system under investigation? Are some regions of conformational space much more probable than others?

2.1 Eliminating Fast Fluctuations

In molecular dynamics simulations the input parameter of main importance is the time step. This is the key parameter in solving the equations of motion numerically based on finite difference algorithms. Clearly this discrete unit of time should be considerably shorter than the period of the fastest dynamical fluctuations to accurately determine the trajectories for all the particles in the simulated system. On the other hand, considering the whole simulation and the time scale covered by the simulation, the time step is the main limiting factor. In simulations of excited state properties of a chromophore the time scales of interest may be within a few hundreds of femtoseconds, while the time step is a fraction of a femtosecond. Conformational changes in large biomolecules take place during a few hundreds of nanoseconds, however. Thus, one wants to use as long time steps as possible to reach these long time scales. To increase the size of the time step one has to get rid of the fast (but interesting) motional degrees of freedom. Below we give a short overview of techniques to do it and thereafter present our own hierarchical method.

In real quantum mechanical simulations one should preferably treat all particles as quantum particles, even the nuclei. Unfortunately no rigorous methods exist able to go beyond the Born–Oppenheimer approximation due to the great complexity of the underlying theory. Currently there are semi-classical wave packet methods and path integral approaches, while all so-called quantum simulation methods are based on the Born–Oppenheimer scheme.

In classical atomistic computer simulations the electronic degrees of freedom have been replaced with point charges as discussed above. This leaves the vibrational fluctuations of bond stretching as the fastest fluctuations. For all-atom simulations with all the hydrogen atoms included it would require a time step of about $0.1 - 0.2$ femtoseconds – a painfully short time step for simulations of biological systems. Normally, these bond vibrations are not of interest, especially when harmonic potentials are used. This idea has lead to approaches where bond stretching has been frozen. This was done first by Ryckaert et al. [22] who introduced the so-called SHAKE algorithm to constrain the bond lengths iteratively while keeping the center-of-mass fixed. In constrained dynamics the time step can be increased up to one order-of-magnitude. Another approach to deal with the same problem is to treat the fast and slow motions separately. This can be done by multiple time step methods, like the one introduced by Tuckerman et al. [23]. Then the bond stretching and angle bending, for example, can be treated using a short enough time step, while the remaining degrees of freedom are described with a long time step (comparable to what is used in constrained dynamics). In the computing time this method is competitive with constrained dynamics, while at the same time one can still describe the vibrational fluctuations and keep all the internal degrees of freedom (in constrained dynamics there are no vibrational contributions from bond stretching). Sometimes some degrees of freedom are restrained to certain intervals, such as bond stretching or angle bending but also other non-bonded interatomic distances such as the proton-proton distances from NMR–NOE (nuclear magnetic resonance nuclear Over-hauser enhancement) measurements. This is done by adding penalty terms to the force field. Note that there are still contributions to the thermal energy from these fluctuations.

Next step would be to keep the molecules completely rigid. This was done already by Rahman and Stillinger in their pioneering MD simulation of water [5]. In fact, it is a standard method to treat small and effectively rigid molecules. In a rigid molecule MD, the equations of translational motion are solved for the molecular center-of-masses. Simultaneously, the Eulerian equations are solved for the rotational motion, calculated using the torques. At the beginning of a new time step the rotated molecules are placed on their center-of-mass positions to provide an updated configuration. Obviously the use of this method cannot be extended for large and flexible molecules.

2.2 Simplifying Molecular Models

Another way to reduce the complexity of a system is to use so-called united atoms. For example, it is common to describe CH_2 and CH_3 groups as united atoms having the masses of 14 and 15 atomic mass units, respectively. In doing so, an effective, larger Lennard–Jones sphere is put either on the center-of-mass of the whole group or on top of the heavy atom. This has several benefits. It eliminates the light hydrogen atoms and thereby allows one to increase the time step. Further, it also reduces the number of interaction sites and therefore increases the speed of the simulation. This is also a reasonable approximation as the CH_2 and CH_3 groups are inert. Consequently hydrocarbon chains, like the ones in lipid membranes are often described using united atoms. Also there are popular three-site models for methanol, thus eliminating the internal rotation of the methyl group. In some cases the united atom model may work even better than the existing all-atom model. It is also possible to extend the united atom model to larger groups like what was done in the simulation of t-butyl alcohol $(CH_3)_3OH$ by reducing a 15-site model to a 3-site model [24].

In principle the simplification procedure can be taken much further by constructing intermediate and low-resolution models made up of beads or clusters of beads connected with springs, representing functional groups or building blocks like amino acids, etc. These types of models have been used for a long time, and particularly in simulations of protein folding. An illustrative selection of these models is presented by Hall and co-workers [25]. We will return to bead models later when we present our own coarse-grained DNA model (see Sect. 4.2).

2.3 Elimination of Explicit Solvent Molecules

Computer simulations of large biomolecules under physiologically relevant conditions and environment require plenty of solvent water molecules around the macromolecule. First, there should be several layers of water forming the hydration shell structure. In addition, a generous amount of bulk water should be added to fill the whole simulation cell allowing exchange of molecules between hydration and bulk phases. Obviously, water in large amounts is important in most biological systems to stabilize their three-dimensional structure, but also as a medium for ions and other molecules coming close to them.

Single water molecules are also vital for the key functions of biomolecules. However, in these simulations the CPU time goes mainly to compute water-water and water-solute interactions. In the first simulations of proteins [26] and many others reported thereafter the water molecules were simply omitted. The solvent was represented as a dielectric continuum. So was done in the first simulation of DNA [27], where even the Coulombic interactions were switched off to prevent the breakdown of the double-helix structure due the repulsion of the anionic phosphate groups. Now, replacing the aqueous solvent of explicit water molecules by a dielectric constant for liquid water is a very crude approximation, particularly in the regions close to the macromolecule where solute-solvent interactions can be very specific on atomic scale. However, various forms of polarizable continuum models can sometimes do the trick. For an interesting method to link together the continuum and microscopic regimes, see [28]. Again, we return to this problem to eliminate the solvent molecules as we present the method of Inverse Monte Carlo in Sect. 3.

3 Effective Potentials and the Inverse Monte Carlo Method

3.1 Inverse Problem in Statistical Mechanics

In statistical mechanical modelling of condensed matter systems, one needs a set of interaction potentials or the Hamiltonian to describe the model system. After the Hamiltonian is specified, analytical or numerical methods within the pairwise approximation can be applied, including computer simulation techniques. The aim is to compute canonical averages of physical quantities and compare them with experiments if these are available. In particular, we can calculate various thermodynamic and structural properties in terms of distribution functions. In case of dynamical simulations we can calculate time-dependent transport properties. Realistic simulations of large enough systems with long time scales are straightforward but computationally expensive.

An interesting and fundamental problem in statistical physics is the "Inverse Problem". Namely, how to deduce the interaction potential (or reconstruct the Hamiltonian) from canonical averages. These averages can be known, for example, from experiments. One such property is the radial distribution function obtained through a Fourier transform of the structure factor from X-ray or neutron scattering experiments of condensed phase samples [29].

In 1974, Henderson presented a theorem [30] of a unique, point-to-point correspondence between pair potentials and radial distribution functions. For a given system under given conditions of temperature and density, two pair potentials, giving rise to the same RDF $g(r)$, do not differ from each other by more than a single constant. This constant can further be defined from the condition that the interaction potential vanishes at the infinite distance, making the potential uniquely defined.

Not surprisingly, the solution of the inverse problem is not as straightforward in practice as a calculation of the RDF from a known interaction potential. It is not possible to write a formal expression for how to compute the interaction potential from the RDF. Some special techniques are required to solve the inverse problem as we will show later after discussing a few related ideas.

In 1988, McGreevy and Pusztai suggested the so-called Reverse Monte Carlo (RMC) method where the starting point for a simulation was the radial distribution function [31]. Within this technique, a MC simulation can be carried out in a consistent way to retain the input RDF. However, this task is accomplished without any knowledge of the interaction potential. The set of configurations produced by the RMC can be used for computation of further structural information, e.g. calculation of 3D spatial or orientational correlations. However, it should be stressed that because the interaction potential is not reconstructed, the inverse problem is not solved. It is not possible to calculate thermodynamical or dynamical properties of the system using this approach. Naturally, it cannot be used for any type of coarse-grained simulations.

In a related approach, Reatto and coworkers [32] have used the Hypernetted Chain (HNC) approximation to solve the inverse problem. This or any similar approach, based on some closure of the Ornstein–Zernike equation, were also applied in several works during the last decade to produce interaction potentials from radial distribution functions [33,34]. Again, it should be stressed that computations within the HNC theory are only feasible for relatively simple models. Although it gives sometimes accurate results [35], the HNC theory is not an exact mathematical solution to a statistical-mechanical problem. Therefore its accuracy should be validated in every specific case. We should also mention a few other works devoted to the inverse problem [36–38].

A practical way to solve the inverse problem completely is to reconstruct the interaction potential from the distribution function along the lines suggested in our paper [39]. This method together with its applications to compute various effective potentials on different levels of description of condensed matter, will be discussed in more details in the following sections.

3.2   The Method

Consider first a simple system of identical particles interacting through a pair potential. The corresponding interaction Hamiltonian is given as

$$H = \sum_{i,j} V(r_{ij}), \tag{1}$$

where $V(r_{ij})$ is a pair potential, and r_{ij} the distance between the particles i and j. Assume further that the radial distribution function $g(r)$ is known so we can construct the corresponding interaction potential $V(r)$.

Let us first apply a grid approximation to digitalize the Hamiltonian:

$$\widetilde{V}(r) = V(r_\alpha) \equiv V_\alpha$$

for

$$r_\alpha - \frac{1}{2M} < r < r_\alpha + \frac{1}{2M};$$
$$r_\alpha = (\alpha - 0.5)r_{\text{cut}}/M;$$
$$\alpha = 1, \dots, M,$$

where r_{cut} is a cutoff distance and M is the number of grid points within the interval $[0, r_{\text{cut}}]$. Then we can rewrite the Hamiltonian (1) as:

$$H = \sum_\alpha V_\alpha S_\alpha , \tag{2}$$

where S_α is the number of pairs of particles whose mutual distance is inside the α-slice, and serves as an estimator of the radial distribution function:

$$\langle S_\alpha \rangle = \frac{4\pi r^2\, g(r) N^2}{2V}.$$

The average values of S_α are now some functions of the potential V_α, and one can write the expansion

$$\Delta \langle S_\alpha \rangle = \sum_\gamma \frac{\partial \langle S_\alpha \rangle}{\partial V_\gamma} \Delta V_\gamma + \mathcal{O}(\Delta V^2) , \tag{3}$$

where the derivatives $\partial \langle S_\alpha \rangle / \partial V_\gamma$ can be expressed using exact relationships based on statistical mechanics [39]:

$$\begin{aligned}
\frac{\partial \langle S_\alpha \rangle}{\partial V_\gamma} &= \frac{\partial}{\partial V_\gamma} \frac{\int dq S_\alpha(q) \exp(-\beta \sum_\lambda V_\lambda S_\lambda(q))}{\int dq \exp(-\beta \sum_\lambda V_\lambda S_\lambda(q))} \\
&= -\frac{\langle S_\alpha S_\gamma \rangle - \langle S_\alpha \rangle \langle S_\gamma \rangle}{k_{\text{B}} T}.
\end{aligned} \tag{4}$$

Equations (3) and (4) give us the interaction potential V_α from radial distribution functions $\langle S_\alpha \rangle$ iteratively. Let $V_\alpha^{(0)}$ be some trial potential such as the corresponding potential of mean force,

$$V_\alpha^{(0)} = -k_{\text{B}} T \ln g(r_\alpha).$$

Then, using standard Monte Carlo simulations, one can evaluate the averages $\langle S_\alpha \rangle$ and their deviations from the reference values S_α^*, defined from the given RDF:

$$\Delta \langle S_\alpha \rangle^{(0)} = \langle S_\alpha \rangle^{(0)} - S_\alpha^*.$$

By solving the system of linear equations (3) with the coefficients defined by (4), and omitting the terms $\mathcal{O}(\Delta V^2)$, we can obtain corrections to the potential $\Delta V_\alpha^{(0)}$. Then the procedure is repeated with the new potential

$$V_\alpha^{(1)} = V_\alpha^{(0)} + \Delta V_\alpha^{(0)},$$

until convergence is achieved. The whole procedure is similar to a solution of a multi-dimensional non-linear equation using the Newton–Rhapson method.

If the initial approximation of the potential is poor, some type of regularization of the iteration procedure is needed. In this case, we multiply the required change of the RDF by a factor between 0 and 1. By doing so, the term $\mathcal{O}(\Delta V^2)$ in (3) can be made small enough to guarantee convergence of the whole procedure at the cost that the number of iterations will increase in this case.

The above algorithm also gives us a possibility to evaluate the uncertainty of the inverse procedure. The input RDF (it can be either an experimental RDF or a RDF obtained in an accurate computer simulation) contains normally some uncertainty. An analysis of the eigen values and eigen vectors of the matrix (4) gives us information about how changes in a certain part of the RDF will change the effective potential. For example, eigen vectors with eigen values close to zero correspond to changes in the potential hardly affecting the RDF. The presence of eigen values close to zero makes the inverse problem less well-defined, which in some cases (for example, liquid water) may cause some serious problems in the RDF inversion [40].

It should be mentioned that our approach is related to the renormalization group Monte Carlo method [41,42], used previously to study phase transitions in lattice models (e.g., Ising models for ferromagnets, polymer models, etc.) as well as in the quantum field theory. The renormalization procedure represents, in fact, a change of scale in representating the given system, during which one consecutively moves towards more and more coarse-grained descriptions of the system. The set of (3)–(4) has been used in references [41,42] to describe how the parameters of the Hamiltonian change during the renormalization procedure. The applications of this method are clearly more general than the lattice systems near the phase transition point, and cover even soft-matter problems, allowing us to "renormalize" the Hamiltonian of a molecular system in such a way that only those degrees of freedom we are interested in are kept.

3.3   How to Implement the Method for More General Systems

The method described above for a mono-component system can easily be extended for more general systems, including mixtures of molecules, presence of long-range electrostatic forces, etc.

In the case of a multicomponent system (assume K different species), we have a set of $K(K+1)/2$ pair potential functions and their corresponding RDFs. Consequently, we will have $M \times K(K+1)/2$ terms in the sum (2) describing the Hamiltonian in the grid approximation, and the same number of lines in the linear system equation (3). In other respects, the algorithm is a complete analogue of the one-component case. Similarly, the method can be generalized to provide effective potentials between molecules, being presented in this case as a sum of site-site interactions.

In calculations of potentials between mobile particles (ions) and more static sites of a large (fixed) macromolecule, we have a set of $K(K + 1)/2$ potential functions, describing the interactions between K species of mobile particles, and $K \times K_M$ potential functions, describing interactions between the mobile particles and K_M different sites on the macromolecule. Naturally, in order to calculate this set of pair potentials, one needs the corresponding set of radial distribution functions.

In a typical case the RDF between two atoms is zero at short distances. In the numerical treatment, we can put the potential at the corresponding distances to $+\infty$ (hard core). This again leads to zero RDF intensities at these distances and zero coefficients in corresponding lines of the linear equation system (3). These zero lines can simply be omitted from the group of equations by assuming that if $g(r)$ is zero for some inter-particle distance, then the probability should be zero for any pair of particles approaching the distance, and the corresponding potential can be therefore set to infinity. The true potential can still assume a finite value, but the absolute value is not important.

In the case of charged particles (ions) special care should be taken of the long-range part of the electrostatic interactions. The inverse procedure yields the effective potential for the range of distances for which the input RDF is defined. Therefore we need some assumptions about the behaviour of the effective potential at distances outside the cut-off range. A reasonable assumption is that outside the cut-off distance the effective potential is made equal to the Coulombic potential with a proper dielectric constant. Our previous studies of ionic solutions [43] show that this assumption is valid already at distances of 10 Å and beyond.

In practical calculations for systems involving charged particles, we can divide the effective pair potential in the whole range $[0, \infty]$ into two parts: the short-range part up to the cut-off radius and the Coulombic potential with some dielectric constant. As the Coulombic part is not changed during the simulations it can be calculated using the Ewald method, while the short-range part changes according to (3) and (4) as it is fitted to the input RDF.

The inverse Monte Carlo algorithm may also easily be generalized to other, non-pairwise types of interactions. For example, intramolecular covalent angles and torsional angle potentials can be obtained, in a similar manner, from the population distributions of these angles. Also, three-body potentials can, in principle, be calculated from three-body correlation functions, however such computations would require very large computer memories to store the matrix (4) for a three-dimensional grid.

4   Some Examples and Discussion

4.1   *Ab–Initio* Effective Potentials

Liquid Water. As discussed in the Introduction, *ab-initio* parameterization of force field parameters from a large number of dimer configurations is very complicated and behind a tedious work. In our recent work [40] we proposed a novel approach to construct site-site interaction potentials from *ab-initio* simulations both accurately and efficiently as well as without any pain. The main idea behind this approach is to use detailed high quality *ab initio* liquid simulations to calculate a set of RDFs between all the different pairs of atoms (or arbitrary sites) in the system. The RDFs will then be used to construct a set of effective pair potentials between the same sites using the Inverse Monte Carlo procedure described in Sect. 3.2.

The suggested method provides us with an automatic procedure to construct a complete set of pair interaction potentials from *ab-initio* (for example, Car–Parrinello) simulations. Unlike, while doing the tedious work manually, we do not need to worry about the sampling of the configurational space because all the relevant parts of the phase space become properly weighted as the *ab-initio* simulation proceeds at ambient temperatures. Most importantly and in contrast to previous *ab initio* potentials like the MCY model [44], which is by definition a pure pair potential, our pair potential is a true effective potential since it is calculated at the proper liquid state conditions at finite temperatures rather than in vacuum at 0 K. An effective pair potential includes all high-order interactions averaged.

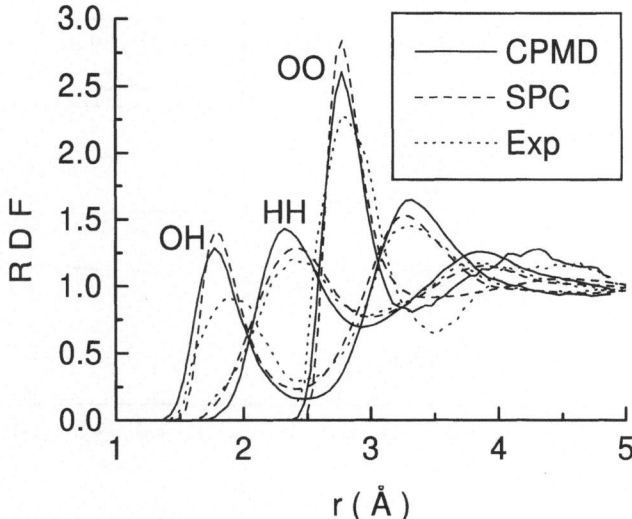

Fig. 1. Radial distribution functions of water obtained in the Car–Parrinello simulations. Results of classical MD for the SPC model and experimental neutron scattering data [46] are also shown

The first application of this approach was the reconstruction of pair potential for water at ambient conditions [40]. Liquid water is an obvious choice to test new theoretical methods. Water is the most studied system using computer simulations, and for water there exist both relatively reliable potential models and experimental data. Most water models are empirically parameterized to reproduce some set of available experimental data.

The Car–Parrinello molecular dynamics simulations [45] were carried out for 32 water molecules in a cubic box of linear size 9.855 Å (corresponding to a density of $1\,g/cm^3$) with periodic boundary conditions. The time step was set to 0.15 fs. The initial configuration was taken from the molecular dynamics simulations of the SPC water model after a proper equilibration at a temperature of 300 K. The system was then further equilibrated for 4 ps using the Car–Parrinello MD method, followed by 10 ps of averaging. The temperature was controlled by scaling the velocity if the kinetic temperature deviated from the target value (300 K) more than 50 K. As a result, the average temperature during the production run was 297 K. More detailed simulation details can be found in [40].

The resulting RDFs for the different pairs of atoms are shown in Fig. 1, and the effective interaction potentials obtained from the RDFs are shown in Fig. 2. These effective potentials reproduce the very same RDFs as were obtained from Car–Parrinello MD simulations, within a precision of at least 0.01 for all distances. For comparison, the potential curves of a popular, empirically parameterized three-site SPC water model [47] are also shown in Fig. 2. It is interesting to observe that the *ab initio* effective potentials are almost superimposed on the SPC model potentials at distances exceeding 3.5 Å. At shorter distances (2.8–3.5 Å for O–O and 1.6–3 Å for O–H pairs) the *ab initio* O–O and O–H potentials rise somewhat less steeply, while at very short distances they are noticeably steeper than the corresponding SPC potential curves.

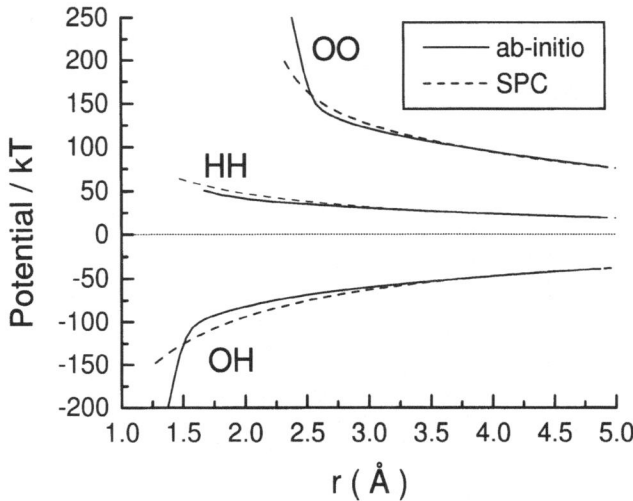

Fig. 2. The effective potentials for water, calculated by the inverse MC method from *ab initio* RDFs. The SPC model potentials are given for comparison

We would like to stress that the main purpose of this example was not to suggest yet another water model, but rather to present and test a pure *"ab initio"* way to derive a set of effective atom-atom potentials for classical MD simulations. We have constructed them from the structure of liquid water through Car–Parrinello simulations. At the moment the effective interactions yield, in a few minutes, the very same set of RDFs as the Car–Parrinello calculations that require a few months of computing time on the same computer. Of course, the Car–Parrinello method does not represent the ultimate high-end simulation tool. Other, more accurate, quantum chemical approaches could be used to calculate the forces during the *ab-initio* molecular dynamics run. It can also be mentioned that *ab initio* simulations, such as the Car–Parrinello technique, are carried out by solving the classical equations of motion on the Born–Oppenheimer surface. In truly quantum mechanical simulations, even the motion of the light atoms should be treated quantum mechanically. The now obtained raw *"ab initio"* effective potential has to be tuned further to give a variety of liquid state properties, both thermodynamical and dynamical, in agreement with the experiment, making it finally a new water model.

Li$^+$ – Water Potential. In another application, the pair interaction potential between a Li$^+$ ion and water molecules has been derived in a similar way. The problem to make a right choice among all the existing empirical ion-water potentials is rather severe: for example, for a Li$^+$ ion, described by the additive pair-potential of Lennard–Jones type, the effective ion diameter σ varies as much as from 0.9 to 2.8 Å in different proposed potential models (see [48–51] and references therein). Such a big change in the effective diameter leads, for example, to a change of the hydration number from 4 to 6 (see [50]), and correspondingly from tetrahedral water coordination to octahedral coordination in the first hydration shell. Other types of ion-water potentials, with an exponential short-range repulsion [52], or the so-called 4–7 potential (in terms of powers of r) [53] and

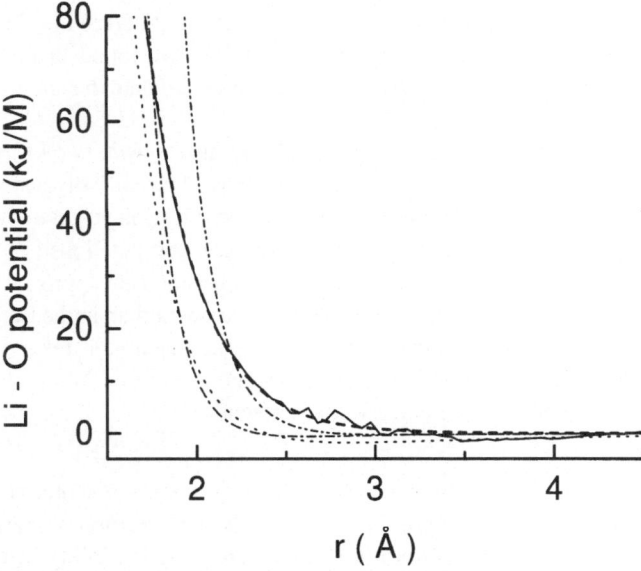

Fig. 3. Short-range part of the effective Li–O potential. The potential derived from *ab initio* simulation using the Inverse Monte Carlo (*solid line*), an exponential fit (*bold dashed line*), two potentials of the Lennard–Jones type from [48] (*dash-dotted line with one dot*) and from [49] (*dash-dotted line with two dots*), and the potential of 4–7 type from [53] (*dot line*)

potentials with more complicated functional forms have been suggested [54,55]. Obviously, different properties of the Li^+ hydration shell as well as the whole electrolyte solution are very sensitive to the chosen potential model.

The underlying Car–Parrinello simulation has been carried out for a single Li^+ ion and 32 water molecules [51]. The simulation setup was mainly the same as in the case of the liquid water described above (for details see the original paper [51]). The simulation time was extended to 20 ps after 5 ps of equilibration.

In principle, it would be possible to invert the whole set of radial distribution functions, obtained in this Car–Parrinello simulation and reconstruct both Li^+-water and water-water potentials. However, for liquid water there already exist several simple models which describe liquid water more or less well under ambient conditions. Therefore we constructed only the Li-water potential, describing water-water interactions by the well-known SPC model. Moreover, since the hydrogens in the SPC model have only electrostatic potential, we kept the same feature in the Li^+-water interactions. Thus, we only vary the Li–O potential, while employing pure electrostatic interactions between water hydrogens and the Li^+ cation.

The result for the short-range part of the Li–O potential (electrostatic part subtracted) is shown in Fig. 3. This potential provides exactly the same Li–O RDF as the one generated in the original *ab initio* simulation. While this potential was fitted to a number of different functional forms, we found that the simple exponential form

$$V_{eff}(r) = A \exp(-Br)$$

with $A = 37\,380\,\text{kJ}/\text{M}$ and $B = 3.63\,\text{Å}^{-1}$ provided the best fit (see Fig. 3). In attempts to fit the potential to different power functions of $1/r$, we found that $1/r^6$ or $1/r^7$ repulsive potentials provided the best fit, but were less good than the above exponential form.

For comparison three other Li^+–O potentials are also shown: two of the Lennard–Jones type from [48] and [49] and a "4–7" type of potential recently suggested by Periole et al. [53]. Clearly, the Lennard–Jones cases have a much stronger repulsive core. It is also interesting that all potentials based on the *ab initio* computations [52,56,57] have a less stiff repulsive part – either exponential (which do not even approach infinity at a zero distance) or a power-law decay with a smaller exponent than in the Lennard–Jones potential. The longer-range repulsive part of the derived potential is responsible for the lower first minimum and correct position of the second maxima of the Li–O RDF, which are impossible to satisfy with a Lennard–Jones potential.

Reliability of the *Ab–Initio* Effective Potentials. The effective potentials calculated by the Inverse Monte Carlo method provide exactly the same structural properties as those obtained through the underlying *ab-initio* simulations. The feasibility of the effective potentials to describe thermodynamical and dynamical properties is still a matter of further research. For the liquid water potential, derived in [40], the internal energy was shown to be $-31\,\text{kJ}/\text{M}$, compared with the experimental value of $-41\,\text{kJ}/\text{M}$. The Li^+–water potential obtained in [51] yields the solvation free energy $-419\,\text{kJ}/\text{M}$, while the experimental value is $-475\,\text{kJ}/\text{M}$. While it is not possible to speak about a perfect agreement, the results may still be considered as good when we keep in mind that no adjustable parameters were used in the derivation of the effective potentials.

There are several approximations affecting the accuracy of the obtained effective potentials. The most significant of these are:

1. several approximations within the Car–Parrinello method itself and in the underlying density functional theory
2. statistical uncertainty of the RDFs, calculated in the Car–Parrinello simulations during a relatively short $(10-20\,\text{ps})$ simulation time
3. small system size
4. limitations of the rigid water model
5. pair potential approximation

With these approximations in mind, the agreement with the experimental results can be considered as good. For the solvation free energy of a Li^+ ion, this method yields results of the same level of precision as those available from empirical models [51]. It is clear that further improvement of quantum-chemical simulation techniques (beyond the Born–Oppenheimer limit) and a substantial increase in computing power would make this method a powerful instrument to derive reliable atom-atom interaction potentials.

4.2 Effective Solvent-Mediated Potentials

Ion-Ion Potentials for Simple Electrolytes. The classical molecular dynamics computer simulations with site-site pair potentials can now be run for molecular systems consisting of the order of 10^4 atoms. This corresponds to a linear system size of 50–60 Å. This size may be large enough for simulations of isotropic liquids of simple molecules. Other systems such as biological macromolecules, however, often require even larger simulation cells. The same can be said about simulations near phase transition points, or about studies of hydrodynamic phenomena. On the other hand, these large-scale simulations do not typically require atomistic resolution, and could be carried out within some simplified, or coarse-grained models.

One of the typical simplifications used in description of macromolecules is to substitute the solvent molecules by a continuum medium. For example, in the primitive electrolyte model the ions in the water solution are substituted by charged spheres moving in a dielectric medium with a proper dielectric constant. Evidently, this is a serious simplification at short distances (a few ångströms) around the ions where it is not possible to define a dielectric constant. Sometimes, good agreement between the experiment can be obtained using the primitive electrolyte model after adjusting the ionic radius. However, the adjusted parameter does not have any physical significance.

A reliable model for effective ion-ion interactions in aqueous solution must take into account the solvation structure around the ions. Such effective solvent-mediated ion-ion potentials can be constructed from the ion-ion radial distribution functions, generated in high-quality all-atomic molecular dynamics simulations, using the Inverse Monte Carlo method. This approach has been already suggested in our original paper describing the inverse Monte Carlo technique [39]. In subsequent papers [35,43], the effective solvent-mediated potentials for NaCl aqueous solution have been calculated with a greater precision and for a number of concentrations.

An example of solvent-mediated ion-ion potentials is presented in Fig 4. The underlying molecular dynamics simulations have been performed using the flexible SPC water model [17] and Smith–Dang parameters for Na^+ and Cl^- ions [58]. Simulation details are described in the original papers [39,43]. Observe that the effective potential makes 1–2 oscillations, thereby reflecting the molecular nature of the solvent, and then finally approaches the primitive model potential with a dielectric constant close to 80. At distances longer than 10 Å, the effective potential almost perfectly coincides with the Coulombic potential. This can be seen from Fig. 5, where the Coulombic part has been removed and only the short-range part of the effective potential is plotted.

Note also that the effective potentials differ from the potential of mean force (PMF) $\psi_{pmf} = -k_B T \ln g(r)$, which corresponds to the Kirkwood approximation for the N-particle correlation function. The potential of mean force is screened by the other ions in the system and decays as $(1/r) \exp(-r/r_D)$ rather than via a Coulombic potential. The potential of mean force is therefore not a suitable choice to present the ion-ion interactions within a continuum solvent model at finite ion concentrations.

Figure 5 shows the effective potentials obtained for 0.5 M and 1 M ion solution after 10 ns MD simulations in a box of linear size 39 Å. One can clearly see that the differences between the two concentrations are very small. This means that the effective potentials depend very little on the ion concentration, and one can use potentials calculated at one specific concentration to describe properties of ion solutions at other concentrations.

Recently, the ion-ion potentials have been successfully used in dynamical mesoscopic simulations through the Dissipative Particle Dynamics (DPD) method [59,60]. It

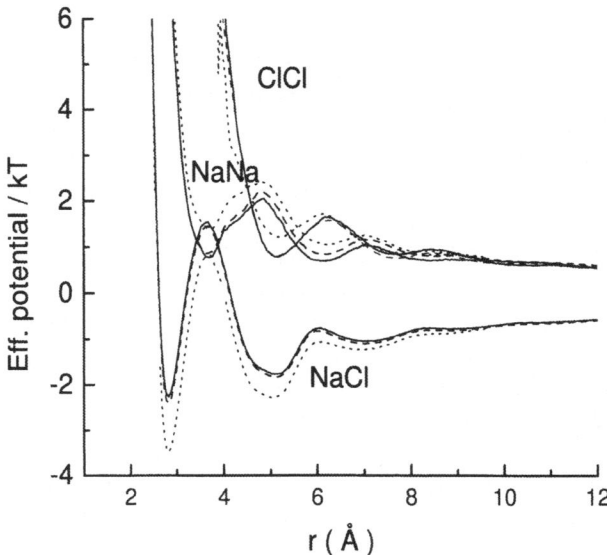

Fig. 4. Ion-ion effective potentials for NaCl ion solution. Ion concentrations: 0.5 (*solid line*), 1 (*dashed line*) and 4 M (*dotted line*)

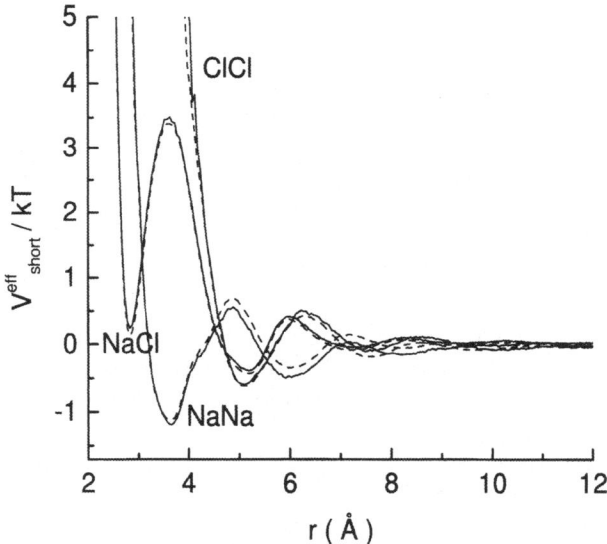

Fig. 5. Short-range part of the effective potential between ion-ion pairs in an aqueous NaCl electrolyte. The cases shown are 0.5 M (*continuous line*) and 1 M (*dashed line*). In subtracting the Coulombic potential we assumed a dielectric constant of 79 and 78 for 0.5 M and 1 M solutions, respectively

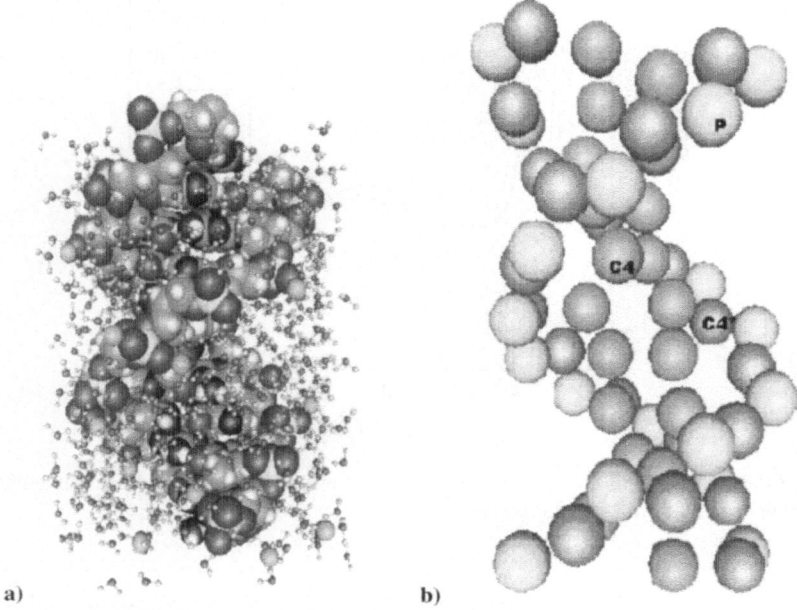

a)   b)

Fig. 6. (a) All-atom model of DNA. **(b)** Coarse-grained model of DNA in aqueous electrolyte solution

was found that the results for structural properties are reproduced accurately, and that transport properties like diffusion coefficients follow the correct temperature dependence very nicely. We are planning to extend the use of DPD simulations with effective potentials to more complex systems such as lipid bilayers. DPD is the obvious choice when effective potentials are used to calculate time-dependent properties.

Ion–DNA Effective Potentials. In the same way as for ionic solutions, one can introduce effective potentials between ions and a macromolecule, for example DNA. Such computations are described in our recent paper [61]. First, MD simulations of a (pseudo) infinitely long DNA model, presented as a periodically repeated fragment (10 base pairs $d(ATGCAGTCAG)_2$) of double-helix B–DNA, surrounded by water and different mono-valent ions in a rectangular box with periodic boundary conditions in all directions, have been carried out. Then, a simplified (coarse-grained) DNA model have been constructed in which three sites on each nucleotide monomer where selected (see Fig. 6) to describe ion-DNA interactions. These sites are the effective middle points of three building blocks of DNA: phosphorus atoms, $C4$ atoms of the bases and $C4'$ atoms of the sugars.

The phosphorus atom is a natural choice since the negative DNA charges are located on the phosphate groups. The choice of other sites is, however, somewhat arbitrary. It is expected that these sites should represent impenetrability of the DNA body for the ions. It is possible to choose additional sites on DNA to calculate RDFs and the effective potentials, in order to provide a more detailed description of the ion distribution within

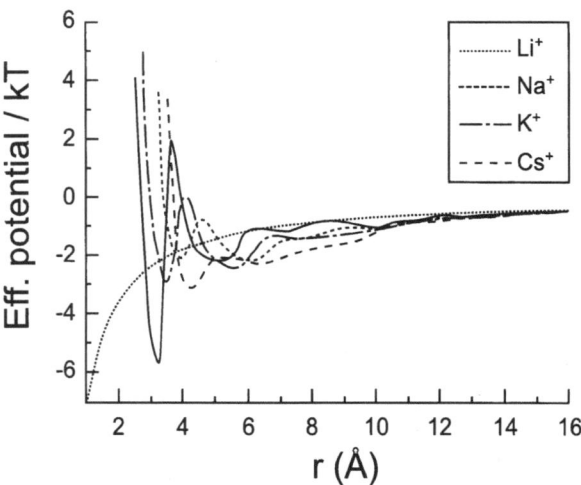

Fig. 7. Effective potentials between counter-ions and phosphorous atoms of DNA, including the Coulombic potential with $\varepsilon = 78$ (*thin dotted line*)

the simplified model. This specific choice of sites to construct the coarse-grained model can be considered as a minimal set of sites, still enough to reproduce the general features of ion binding properties and ion distribution around DNA.

A selected set of effective potentials for ion-DNA interactions are shown in Fig. 7. It can be seen that the effective potentials, as in the case of simple electrolyte solutions, have an oscillatory character representing the molecular nature of the solvent. We have seen that the effective potentials between the charged particles (or sites) approach the Coulombic potential at distances $r > 10$ Å. The effective potentials between the un-charged DNA groups and ions tend to go to zero faster, reaching it already at distances around 8 Å. Also, the potentials have a less regular shape than the corresponding poten-tials between the ions and charged phosphates. Again, we want to stress that the displayed effective potentials should not be interpreted as potentials between the ions and specific DNA atoms, but rather as potentials between the ions and the whole functional group of DNA (phosphate group, sugar ring, nucleotide base), represented by the corresponding interaction center.

One important observation made in [61] was that the presence of the bulky DNA has only a weak influence on the effective potential between the ions. The influence was about the same order of magnitude as was caused by an increase of the salt concentration in pure electrolyte solutions. This observation shows a reasonable transferability among the effective potentials. The effective potentials, calculated at some conditions, can to a good approximation be used to describe other situations, e.g at different concentrations or in the presence of other components or solutes.

As an application of the ion-DNA effective potentials, we have calculated the relative binding affinities of different alkali ions to DNA. The problem of ion binding and ion association to DNA is biologically highly important because of the strong effect different ions exert on DNA. A problem of general interest could be formulated by asking how many ions of a given species can be found near the DNA surface, say, within 5 Å distance (and be able to take part in specific ion-DNA interactions or reactions) at a given bulk ion concentration? All polyelectrolyte theories, calculating ion distributions and electrostatic potentials around polyions, are trying to provide the answer to this problem. To find the answer based on data from computer simulations, we must simulate both the region around the DNA and also the bulk solution. Moreover, the thermodynamic equilibrium between these two regions has to be reached. For physiological ion concentrations of the order of $0.1 - 0.2$ M, the "bulk" conditions, i.e., a constant ion distribution and zero electrostatic potential are typically reached at distances of about 50 Å from the DNA; for lower salt concentrations this region should be even greater. In typical molecular dynamics simulations of DNA, the distance from the DNA surface to the border of the simulation box is on the scale of few ångströms. Therefore, such MD simulations allow one to obtain, for example, the answer to the question: where are the binding sites of different ions. However, the MD simulations would not allow one to address the question: what is the total amount of bound DNA ions, related to the ion concentration in the bulk. Currently, it would not be realistic to perform full-atomic MD simulations with a box length of the order of 100 Å, especially since just the equilibration of such large systems may well exceed the nanosecond time scale.

A coarse-grained model, based on our effective potentials could give us the answer to all the questions mentioned above. Monte Carlo simulations within the coarse-grained model have been carried out in a rectangular box of size $100 \times 100 \times 68$ Å^3. Cylindrically averaged density profiles of different alkali ions around DNA were calculated. They are shown in Figs. 8(a) and (b). The binding affinities of alkali ions to DNA, as determined in the present work, follow the order: $Cs^+ > Li^+ > Na^+ \gtrsim K^+$. This order of binding affinities is in perfect agreement with experimental measurements performed by techniques of very different nature [61].

4.3   Effective Potentials for Macromolecules

As an additional example of coarse graining we can describe the computation of effective potentials between charged colloidal particles. This system is typically simulated as an "asymmetric electrolyte model" where the colloid particles are represented as highly charged and large spheres surrounded by small counterions and eventually coions, all interacting by a Coulombic potential with a suitable dielectric constant. As a step of coarse-graining, the ion component is removed and substituted by an effective interaction between colloid particles.

Such effective potentials for different concentrations of ions have been calculated in [62]. The radial distribution function between colloid particles were determined in Monte Carlo calculations of 80 spherical macroions of charge -60 e and radius 40 Å, and few thousands of small ions. Then the effective potentials were determined in the inverse MC procedure. It was found that in the case of monovalent ions the effective potentials could be nicely fitted to screened Coulomb (Yukawa) potentials, corresponding to the well-known DLVO (Derjaguin–Landau–Verley–Overbeek) theory of colloidal systems. However, in the case of divalent ions, a deep minimum in the effective potential appears at small distances between colloids (see Fig. 9). This minimum corresponds to an effective attraction between the macroions, which in some cases may cause a spontaneous aggregation of colloid particles.

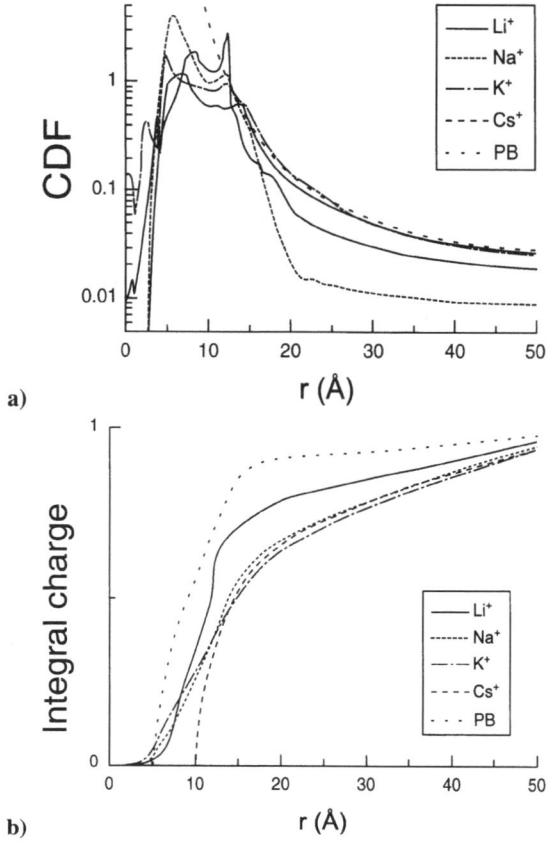

Fig. 8. (a) Cylindrically averaged density profiles of counter-ions around DNA for the case of salt-free solution, obtained in MC simulations with effective potentials. The solution of the Poisson–Boltzmann equation is included for comparison (*thin dotted line*). **(b)** The integrated charge around DNA, calculated from the cylindrical distribution functions

5   Summary

To summarize, the results from the above examples show how one can successively reduce molecular degrees of freedom in molecular computer simulations, going from the *ab-initio* description of electron structure to assemblies of macromolecules. Although we did not demonstrate it all the way for any of the examples, it could have been possible using the method of Inverse Monte Carlo. *Ab initio* MD simulations for DNA are still too expensive at the moment, although in principle they are possible. We are confident that with fine-tuning together with more powerful computers, it will be the ultimate tool for coarse-graining.

Acknowledgements

This work has been supported by the Swedish Science Council (VR).

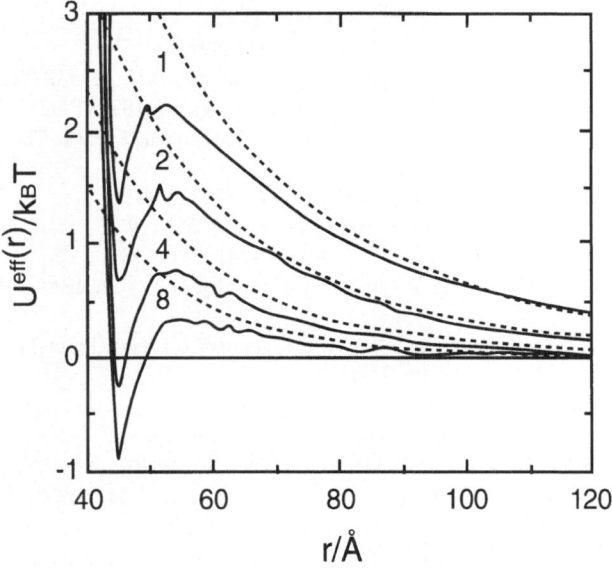

Fig. 9. Effective potentials between spherical macroions in the presence of divalent counterions at different polyion number densities in units of 10^{-6}Å^{-3}. For comparison, Yukawa fits to the long-range repulsive tail are also shown (*dotted lines*)

References

1. J.D. van der Waals: *Over de continuïteit van den Gas – en Vloeistoftoestand*. PhD thesis, University of Leiden, The Netherlands (1873)
2. M. Rigby, E.B. Smith, W.A. Wakeham, G.C. Maitland: *Intermolecular Forces, Their Origin and Determination* (Clarendon Press, Oxford 1981)
3. B.J. Alder, T.E. Wainwright: J. Chem. Phys. **27**, 1208 (1957)
4. A. Rahman: Phys. Rev. **136**A, 405 (1964)
5. A. Rahman, F.H. Stillinger: J. Chem. Phys. **55**, 3336 (1971)
6. J.A. McCammon, B.R. Gelin, M. Karplus: Nature **267**, 585 (1976)
7. P.K. Weiner, P.A. Kollman: J. Comp. Chem. **2**, 287 (1981)
8. B.R.B. Brooks, R.E. Bruccoleri, B.D. Olafson, D.J. States, S. Swaminathan, M. Karplus: J. Comp. Chem. **4**, 187 (1983)
9. W.F. van Gunsteren, H.J.C. Berendsen: *GROMOS Library Manual* (Biomos Nijenborgh 16, Groningen 1987)
10. W.L. Jorgensen, J. Tirado-Rives: J. Am. Chem. Soc. **110**, 1657 (1988)
11. T.L. Hill: J. Chem. Phys. **14**, 465 (1946)
12. J.D. Kemp, K.S. Pitzer: J. Chem. Phys. **4**, 749 (1936)
13. J.E. Lennard-Jones: Proc. R. Soc. London A **106**, 463 (1924)
14. L. Pauling: J. Am. Chem. Soc. **54**, 3570 (1932)
15. A.I. Kitaigorodskii: Tetrahedron **14**, 230 (1961)
16. P.M. Morse: Phys. Rev. **34**, 57 (1929)
17. K. Toukan, A. Rahman: Phys. Rev. B **31**, 2643 (1985)
18. R.A. Buckingham: Proc. R. Soc. London A **163**, 57 (1938)
19. A.D. Leach: *Molecular Modelling – Principles and Applications* (Addison–Wesley Longman, Edinburgh 1996)

20. A. Laaksonen, E. Clementi: Mol. Phys. **56**, 495 (1985)
21. A. Laaksonen, P.-O. Westlund: Mol. Phys. **73**, 663 (1991)
22. J.P. Ryckaert, G. Ciccotti, H.J.C. Berendsen: J. Comp. Phys. **23**, 327 (1977)
23. M. Tuckerman, B.J. Berne, G.J. Martyna: J. Chem. Phys. **97**, 1990 (1992)
24. P.G. Kusalik, A.P. Lyubartsev, D.L. Bergman, A. Laaksonen: J. Phys. Chem. B **104**, 9526 (2000)
25. A. Voegler-Smith, C.K. Hall: J. Chem. Phys. **133**, 9331 (2000)
26. J.A. McCammon, B.R. Gelin, M. Karplus: Nature **267**, 267 (1977)
27. M. Levitt: Cold Spring Harbor Symp. Quant. Biol. **47**, 251 (1983)
28. G.S.D. Ayton, S. Barderhagen, P. McMurtry, D. Sulsky, G.A. Voth: IBM J. Res. & Dev. **45**, 417 (2001)
29. J.P. Hansen, I.R. McDonald: *Theory of Simple Liquids* (Academic Press, London 1986)
30. R.L. Henderson: Phys. Lett. A**49**, 197 (1974)
31. R.L. McGreevy, L. Pusztai: Mol. Sim. **1**, 359 (1988)
32. L. Reatto, D. Levesque, J.J. Weis: Phys. Rev. A **33**, 3451 (1986)
33. Y. Rosenfeld, G. Kahl: J. Phys.: Condens. Matter **9**, L89 (1997)
34. P. Gonzalez-Mozuelos, M.D. Carbajal-Tinoco: J. Chem. Phys. **24**, 11074 (1998)
35. A.P. Lyubartsev, S. Marcelja: Phys. Rev. E **65**, 041202 (2002)
36. M. Ostheimer, H. Bertagnolli: Mol. Sim. **3**, 227 (1989)
37. A.K. Soper: Chem. Phys. **202**, 295 (1996)
38. G. Toth, A. Baranyai: J. Mol. Liquids **85**, 3 (2000)
39. A.P. Lyubartsev, A. Laaksonen: Phys. Rev. E **52**, 3730 (1995)
40. A.P. Lyubartsev, A. Laaksonen: Chem. Phys. Lett. **325**, 15 (2000)
41. R.H. Swendsen: Phys. Rev. Lett. **42**, 859 (1979)
42. G.S. Pawley, R.H. Swendsen, D.J. Wallace, K.G. Wilson: Phys. Rev. B **29**, 4030 (1984)
43. A.P. Lyubartsev, A. Laaksonen: Phys. Rev. E **55**, 5689 (1997)
44. O. Matsuoka, E. Clementi, M. Yoshimine: J. Chem. Phys. **64**, 1351 (1976)
45. R. Car, M. Parrinello: Phys. Rev. Lett. **55**, 2471 (1985)
46. A.K. Soper, F. Bruni, M.A. Ricci: J. Chem. Phys. **106**, 247 (1997)
47. H.J.C. Berendsen, J.P.M. Postma, W.F. van Gunsteren, J. Hermans: 'Interaction Models for Water in Relation to Protein Hydration'. In: *Jerusalem Symposia on Quantum Chemistry and Biochemistry*, Vol. 14, ed. by B. Pullman (Reidel, Dordrecht, Holland 1981) pp. 331–342
48. L.X. Dang: J. Chem. Phys. **96**, 6970 (1992)
49. K. Heinzinger: Physica B & C **131**, 196 (1985)
50. Y. Suwannachot, S. Hannongbua, B.M. Rode: J. Chem. Phys. **102**, 7602 (1995)
51. A.P. Lyubartsev, K. Laasonen, A. Laaksonen: J. Chem. Phys. **114**, 3120 (2001)
52. H. Kistenmasher, H. Porkie, E. Clementi: J. Chem. Phys. **59**, 5842 (1973)
53. X. Periole, D. Allouche, A. Ramirez-Solis, I. Ortega-Blake, J.P. Daudey, Y.H. Sanejouand: J. Phys. Chem. B **102**, 8579 (1998)
54. D.G. Bounds: Mol. Phys. **54**, 1335 (1985)
55. B.T. Gowda, S.W. Benson: J. Chem. Phys. **79**, 1235 (1983)
56. M. Migliore, G. Corongie, E. Clementi, G.C. Lee: J. Chem. Phys. **88**, 7766 (1988)
57. X. Periole, D. Allouche, J.P. Daudey, Y.H. Sanejouand: J. Phys. Chem. B **101**, 5018 (1997)
58. D.E. Smith, L.X. Dang: J. Chem. Phys. **100**, 3757 (1994)
59. A.P. Lyubartsev, M. Karttunen, I. Vattulainen, A. Laaksonen: Soft Materials **1**, 121, (2002)
60. M. Karttunen, A. Laaksonen, A.P. Lyubartsev, I. Vattulainen: unpublished
61. A.P. Lyubartsev, A. Laaksonen: J. Chem. Phys. **109** 11207 (1999)
62. V. Lobaskin, A.P. Lyubartsev, P. Linse: Phys. Rev. E **63**, 020401 (2001)

Computer Simulations of the Electric Double Layer

André G. Moreira[1] and Roland R. Netz[2]

[1] Max-Planck-Institut für Kolloid- und Grenzflächenforschung, 14424 Potsdam, Germany, and
   Materials Research Laboratory, UCSB - Santa Barbara, CA. 93106, USA
[2] Sektion Physik, LMU - Theresienstr. 37, 80333 München, Germany

Abstract. We describe the Lekner–Sperb summation technique used to calculate the Coulomb interaction in 2D periodic systems, and discuss in detail the methods used to perform computer simulations of counterions close to charged objects (electric double layer). We focus on three situations where the double layer is formed: (1) when a wall with smeared out surface charge is in the presence of its counterions, (2) when the surface charge at the wall is modulated (e.g. formed by discrete charges), and (3) when two simple double layers interact with each other (i.e., the counterions are confined between two walls). Using Monte Carlo (MC) simulations, we obtain counterion density profiles around the charged objects and use them to test both Poisson–Boltzmann (asymptotically exact in the limit of low surface charge and/or low counterion valence) and the strong coupling theory, which becomes exact in the opposite limit of high surface charge and/or high counterion valence. For the case of single wall with smeared out surface charge (system 1), we also study the the counterion pair correlation function, which indicates a behavioral change from a three-dimensional, weakly correlated counterion distribution (at low coupling) to a two-dimensional, strongly correlated counterion distribution (at high coupling), which is paralleled by the specific heat capacity which displays a rounded hump at intermediate coupling strengths. For the charge-modulated case (system 2), we show (for high surface charge and/or multivalent counterions) that the counterions tend to stay in the close vicinity of the surface and are laterally correlated with the surface charges when the minimum approach distance between the latter and counterions is smaller than the distance between surface charges. We obtain a set of parameters for which the average counterion density profiles are very different from the ones obtained with smeared out charged surfaces, and show that in the regime where the classical Poisson–Boltzmann theory is expected to fail, the strong coupling theory agrees very well with Monte Carlo simulations. Finally, for the case of counterions confined between two equally charged walls (system 3), we analyze the inter-wall pressure and establish the complete phase diagram, featuring attraction between the walls for large enough coupling strength and at intermediate wall separation.

1 Introduction

One of the rules-of-thumb of colloid and surface science is that most surfaces are charged when in contact with water. The charges at the surface may be chemically bounded or adsorbed. In soft systems, there are usually two interactions between the particles that are important, viz. the electrostatic and van der Waals interactions. However, if the solutions are dilute, van der Waals interactions become generally unimportant in comparison to the Coulomb interaction – that is, in many situations, it is a good approximation to study only the electrostatics. One can find examples of soft systems where electrostatics is very important in many areas: in biology, ions play a crucial role in regulating the cell behavior. Another biological example is the DNA, which is a highly charged molecule, and forms with the (oppositely charged) histones a complex that is very important concerning

A.G. Moreira and R.R. Netz, Computer Simulations of the Electric Double Layer, Lect. Notes Phys. **640**, 245–278 (2004)

the packing of DNA in cells [1]. Finally, the fabrication of hollow nanoparticles using polyelectrolyte adsorption on charged colloids has recently attracted a lot of attention, since these can potentially have a great technological impact (e.g. targeted drug delivery or chemical microreactors [2]).

Charged colloidal particles have typically surface charge densities of the order of $1e/100 \, \text{Å}^2$ (where e is the elementary charge unit). The asymmetry in size and charge between the colloidal particles and their counterions is very large: in many systems, the surface of a colloidal particle can be approximated to an infinite charged plane, and the counterions (i.e. the ions that compensate for the charge on the surface, making the system globally neutral) can be assumed to be point-like. Generally, this latter assumption cannot be used if the system contains free charges of both signs, since oppositely charged particles would collapse onto each other. In the simulations we will discuss here, however, we study systems where only counterions are present.

The pioneering work in the field of charged fluids and colloidal physics done by Faraday, Debye and Hückel, Gouy and Chapman, Bjerrum, Meyer, Verwey and Overbeek, just to mention a few, brought our understanding of these systems to a relatively high standard. However, there are still many open questions which makes this an interesting area for research. In this article, we summarize some of the computer simulations and analytical results obtained by us in the past few years for some simple models of charged systems. It is organized as follows. In Sect. 2 we describe in detail the Lekner and Sperb summations for periodic Coulomb systems, the two techniques that we extensively used in our computer simulations. In the next three sections we summarize some of our previous work for three specific systems with charges: the simple double layer [3] (Sect. 3), the double layer with a charge-modulated substrate [4] (Sect. 4) and finally the interaction between two simple double layers [5] (Sect. 5). More results can be found in [6] (theory) and [7] (simulations). Section 6 contains some concluding remarks.

2 Numerics of the Coulomb Interaction

Assume a collection of N counterions, each with charge valence q, confined to the half-space $z > 0$ in the presence of an oppositely charged hard-wall located at $z = 0$ with surface charge σ_s. (Without loss of generality, we assume the counterions to be positively charged.) The generalization to other cases (e.g. discrete charges at the surface) is usually straightforward and will be done later. For a given configuration of the counterions, the Hamiltonian of the system is given by

$$\frac{\mathcal{H}}{k_B T} = \sum_{j=1}^{N-1} \sum_{k=j+1}^{N} \frac{q^2 \ell_B}{|\mathbf{r}_j - \mathbf{r}_k|} + 2\pi q \ell_B \sigma_s \sum_{j=1}^{N} z_j, \tag{1}$$

where

$$\ell_B \equiv \frac{e^2}{4\pi \varepsilon_0 \varepsilon_r k_B T}$$

is the Bjerrum length (the distance at which two elementary charges interact with the same strength as the thermal energy, $k_B T$) and $\mathbf{r}_j$ is the position of counterion j (ε_0 is the dielectric constant of vacuum and ε_r is the relative dielectric constant of the solvent). The first sum in (1) corresponds to the interaction between counterions in solution, while the second one is the interaction between each counterion and the charged wall. The system is globally neutral, i.e. $\sigma_s = q\,N/A$, where A is the area of the wall.

If one rescales all lengths to be in units of the Gouy–Chapman length

$$\mu \equiv \frac{1}{2\pi q\ell_B\sigma_s} \tag{2}$$

the Hamiltonian of the system can be then rewritten as

$$\frac{\mathcal{H}}{k_B T} = \sum_{j=1}^{N-1}\sum_{k=j+1}^{N}\frac{\Xi}{|\widetilde{\mathbf{r}}_j - \widetilde{\mathbf{r}}_k|} + \sum_{j=1}^{N}\widetilde{z}_j\,, \tag{3}$$

where $\widetilde{z} \equiv z/\mu$ and

$$\Xi \equiv \frac{q^2\,\ell_B}{\mu} = 2\pi q^3\ell_B^2\sigma_s$$

is the *coupling parameter* of the system. Note that now the Hamiltonian depends explicitly only on this parameter.

The system described above is the simplest model one can imagine for the electric double layer. Having the energy (Hamiltonian) of a system is generally the starting point for a computer simulation – in our case, we used the Monte Carlo (MC) method with Metropolis algorithm [8,9]. However, there is one important constraint: one wishes to simulate systems in the thermodynamic limit. Naturally, one cannot expect (at least with today's computers!) to be able to simulate a system with 10^{23} particles. In other words, unless one devises some way to obtain the thermodynamic limit, computer simulations will be useful only for systems with finite size.

One of the most widely used techniques to simulate thermodynamic systems is the use of periodic boundary conditions implemented with minimal image scheme. The periodicity basically imposes that when a particle crosses the edge of the simulation box, it reappears at the opposite side of it; the minimal image imposes that each particle has always the "feeling" that it is at the center of the simulation box. This means that if one calculates, say, the force exerted by particle 2 on particle 1, the distance between the two particles is in fact the distance between particle 1 and the closest replica of particle 2. The use of periodic boundary conditions (in the context of charged systems) received some critique in the past [10,11], but was later on established [12] to give very good results, provided that a suitable method to calculate the energy is used.

With the periodic boundary conditions with minimal image, each particle has the impression to be immersed in a bulk of particles, even though this bulk is formed by replicas of the original simulation box. It is nevertheless important to check for finite-size effects by comparing the simulation results between runs with different number of particles. Generally, there is a threshold in the number of particles above which the results become independent of it: this means that above this threshold, all simulations will give the same result; by induction, the simulations become equivalent to a system with an infinite number of particles.

For the simulation of the double layer, the intrinsic symmetry of the problem suggests that one should use a simulation box that is periodic in the directions parallel to the wall and non-periodic in the direction perpendicular to the wall. As we will see next, one has to account for the periodicity in the Coulomb potential.

2.1 Electrostatic Energy for Periodic Systems

In a periodically replicated system, when calculating the force (or the potential) exerted by, say, particle j on particle i, one has to take into account not only the contribution coming from the pair in the central box (cf. Fig. 1), but also the contributions coming from all the replicas of particle j in the "virtual" simulation boxes (in principle infinite in number) around the central box. The replicas of particle i do not contribute to this, provided that the interaction potential between the particles is spherically symmetric.

For most short-ranged potentials there is no need to account for the contributions of the neighboring boxes (except for particles close to the edges of the central box) since, in general, it is enough to have a box size a few times larger than the range of the interaction in order to avoid finite-size effects. For ionic systems, the contributions coming from the replicas in the neighboring boxes cannot be neglected. The Coulomb potential $\sim 1/r$ is long-ranged and in the thermodynamic limit the coupling between parts of the system at large distances contribute non-vanishingly to the total energy. Nevertheless, as long as electroneutrality holds, the energy per particle will always be bounded from below.

In a two-dimensional (2D) periodic system the electrostatic energy is given by

$$\frac{E_{\text{el}}}{k_{\text{B}}T} = \sum_{i \neq j} q_i q_j \frac{\ell_{\text{B}}}{L} v_{\text{L}}(\mathbf{r}_i - \mathbf{r}_j), \tag{4}$$

where the sum over $i \neq j$ refers to pairs of particles in the simulation box (located at $\mathbf{r}_i$ and at $\mathbf{r}_j$) and v_{L} is given by

$$v_{\text{L}}(\mathbf{r}_i, \mathbf{r}_j) = \sum_{l,m=-\infty}^{+\infty} \frac{1}{\sqrt{[\xi + l]^2 + [\eta + m]^2 + \zeta^2}}, \tag{5}$$

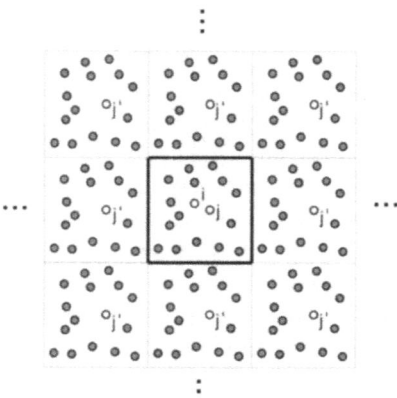

Fig. 1. Schematic top view of the two-dimensional periodic boundary conditions used in the simulations to calculate the contribution to the electrostatic energy due to the interaction between ion j (and all its replicas) and ion i

where we defined

$$\xi = \frac{|x_i - x_j|}{L}, \quad \eta = \frac{|y_i - y_j|}{L}, \quad \zeta = \frac{|z_i - z_j|}{L}. \tag{6}$$

Although the system is three-dimensional, we are assuming that it is only periodic in the x and y directions.

The sum (4) includes *all* charged particles in the system, i.e., it is extended to a region that is electrically neutral, and is finite. However, notice that the series in (5) is a non-convergent one. Although this could cause some concern, one should realize that this is related to the fact that the energy of a system is always defined with respect to a reference state. This state can be chosen arbitrarily (but be the same for all realizations of the system), and all observables, e.g. the force acting on a particle, are not affected by this choice. The sum (5) may diverge, but the *term-by-term* difference between two such sums converges. Let us then redefine v_L as

$$v_{\mathrm{L}}(\mathbf{r}_i, \mathbf{r}_j) = \sum_{l,m=-\infty}^{+\infty} \left[\frac{1}{\sqrt{[\xi + l]^2 + [\eta + m]^2 + \zeta^2}} - \frac{1}{\sqrt{[\xi_0 + l]^2 + [\eta_0 + m]^2 + \zeta_0^2}} \right], \tag{7}$$

where

$$\xi_0 = \frac{|x_i^0 - x_j^0|}{L}, \quad \eta_0 = \frac{|y_i^0 - y_j^0|}{L}, \quad \zeta_0 = \frac{|z_i^0 - z_j^0|}{L}.$$

Here $\mathbf{r}_i^0$ is the position of particle i in the reference state. From now on, the electrostatic energy by (4) will be always given with respect to this arbitrary reference state.

The specification of a reference state solves the convergence problem, but for the simulations there is still one important issue: for computer simulations, this sum has to be at some point truncated. It is then desirable to transform it into a fast converging series, so that the number of terms to be summed before a reliable truncation is not too big and the computations can be performed reasonably fast and accurate. Figure 2 makes the point in this respect: a naive summation using (7) still does not converge after summing over 10^4 (100×100) terms.

A popular way to achieve faster convergence is known as the Ewald summation [9,13,14] first introduced in the context of ionic crystals. Here we use a different method introduced by Lekner [15–17] and Sperb [18]. The main reason is that the use of the Ewald summation for systems with two-dimensional periodicity (in contrast to crystals, with periodicity in all three dimensions) requires some modifications to the original method [17,19], and only works if the ions are confined to a thin layer in the z direction. The Lekner–Sperb method is general, and is particularly suited for systems with 2D symmetry [20]. A comparison between the two methods in a system with 3D symmetry can be found in [21], and for a system with 2D symmetry in [17].

In what follows, we quickly derive the Lekner summation formula for the electrostatic energy in a 2D periodic system, based on the original derivation by Lekner [16]. To obtain a fast convergent form for the sums, we first apply to (7) the Euler transformation (valid for $\nu > 0$)

$$\frac{1}{x^\nu} = \frac{1}{\Gamma(\nu)} \int_0^{+\infty} dt\, t^{\nu-1} e^{-xt}$$

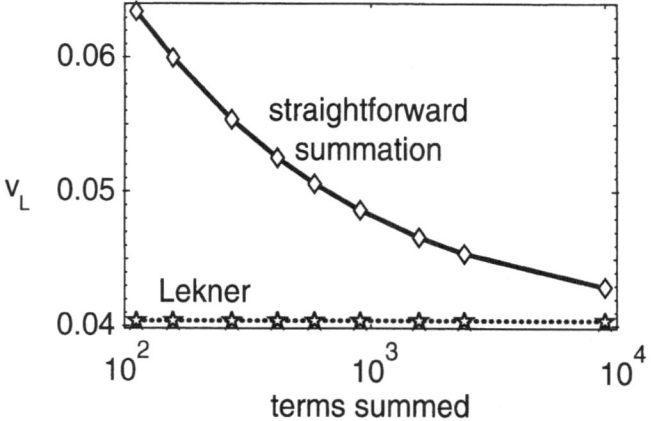

Fig. 2. Convergence of the straightforward (*full line*) and Lekner (*dashed line*) summations of (7)

followed by the Poisson–Jacobi identity

$$\sum_{m=-\infty}^{+\infty} e^{-[u+m]^2 t} = \sqrt{\frac{\pi}{t}} \sum_{m=-\infty}^{+\infty} e^{-\pi^2 m^2/t} \cos(2\pi\, m\, u)\,,$$

which leads to

$$v_L(\mathbf{r}_i,\mathbf{r}_j) = \sum_{l,m=-\infty}^{+\infty} \left\{ \int_0^\infty dt\, \frac{1}{t} e^{-t\varrho^2 - \pi^2 l^2/t} \cos(2\pi l\xi) - \int_0^\infty dt\, \frac{1}{t} e^{-t\bar\varrho^2 - \pi^2 l^2/t} \cos(2\pi l\xi_0) \right\},$$

where $\varrho^2 = [\eta + m]^2 + \zeta^2$ and $\bar\varrho^2 = [\eta_0 + m]^2 + \zeta_0^2$. We now isolate the terms with $l = 0$, and use the integral representation of the modified Bessel function of second kind, viz.

$$\int_0^{+\infty} dt\, t^{\nu-1} e^{-t\varrho^2 - \pi^2 l^2/t} = 2 \left[\pi \frac{|l|}{|\varrho|}\right]^\nu K_\nu(2\pi|l|\,\varrho|),$$

on the terms with $l \neq 0$ and obtain

$$v_L(\mathbf{r}_i,\mathbf{r}_j) = \sum_{m=-\infty}^{+\infty} \int_0^{+\infty} dt\, \frac{1}{t} \left[e^{-t\varrho^2} - e^{-t\bar\varrho^2} \right]$$

$$+ 4 \sum_{l=1}^{+\infty} \cos(2\pi l\xi) \sum_{m=-\infty}^{+\infty} K_0(2\pi l\varrho)$$

$$- 4 \sum_{l=1}^{+\infty} \cos(2\pi l\xi_0) \sum_{m=-\infty}^{+\infty} K_0(2\pi l\bar\varrho).$$

The term corresponding to $l = 0$ (the second term on the right hand side) leads to

$$\sum_{m=-\infty}^{+\infty} \int_0^{+\infty} dt\, \frac{1}{t} \left[e^{-t\varrho^2} - e^{-t\bar\varrho^2} \right] = -\ln\left(\frac{\cosh(2\pi\zeta) - \cos(2\pi\eta)}{\cosh(2\pi\zeta_0) - \cos(2\pi\eta_0)} \right). \tag{8}$$

Notice that the existence of the reference state ensures that the integral in (8) is well behaved.

We finally arrive to Lekner's summation formula for v_{L}, which reads

$$v_{\mathrm{L}}(\mathbf{r}_i, \mathbf{r}_j) = C - \ln\Big(\cosh(2\pi\zeta) - \cos(2\pi\eta)\Big) + s(\xi, \eta, \zeta), \qquad (9)$$

where s is the series

$$s(\xi, \eta, \zeta) = 4\sum_{l=1}^{+\infty}\cos(2\pi l\xi)\sum_{m=-\infty}^{+\infty}K_0\Big(2\pi l\sqrt{[\eta + m]^2 + \zeta^2}\Big) \qquad (10)$$

and C is

$$C = \ln\Big(\cosh(2\pi\zeta_0) - cos(2\pi\xi_0)\Big) - s(\xi_0, \eta_0, \zeta_0).$$

This formula has two major advantages over the original formula (7). On one hand, it is much faster in convergence, i.e., one can truncate the sum after a few terms in l and m (cf. Fig. 2). On the other hand, one does not have to specify a reference state (given by C in (9)), since the sums involved in (9) are finite. In other words, by doing the transformations to obtain the Lekner formula, (7) has been regularized; the existence of the reference state only ensures that the mathematical steps taken between (7) and (9) are well defined. This means that, during the simulations, one can choose any value for C.

The Bessel function diverges as its argument becomes small, and the sum $s(\xi, \eta, \zeta)$ can become very inefficient for small $\sqrt{\eta^2 + \zeta^2}$. To solve this, Sperb [18] proposes the following transformation

$$4\sum_{l=1}^{+\infty}\cos(2\pi l\xi)K_0(2\pi l\varrho) = -1.386294 + 2\ln(\varrho) + \frac{1}{\sqrt{\xi^2 + \varrho^2}} - \Psi(1 + \xi)$$

$$-\Psi(1 - \xi) + \sum_{l=1}^{+\infty}\binom{-1/2}{l}\varrho^{2l}\big[Z(2l + 1, 1 + \xi)$$

$$+Z(2l + 1, 1 - \xi)\big], \qquad (11)$$

where Ψ is the digamma function, Z is the commonly known Hurwitz zeta function [22] and $\varrho = \sqrt{[\eta + m]^2 + \zeta^2}$. This series converges very quickly if ϱ is small, provided that $\xi < 1/2$.

Given (9), we can write down the electrostatic energy of any 2D periodic system. The energy of the system composed of a charged wall with (smeared out) charge density σ_{s} and counterions with charge valence q is given by

$$\frac{E_{\mathrm{el}}}{k_{\mathrm{B}}T} = E_0 + \frac{\ell_{\mathrm{B}}}{2L}\int d\mathbf{r}\,d\mathbf{r}'\,\hat{\varrho}_{\mathrm{c}}(\mathbf{r})v_{\mathrm{L}}(\mathbf{r}, \mathbf{r}')\hat{\varrho}_{\mathrm{c}}(\mathbf{r}'), \qquad (12)$$

where

$$E_0 = -\frac{\ell_{\mathrm{B}}}{2L}\sum_i q_i^2 v_{\mathrm{L}}(\mathbf{r}_i, \mathbf{r}_i)$$

is the self-energy of each particle in the system and

$$\hat{\varrho}_c(\mathbf{r}) = q \sum_{i=1}^{N} \delta(\mathbf{r} - \mathbf{r}_i) - \sigma_s \delta(z) \Theta_0^L(x) \Theta_0^L(y) \tag{13}$$

is the charge distribution within the simulation box and $\Theta_0^L(x) = 1$ if $0 < x < L$, and zero otherwise. The electroneutrality condition implies that $\int d\mathbf{r} \, \hat{\varrho}_c(\mathbf{r}) = 0$, i.e., $\sigma_s = qN/L^2$. Putting (13) and (9) in the electrostatic energy (12) one obtains, after some algebra, the result

$$\frac{E_{el}}{k_B T} = \frac{q^2 \ell_B}{L} \sum_{i \neq j} v_L(\xi, \eta, \zeta) + 2\pi q \ell_B \sigma_s \sum_{i=1}^{N} z_i - \frac{\ell_B q^2 N^2}{2L} \int_0^1 dx \, dy \, v_L(x, y, 0),$$

$$\tag{14}$$

where ξ, η and ζ are defined according to (6). The integral over v_L in (14) is

$$\int_0^1 dx \, dy \, v_L(x, y, 0) = \ln(2)$$

and after rescaling all lengths with μ (the Gouy–Chapman length), one finally arrives to

$$\frac{E_{el}}{k_B T} = \frac{\Xi}{\tilde{L}} \sum_{i \neq j} v_L(\xi, \eta, \zeta) + \sum_{i=1}^{N} \tilde{z}_i - \frac{\ln(2)}{2\sqrt{2\pi}} \sqrt{\Xi} N^{3/2}. \tag{15}$$

This corresponds to the energy given by (3) for the system with periodic boundary condition. $\tilde{L}$ is the box size in units of Gouy–Chapman length given by

$$\tilde{L} = \frac{L}{\mu} = \sqrt{2\pi N \Xi}.$$

Then, electroneutrality in the simulation box is fulfilled. $\Xi = 2\pi q^3 \ell_B^2 \sigma_s$ is the coupling parameter, as previously defined. Notice that we neglected the self-energy of the particles, which corresponds to a constant term directly proportional to N that can be added to E_{el} without any change of the thermodynamic properties of the system. This is the definition we use to calculate the electrostatic energy for each trial configuration in the computer simulations. The first term corresponds to the interaction between the counterions, and the second term to the interaction between the counterions and the charged wall. The third term in the right hand side of (15) is the self-energy of a $\tilde{L} \times \tilde{L}$ periodically repeated wall.

2.2 Precision of the Lekner Summation

When calculating (10) in our simulations, we have typically included seven terms in the sum in m ($m = -3, \ldots, 3$), and made the number of terms to be summed in the index l to depend on the value of $\varrho = \sqrt{[\eta + m]^2 + \zeta^2}$:

a   if $\varrho \geq 3$, truncate the Lekner sum at $l = 3$;

b if $1/3 < \varrho < 3$, truncate the Lekner sum at $l = 2 + \text{Integer}[3/\varrho]$;

c if $\varrho < 1/3$ and $\xi < 1/2$, use Sperb's formula (11) truncated at $l = 8$; if $\xi > 1/2$, use Lekner sum as above.

This scheme is arbitrary, but it proves to be quite accurate and fast. In order to increase the efficiency of the method, one should use the 2D symmetry and exchange ξ and η such that the former becomes the smallest of the two. This also ensures, given the procedure above, that $\xi < 1/2$, a necessary condition to use Sperb's formula.

It is instructive to make a quick estimate on the truncation error in the Lekner summation with the recipe given above. Define

$$e = 4 \sum_{l=l_{\max}+1}^{+\infty} \cos(2\pi l \xi) \sum_{m=m_{\max}+1}^{m=+\infty} \left[K_0(2\pi l \varrho_m^+) + K_0(2\pi l \varrho_m^-) \right]$$

where $\varrho_m^+ = \sqrt{[\eta + m]^2 + \zeta^2}$ and $\varrho_m^- = \sqrt{[\eta - m]^2 + \zeta^2}$. This gives the missing part of v_{L} due to truncation in the sums over l and m. Notice that $\varrho_m^+ \geq m$ and $\varrho_m^- \geq \sqrt{m^2 - 2m}$ (for $m \geq 2$), which implies that

$$K_0(2\pi l \varrho_m^+) \leq K_0(2\pi \, l \, m) \qquad \text{and} \qquad K_0(2\pi l \varrho_m^-) \leq K_0\left(2\pi \, l \, \sqrt{m^2 - 2m}\right).$$

Using the fact that $\cos(x) \leq 1$, we conclude that $e \leq e_{\max}$, with

$$e_{\max} = 4 \sum_{\substack{l=l_{\max}+1 \\ m=m_{\max}+1}}^{+\infty} \left[K_0(2\pi l m) + K_0(2\pi l \sqrt{m^2 - 2m}) \right].$$

For $m_{\max} = 3$ and $l_{\max} = 3$, $e_{\max} \sim 8 \times 10^{-32}$. This means that any error involved in the Lekner summation comes from the calculation of the functions involved (Bessel, cosine, etc.), since the truncation error, using the recipe given above, is negligible. For instance, we used the scheme described in Abramowitz [22] to calculate the Bessel function, which has an absolute error of $\sim 10^{-8}$.

We now turn our attention to the computer simulation results obtained using the techniques described above. The first system we present is the simple double layer problem.

3 Counterions Close to a Single Charged Wall

The ionic distribution close to a charged wall (electric double layer) is an old problem which has attracted great attention in recent years. The basic system consists of an impenetrable wall with a smeared out charge density σ_s in the presence of its counterions (with charge valence q) immersed in a solvent characterized by a certain dielectric constant. Some examples where the double layer problem is relevant can be found in biology (e.g. cell membranes are charged due to the phospholipids that compose it), in colloidal chemistry (the stability problem of lyophobic colloids [23]) or in interface physics [24].

The first approximate solution to this model was obtained by Gouy [25] and Chapman [26] using the Poisson–Boltzmann (PB) equation and point-like counterions, in a way similar to the one later used by Debye and Hückel to treat strong electrolytes [27]. This solution is asymptotically exact in the limit of weakly charged systems or at high temperatures, where the correlations between the counterions become less important, and a mean-field theory can be used to describe the system.

Since then, there have been various attempts to improve or correct the Poisson-Boltzmann solution [28,29]. One example is the assumption of a layer of condensed counterions (a Stern layer [30] with approximately 2 Å of thickness) in coexistence with non-condensed counterions, which would be described by the Poisson–Boltzmann equation. Liquid state theory has also been applied to these systems, where the hypernetted chain (HNC) closure relation was proved to be a very powerful technique for charged systems [31]. When applied to the double layer problem [32–35], it yields numerical results that are in very good agreement with computer simulations [36–39]. However, integral equation theories sometimes rely on numerical work and fail to provide an intuitive insight into the problem. More recently, there has been some interest in applying field-theoretic methods to charged systems. Perturbative field theory has been applied to study different aspects of these systems, ranging from the critical behavior of the Restricted Primitive Model [40–42] to the equation of state of a One-Component Plasma [43,44]. Also the double layer has been tackled with these methods, as for instance to study image-charge effects [45,46], or to obtain corrections to Poisson–Boltzmann [47,48].

In what follows we summarize the field-theoretic derivation [3,6] of an analytic expression for the counterion distribution which is asymptotically exact for systems with highly valent counterions and highly charged surfaces (strong coupling (SC) limit), complementing the Poisson–Boltzmann theory, which is valid in the opposite (weak coupling) limit. We also present in detail the computer simulations performed to study the double layer and used to test the analytical results.

The starting point is the Hamiltonian given in (3), viz.

$$\frac{\mathcal{H}}{k_B T} = \sum_{j=1}^{N-1} \sum_{k=j+1}^{N} \frac{\Xi}{|\widetilde{\mathbf{r}}_j - \widetilde{\mathbf{r}}_k|} + \sum_{j=1}^{N} \widetilde{z}_j . \tag{16}$$

We remind that all lengths are given in units of Gouy–Chapman length (2), and that the coupling parameter is defined as

$$\Xi \equiv \frac{q^2 \ell_B}{\mu} = 2\pi q^3 \ell_B^2 \sigma_s. \tag{17}$$

The Hamiltonian above can give us some useful information about this system. The Monte Carlo simulations that we will present show that the typical distance between an ion and the charged wall is of the order of one Gouy–Chapman length, meaning that the binding energy per ion (cf. the second term on the right hand side of (16)) is of the order of unity. In Fig. 3 we show snapshots of Monte Carlo simulations for three different values of the coupling, viz. $\Xi = 0.1$, 10 and 10^4. For low coupling (Fig. 3(a)), the counterions distribute themselves in a diffuse cloud, and each counterion is surrounded

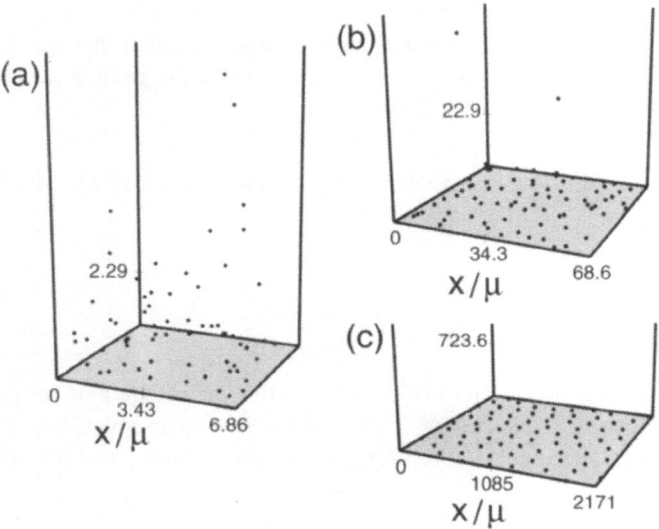

Fig. 3. Snapshots of counterion distributions containing 75 particles for different values of the coupling constant Ξ: (a) weak coupling regime ($\Xi = 0.1$) where the Poisson–Boltzmann prediction is accurate; (b) intermediate coupling regime ($\Xi = 10$); (c) strong coupling regime ($\Xi = 10^4$). Notice that in (a) there are 8 particles located far away from the wall (particles not shown here)

in all directions by other counterions – a usual situation where mean-field theory yields good results. The typical distance between the counterions scales like $\Xi^{1/3}$, and the repulsion energy per pair goes like $\Xi^{2/3}$ (cf. the first term on the right hand side of (16)), i.e., the repulsion between counterions becomes small (compared to the binding energy per ion), and the counterion distribution should be quite disordered. In the other limit, (Fig. 3(c)), the counterions organize themselves in a two-dimensional layer close to the wall. Each counterion is laterally surrounded by other counterions, typically at a distance $\Xi^{1/2}$ away from each other. The repulsion energy per pair scales like $\Xi^{1/2}$, that is, it is now the dominant term in the energy. Note that here mean-field theory is expected to break down, since the ions are almost independent to move in the direction perpendicular to the wall, and are only confined by the linear potential due to the fixed smeared out charge distribution on the wall. In fact, the latter suggests that one should expect the counterion distribution perpendicular to the wall to decay like an exponential law (cf. barometric law) for large Ξ.

The partition function of a system of N counterions interacting through the Coulomb potential $v_c(\mathbf{r}) = \ell_B/r$ (in units of $k_B T$) with each other and with a fixed charge distribution $\sigma(\mathbf{r})$ is given by

$$\mathcal{Z} = \frac{1}{N!} \int \prod_{i=1}^{N} d\mathbf{r}_i \, \exp\{-\mathcal{H}\}, \tag{18}$$

where the integration is done over the positions of all particles and $\mathcal{H}$ is given by (16). In order to get a field-theoretic description of the system, one performs a Hubbard–

Stratonovich on (18). Another simplifying step is to Legendre-transform the resulting field-theoretic description, obtaining a grand partition function that depends on the chemical potential of the counterions. A detailed derivation is given in [6]. After some algebra,one obtains the grand partition function

$$Q = \int \frac{\mathcal{D}\bar{\phi}}{\mathcal{Z}_v} \exp\left(-\frac{1}{8\pi\Xi} \int d\tilde{\mathbf{r}} \left[[\nabla\bar{\phi}(\tilde{\mathbf{r}})]^2 - 4i\delta(\tilde{z})\bar{\phi}(\tilde{\mathbf{r}}) - 4\Lambda\theta(\tilde{z})e^{h(\tilde{\mathbf{r}})-i\bar{\phi}(\tilde{\mathbf{r}})} \right] \right),$$

(19)

where

$$\theta(\tilde{z}) = \begin{cases} 1 & \text{if } \tilde{z} > 0 \\ 0 & \text{otherwise} \end{cases}$$

is the Heavyside function (which basically states that the counterions are confined to the positive half-space), Λ is the fugacity (exponential of the chemical potential) of the counterions, and $\mathcal{Z}_v = \sqrt{\det v_c}$. The average counterion density, $\varrho(\tilde{\mathbf{r}})$ is given by

$$\frac{\varrho(\tilde{\mathbf{r}})}{2\pi\ell_B\sigma_s^2} = \Lambda\langle e^{-i\bar{\phi}(\tilde{z})}\rangle.$$

(20)

The normalization condition for the counterion distribution, $\mu \int d\tilde{z}\varrho(\tilde{z}) = \sigma_s/q$, leads to

$$\Lambda \int_0^\infty d\tilde{z}\langle e^{-i\bar{\phi}(\tilde{z})}\rangle = 1$$

(21)

showing that the expectation value of the fugacity term in (19) is bounded and of the order of unity per unit area.

Given the structure of the action in (19), the saddle-point analysis [47] should be valid for $\Xi \ll 1$. This leads to the equation

$$\frac{d^2\bar{\phi}(\tilde{z})}{d\tilde{z}^2} = 2i\Lambda e^{-i\bar{\phi}(\tilde{z})}$$

with the boundary condition $d\bar{\phi}(\tilde{z})/d\tilde{z} = -2i$ at $\tilde{z} = 0$. The solution of this differential equation is

$$i\bar{\phi}(\tilde{z}) = 2\ln\left(1 + \Lambda^{1/2}\tilde{z}\right),$$

(22)

while the boundary condition leads to $\Lambda = 1$, which shows that the saddle-point approximation is indeed valid in the limit $\Xi \ll 1$. Combining (20) and (22), the density distribution of counterions is given by the well-known Poisson–Boltzmann prediction [23,49]

$$\frac{\varrho(\tilde{z})}{2\pi\ell_B\sigma_s^2} = \frac{1}{[1+\tilde{z}]^2}.$$

(23)

This result is exact in the limit of vanishing Ξ.

Let us now consider the opposite limit, when the coupling constant Ξ is large [6]. In this case, the saddle-point approximation breaks down, since the prefactor in front of the action in (19) becomes small. However, from the field-theoretic partition function (19), it is clear what has to be done in this limit. Since the fugacity term is bounded, as evidenced by (21), one can expand the partition function (and also all expectation values) in powers of Λ/Ξ. Upon Legendre transformation to the canonical ensemble, this gives the standard virial expansion. The normalization condition (21) can be solved by an expansion of the fugacity as $\Lambda = \Lambda_0 + \Lambda_1/\Xi + \cdots$, which leads to an expansion of the density profile with the small parameter $1/\Xi$. While the standard virial expansion fails for homogeneous bulk charged systems because of infra-red divergences, these divergences are renormalized for the present case of inhomogeneous distribution functions via the normalization condition (21). To leading order in this expansion, the rescaled density is

$$\frac{\varrho(\widetilde{\mathbf{r}})}{2\pi\ell_{\mathrm{B}}\sigma_{\mathrm{s}}^2} = \Lambda \exp\left(-\frac{\Xi}{2}v_{\mathrm{c}}(0) + \frac{1}{2\pi}\int d\widetilde{\mathbf{r}}'\,\delta(\widetilde{z})\frac{1}{|\widetilde{\mathbf{r}}' - \widetilde{\mathbf{r}}|}\right),$$

where all lengths have been rescaled by μ. From the normalization condition (21) we obtain

$$\Lambda_0 = \exp\left(\frac{\Xi}{2}v_{\mathrm{c}}(0) - \frac{1}{2\pi}\int d\widetilde{\mathbf{r}}'\,\delta(z')\frac{1}{\widetilde{r}'}\right),$$

and thus to leading order the density distribution is given by

$$\frac{\varrho(\widetilde{z})}{2\pi\ell_{\mathrm{B}}\sigma_{\mathrm{s}}^2} = \exp(-\widetilde{z}). \tag{24}$$

This is the exponential decay suggested by the previous scaling analysis of the Hamiltonian. Notice that this result is *exact* in the limit $\Xi \to \infty$. An exponential density profile (however with a different prefactor) has also been obtained by Shklovskii [50] using an heuristic model for a highly charged surface, where counterions bound to the wall are in chemical equilibrium with free ions.

We mention in passing that a similar exponential decay would also follow from the linearization of the Poisson–Boltzmann equation, very much in the spirit of Debye and Hückel [27]. The density profile in this case is given by

$$\frac{\varrho(\widetilde{z})}{2\pi\ell_{\mathrm{B}}\sigma_{\mathrm{s}}^2} = 2\exp(-2\widetilde{z}). \tag{25}$$

However, this should be regarded as a coincidence: The Poisson–Boltzmann equation is only valid when Ξ is small; a linear approximation of the equation cannot describe better the behavior of the system than the full equation itself, especially for values of Ξ where the equation in expected to break-down.

3.1 Counterion Density Profile

In Fig. 4 we show the Monte Carlo results for the average counterion density distribution for various values of Ξ, as well as the predictions from Poisson–Boltzmann and strong coupling. Note the very good agreement between Poisson–Boltzmann and strong coupling theory (main graph) with the simulation results at low and high coupling, respectively. The inset in Fig. 4 shows Monte Carlos results for $\Xi = 10^5, 10^4, 100, 10, 1,$

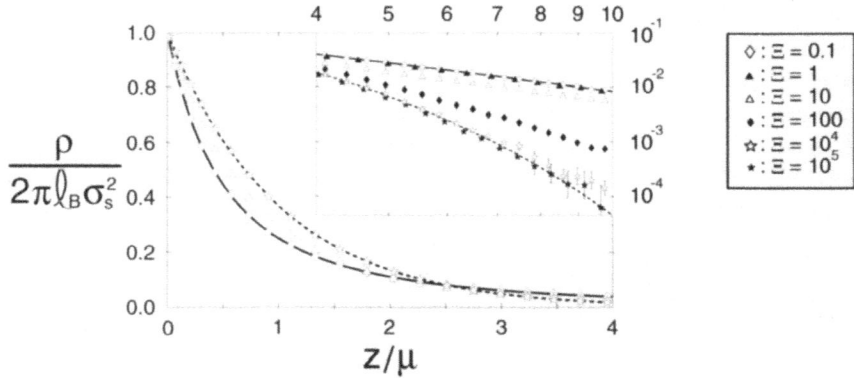

Fig. 4. Rescaled counterion density distribution $\varrho/2\pi\ell_B\sigma_s^2$ as a function of the rescaled distance $\tilde{z} = z/\mu$ from the charged wall. The main figure shows Monte Carlo results for $\Xi = 0.1$ (*open diamonds*), $\Xi = 10$ (*open triangles*) and $\Xi = 10^4$ (*open stars*). The *solid* and the *dashed* lines denote the Poisson–Boltzmann and the strong coupling theory predictions, equations (23) and (24) respectively. The inset shows a log-log plot of the density distribution at distances larger than $\tilde{z} = 4$ for $\Xi = 10^5, 10^4, 100, 10, 1$ and 0.1 (from bottom to top). All simulations were performed with 75 particles and 10^6 Monte Carlo steps (MCS) per particle, except the data for $\Xi = 0.1$ where 600 particles where simulated. Unless when explicitly shown, the error bars are comparable to or smaller than the size of the symbols

and 0.1 (from bottom to top) at larger distances from the wall (between 4 and 10 Gouy–Chapman lengths).

Previous computer simulations [51,52] have also confirmed the validity of PB at the weak coupling regime. In fact, a picture that has emerged from previous studies is that PB is a good description for systems with monovalent counterions, but a poor description for systems with counterions with higher valence; deviations from the PB behavior can even lead to attraction between similarly charged plates [36] (see also Sect. 5). Equation (17) gives us the clue to this: q has a power 3 in the expression for the coupling; if a system with monovalent ions has $\Xi \lesssim 1$, the same system with divalent ions will have a coupling 8 times larger! As is already clear from the simulation results in Fig. 4, a system with $\Xi = 10$ clearly deviates from the PB curve in the vicinity of the wall.

The computer simulations done at high Ξ confirm the novel strong coupling limit as the correct asymptotic limit at infinite coupling. With this in mind, we can say that Fig. 4 presents a unified picture of the simple double layer problem (when only counterions are present): PB is the asymptotically correct as $\Xi \rightarrow 0$, SC is asymptotically correct as $\Xi \rightarrow \infty$, and any system with a Ξ between those limits will present a density profile that is between the power-law (PB) and the exponential decay (SC). Experimentally, a coupling of $\Xi = 100$ – which is already quite close to the strong coupling regime – can be reached with divalent counterions for a surface charge density $\sigma_s \simeq 3.9e \, \text{nm}^{-2}$, which is feasible with compressed charged monolayers, and with trivalent counterions with $\sigma_s \simeq 1.2e \, \text{nm}^{-2}$, which is a typical value.

Finally, note that the rescaled densities as shown in Fig. 4 always fulfill

$$\frac{\varrho(0)}{2\pi\ell_{\mathrm{B}}\sigma_{\mathrm{s}}^2} = 1. \tag{26}$$

This is a trivial consequence of the contact-value theorem [36,47], which states that the value of the counterion density at contact with the wall $\varrho(0)$ is related to the pressure P acting on the wall through

$$\frac{P}{2\pi\ell_{\mathrm{B}}\sigma_{\mathrm{s}}^2} = 1 - \frac{\varrho(0)}{2\pi\ell_{\mathrm{B}}\sigma_{\mathrm{s}}^2}.$$

When in equilibrium, $P = 0$ and the result (26) follows. Incidentally, we remind that in (25) the density profile obtained through the linearized PB equation leads to a value of two for the rescaled density at contact, violating this theorem.

3.2   Two-Dimensional Liquid and Crystal

At very high couplings, one should expect a crystallization of the counterions into a Wigner crystal [53,54]. A two-dimensional one-component plasma is known to crystallize at values of the plasma parameter $\Gamma \simeq 125$ [31], which can be related to our coupling constant Ξ through $\Xi = 2\,\Gamma^2$, meaning that the Wigner crystal should be visible in the simulations at around $\Xi \simeq 31\,000$. In Fig. 5 we show the top view snapshots of two systems with high Ξ, one with $\Xi = 10^4$ (below crystallization) and another with $\Xi = 10^5$ (above crystallization). The latter shows, in contrast to the former, a clear two-dimensional ordering.

In the strong coupling regime, the average lateral distance between neighboring counterions is approximately given by

$$2\tilde{a}_\perp = 2\sqrt{2\,\Xi},$$

which is larger than the typical distance between the counterions and the charged wall ($\tilde{z} \sim 1$). For very large coupling, $\tilde{a}_\perp \gg 1$ and the counterion layer is essentially two-dimensional. This effect is visible in the snapshots shown in Fig. 3: while for $\Xi = 0.1$

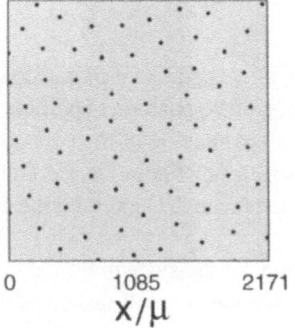
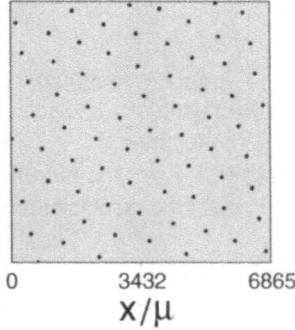

| 0 | 1085 | 2171 |

x/μ

| 0 | 3432 | 6865 |

x/μ

Fig. 5. Top view snapshots of a system with $\Xi = 10^4$ (left: below the expected crystallization transition) and $\Xi = 10^5$ (right: above the transition) (see Fig. 3(a)) the counterions form a three-dimensional diffuse cloud, for $\Xi = 10^4$ (Fig. 3(c)) they form a clearly two-dimensional layer. This means that, although the perpendicular density profiles shown in Fig. 4 are relatively close to each other, the physics behind each of those curves is fundamentally different. This becomes evident from the heat capacity and the 2D pair distribution function, as we now demonstrate.

The heat capacity is related to the fluctuations of the energy around its average through [9]

$$C_{\mathrm{v}} = k_{\mathrm{B}} \left\langle \left(\frac{\delta \mathcal{H}}{k_{\mathrm{B}} T} \right)^2 \right\rangle,$$

where the brackets denote the average over trial configurations, and $\delta \mathcal{H}$ is the instantaneous deviation of the total electrostatic energy from its mean,

$$\delta \mathcal{H} = \mathcal{H} - \langle \mathcal{H} \rangle.$$

In Fig. 6 we show the excess heat capacity per particle (the total C_{v} includes the kinetic contribution $3/2$) for systems with a coupling constant between $\Xi = 0.1$ and $\Xi = 2 \times 10^4$. In the limit $\Xi \to 0$, the excess C_{v} tends to 1, which is the value predicted by PB theory. At high coupling constants in the range $200 < \Xi < 10^4$, the specific heat increases logarithmically with the coupling, following approximately the functional form $C_{\mathrm{v}}/(N k_{\mathrm{B}}) = 1.04 + 0.225 \log_{10} \Xi$ over two orders of magnitudes, which is denoted by a full line in Fig. 6. The reason for this behavior is at present not clear. The broad hump that appears in the range $10 < \Xi < 100$ can be associated with the freezing out of lateral degrees of freedom of the counterions, since in this range the counterion distribution changes from being three-dimensional and rather weakly correlated (for small coupling), to being two-dimensional and rather strongly correlated in the lateral directions. The hump does not change its size or shape as the system size is varied (cf. Fig. 9(b)), suggesting that it is not caused by a phase transition but merely reflects the change in the short-ranged correlations. However, we cannot rule out a singular behavior with a negative heat-capacity exponent.

The behavior of the heat capacity becomes clearer if one looks at the two-dimensional pair distribution function [9], defined as

$$g_{\mathrm{2D}}(\tilde{r}_{xy}) = \frac{A}{N^2} \left\langle \sum_{\langle ij \rangle} \delta(\tilde{\mathbf{r}}_{xy} - \tilde{\mathbf{r}}_{xy,i} + \tilde{\mathbf{r}}_{xy,j}) \right\rangle,$$

where $\sum_{\langle ij \rangle}$ denotes a sum over pairs of particles, $\tilde{\mathbf{r}}_{xy}$ is the two-dimensional vector $(\tilde{x}, \tilde{y})$ (with magnitude $\tilde{r}_{xy}$) and $\tilde{\mathbf{r}}_{xy,i}$ is the projection of the position of particle i into the xy plane. The function $g_{\mathrm{2D}}(\tilde{r}_{xy})$ is the two-dimensional analog of the pair distribution function [49]. It gives the ratio between the probability of finding two counterions at distance $\tilde{r}_{xy}$ and the expected probability for a homogeneous 2D gas with the same bulk density.

As one can see in Fig. 7, the correlation function for the system with $\Xi = 1$ (filled triangles) only shows a very short-ranged depletion zone at small distances, while the correlation function for $\Xi = 100$ (filled diamonds) exhibits a pronounced correlation hole at small separations where the correlation function is essentially zero over a finite range. The position of the hump in the specific heat, see Fig. 6, coincides with the region

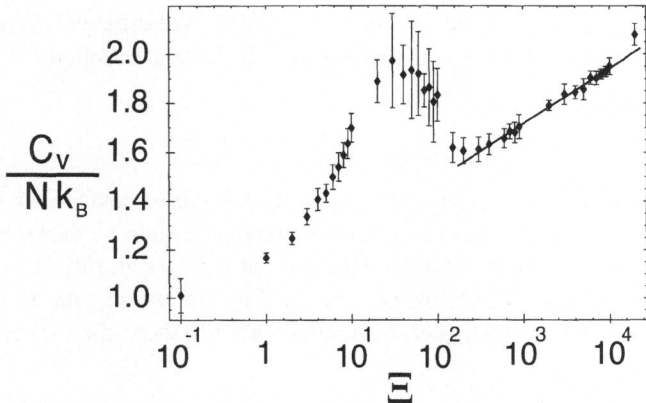

Fig. 6. The specific heat capacity per counterion as a function of the coupling constant. The hump at intermediate coupling constants $10 < \Xi < 100$ indicates a structural change from a weakly correlated (for small coupling) to a strongly correlated counterion layer (for large coupling). The *solid line*, describing the data for coupling constants $\Xi > 100$, is a fit given by $C_v/(Nk_B) = 1.04 + 0.225 \log_{10} \Xi$

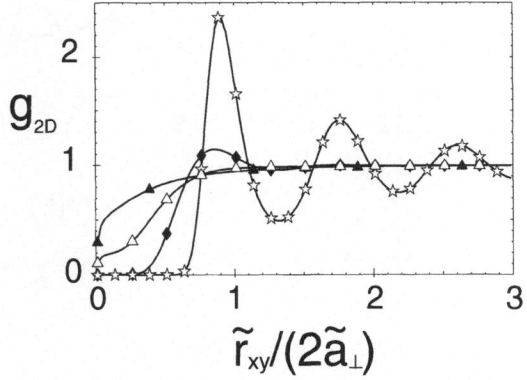

Fig. 7. Two-dimensional pair correlation function for a system with $\Xi = 1$ (*filled triangles*), 10 (*open triangles*), 100 (*filled diamonds*), and 10^4 (*open stars*), as a function of the xy projected distance between the ions $\tilde{r}_{xy} = r_{xy}/\mu$; $2\tilde{a}_\perp = 2/\sqrt{\pi N/A}$ is the distance expected between the particles for a homogeneous two-dimensional gas with N/A particles per unit area. The error bars are comparable to or smaller than the symbols

where the two-dimensional correlation hole is created around each of the counterions, namely the range of coupling constants $10 < \Xi < 100$.

At the highest coupling constant shown in Fig. 7, $\Xi = 10^4$ (open stars), the oscillatory behavior of $g_{2D}(\tilde{r}_{xy})$ indicates that the counterions behave like a strongly correlated two-dimensional liquid. The first maximum in the pair correlation function appears at a distance of $\tilde{r}_{xy}/(2\tilde{a}_\perp) \approx 0.9$, which can be compared with the expected peak in a perfect hexagonal crystal, $\tilde{r}_{xy}/(2\tilde{a}_\perp) = (\pi/(2\sqrt{3}))^{1/2} \approx 0.95$, and in a perfect square crystal, $\tilde{r}_{xy}/(2\tilde{a}_\perp) = \sqrt{\pi}/2 \approx 0.89$. Clearly, the structure seen in the simulation at

$\Xi = 10^4$ is not a perfect crystal, but the position of the nearest-neighbor maximum is what one would expect for a system close to a crystallization transition.

3.3 Finite-Size Effects

The density distribution as a function of N, the number of counterions used in the simulation, converges very quickly to its thermodynamic limit as shown in Fig. 8(a): for a coupling of $\Xi = 0.1$, a system with only 35 particles already shows a density distribution that is very close to the one exhibited by the system with $N = 200$. This convergence is even faster for higher couplings. Figure 8(b) shows the difference between

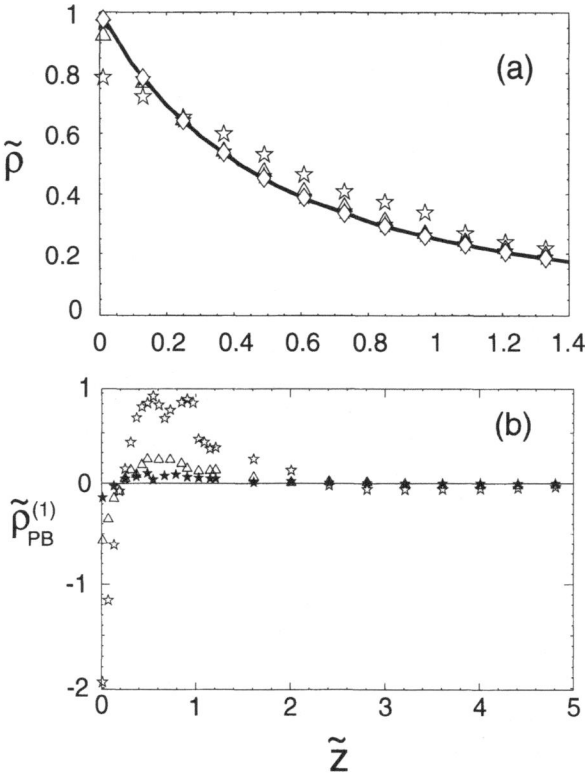

Fig. 8. Plot (**a**): results for the rescaled counterion density distribution as a function of the rescaled distance from the wall for $\Xi = 0.1$ and 5 particles and 10^8 MCS (*open stars*), 15 particles and 10^7 MCS (*open triangles*), 35 particles and 10^7 MCS (*filled stars*) and 200 particles and 2×10^6 MCS (*open diamonds*). The *full line* denotes the PB result. Notice that finite-size effects are almost negligible for 35 particles in comparison to $N = 200$ (the filled stars are essentially under the open diamonds), and that the system with 15 particles is already quite close to the large-N behavior. For systems with larger Ξ, finite-size effects become negligible with even less particles. The error bars are smaller than the symbols. Plot (**b**): Difference between the simulation result and the PB result normalized to the coupling parameter according to $\widetilde{\varrho}_{PB}^{(1)}(\widetilde{z}) = \left(\widetilde{\varrho}(\widetilde{z}) - \widetilde{\varrho}_{PB}(\widetilde{z})\right)/\Xi$

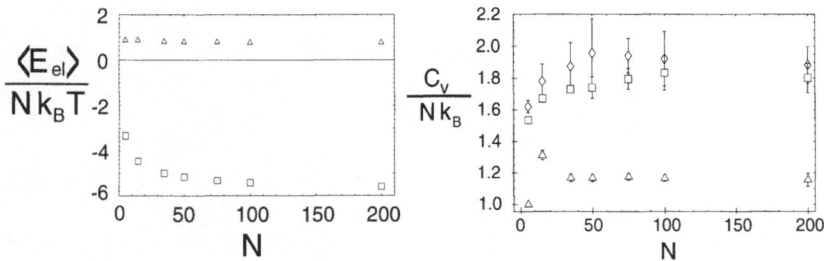

Fig. 9. Finite-size effects on the energy and heat capacity. Left: average energy per particle for $\Xi = 1$ (*open triangles*) and $\Xi = 100$ (*open squares*) as a function of N, the number of counterions in the simulation. Right: heat capacity per particle for $\Xi = 1$ (*open triangles*), $\Xi = 60$ (*open diamonds*) and $\Xi = 100$ (*open squares*) as a function of N, the number of counterions in the simulation. Unless when explicitly shown, the error bars are smaller than the symbols

the simulations and the Poisson–Boltzmann result (normalized to the coupling parameter, which can be related to the first loop correction in the PB theory [7,47]). This emphasizes the fact that the convergence to a final curve is quite fast using the Lekner summation with the periodic boundary condition and minimal image. We will see later that other boundary conditions do not behave so well when this comparison is done.

Figure 9(a) shows the average energy per particle obtained from Monte Carlo simulations according to (15). The electrostatic energy per particle is plotted for coupling $\Xi = 1$ and $\Xi = 100$ as a function of the number of counterions N. This is the energy given by (15) (per particle) averaged over the trial configurations used in the course of the simulations. As expected, after a certain critical size, the energy per particle becomes constant. Notice that the thermodynamic energy per particle is equal to the values obtained from the simulation plus a constant (independent of N or Ξ) that has been neglected.

In Fig. 9(b) we show $C_{\mathrm{v}}/k_{\mathrm{B}}N$ (without the ideal gas contribution) for simulations with $\Xi = 1, 60$ and 100 as a function of N. As expected, the heat capacity also becomes independent of N after a certain size of the system. Notice that this is also true for $\Xi = 60$, showing that the shoulder in the heat capacity per particle shown in Fig. 6 indeed does not change its shape or size as the number of particles in the simulation changes.

3.4 Other Boundary Conditions

In all numerical results presented up to this point, we used periodic boundary conditions implemented with the Lekner–Sperb summation. We now investigate how other boundary conditions perform under a finite-size scaling analysis.

If one first substitutes the Lekner summation by a pure $1/r$ potential with the minimal image scheme (that is, each particle is surrounded by a symmetric shell of neighboring particles), one obtains the density profiles shown in the plots (a) and (b) of Fig. 10 (for $\Xi = 0.1$). These profiles can be directly compared with the results for periodic boundary conditions in Fig. 8, which are obtained for the same parameter values. Notice that, although the results for the bare density profile in Fig. 10(a) is comparable to the

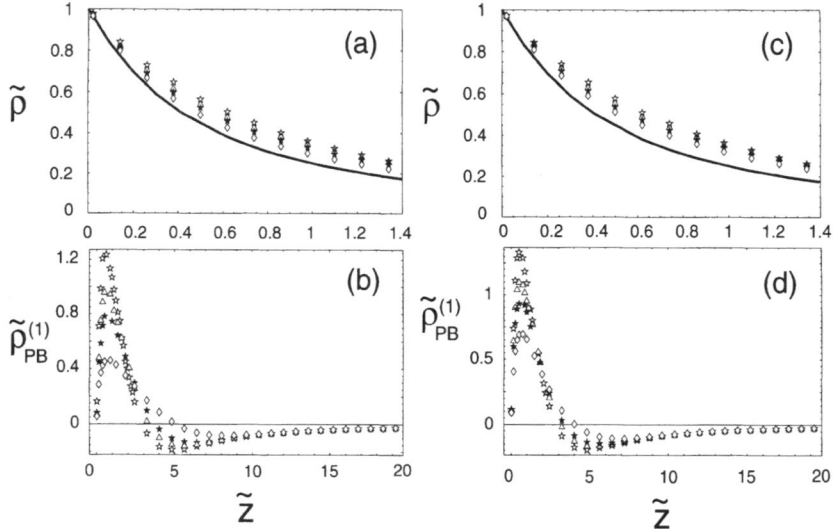

Fig. 10. Finite-size effects on a system with minimal image and $1/r$ potential (left) and with $1/r$ potential but without minimal image (right). The symbols denote MC results for $\Xi = 0.1$ and $N = 5$ (*open stars*), 15 (*open triangles*), 35 (*filled stars*) and 200 (*open diamonds*), all data with 10^6 MC steps per particle. The full line correspond to the PB result. Plots **(b)** and **(d)** show the difference between the simulation result and the PB result normalized to the coupling parameter according to $\widetilde{\varrho}_{PB}^{(1)}(\widetilde{z}) = \left(\widetilde{\varrho}(\widetilde{z}) - \widetilde{\varrho}_{PB}(\widetilde{z})\right)/\Xi$

results with periodic boundary conditions, Fig. 8(a), the plot of the differences to the PB result in Fig. 10(b) reveals that finite-size effects are much more pronounced with the minimal image condition. As a side remark, we note that the data for very small system sizes ($N = 5$) are with minimal image boundary conditions somewhat better than the periodic-boundary results, which is due to artifacts introduced by enforcing periodicity for too small particle numbers.

Similar conclusions apply to a system with open boundaries (i.e., $1/r$ potential without minimal image scheme), as shown in plots (c) and (d) of Fig. 10. Surprisingly, the differences between the systems with open boundaries and minimal image conditions are rather small, and both show pronounced deviations from the periodic boundary system. This is rather unexpected, since with open boundary conditions, the lateral distribution of counterions shows a pronounced increase at the boundaries, as shown in Fig. 11. This effect, which will have important consequences for laterally finite systems (such as small charged crystallites) is due to the mutual repulsion between counterions.

As a summary of our investigation of different boundary conditions, we note that open boundaries and minimal-image boundary conditions give roughly the same results. Open boundaries lead to an accumulation of counterions at the system edge, which is in most cases an unwanted phenomenon. Since the computational effort of implementing minimal image boundary conditions is minute, this is always preferred over open boundaries (unless one is specifically interested in edge effects). On the other hand, the finite-size scaling behavior of the periodic boundary condition is superior to the

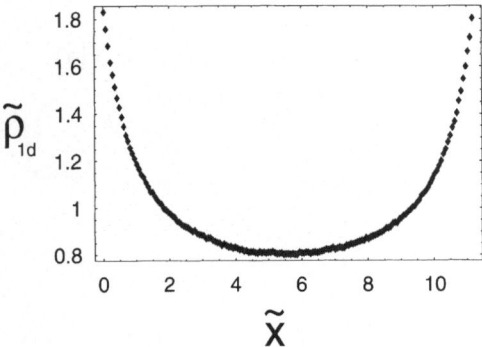

Fig. 11. Lateral density distribution $\tilde{\varrho}_{yz}$ (integrated in the zy-plane) of a system with open boundary conditions as a function of the lateral coordinate $\tilde{x}$ for $\Xi = 0.1$ and $N = 200$. The counterions accumulate at the edge of the simulation box, which is due to the mutual repulsion between counterions

non-periodic ones. If one is therefore interested in the thermodynamic behavior of a quantity which sensitively depends on the system size, one is forced to implement periodic boundary conditions, since increasing the system size by too much is prohibitive from the numerical point of view.

4 Charge-Modulated Substrate

Until now we have only examined the case where the surface charge is smoothly smeared out over the wall. What is the impact of this assumption, and how does the counterion distribution change when the substrate has a charge modulation?

The influence of non-uniform surface charge distributions has been experimentally established, for instance, in colloidal flocculation [55] and deposition [56]. Theoretically, this has been studied either using the Poisson–Boltzmann (PB) equation [57–62], liquid state theory [63] or computer simulations [64–66]. We have recently performed a systematic study [4] where we demonstrated under which conditions one can assume a smeared out charge density. As it turns out, this latter assumption is a poor one when the charge modulation is pronounced, but also for moderate charge-modulation when the electrostatic coupling is large (i.e., with highly charged surfaces or with multivalent counterions). Under these conditions, the counterions become highly correlated with the fixed charges and tend to form a two-dimensional layer close to the surface, a fact that is reflected in the lateral (Fig. 12(b)) and in the normal (Fig. 13) density distribution of counterions.

A convenient measure for the effects of substrate charge modulation is the counterion contact density, i.e., the laterally averaged counterion density at the substrate surface, since it is known exactly in the smeared-out case and can be easily determined from simulations. We find that the contact density for charge-modulated substrates can be much larger than for the smeared-out case, which is in agreement with recent experimental measurements on highly charged surfactant monolayers [67]. Quite surprisingly,

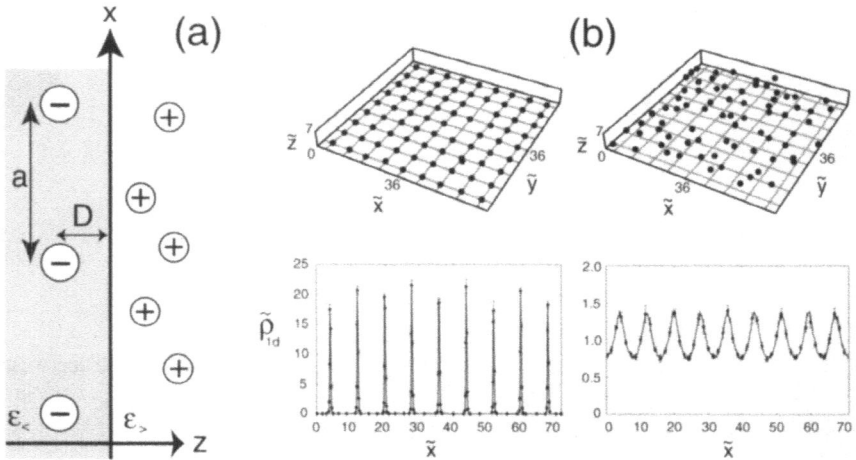

Fig. 12. Plot (**a**): Side view of the model studied here. We assume $\varepsilon_> = \varepsilon_<$. Plot (**b**): simulation snapshots and lateral density profile ($\tilde{\varrho}_{1d}$, integrated in the yz-plane) for $D/a \simeq 0.06$ (left) and $D/a \simeq 0.24$ (right), both with coupling $\Xi = 10$. In the snapshots, the surface charges are located at the line crossings

substrate charge modulation tends to have a more drastic effect on the counterion distribution than fluctuations and correlations, which have been the subject of numerous recent studies (see [7] and references therein). In contrast to the mean-field Poisson–Boltzmann approach, the strong coupling theory describes quantitatively the counterion distributions (obtained through Monte Carlo simulations) in the SC limit and in particular in the limit of pronounced surface-charge modulation.

The charge modulation in the substrate is modeled, for simplicity, by a square lattice with lattice constant a (Fig. 12(a)). The point-like oppositely charged counterions are confined to the half-space $z > 0$, and the minimal distance of approach between the counterions and the fixed ions is given by D. In the case treated here the dielectric constant is assumed to be same everywhere. The system is globally neutral, and only the fixed surface charges and counterions are present, i.e., no salt is added. In principle, one could interpret D as the sum of the surface-ion and counterion hard-core radii. In our model, however, we use the ratio D/a more generally to control the degree of substrate-charge modulation, which allows to describe a variety of experimental situations and systems by a single parameter. The limit $D/a \to \infty$ is equivalent to a smeared-out surface-charge distribution, while $D/a \to 0$ corresponds to a delta-peaked distribution.

As before, we define the Gouy–Chapman length though $\mu = 1/(2\pi q \ell_B \sigma_s)$, where q is the valence of the counterions, σ_s is the number charge density at the wall and $\ell_B = e^2/4\pi\varepsilon_>\varepsilon_0 k_B T$ is the Bjerrum length (the distance at which two elementary charges interact with the same strength as the thermal energy, $k_B T$). As done previously, all lengths will be rescaled with μ according to $\tilde{r} \equiv r/\mu$. For the simple double layer (smeared out charges at the surface) the coupling parameter $\Xi = q^2 \ell_B/\mu = 2\pi q^3 \ell_B^2 \sigma_s$ is the only explicit parameter in the problem. In the limit $\Xi \to 0$ (weakly charged surfaces and/or low-valence counterions) the Poisson–Boltzmann theory is asymptotically exact, while in the opposite limit $\Xi \to \infty$ (highly charged surfaces and/or high-valence counterions) the strong coupling theory which is exact [3]. The present model is in addition characterized by D/a and the valence ration q/Q between counterions and surface charges.

For a given configuration of the counterions, the contribution to the electrostatic energy due to the interaction between counterions and between counterions and surface charges reads

$$\frac{\mathcal{H}}{k_\mathrm{B} T} = \Xi \left\{ \sum_{i \neq j} \frac{1}{|\widetilde{\mathbf{r}}_j - \widetilde{\mathbf{r}}_k|} - \frac{Q}{q} \sum_{i,\alpha} \frac{1}{|\widetilde{\mathbf{r}}_i - \widetilde{\mathbf{R}}_\alpha|} \right\}. \tag{27}$$

The first sum corresponds to the interaction between pairs of counterions, while the second sum is the interaction between the counterions and the fixed surface charges (the sum over α corresponds to all lattice vectors $\widetilde{\mathbf{R}}_\alpha$). We use (27), together with the Lekner–Sperb technique [16,18] (cf. Sect. 2), to calculate the electrostatic energy in our computer simulations. These were typically performed with N between 81 and 100 counterions and periodic boundary conditions in order to minimize finite size effects; the results reported here are always taken from runs with 10^6 Monte Carlo steps per particle. In Fig. 12(b) we show the effects of varying surface-charge modulation: for a ratio $D/a = 0.24$ (to the right) the counterion snapshot shows a rather irregular configuration, with no or little visible correlation between surface charges (located beneath the nodes of the square lattice) and counterions. For a ratio $D/a = 0.06$ (to the left), on the other hand, the counterions are strongly correlated with the square lattice of the surface ions (two counterions obviously have, driven by thermal fluctuations, escaped from their assigned lattice positions). This is reflected by the normalized lateral counterion distribution $\widetilde{\varrho}_{yz}(\widetilde{x})$, which for $D/a = 0.24$ oscillates weakly around its mean value of unity, while for $D/a = 0.06$ this distribution function is strongly peaked at the positions of the surface ions.

At leading order in inverse powers of the coupling Ξ, the strong coupling counterion density profile is given by [6]

$$\widetilde{\varrho}(\widetilde{\mathbf{r}}) = \frac{\varrho(\widetilde{\mathbf{r}})}{2\pi \ell_\mathrm{B} \sigma_\mathrm{s}^2} = \Lambda\, e^{-\widetilde{u}(\widetilde{\mathbf{r}})} + \mathcal{O}(\Xi^{-1}). \tag{28}$$

where $\widetilde{u}$ is is the potential created at $\widetilde{\mathbf{r}}$ by the surface ions at lattice positions $\widetilde{\mathbf{R}}_\alpha$ (with $\widetilde{z}_\alpha = -\widetilde{D}$) and is given by

$$\widetilde{u}(\widetilde{\mathbf{r}}) = -\Xi \frac{Q}{q} \sum_\alpha \frac{1}{|\widetilde{\mathbf{r}} - \widetilde{\mathbf{R}}_\alpha|}. \tag{29}$$

The factor Λ is the fugacity of the counterions, and is determined by the normalization $\int_0^\infty \mathrm{d}\widetilde{z}\, \varrho_0(\widetilde{\mathbf{r}})/(2\pi \ell_\mathrm{B} \sigma_\mathrm{s}^2) = 1$, which is equivalent to the condition of global electroneutrality. The discreteness of surface charges tends to decouple different counterions from each other, as witnessed by the snapshots shown in Fig. 12(b). One would therefore expect corrections to the leading term of the systematic expansion in (28), which come from correlations between counterions, to be weakened and the SC approach to perform

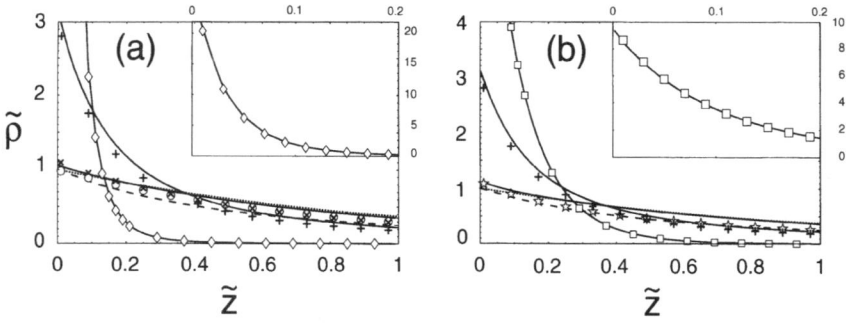

Fig. 13. Counterion density profile $\tilde{\varrho} \equiv \varrho/(2\pi\ell_B\sigma_s^2)$ (integrated in the xy-plane) as a function of $\tilde{z} \equiv z/\mu$. Plot **(a)**: MC results (symbols) and SC predictions (*full lines*) for a system with fixed $\Xi = 10$ ($q = Q$) and $D/a \simeq 0.06$ ($\tilde{D} = 0.5$, *diamonds and inset*), $D/a \simeq 0.12$ ($\tilde{D} = 1$, *plus symbols*), $D/a \simeq 0.24$ ($\tilde{D} = 2$, *crosses*) and $D/a \to \infty$ (*circles*). Plot **(b)**: same as (a) for a system with fixed $D/a \simeq 0.12$ and $\Xi = 1$ (*stars*), $\Xi = 10$ (*plus symbols*) and $\Xi = 100$ (*squares and inset*). The *dashed* and *dotted* lines denote, respectively, the PB and SC solutions at $D/a \to \infty$. Error bars are smaller than the size of the symbols

even better in the presence of modulated surface charges, as indeed confirmed by our data. In the limit $D/a \to \infty$, (29) reduces to $\tilde{u}(\tilde{\mathbf{r}}) = \tilde{z}$, i.e. the smeared out case treated in Sect. 3 is recovered.

We now discuss our results for the case where the surface ions and the counterions have the same charge valence ($Q = q$). In Fig. 13(a) we fix $\Xi = 10$ while D/a is varied; in Fig. 13(b) we fix $D/a = 0.12$ while Ξ is varied. Clearly, the counterion density profiles $\tilde{\varrho}_{xy}(\tilde{z})$ (which are averaged in the xy-plane) are very sensitive to both the coupling constant Ξ and the ratio D/a. For $\Xi = 10$ and $D/a = 0.24$ (crosses in Fig. 13(a)) the discretization has a small effect, as the data almost coincide with the MC results for the smeared-out case $D/a \to \infty$ (circles). With the rescaling of Fig. 13(a), the difference between the smeared-out SC and PB profiles (dotted and dashed lines, respectively) is in fact rather small compared to the effects of surface-charge modulation, and the smeared-out data (circles) are somewhat in between the SC and PB predictions, demonstrating that $\Xi = 10$ is in the crossover regime between strong and weak coupling [3]. For $D/a = \infty$ the contact-value theorem predicts a rescaled contact density of unity, $\tilde{\varrho}_{xy}(\tilde{z} = 0) = 1$, as confirmed by simulations at various values of Ξ [3]. However, as D/a becomes smaller there is a greater accumulation of counterions in the immediate vicinity of the charged surface, and the laterally averaged contact density can be several times larger than unity. This is reflected by a high lateral correlation between counterions and the surface charges, see Fig. 12(b). As D/a becomes smaller, the SC predictions as defined in (28) and (29) (solid lines in Fig. 13(a)) show progressively better agreement with the MC results. Similarly, in Fig. 13(b) for fixed $D/a = 0.12$, the contact density rises while the agreement between MC data and SC predictions becomes quantitative as Ξ is increased.

These results are summarized in Fig. 14, where contour plots of constant contact density $\tilde{\varrho}_{xy}(\tilde{z} = 0)$ are shown as a function of Ξ and D/a. The lines were obtained from SC theory via (28) and agree well with the MC results (open circles), obtained

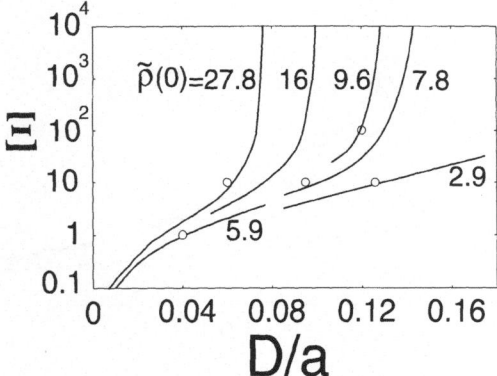

Fig. 14. Density at contact $\widetilde{\varrho}(0) \equiv \varrho(\widetilde{z} = 0)/(2\pi\ell_B\sigma_s^2)$ (integrated in the xy-plane) as a function of both the coupling Ξ and the ratio D/a ($Q = q$). The *full curves* with constant $\widetilde{\varrho}(0)$ were obtained with the SC theory (see (28)), the *open circles* show the MC results with $\varrho(0) = 2.9$, 5.9, 7.8, 9.6 and 27.8. At large D/a, the smeared-out assumption for the surface charge density (which leads to a $\widetilde{\varrho}(0) = 1$, independently of Ξ) is again valid

from density profiles via extrapolation. In the limit $D/a \to 0$ and finite coupling Ξ the counterions collapse onto the surface charges and the density at contact diverges, while for fixed D/a and in the limit $\Xi \to \infty$ the contact density saturates at a finite value. As D/a grows (or Ξ becomes smaller), the smeared-out assumption becomes valid and $\widetilde{\varrho}_{xy}(0)$ approaches unity.

We finally analyze the case where the restriction $q = Q$ is relaxed. In Fig. 15 we show density profiles for $\Xi Q^3/q^3 = 10$ (which corresponds to keeping the surface charge density constant) and $D/a = 0.12$, with varying valence ratios of $q/Q = 1$ (plus symbols, already featured in Fig. 13), $q/Q = 2$ (stars), $q/Q = 4$ (triangles) and $q/Q = 16$ (squares). The inhomogeneity of the surface charge distribution becomes more important as the counterion valence increases. The SC results (solid lines) capture this trend and quantitatively agree with simulation data. The contact density $\widetilde{\varrho}_{xy}(0)$ saturates at a constant value as $q/Q \to \infty$ (see the inset). As one would expect, the modulation of the surface charge becomes increasingly important when multivalent counterions are present.

In summary, the charge inhomogeneity affects the counterion distribution strongly for pronounced charge modulation as well as when the coupling constant Ξ is large. As shown in Fig. 14, the laterally averaged counterion contact density is much larger than the smeared-out value at decreasing D/a (equivalent to a strongly charge-modulated substrate) and at increasing coupling constant Ξ, and counterions become strongly correlated with the surface charges. The same effect is reflected in the counterion density profiles, which under such conditions decay faster to zero as one moves away from the substrate.

This has direct experimental consequences: very recently the counterion density profile (measured using ellipsometry) at a surfactant monolayer with a high surface charge density (determined via second-harmonic generation) could be described by PB

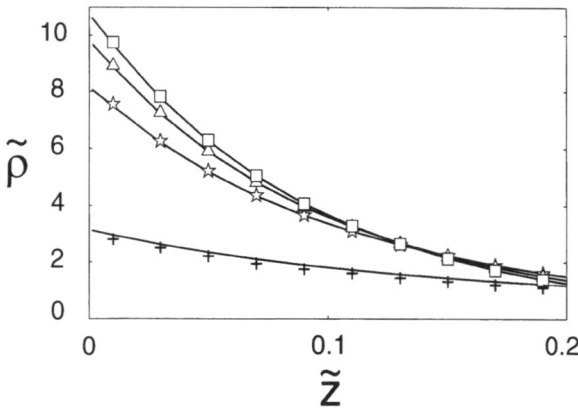

Fig. 15. Effect of the ionic valence: the plot shows the integrated counterion profile for fixed $\Xi Q^3/q^3 = 10$ (corresponding to a constant surface charge density), $D/a \simeq 0.12$ and different values of q. *Full lines* correspond to the SC predictions and symbols to MC results for $q/Q = 1$ (*plus symbols*), $q/Q = 2$ (*stars*), $q/Q = 4$ (*triangles*) and $q/Q = 16$ (*squares*)

profiles only by assuming a fraction of counterions to be bound to the monolayer [67], similar to what we find. Our results suggest to reconsider the traditional picture of a Stern layer, where the counterion surface concentration is subject to electrostatic and specific surface attraction and packing constraints, but where the influence of lateral surface charge modulation (which is always present experimentally and leads to sizable effects as shown here) is typically neglected. One should emphasize the experimental relevance of our parameters: for a system with $\sigma_s \simeq 1/257 \,\text{Å}^{-2}$ and divalent counterions in water (which at room temperature leads to $\Xi = 10$), the surface ions are on average at a distance $a \simeq 16 \,\text{Å}$ from each other. For a minimal distance between surface ions and counterions of $D \simeq 2 \,\text{Å}$ (noting that at close contact the hydrated water shell is stripped off) the ratio $D/a = 0.12$ is obtained; as our results demonstrate, the inhomogeneous character of the surface charges leads to pronounced deviations from the smeared-out case.

Although the model used here is quite crude and neglects a number of important effects, it has the advantage that its simplicity allows for a global analysis, encompassing different and limiting values of the coupling strength and the degree of substrate charge modulation. In a complementary study, similar numerical calculations have been recently analyzed within a modified PB approach [68].

5   Interacting Double Layers

A great deal of work has been devoted in the past twenty years to the interaction between two double layers. Specifically, it has been known for some time that two similarly and strongly charged plates can attract each other in the presence of multivalent counterions. This has been seen in Monte Carlo simulations [36,37], observed experimentally with the surface force apparatus [69] and also deduced from the phase diagrams of charged lamellar systems [70–72]. This result is quite relevant concerning the stability of colloidal solutions, since it means that the stabilization of colloids with charges can fail if the surfaces are too highly charged. Such behavior strongly contradicts the Poisson–Boltzmann theory, which predicts that the electrostatic interaction between similarly charged surfaces is always repulsive. Most theoretical approaches (beyond PB) tried to include the correlations between counterions, which were thought to be the reason for the discrepancy between the mean-field and the experimental/simulation results [73,74]. The first theoretical approach that demonstrated the existence of attraction between equally charged plates (with electrostatic origin) is due to Kjellander and Marčelja [32], which used a sophisticated integral-equation theory (with the hypernetted chain closure) and obtained results that compared very well with simulations [32,36,39]. Also perturbative expansions around the PB solution [35,48] and density-functional theory [75,76] were used, and predicted as well the existence of an attractive interaction. For plates far away from each other, i.e., at distances such that the two double layers weakly overlap, the attractive force was obtained within the approximation of two-dimensional counterion layers by including in-plane Gaussian fluctuations [77–79] and, more recently, plasmon fluctuations at zero temperature [80] and at non-zero temperatures [81].

In this section we focus on the mechanism for electrostatic attraction between similarly charged plates and, in particular, on the bound state, which occurs for finite plate separations [5,7]. Like the systems treated in the previous sections, the strong coupling theory is valid in the limit when PB breaks down and, as we will show, naturally yields electrostatic attraction between similarly charged plates. To check these theoretical results, we have performed extensive MC simulations in the complete parameter space. In the SC limit we obtain quantitative agreement of MC density and pressure profiles with our SC theory, whereas PB theory describes the numerical data well in the opposite limit of weak coupling.

The leading term of our SC theory is the first virial term and thus corresponds to the partition function of a single counterion sandwiched between two charged plates, which we now explicitly evaluate, thereby providing some physical insight into the mechanism for electrostatic attraction. That this simple calculation in fact is exact in the SC limit will be demonstrated below in a formal field-theoretic calculation and by comparison with MC results. Denoting the distance between the counterion and the plates (of area A) as x and $d - x$, respectively, we obtain for the electrostatic interaction between the ion and the plates (note that all energies and forces are given in units of k_BT) for $d \ll \sqrt{A}$ the results $W_1 = 2\pi\ell_B q\sigma x$ and $W_2 = 2\pi\ell_B q\sigma(d-x)$, respectively, as follows from the potential at an infinite charged wall and omitting constant terms. The Bjerrum length $\ell_B = e^2/4\pi\varepsilon k_BT$ is the distance at which two unit charges interact with k_BT. The sum of the two interactions is $W_{1+2} = W_1 + W_2 = 2\pi\ell_B q\sigma d$ which shows that i) no pressure is acting on the counterion since the forces exerted by the two plates exactly cancel and ii) that the counterion mediates an effective attraction between the two plates. The interaction between the two plates is proportional to the total charge on one plate, $A\sigma$, and for $d \ll \sqrt{A}$ given by $W_{12} = -2\pi A\ell_B\sigma^2 d$. Since the system is electroneutral, $q = 2A\sigma$, the total energy is $W = W_{12} + W_1 + W_2 = 2\pi A\ell_B\sigma^2 d$, leading to an electrostatic pressure $P_{el} = -\partial(W/A)/\partial d = -2\pi\ell_B\sigma^2$ per unit area. *The two plates attract each other!* The entropic pressure due to counterion confinement is $P_{en} = 1/Ad = 2\sigma/qd$. The equilibrium plate separation is characterized by zero total pressure, $P_{\text{tot}} = P_{\text{el}} + P_{\text{en}} = 0$, leading to an equilibrium plate separation $d^* = 1/\pi \ell_{\text{B}} q \sigma$. In fact, this simple one-particle derivation for the attraction between charged plates is conceptually simpler than the PB result of repulsion, because the latter case involves many-body effects. In the following we will in fact show that the results for P_{tot} and d^* become exact in the SC limit.

We now quickly recapitulate the Poisson–Boltzmann and strong coupling results for this system. The saddle-point analysis, valid for small Ξ, leads to the well-known PB result

$$\tilde{\varrho}(\tilde{z}) = 1/\cos^2\left(\Lambda^{1/2}[\tilde{z} - \tilde{d}/2]\right) + \mathcal{O}(\Xi) \tag{30}$$

with corrections proportional to the coupling constant Ξ [47]. In the opposite limit of high couplings the SC theory leads to the density profile

$$\tilde{\varrho}(\tilde{z}) = \frac{2}{\tilde{d}} + \frac{2}{\tilde{d}\Xi}\left[\left(\tilde{z} - \tilde{d}/2\right)^2 - \tilde{d}^2/12\right]. \tag{31}$$

The leading term, $\tilde{\varrho}_0 = 2/\tilde{d}$, is the first virial contribution, which originates from the one-particle partition function, and therefore coincides with the scaling result obtained in the beginning. The second leading term gives a contribution of maximal magnitude $\tilde{d}/3\Xi$ and therefore dominates the leading term for $\tilde{d}/\Xi^{1/2} = d/a > 1$, where the mean lateral distance between ions, a, is defined by $2\sigma/q = 1/(\pi a^2)$. This shows that the virial expansion, and in particular the SC result, should be valid as long as the plate separation d is smaller than the lateral distance between ions a, or, in rescaled units, for $\tilde{d} < \Xi^{1/2}$. In Fig. 16(a) we show density profiles obtained from MC simulations for small coupling parameter $\Xi = 0.5$ for various plate distances, which are well described by the PB profiles (30) shown as solid lines. Figure 16(b) shows that for $\Xi = 100$ PB (thin solid lines) is inadequate. For $\tilde{d} = 1.5$ (open diamonds) we have $d/a = 0.15$, and the leading term of (31) is indeed accurate. For $\tilde{d} = 10$ (open stars) we find $d/a = 1$, the density profile is neither described by (31) nor (30). Finally, for $\tilde{d} = 30$ (open stars) we find $d/a = 3$, the two layers are decoupled and the density profile is described by a double exponential $\tilde{\varrho}(\tilde{z}) = (\exp(-\tilde{z}) + \exp(\tilde{z} - \tilde{d}))/(1 - \exp(-\tilde{d}))$ (dashed-dotted line), which is the superposition of the density profiles of two isolated charged surfaces in the SC limit. The crossover from PB to SC is demonstrated in Fig. 17, where we plot density profiles for fixed separation $\tilde{d} = 2$ for various coupling parameters Ξ.

Using the contact-value theorem, the pressure P between the two plates is related to the counterion density at a plate, $\tilde{\varrho}(\tilde{d})$, by [36,47]

$$\tilde{P} = P/(2\pi \ell_{\text{B}} \sigma^2) = \tilde{\varrho}(\tilde{d}) - 1. \tag{32}$$

Numerically, the contact ion density $\tilde{\varrho}(\tilde{d})$ is obtained from the density profiles by extrapolation. In Fig. 18 we show numerical pressure data for selected values of Ξ. Attraction (negative pressure) is obtained for $\Xi > 12$ and intermediate distances only, as evidenced by the inset where the pressure profile for $\Xi = 17$ is shown. The numerical pressure for $\Xi = 0.5$ (open diamonds) agrees well with the PB prediction (thick solid line), which from (32) and (30) is given by $\tilde{P} = \Lambda$ with Λ determined by $\Lambda^{1/2} \tan[\tilde{d}\Lambda^{1/2}/2] = 1$. The SC prediction for $\tilde{P}$ is obtained by combining the leading term of (31) and (32),

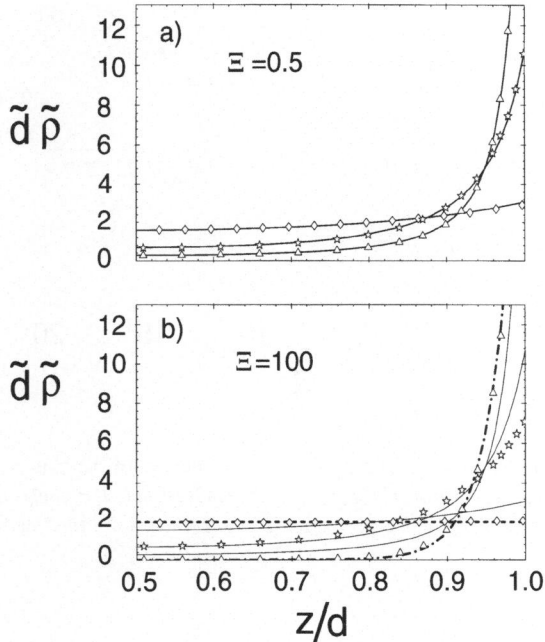

Fig. 16. MC results for the rescaled counterion density $\tilde{d}\tilde{\varrho}$ as a function of the rescaled distance from the wall z/d in the **(a)** PB limit for $\Xi = 0.5$ and in the **(b)** SC limit for $\Xi = 100$ for various plate separations $\tilde{d} = d/\mu = 1.5$ (*open diamonds*), $\tilde{d} = 10$ (*open stars*), and $\tilde{d} = 30$ (*open triangles*). In (a) MC results agree well with the corresponding PB predictions (see (30), *solid lines*), whereas in b) results for $\tilde{d} = 1.5$ agree with the SC prediction $\tilde{\varrho} = 2/\tilde{d}$ (*dashed line*) and for $\tilde{d} = 30$ with a double-exponential curve (*dashed-dotted line*, see text)

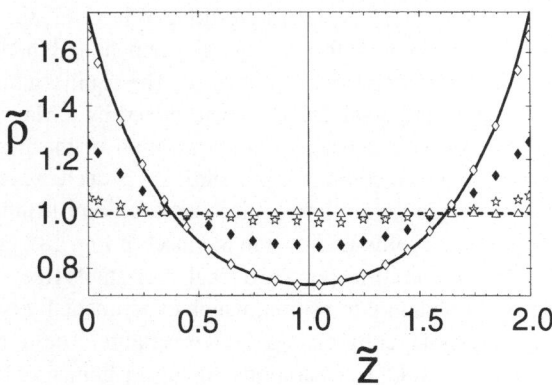

Fig. 17. MC results for the rescaled counterion density profiles $\tilde{\varrho} = \varrho/2\pi\ell_{\mathrm{B}}\sigma^2$ for fixed plate separation $\tilde{d} = d/\mu = 2$ as a function of the rescaled distance $\tilde{z} = z/\mu$ from one wall. Symbols correspond to coupling parameters $\Xi = 0.5$ (*open diamonds*), $\Xi = 10$ (*filled diamonds*), $\Xi = 100$ (*open stars*), and $\Xi = 10^5$ (*open triangles*), exhibiting clearly the crossover from the PB prediction (*solid line*, (30)) to the SC prediction $\tilde{\varrho} = 2/\tilde{d}$ (*dashed line*, (31))

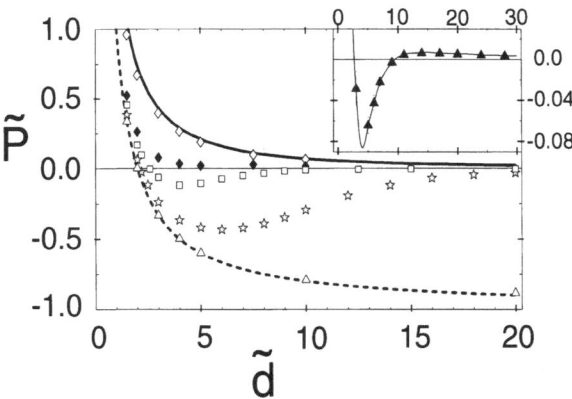

Fig. 18. MC results for the rescaled pressure $\tilde{P}$ as a function of the plate separation $\tilde{d}$ for the same parameter values as in Fig. 17 (and $\Xi = 17$ and 20, *filled triangles* and *open squares*, respectively), compared with the PB prediction $\tilde{P} = \Lambda$ (*thick solid line*) and the SC prediction $\tilde{P} = 2/\tilde{d} - 1$ (*dashed line*)

yielding $\tilde{P} = 2/\tilde{d} - 1$, from which the equilibrium separation, determined by $\tilde{P} = 0$, follows as $\tilde{d}^* = 2$. Incidentally, this is exactly the scaling prediction for the pressure derived in the beginning of this section. The small distance range of most data, and the complete pressure data for $\Xi = 10^5$ (open triangles) are well described by the SC prediction (dashed line), demonstrating again that the SC result is valid for $\tilde{d} < \Xi^{1/2}$.

Finally, combining all pressure data, we obtain the global phase diagram shown in Fig. 19, featuring attractive (negative) inter-plate pressure at intermediate distances and above a threshold coupling of $\Xi^* \approx 12$. In the limit of large Ξ, the phase boundary saturates at $\tilde{d}^* = 2$, in agreement with our scaling argument and the leading SC term. For large separation re-entrant repulsion is observed. The equilibrium plate separation (denoted by filled symbols and a solid line) is determined by a Maxwell construction on the pressure profile, or, equivalently, by minimization of the free energy (which numerically is determined by integrating the pressure). As Ξ decreases from large values this equilibrium separation grows and shows a discontinuous jump to infinity at $\Xi^{**} \approx 17$ (the corresponding pressure profile is shown in the inset of Fig. 18). For $\Xi < \Xi^{**}$, the lower branch of the dashed line correspond to a local, metastable free-energy minimum. This constitutes a novel unbinding transition, which experimentally is observable with charged lamellar or clay systems by raising the temperature. The re-entrant transition from attraction to repulsion at large separations, the upper branch of the dashed curve, is expected to scale as $\tilde{d} \sim \Xi$ (plus logarithmic corrections) [35,48], as denoted by the straight dashed line and in agreement with the MC data.

The range of validity of our novel SC theory is $\tilde{d} < \sqrt{\Xi}$ which includes most of the equilibrium-plate-separation line in Fig. 19. It follows that the bound state is i) well characterized by our SC theory (as demonstrated by the fact that the equilibrium plate separation is for the most part close to $\tilde{d}^* = 2$) and ii) the counterion distribution is indeed two-dimensional [82], $d < a$, though Wigner crystallization (which occurs

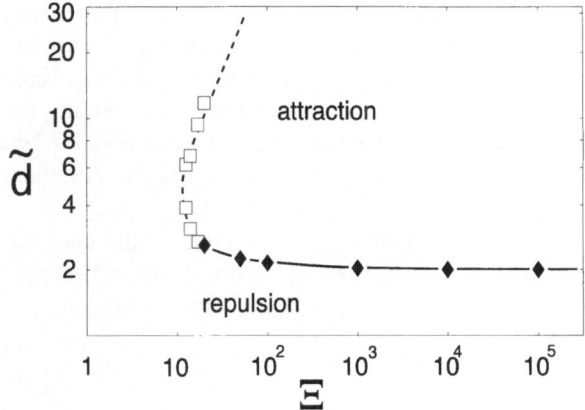

Fig. 19. MC phase diagram showing regions of attractive and repulsive pressure as a function of plate separation $\tilde{d}$ and coupling strength $\varXi$. Attraction only occurs for intermediate distances and $\varXi > 12$. The equilibrium plate separation (determined by minimization of the free energy, *solid line*) exhibits a discontinuous unbinding to infinity at $\varXi^{**} \approx 17$ and saturates at $\tilde{d}^* = 2$ for $\varXi \to \infty$

at $\varXi \simeq 15\,600$ in the limit $\tilde{d} \to 0$) is not required to find attraction. Correlations between counter ions, except the lateral exclusion correlation which keeps ions apart, are unimportant in the bound state for large $\varXi$, but of course play an important role for intermediate values of $\varXi$. The range of validity of PB is $\tilde{d} > \varXi \ln \varXi$ for $\varXi > 1$ and $\tilde{d} > 1$ for $\varXi < 1$, as follows by comparison of the PB and one-loop-correction pressures [35,48]. Between the PB and SC ranges of validity is therefore only a small region where none of these asymptotic theories is valid.

6 Concluding Remarks

We described in detail the methods used in our Monte Carlo simulations of counterions close to charged objects. We discussed the Lekner–Sperb summation technique for 2D periodic systems, finite-size effects, and the consequences of using different boundary conditions. We compared the simulation results with the Poisson–Boltzmann (PB) and the strong coupling (SC) theories, establishing the regimes of validity of each of the theories. As we demonstrated, the PB theory is in general valid when the surface charges and/or the valence of the counterions is low, or when the temperatures are high. The SC theory describes the opposite limit of high surface charges and/or the counterion valence, or low temperatures.

For the case of single wall with smeared out surface charge, we also studied the counterion pair correlation function, which indicates a behavioral change from a three-dimensional, weakly correlated counterion distribution (at low coupling) to a two-dimensional, strongly correlated counterion distribution (at high coupling). As we showed, the specific heat capacity displays a rounded hump at the intermediate coupling strengths where this crossover occurs.

For the charge-modulated system, we showed that the counterions are laterally highly correlated with the surface charges when the minimum approach distance between the latter and counterions is smaller than the distance between surface charges. We obtained the set of parameters where the average counterion density profiles are very different from the ones obtained with smeared out charged surfaces, and showed that in the regime where the classical Poisson–Boltzmann theory is expected to fail, the strong coupling theory agrees very well with Monte Carlo simulations.

Finally, for the interacting double layers, we studied the inter-wall pressure and obtained the complete phase diagram, which shows attraction between the walls for a large enough coupling strength and at intermediate wall separation.

References

1. K.K. Kunze, R.R. Netz: Phys. Rev. Lett. **85**, 4389 (2000)
2. G.B. Sukhorukov, E. Donath, S. Davis, H. Lichtenfeld, F. Caruso, V.I. Popov, H. Möhwald: Polym. Adv. Technol. **9**, 759 (1998)
3. A.G. Moreira, R.R. Netz: Europhys. Lett. **52**, 705 (2000)
4. A.G. Moreira, R.R. Netz: Europhys. Lett. **57**, 911 (2002)
5. A.G. Moreira, R.R. Netz: Phys. Rev. Lett. **87**, 078301 (2001)
6. R.R. Netz: Eur. Phys. J. E **5**, 557 (2001)
7. A.G. Moreira, R.R. Netz: Eur. Phys. J. E **8**, 33 (2002)
8. N. Metropolis, A.W. Rosenbluth, M.N. Rosenbluth, A.H. Teller, E. Teller: J. Chem. Phys. **21**, 1087 (1953)
9. M.P. Allen, D.J. Tildesley: *Computer Simulations of Liquids* (Clarendon Press, Oxford 1987)
10. J.P. Valleau. In: *The Problem of Long-Range Forces in the Computer Simulation of Condensed Media*, ed. by D. Ceperley. NRCC workshop proceedings **9**, 3 (1980)
11. J.P. Valleau, S.G. Whittington. In: *Modern Theoretical Chemistry*, ed. by B. Berne (Plenum, NY 1977)
12. D.J. Adams: J. Chem. Phys. **78**, 2585 (1983)
13. P.P. Ewald: Ann. Phys. **64**, 253 (1921)
14. S.W. de Leeuw, J.W. Perram, E.R. Smith: Proc. R. Soc. Lond. A **373**, 27 (1980)
15. J. Lekner: Phys. A **157**, 826 (1989)
16. J. Lekner: Phys. A **176**, 485 (1991)
17. D.J. Tildesley. In: *Computer Simulations in Chemical Physics*, ed. by M.P. Allen, D.J. Tildesley (Kluwer Academic Publishers, Dordrecht 1993)
18. R. Sperb: Mol. Simul.**20**, 179 (1998)
19. J. Hautman, M.L. Klein: Mol. Phys. **75**, 379 (1992)
20. F.S. Csajka, C. Seidel: Macromolecules **33**, 2728 (2000)
21. G. Landhäußer: Ph. D. Thesis, Technical University of Berlin (1997)
22. *Handbook of Mathematical Functions*, ed. by M. Abramowitz, I.A. Stegun (Dover, NY 1965)
23. E.J.W. Verwey, J.T.G. Overbeek: *Theory of the Stability of Lyophobic Colloids* (Elsevier, NY 1948)
24. J. Israelachvili: *Intermolecular and Surface Forces*, 2nd edn. (Academic Press, London 1991)
25. G. Gouy: J. de Phys. **IX**, 457 (1910)
26. D.L. Chapman: Phil. Mag. **25**, 475 (1913)
27. P. Debye, E. Hückel: Physik. Z. **24**, 185 (1923)
28. D. Andelman. In: *Handbook of Biological Physics*, ed. by R. Lipowsky, E. Sackmann (Elsevier, Amsterdam 1995)

29. I. Borukhov, D. Andelman, H. Orland: Phys. Rev. Lett. **79**, 435 (1997)
30. O. Stern: Z. Elektrochem. **30**, 508 (1924)
31. M. Baus, J. Hansen: Phys. Rep. **59**, 1 (1980)
32. R. Kjellander, S. Marčelja: Chem. Phys. Lett. **112**, 49 (1984)
33. R. Kjellander, S. Marčelja: J. Chem. Phys. **82**, 2122 (1985)
34. P. Nielaba, T. Alts, B. D'Aguanno, F. Forstmann: Phys. Rev. A **34**, 1505 (1986)
35. P. Attard, D.J. Mitchell, B.W. Ninham: J. Chem. Phys. **88**, 4987 (1988)
36. L. Guldbrand, B. Jönson, H. Wennerström, P. Linse: J. Chem. Phys. **80**, 2221 (1984)
37. D. Bratko, B. Jönsson, H. Wennerström: Chem. Phys. Lett. **128**, 449 (1986)
38. B. D'Aguanno, P. Nielaba, T. Alts, F. Forstmann: J. Chem. Phys. **85**, 3476 (1986)
39. R. Kjellander, T. Åkesson, B. Jönsson, S. Marčelja: J. Chem. Phys. **97**, 1424 (1992)
40. A.L. Kholodenko, A.L. Beyerlein: Phys. Rev. A **34**, 3309 (1986)
41. N.V. Brilliantov, C. Bagnuls, C. Bervillier: Phys. Lett. A **245**, 274 (1998)
42. A.G. Moreira, M.M. Telo da Gama, M.E. Fisher: J. Chem. Phys. **110**, 10058 (1999)
43. N.V. Brilliantov: Contrib. Plasma Phys. **38**, 489 (1998)
44. A.G. Moreira, R.R. Netz: Eur. Phys. J. D **8**, 145 (2000)
45. R.R. Netz: Phys. Rev. E **60**, 3174 (1999)
46. R. Podgornik, B. Žekš: J. Chem. Soc. Faraday Trans. 2 **84**, 611 (1988)
47. R.R. Netz, H. Orland: Eur. Phys. J. E **1**, 203 (2000)
48. R. Podgornik: J. Phys. A: Math. Gen. **23**, 275 (1990)
49. D.A. McQuarrie: *Statistical Mechanics* (Harper Collins, NY 1976)
50. B.I. Shklovkii: Phys. Rev. E **60**, 5802 (1999)
51. G.M. Torrie, J.P. Valleau: Chem. Phys. Lett. **65**, 343 (1979)
52. B. Jönson, H. Wennerström, B. Halle: J. Phys. Chem. **84**, 2179 (1980)
53. R. Morf. In: *Physics of Intercalation Compounds*, Vol. 38, ed. by L. Pietronero, E. Tosati (Springer, Berlin 1981)
54. D.R. Nelson, M. Rubinstein: Phil. Mag.A **46**, 105 (1982)
55. H. Kihira, N. Ryde, E. Matijević: J. Chem. Soc. Faraday Trans. **88**, 2379 (1992)
56. G.M. Litton, T.M. Olson: J. Colloid Interface Sci. **165**, 522 (1994)
57. P. Richmond: J. Chem. Soc. Faraday Trans. II **71**, 1154 (1975)
58. D.Y. Chan, D.J. Mitchell, B.W. Ninham: J. Chem. Phys. **72**, 5159 (1980)
59. M. Kostoglou, A.J. Karabelas: J. Colloid Interface Sci. **151**, 534 (1992)
60. R.M. Peitzsch, M. Eisenberg, K.A. Sharp, S. McLaughlin: Biophys. J. **68**, 729 (1995)
61. J.Y. Walz: Adv. Colloid Interface Sci. **74**, 119 (1998)
62. T.T. Nguyen, A.Y. Grosberg, B.I. Shklovkii: J. Chem. Phys. **113**, 1110 (2000)
63. R. Kjellander, S. Marčelja: J. Chem. Phys. **88**, 7138 (1988)
64. W. van Megen, I. Snook: J. Chem. Phys. **73**, 4656 (1980)
65. T. Åkesson, B. Jönson: J. Phys. Chem. **89**, 2401 (1985); H. Wennerström, B. Jönsson: J. Phys. France **49**, 1033 (1988)
66. R. Messina, C. Holm, K. Kremer: Eur. Phys. J. E **4**, 363 (2001)
67. R. Teppner, K. Haage, D. Wantke, H. Motschmann: J. Phys. Chem. B **104**, 11489 (2000)
68. D.B. Lukatsky, S.A. Safran, A.W.C. Lau, P. Pincus: Europhys. Lett. **58**, 785 (2002)
69. P. Kékicheff, S. Marčelja, T.J. Senden, V.E. Shubin: J. Chem. Phys. **99**, 6098 (1993)
70. A. Khan, B. Jonsson, H. Wennerström: J. Phys. Chem. **89**, 5180 (1985)
71. H. Wennerström, A. Khan, B. Lindman: Adv. Colloid Interface Sci. **34**, 433 (1991)
72. M. Dubois, T. Zemb, N. Fuller, R.P. Rand, V.A. Parsegian: J. Chem. Phys. **108**, 7855 (1998)
73. R. Kjellander: Ber. Bunsenges. Phys. Chem. **100**, 894 (1996)
74. P. Attard. In: *Advances in Chemical Physics*, Vol. XCII, ed. by I. Prigogine, S. A. Rice (John Wiley and Sons, NY 1996)
75. M.J. Stevens, M.O. Robbins: Europhys. Lett. **12**, 81 (1990)

76. A. Diehl, M.N. Tamashiro, M.C. Barbosa, Y. Levin: Physica A **274**, 433 (1999)
77. P. Attard, R.K.D.J. Mitchell: Chem. Phys. Lett. **139**, 219 (1987)
78. P.A. Pincus, S.A. Safran: Europhys. Lett. **42**, 103 (1998)
79. D.B. Lukastky, S.A. Safran: Phys. Rev. E **60**, 5848 (1999)
80. A.W.C. Lau, D. Levine, P. Pincus: Phys. Rev. Lett. **84**, 4116 (2000)
81. A.W.C. Lau, P. Pincus, D. Levine, H.A. Fertig: Phys. Rev. E **63**, 051604 (2001)
82. I. Rouzina, V.A. Bloomfield: J. Phys. Chem. **100**, 9977 (1996)

Lattice Boltzmann Modeling of Complex Fluids: Colloidal Suspensions and Fluid Mixtures

Ignacio Pagonabarraga

Departament de Física Fonamental, Universitat de Barcelona, C. Martí i Franqués, 1, 08028-Barcelona, Spain

Abstract. The study of complex fluid dynamics requires development of numerical tools that capture the essentials of the dynamic coupling among the different particles that characterize these materials. At the same time these techniques should operate on mesoscopic time and length scales, such that the relevant phenomena can be addressed in detail. Lattice Boltzmann (LB) is a simulation procedure that, although initially introduced to address problems related to fluid flows at high Reynolds numbers, has proven to be a very flexible technique to study complex fluids in generic geometries. I will address the fundamentals of the method and describe in detail two different perspectives to model complex fluids: Colloidal suspensions, where the mesoscopic particles are described in full detail, and non-ideal fluid mixtures, where there exists no clear length scale separation. In this case, a coarse-grained description of the interactions is enough to capture the essentials of the collective dynamics. In both cases I analyze the basic formulation, emphasizing their similarities, and I will afterward discuss new results that have been obtained using LB in these systems.

1   Introduction

The study of fluids on different characteristic length scales poses a number of challenges both for the description of their equilibrium and non-equilibrium behavior. These systems range from heterogeneous mixtures, which develop domains of different sizes and textures, to fluids in which heterogeneities appear at mesoscopic scales [1]. The latter are characterized by elastic interactions among a fraction of their degrees of freedom, e.g. the elastic constraints between monomers in polymer chains, or the rigidity of colloidal particles [2]. These degrees of freedom are in equilibrium with the thermal fluctuations of the molecules that constitute the solvent in which these objects are suspended. Such equilibrium defines the typical size of these objects (varying between nano and micrometers). Hence, they introduce length scales much larger than the one characterizing the molecular solvent (∼ Ångström). As a result both the equilibrium and out of equilibrium behavior of these materials have qualitative differences with respect to the behavior of molecular fluids. When heterogeneities take place at larger length scales, there still exist the question of how to describe them in terms of the basic interactions that control the fluid.

   The complexity in understanding these systems and their rich phenomenology makes it necessary to resort to numerical simulations in order to explore their behavior. However, the existence of separate relevant length scales poses a number of challenges. It is clear that a first principles approach (based on the study of the behavior at the level of the interactions and dynamics of the constitutive molecules) is impractical. Hence, new approaches that focus on the mesoscopic level are needed. This level corresponds to the

I. Pagonabarraga, Lattice Boltzmann Modeling of Complex Fluids: Colloidal Suspensions and Fluid Mixtures, Lect. Notes Phys. **640**, 279–309 (2004)
http://www.springerlink.com/

scale defined by the solute objects which capture the essentials of the collective behavior controlling domain and formation of heterogeneities. To this end, a great deal of effort has been invested in understanding the effective interactions in these systems at this coarse grained level. Such studies have shown the differences between coarse effective interactions and molecular ones, although a lot of work is required in order to consistently link the different length and time scales. Nonetheless, in equilibrium, concepts such as depletion potentials, or entropy-induced potentials are now well-established and used as tools in predicting the phase behavior of these materials [3]. Once the effective interactions are known, Monte Carlo and Molecular Dynamics schemes are applicable to these systems. In addition, hybrid methods such as the inverse Monte Carlo [4,5] and variants offer alternative procedures to obtain such potentials and study their phase behavior. Such inverse methods offer new possibilities when combined with mesoscopic approaches, such as dissipative particle dynamics (DPD) [6].

The situation is analogous from the point of view of the dynamics, although it is less developed than its equilibrium counterpart. The central idea in many of the numerical approaches has been to coarse grain the solvent molecules to obtain an effective description at the level at which heterogeneities develop [7]. While in equilibrium the effect of solvent reduces to inducing effective interactions among the solute particles, the dynamics is more involved because one has to describe the dynamic evolution of the relevant collective modes of the solvent, and also those which will interact dynamically when mesoscopic particles are present. Different strategies have been put forward to tackle these problems. Some of them, such as DPD [8], borrow ideas from molecular dynamics. Some others try to build on kinetic theory concepts. These methods include lattice gas cellular automata (LGCA) [9], lattice Boltzmann (LB) [10] or the Malevanets–Kapral method [11,12]. In addition the the above, other methods start from a macroscopic approach using ideas from non-equilibrium thermodynamics [13].

In this contribution, I will concentrate on the description of the fundamentals and some of the applications of the lattice Boltzmann method which has shown a great potential in modeling of complex fluids. In the next section I will describe the fundamentals of the method. I will then discuss how generic boundaries are introduced and how they can be used to simulate mesoscopic suspensions, and subsequently describe some of the results that have been obtained with this approach. In Sect. 4, I will analyze how to describe generic non-ideal mixtures within LB, and discuss some recent physical results for binary mixtures that demonstrate the potential of this technique. I will finish with a concluding section.

2   Lattice Boltzmann: The Model

In order to overcome the computational limitations in the study of fluids at high Reynolds numbers, lattice gas cellular automaton (LGCA) was introduced in the late eighties [9]. At discrete time steps particles move from one lattice node to a neighboring node according to a prescribed set of velocities (which is finite), leading to a discretized dynamics both in space and time. The set of allowed velocities is defined by the set of vectors $\{c_k\}$ which joins the neighboring nodes where a particle can jump in one time step. In addition, the occupation number of each given node is discretized; in fact, only single occupancy is allowed for each possible velocity at a given node. Accordingly, the dynamics has two steps: A propagation step, in which particles move from their current node to a neighboring one according to their velocity. The move is accepted only if the final node is empty for that velocity. The second one is a collision step, in which the velocities of a given node are rearranged with the only constraint being that number of particles and momentum are conserved. If the lattice and the velocity space are chosen carefully, the hydrodynamic behavior of LGCA recovers Navier–Stokes [14]. Despite the potential of the method (unconditionally stable, easily parallelizable), such models do not allow as large Reynolds numbers as it was initially thought. In addition, they suffer from a number of physical limitations, e.g. Galilean invariance is lost.

In the context of LGCA, the lattice Boltzmann model (LB) was introduced as a preaveraged version of LGCA [10]. It inherited the discretized lattice dynamics characterized by the collision and propagation steps. However, instead of following particle dynamics, it assumed that the relevant dynamic variable was the one-particle distribution function. Hence, the occupation of the nodes is no longer described by a Boolean variable, but by a real number that characterizes the density of particles moving at a given node with a particular velocity. At the first stage, the collision matrix in LB was computed by preaveraging the collision table of the corresponding underlying LGCA [10][1]. However, in the second stage, this collision mechanism is substituted by a linearized version of the collision matrix in which the distribution function relaxes toward a local equilibrium distribution [15]. The simplicity of the dynamics when considered from this second perspective has made LB a powerful tool. LGCA is intrinsically stable as a particle based method; such intrinsic stability is lost in LB. Despite this drawback, the fact that thermal noise is not present makes LB much more efficient in comparison to LGCA for hydrodynamic problems. Even though some of the LB models suffer from Galilean invariance problems, it is easier to correct these limitations in LB as compared with LGCA.

As I have pointed out, in LB the basic dynamic variable is the one-particle distribution function, $f_j(\mathbf{r}_i, t)$, which gives the density of particles at node $\mathbf{r}_i$ at time t moving with velocity $\mathbf{c}_j$. The lattice in which this density moves is characterized by both the sets of nodes that make it up and the velocity subspace $\{\mathbf{c}_k\}$ which determines the neighboring nodes to which a given density will be able to move to in one time step. For example, Fig. 1 shows two examples of lattices corresponding to cubic three-dimensional lattices. The lattice on the left hand side defines 15 allowed velocities (a particle may have zero velocity), while in the second cubic lattice there are 19 allowed velocities [2]. Although there is a certain freedom in the choice of the lattice, its symmetry, as well as the minimum allowed set of velocities should conform to a minimum set of symmetry properties ensuring that the underlying anisotropy of the lattice does not affect the behavior of the system at the hydrodynamic level [14].

---

[1] In LGCA the collision step is determined by a table of all possible particle-velocity rearrangements at a node consistent with mass and momentum conservation.

[2] If no rest densities are allowed, the velocity sets reduce to 14 and 18 velocities respectively.

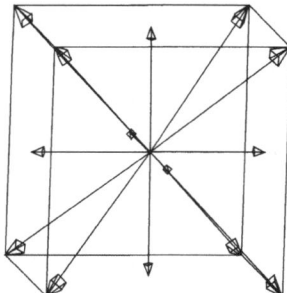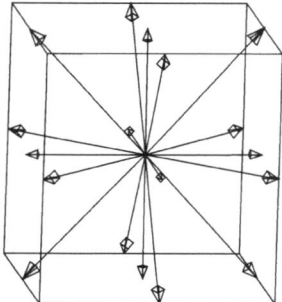

Fig. 1. Two different usual cubic lattices where LB can be defined. Left: 14 or 15 velocities (if particles at rest are allowed) are defined (giving rise to the so called, D3Q14 and D3Q15, respectively). Right: Nodes are still defined on a cubic lattice but the number of allowed velocities is now 18 or 19 (defining the models D3Q18 and D3Q19, respectively)

2.1 Elementary Variables

The densities $f_j(\mathbf{r}_i, t)$ are the elementary dynamical variables in LB. The connection to the relevant collective variables of the fluid is done by taking moments of these elementary densities in the same way that hydrodynamic variables are obtained from the one-particle distribution function in kinetic theory. Hence, the local density $\rho(\mathbf{r}_i, t)$, momentum $\rho\mathbf{v}(\mathbf{r}_i, t)$ and momentum flux $\mathbf{P}(\mathbf{r}_i, t)$ are obtained from

$$\rho(\mathbf{r}_i, t) \qquad = \sum_k f_k(\mathbf{r}_i, t) \qquad (1)$$

$$\rho(\mathbf{r}_i, t)\mathbf{v}(\mathbf{r}_i, t) = \sum_k \mathbf{c}_k f_k(\mathbf{r}_i, t) \equiv \mathbf{j}(\mathbf{r}_i, t) \qquad (2)$$

$$\mathbf{P}(\mathbf{r}_i, t) \qquad = \sum_k \mathbf{c}_k \mathbf{c}_k f_k(\mathbf{r}_i, t), \qquad (3)$$

where the index k runs over all allowed velocities.

2.2 Time Evolution

As in LGCA, the distribution function dynamics is composed of two steps: Collision and propagation. During the collision step, the distribution function is relaxed locally toward the corresponding equilibrium distribution. This is a purely local step. At the propagation step, the densities will move to a neighboring node according to their velocities. These two steps define the elementary time update, and hence we can write the LB dynamics in condensed notation,

$$f_j(\mathbf{r}_i + \mathbf{c}, t + 1) = f_j(\mathbf{r}_i, t) + \sum_k \Lambda_{jk} \left(f_k(\mathbf{r}_i, t) - f_k^{\text{eq}}(\mathbf{r}_i, t) \right), \qquad (4)$$

where the index k spans the velocity subspace, $f_j^{\text{eq}}(\mathbf{r}_i, t)$ is the equilibrium distribution function, and Λ_{jk} is a matrix that mixes the densities with different velocities at the corresponding node. This collision matrix is intimately related to the viscosities of the fluid. In the simplest model it is taken to be diagonal. In this case, i.e., $\Lambda_{jk} = -\lambda\delta_{jk}$, the resulting LB is the lattice equivalent of Bhatnagar, Gross and Krook (BGK) kinetic model , also referred to as exponential relaxation time model [16] (ERT).

The initial problem in LGCA, i.e., determining the collision tables that mix the particle velocities at each node (ensuring particle and momentum conservation) is now substituted by the problem of prescribing the properties of the equilibrium distribution in the linearized collision model giving rise to the desired physical behavior, i.e. that they still ensure the appropriate conservation laws. Here, I will concentrate on the simplest model where mass and momentum is conserved. In addition, it is also possible to enforce energy conservation [17].

2.3   The Equilibrium Distribution

Equilibrium distribution is determined by requiring that the collision operator conserves mass and momentum, namely

$$\sum_k f_k^{\text{eq}}(\mathbf{r}_i, t) \quad = \sum_k f_k(\mathbf{r}_i, t) = \rho(\mathbf{r}_i, t)$$

$$\sum_k \mathbf{c}_k f_k^{\text{eq}}(\mathbf{r}_i, t) = \sum_k \mathbf{c}_k f_k(\mathbf{r}_i, t) = \rho(\mathbf{r}_i, t)\mathbf{v}(\mathbf{r}_i, t). \tag{5}$$

These two constraints, together with the relation between the second moment of f_i^{eq} and the momentum flux,

$$\sum_k \mathbf{c}_k\mathbf{c}_k f_k^{\text{eq}}(\mathbf{r}_i, t) = \rho(\mathbf{r}_i, t)\mathbf{v}(\mathbf{r}_i, t)\mathbf{v}(\mathbf{r}_i, t) + \mathbf{P}^{\text{th}}, \tag{6}$$

give in turn a set of conditions that the equilibrium distribution should satisfy. According to (3), $\mathbf{P}^{\text{th}}$ in the previous equation corresponds to the thermodynamic pressure tensor. In addition, there exist further constraints on the acceptable forms of the equilibrium distribution function to ensure that the underlying lattice does not introduce any spurious anisotropy in the hydrodynamic limit of the LB dynamics (see Sect. 2.4).

As a result of the previous physical requirements, a suitable form for the equilibrium distribution is a quadratic function in velocities (equivalent to a low-velocity expansion of the velocity distribution). It can be expressed as

$$f_i^{\text{eq}} = a^\nu \left[A_\nu + \frac{\mathbf{j} \cdot \mathbf{c}_i}{c_s^2} + \frac{\rho\mathbf{v}\mathbf{v} : (\mathbf{c}_i\mathbf{c}_i - c_s^2\mathbf{1})}{c_s^4} \right], \tag{7}$$

where c_s is the speed of sound, and the values of the coefficients depend both on the specific lattice considered and the particular fluid model. The superscript ν indicates that the corresponding coefficients will have different values depending on the modulus of the velocity $\mathbf{c}_i$ (e.g., in the two lattices of Fig. 1 there are velocities with moduli 1 and $\sqrt{3}$ for the D3Q15 model, and 1 and $\sqrt{2}$ for the D3Q19 model).

Although (6) is expressed generically in terms of the thermodynamic pressure, the way in which non-ideality is incorporated into LB models has been a subject of continuous research. In the original models, the fluid was considered to be ideal; hence, they imposed $\mathbf{P}^{\text{th}} = \rho c_s^2 \mathbf{1}$. This approach is useful when modeling complex fluids where a length scale separation between solvent and solute molecules exists, and hence suspended particles should be dealt with as individual entities (see Sect. 3). In fact, there are a number of alternative proposals to incorporate interactions in LB in the literature. In many of them, $\mathbf{P}^{\text{th}}$ is still assumed to be that of an ideal fluid, and interactions are introduced effectively through a modified collision matrix, or through effective forces at mean field level (see e.g. [18]). However, in this contribution I will concentrate on the approach in which one modifies the equilibrium distribution function (selecting $\mathbf{P}^{\text{th}}$ appropriately) rather than the collision matrix, to ensure a specific thermodynamic behavior for the non-ideal fluid to be modelled (see Sect. 4). In this case, the coefficients of the equilibrium distribution function depend non-linearly on thermodynamic variables to reproduce the appropriate pressure tensor in equilibrium. Since generically one deals with coexisting phases, the pressure tensor must be able to sustain density gradients (see Sect. 4 for details). In this the microscopic origin of the interactions is completely disregarded; it is, therefore, fundamentally different approach to complex fluids from the LGCA approach, where one tries to determine simple interactions among the particles and analyze the free energy they give rise to.

2.4 Macroscopic Dynamics

The hydrodynamic equations to which LB dynamics gives rise to can be obtained from a Chapman–Enskog expansion, analogous to the procedure used in derivation of hydrodynamics starting from the Boltzmann equation.

For simplicity, I will restrict the discussion to the BGK model (see [19] for a more general discussion). The starting point is a Taylor expansion of (4). Both time and space should be expanded taking into account that they are discretized variables defined on a lattice. Hence,

$$\sum_{k=1}^{\infty} \frac{\bar{\epsilon}^k}{k!} \left(\frac{\partial}{\partial t} + \mathbf{c} \cdot \frac{\partial}{\partial \mathbf{r}} \right)^k f_i = -\lambda(f_i - f_i^{\text{eq}}),$$

where $\bar{\epsilon}$ is simply introduced as a label to keep track of the order of the expansion, and where the explicit time and position dependence of the distribution functions is omitted for the ease of the notation. Expanding to the second order in $\bar{\epsilon}$ (analogous to assuming an expansion to second order in gradients), one can rewrite the previous equation as

$$\bar{\epsilon} \left(\frac{\partial}{\partial t} + \mathbf{c}_i \cdot \frac{\partial}{\partial \mathbf{r}} \right) f_i^{\text{eq}} - \bar{\epsilon}^2 \tau \left(\frac{\partial}{\partial t} + \mathbf{c}_i \cdot \frac{\partial}{\partial \mathbf{r}} \right)^2 f_i^{\text{eq}} = -\lambda(f_i - f_i^{\text{eq}}). \qquad (8)$$

The left-hand-side has been written using the order of the expansion in terms of the equilibrium distribution. In the previous equation I have defined

$$\tau = \left(\frac{1}{\lambda} - \frac{1}{2} \right),$$

which, as it will be seen later, is closely related to the LB viscosity. From (8), one can easily derive the macroscopic constitutive equations by making use of the conservation laws (as expressed in (5)). For example, by summing over velocities in (8), we obtain

$$0 = \bar{\epsilon} \left[\frac{\partial \rho}{\partial t} + \nabla \cdot (\rho \mathbf{v}) \right] - \bar{\epsilon}^2 \tau \left\{ \frac{\partial}{\partial t} \left[\frac{\partial \rho}{\partial t} + \nabla \cdot (\rho \mathbf{v}) \right] + \nabla \cdot \left[\frac{\partial \rho \mathbf{v}}{\partial t} + \nabla \cdot (\rho \mathbf{v} \mathbf{v} + \mathbf{P}) \right] \right\} \quad (9)$$

Multiplying (8) with the microscopic velocity $\mathbf{c}_i$, and summing over velocities, one obtains (using momentum conservation),

$$0 = \bar{\epsilon} \left[\frac{\partial \rho \mathbf{v}}{\partial t} + \nabla \cdot (\rho \mathbf{v} \mathbf{v} + \mathbf{P}^{\mathrm{th}}) \right] - \bar{\epsilon}^2 \tau \nabla \cdot \left[\frac{\partial}{\partial t} \left(\rho \mathbf{v} \mathbf{v} + \mathbf{P}^{\mathrm{th}} \right) + \nabla \cdot \sum_i f_i^{\mathrm{eq}} \mathbf{c}_i \mathbf{c}_i \mathbf{c}_i \right] \quad (10)$$

The first term in (9) gives

$$\frac{\partial \rho}{\partial t} + \nabla \cdot (\rho \mathbf{v}) = 0 \quad (11)$$

which is the continuity equation expressing mass conservation. The two terms in square brackets in (8) are in fact already at least linear in $\bar{\epsilon}$. Hence, their contributions are of higher order in $\bar{\epsilon}$ and can be neglected. The absence of additional terms shows that LB recovers the continuity equation in the hydrodynamic limit.

If I consider an ideal gas, to lowest order in $\bar{\epsilon}$, (10) recovers the Euler equation

$$\frac{\partial \rho \mathbf{v}}{\partial t} + \nabla \cdot (\rho \mathbf{v} \mathbf{v}) + c_s^2 \nabla \rho = 0,$$

i.e., the evolution equation for a perfect fluid (for a discussion of generic non-ideal fluids, see e.g. [20]). The terms quadratic in $\bar{\epsilon}$ in (10) give the viscous contribution to the fluid motion. However, looking at the terms in brackets in the right-hand-side of (10), one can recognize that the first term is already of order $\bar{\epsilon}$. Therefore, dissipation is proportional to the relaxation parameter through τ, as had been anticipated. By analyzing this term one also can recognize that in order to ensure an isotropic dissipation, the third moment of the equilibrium distribution must be isotropic. This is the additional constraint that must be imposed on f_i^{eq} when determining the coefficients of (7) (see Sect. 2.3). The ability to impose such a restriction depends on the particular velocity subset considered [14] (e.g. it is not possible in a 2D 9-velocity model [21]). Isotropy requires

$$\sum_i f_i^{\mathrm{eq}} c_{i\alpha} c_{i\beta} c_{i\gamma} = \frac{\rho}{3} \left[v_\alpha \delta_{\beta\gamma} + v_\beta \delta_{\alpha\gamma} + v_\gamma \delta_{\alpha\beta} \right], \quad (12)$$

where $\delta_{\alpha\beta}$ is the Kronecker delta function, and the Greek indices refer to spatial components. Using this expression, and assuming an ideal gas, it is possible to rewrite (10) as

$$\frac{\partial \rho \mathbf{v}}{\partial t} + \nabla \cdot (\rho \mathbf{v} \mathbf{v}) = -\nabla (\rho c_s^2) - \nabla \cdot \Pi - \frac{3\eta}{\rho} \nabla \nabla : (\rho \mathbf{v} \mathbf{v} \mathbf{v}), \quad (13)$$

where the viscous contribution to the stress tensor is given by

$$\Pi = \eta \bar{\bar{\nabla}} \mathbf{v} + \xi \nabla \cdot \mathbf{v} \mathbf{1}, \quad (14)$$

where $\bar{\bar{\nabla}}\mathbf{v}$ denotes the symmetric and traceless contribution from the tensor $\nabla\mathbf{v}$, while $\mathbf{1}$ is the identity tensor. The shear and bulk viscosities, η and ξ, respectively, are given in terms of the relaxation parameter λ as

$$\eta = \frac{1}{6}\left(\frac{2}{\lambda} - 1\right) \quad \text{and} \quad \xi = \frac{2}{3}\eta.$$

The momentum balance equation (13), together with the constitutive relation (14) essentially recover the Navier–Stokes equation, except for the last term. Such a term is of higher order in velocities, and hence can be neglected under the usual conditions (consistent with the low-velocity expansion of f_i^{eq}).

It is possible to study the collective dynamics of LB beyond the hydrodynamic limit, e.g. by analyzing the hydrodynamic modes of LB at finite wave vectors. In the case of an ideal gas, it has been shown that collective modes have an analytic dependence on the wave vector [22].

This fact shows one of the useful aspects of using a kinetic equation to study collective fluid motion. Indeed, even if LB is tailored to reproduce hydrodynamics, its sensible behavior at finite wave vectors ensures that, even in dynamic regimes beyond the strict hydrodynamic limit still the response is meaningful and can be controlled and quantified. For non-ideal fluids this issue may be more controversial, but still there, since the free energy functional is essentially a low wave vector expansion of a more microscopic one (see Sect. 4).

Another important aspect related to the kinetic-equation structure of LB is the fact that the collision matrix can also be used to model the decay of the fast (non-conserved) modes. Although in most models this feature is not used (and such eigenvalues are set to their maximum value to ensure a proper separation of time scales to yield hydrodynamic behavior more easily), modifying them appropriately can be used to one's advantage to model different rheological responses, as for example viscoelasticity [23].

Having described the essentials of LB above, in the next two sections I will discuss in some detail how interfaces can be incorporated and how they can be applied to study colloidal dynamics.

3   Colloidal Suspensions

The characteristic sizes of colloidal particles are orders of magnitude larger than those of the solvent molecules. As a result, colloids will interact effectively with the collective modes of the solvent, rather than with the individual solvent molecules (for a general overview, see [3]). Hence, one can assume that LB describes the solvent at a coarse grained level, and we have to discuss how particles of rigid shape and finite extent, i.e., colloids, will interact with such a fluid.

3.1   Modeling Solid Particles

Since the basic characteristic of colloids is their rigid shape and extension, the first step is to model a generic solid interface in LB – a closed rigid interface (which does not need to coincide with the lattice nodes) defines the shape and size [24] – and to characterize how the fluid densities, f_i, will interact with it. This interaction will depend on how momentum is transferred at the solid surface. The usual situation is that the fluid velocity accommodates to that imposed by the solid boundary. This is referred to as *stick boundary condition*. This situation is completely different from the one encountered at a liquid/liquid interface, in which the absence of an elastic moduli makes it impossible to sustain stress jumps. In this case the stress tensor is continuous across the boundary (*slip boundary conditions*).

Solid Interfaces. Different schemes have been proposed to locate the interface and identify the fluid densities, f_i, that are affected by the presence of the interface. To describe colloids in a simple solvent, the most efficient scheme seems to be the approach in which one identifies the links (which join neighbor lattice nodes) that are crossed by the interface[3]. The boundary is then defined through the sets of densities at both nodes that would cross the interface during the propagation step.

In LB both stick and slip (and in general, a mixture of both) boundary conditions are easily simulated. If one uses the links to define the interface, stick boundary conditions for a solid boundary at rest are satisfied if one bounces back the relevant impinging densities of the corresponding neighbouring nodes – this procedure is referred to as *bounce-back at the links* (BBL). Instead, slip boundary conditions are satisfied if a specular reflection is executed on the impinging densities. A linear combination of both reflections gives rise to mixed boundary conditions. If the solid interface is moving, the same boundary condition is accomplished if such reflections are carried out in the frame of reference in which the interface is at rest [25].

To be specific, let us assume that the interface crosses the link joining nodes $\mathbf{r}$ and $\mathbf{r} + \mathbf{c}_i$. After the collision (and prior to the propagation step, at a time I will refer to as t^+), the densities that would cross the boundary are $f_i(\mathbf{r}, t^+)$ and $f_{i'}(\mathbf{r} + \mathbf{c}_i, t^+)$. If the solid interface is moving at a velocity $\mathbf{v}_w$, BBL can be expressed as

$$f_i(\mathbf{r} + \mathbf{c}_i, t + 1) = f_{i'}(\mathbf{r} + \mathbf{c}_i, t^+) + 2\frac{a^\nu}{c_s^2}\rho\mathbf{v}_w \cdot \mathbf{c}_i$$

$$f_{i'}(\mathbf{r}, t + 1) = f_i(\mathbf{r}, t^+) - 2\frac{a^\nu}{c_s^2}\rho\mathbf{v}_w \cdot \mathbf{c}_i, \tag{15}$$

where I have introduced the notation $\mathbf{c}_i = -\mathbf{c}_{i'}$. The coefficient $2a^\nu/c_s^2$ can be determined by requiring that at equilibrium bounce-back does not disturb the equilibrium distribution.

The description so far, even if put on the perspective of colloids, is completely general for an interface. In fact, the flexibility in defining interfaces is one of the great advantages of LB and it has been used to study suspension and flow problems in porous media and generic geometries [26–28].

Moving Particles. To study suspended particles, one has to determine a closed interface on the basis of information of the colloid position and geometry. For a spherical particle,

---

[3] The real interface is somewhere in between the two nodes. It is usually assumed to be at the middle of the link for simplicity.

knowing its center of mass position (which does not need to coincide with a lattice node) allows one to define the position of the interface. Once the velocity and angular velocity of the colloid, and the location of the interface are known, and the relevant links and corresponding densities have been identified, the BBL rule can be applied.

Strictly speaking, one would have to void the interior nodes from fluid (and hence restrict the bounce-back to the nodes that lie outside the solid interface). However, there are subtleties related to what happens when the colloid moves (how this happens will be discussed later) and how the internal and neighboring nodes exchange their identity. Again, this is a non-trivial issue, and a number of proposals exist in the literature. The most promising ones are those which approach the fluid/solid exchange from a volumetric perspective [19,29], and which, computationally, can be viewed as a sophisticated version of the simpler bounce-back boundary condition I described above. For the particular case of colloids in a simple solvent, it is enough (and simpler) for practical purposes to keep the nodes inside the colloid full of fluid. At low Reynolds number the presence of a closed solid surface decouples the dynamics of the interior of the shell from the exterior one. In this case, the interior fluid decays fast toward a fluid motion equivalent to that of a solid particle. Deviations from the appropriate behavior appear only at very short times (on time scale in which the speed of sound propagates a distance of the order of the colloidal size), and they can be quantified and corrected if necessary [30].

Once the boundary condition is enforced, it remains to determine the hydrodynamic coupling between the colloid and the solvent. The overall momentum change experienced by the fluid when it bounces back at the colloid interface adds up to the total force the solvent exerts on the colloid. Such a force is easily computed for a colloid moving with velocity $\mathbf{v}_c$:

$$\mathbf{F}\left(\mathbf{r}_b, t + \frac{1}{2}\right) = 2\left[f_i(\mathbf{r}, t^+) - f_{i'}(\mathbf{r} + \mathbf{c}_i, t^+) - \frac{2a^\nu}{c_s^2}\rho\mathbf{v}_c \cdot \mathbf{c}_i\right]\mathbf{c}_i, \qquad (16)$$

where $\mathbf{r}_b = \mathbf{r} + \mathbf{c}_i/2$ is the approximate position of the boundary between nodes $\mathbf{r}$ and $\mathbf{r} + \mathbf{c}_i$.

Once this overall force and the corresponding hydrodynamic torque are known, the velocity and position of the colloid can be updated by performing a simple molecular dynamics (MD) step. Although this is true to the first approximation, one has to take into account that the hydrodynamic force (as displayed in (16)) depends on the colloid velocity, $\mathbf{v}_c$. Hence, the new velocity obtained from the MD step should be used when computing the bounce-back to ensure consistency (although it can be achieved in one step, see [31]). If the colloid velocity changes appreciably during one time step (which may be the case for buoyant particles) numerical instabilities may appear. The explicit scheme is stable for dense colloids, where there exists a time scale separation between solvent and particle velocity relaxation.

The overall scheme to deal with suspended rigid particles is thus a mixed algorithm in which LB, determining solvent dynamics, is coupled to an MD scheme for the colloidal particles in the suspension, and where bounce-back is applied (locally) to the solvent to enforce the appropriate hydrodynamic boundary condition provides also the force exerted by the solvent on the colloid (which feeds-back into the solvent dynamics through the bounce-back itself).

A similar procedure has also been developed to model polymer solutions [32]. In this case the polymer beads are treated as solid spheres which are connected by springs. In principle, one could deal with every bead in the same way I have described for the colloids. However, due to its computational cost the beads have smaller radius than the lattice spacing. The lack of spatial resolution is solved by assuming a friction coefficient for the beads that must be fixed numerically.

Finally, in order to simulate an equilibrium state of a suspension, thermal fluctuations of the solvent must be included. Since there exist no intrinsic fluctuations in LB (arising from an averaging of microscopic degrees of freedom), they must be introduced with physical insight. The usual approach is to assume local equilibrium in the fluid, and hence suppose that the relevant fluctuations are the hydrodynamic ones [33]. One can then add a random contribution to the dynamic evolution of the distribution function (4) with the constraint that it only affects the pressure tensor which then has a random component. Its amplitude is tuned to enforce the fluctuation-dissipation theorem in equilibrium. This is the lattice equivalent of the procedure developed by Zwanzig and Bixon to recover fluctuating hydrodynamics from a fluctuating Boltzmann equation [34]. Nonetheless, as I will discuss in the next subsection, a lot of information on the fundamentals of colloidal hydrodynamics can be obtained by performing simulations in the short-time regime, where linear response theory can be used to avoid the use of such fluctuations.

3.2 Colloidal Hydrodynamics

The LB model for suspensions has the advantage that it incorporates all the relevant dynamical couplings between the colloids and the collective modes of the solvent. All the (dissipative) interactions are properly accounted for locally through the boundary conditions. This scheme will then reproduce long-range hydrodynamic interactions, i.e. the fact that solvent hydrodynamics is properly incorporated ensures that the perturbations in the fluid flow caused by the solvent particles will be propagated through the fluid yielding the effective dissipative interactions among the colloids that characterize their dynamics (through a completely local dynamics).

This approach is distinct from Stokesian Dynamics [35], where the solvent is integrated out. In Stokesian Dynamics one works directly with the friction matrices that determine the motion of the colloidal particles. Long-range hydrodynamic interactions yield friction matrices that couple the velocities of all colloidal particles. As a result, to solve for colloidal hydrodynamics, it is necessary to invert the friction matrix which becomes a numerically demanding step, unless specific tricks are considered. LB resolves the time scales at which the solvent evolves building up the hydrodynamic interactions through purely local dynamics. However, in comparison to Stokesian Dynamics, this has the additional computational cost that all relevant hydrodynamic time-scales must be resolved. There may be five orders of magnitude between the time scales related to solvent momentum relaxation and colloid diffusion. Hence, how much of the true colloidal hydrodynamics can be reproduced with LB depends on the possibility of using scaling ideas to narrow down such a time scale gap. Nevertheless, LB has proved useful, for example, to study sedimentation at high Péclet numbers [36].

Extensive studies have quantified the performance of LB in simulations of colloidal hydrodynamics (see, e.g. [33]). Mobility coefficients and viscosities are recovered and shown to be in excellent agreement with theoretical predictions and experiments. Only at high concentrations detailed lubrication effects must be included to obtain quantitative results.

The scheme described above for spherical colloidal suspensions has been extended to consider non-spherical colloidal particles [37] as well. This class of suspensions has actually been modeled both by resolving the individual mesoscopic objects (especially in the dilute regime, hence addressing question related to ordering when subject to external forces of flows) and by using an order parameter (studying e.g. spinodal decomposition or phase transitions out of equilibrium). The latter corresponds to the situation where the nematogen molecules are comparable in size to solvent molecules [38]. This is a good example of the flexibility of the method and the various issues it can address.

3.3 Colloids at High Confinement

The physics of fluid flow at high confinement has been an issue of interest in fluid mechanics and physicochemistry for a long time. During past decades, oil recovery was one of the issues with obvious potential applications. More recently, the technology developed in last decades in the miniaturization of chips has migrated to the design small circuits to control the flow of small quantities of fluids [39]. This is a growing area since such microdevices allow the delivery of small and well controlled amounts of materials, and constitute suitable environments to develop bio-electrical interfaces.

The understanding of the flow of complex fluids under such conditions poses new challenges [40], e.g. how to mix fluids at low Reynolds numbers taking into account that mixing strategies in usual reactors work in the opposite limit, or how to develop structures that control fluid flow or pump it. In this respect, the use of colloids becomes natural, given that their size is smaller than the typical width of these circuits [41]. Hence, it is important to understand the basic mechanisms that characterize colloid motion under these conditions. LB is in fact a good model to study such behavior, as generic boundaries are easily accounted for.

One of the first questions one can address from a fundamental point of view is the way in which confinement affects the diffusion of a colloidal particle. One way to do it is by analyzing the velocity autocorrelation function (*vacf*), $C(t)$, which measures the loss of memory of the velocity of a colloidal suspension. It can be written as

$$C(t) = \sum_i \frac{\langle \mathbf{v}_i(t) \cdot \mathbf{v}_i(0) \rangle}{\langle \mathbf{v}_i(0)^2 \rangle}, \tag{17}$$

where the sum runs over all colloidal particles. Such an analysis is interesting since it provides us with the relevant dynamical information characterizing the hydrodynamical coupling between the suspended colloids and the solvent.

Unbounded Suspension. For an unbounded suspension such a coupling is well understood [42]. The colloids perturb the solvent flow according to their initial velocity generating both density and vorticity perturbations. The former will propagate away from the colloids by sound waves, affecting the colloid dynamics on a time scale $\tau_s \sim R/c_s$

for a colloid of size R ($\sim 10^{-8}$ s for an ordinary colloidal suspension). Due to momentum conservation, vorticity can only diffuse away from the colloid, and it does so on a time scale $\tau_\nu \sim R^2/\nu$ ($\sim 10^{-6}$ s) for a solvent with kinematic viscosity $\nu = \eta/\rho$. This time scale turns out to be of the same order as the time scale in which colloids lose memory of their initial velocity (the inertial time scale, τ_ξ, related to the friction that the solvent exerts on the colloids)[4]. The fact that τ_ν and τ_ξ are of the same order implies that colloidal velocity and vorticity relaxation are coupled [43] and both will control the decay of *vacf*. There exists a third relevant time scale related to the diffusion of the colloids themselves. It can be estimated easily as the time needed for the colloids to diffuse their own size is $\tau_D \sim k_B T/(\rho \nu R)$ ($\sim 10^{-3}$ s). In the absence of this mode-coupling, the velocity of a colloid will decay exponentially due to the friction force the solvent exerts on it.

The dynamic coupling between the particle velocities and the collective modes of the solvent induces a memory effect leading to colloid relaxation slower than exponential. The latter is the expected *vacf* decay if the solvent is assumed to be a passive friction medium. It is possible to gain some physical insight into this coupling – and to understand the velocity relaxation it gives rise to – by assuming that the solvent velocity field is a scalar. Assuming one single colloidal particle for simplicity, the slowly decaying part of the solvent velocity (induced by the colloidal motion itself) diffuses away from the particle. Hence it follows

$$\frac{\partial v}{\partial t} = \nu \nabla^2 v. \tag{18}$$

The initial value of the velocity is provided by the moving colloid and it can be regarded as an initial localized perturbation of the fluid field, at the position where the colloidal particle is, r_0. Hence, at later times the fluid velocity will decay following

$$v(\mathbf{r}, t) = \frac{v_0}{(4\pi\nu t)^{d/2}} e^{-(\mathbf{r}-\mathbf{r}_0)^2/(4\nu t)}, \tag{19}$$

where d refers to the dimensionality of the system. The above equation shows that the amplitude decays algebraically as $1/t^{d/2}$. If we assume that due to mode-coupling the particle velocity will follow its surrounding fluid velocity field, we see that the *vacf* will decay algebraically, i.e., $C(t) \sim 1/t^{d/2}$, instead of exponentially as predicted by a simple Langevin picture where the solvent acts as a passive friction medium.

At finite concentrations the same picture applies; the same algebraic decrease is obtained, although with concentration-dependent amplitudes. This has been quantified by Lowe and Frenkel [44]. In fact, while *vacf* refers to the motion of a single particle in a given environment, the complete hydrodynamic response of a suspension is contained in its generalization for finite-wave vector. This corresponds to the wave-vector longitudinal current-current correlation function defined as

$$J(q,t) = \sum_{j\neq i} \sum_i e^{i\mathbf{k}\cdot\mathbf{r}_{ij}(t)} \hat{\mathbf{k}} \cdot \langle \mathbf{v}_j(t)\mathbf{v}_i(0)\rangle \cdot \hat{\mathbf{k}}, \tag{20}$$

---

[4] τ_ξ is of order $\tau_\xi \sim M/(\nu R\rho)$ with M being the mass of the colloid.

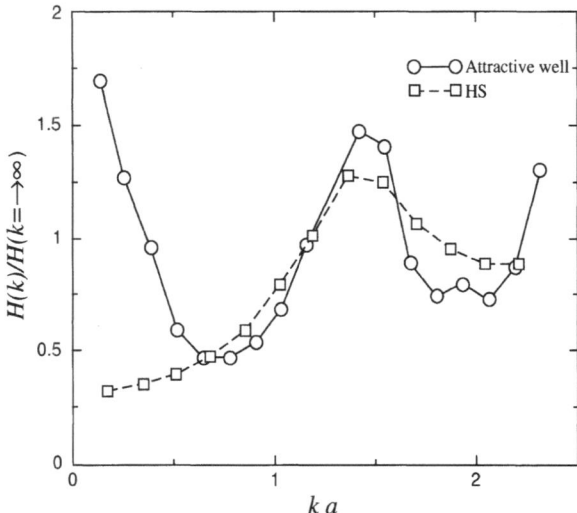

Fig. 2. Time integral of the wave-vector longitudinal current-current correlation function, $H(k)$, as a function of the wave vector for a suspension of colloidal particles at a volume fraction $\phi = 0.3$. The colloids have radius $a = 2.5$ in lattice units. Squares: Colloids interacting with hard core repulsion. Circles: A linear attractive potential is added to the hard core exclusion for distances larger than the particle diameter but smaller than two times the diameter. $H(k)$ is normalized by its value at large wave vectors

where the sum runs over all suspended particles. In the limit of large wave vectors, the previous expression reduces to the *vacf*, while in the limit of vanishing wave vectors it is related to the short time sedimentation velocity. In between, it provides information of the correlated motion induced by hydrodynamics at a characteristic length-scale defined as the inverse of the wave vector. This quantity can be accessed by diffusive wave spectroscopy (DWS) [45] and it is easy to compute using LB. LB has already been used to analyze the behavior of this quantity for colloids interacting through hard core interactions [45,46], and it can be easily generalized to yield dynamic structure factors for colloid suspensions with generic interactions [47]. An example is displayed in Fig. 2.

Colloids in a Tube. In order to address the basic dynamic couplings between a suspension and the solvent in the present of confining walls, I will focus on the *vacf*, rather than in the general current-current correlation function. In order to clarify the analysis, I will restrict myself to the simplest geometry, a cylindrical tube or two parallel plane walls in three dimensions, or two parallel slits in two dimensions.

The presence of confining walls alters significantly the dynamical behavior of the suspension. Contrary to what happens in the unbounded case, now the momentum of the suspension, including the solid boundaries, is no longer a conserved variable. Assuming the picture introduced in the previous subsection, if I consider that solvent vorticity is the only relevant motion of the fluid that couples to the colloid dynamics, then the decay of the *vacf* will be exponential, rather than algebraic. Using again the decay given by

(19), the existence of a characteristic length given by the tube width H will define a characteristic wave length, $k_0 = 2\pi/L$ on which velocity perturbations will decay,

$$v(t) \sim \frac{1}{t^{d/2}} e^{-k_0^2 \nu t}$$

yielding an exponential decay of the velocity. Such a decay arises as a result of vortex damping once it starts to interact with the solid walls.

A more careful mode-coupling analysis confirms the exponential decay of vorticity due to the presence of the rigid walls [48]. Note that this situation at high confinement is different from cases in which only one wall is present, where the decay is still algebraic. In this case, the presence of a rigid wall perturbs the mode-coupling giving rise to different exponents characterizing the algebraic decay [49][5].

The previous description suggests that colloid dynamics between solid interfaces complies to the Langevin theory. However, the *vacf* of a spherical particle in a cylindrical tube moving initially along its axial direction decays as shown in Fig. 3. Its time

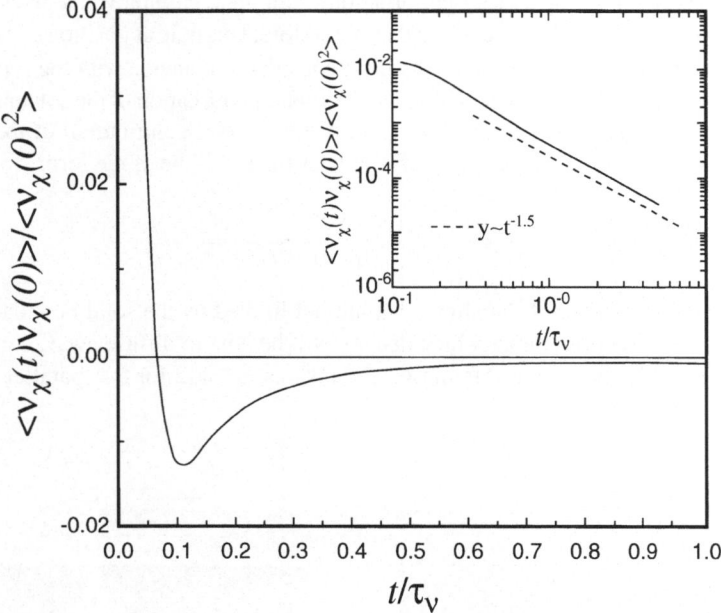

Fig. 3. *Vacf* for a spherical particle of radius $R = 2.5$ lattice spacings in a circular cylinder of width $H = 9$ lattice spacings. The particle is initially moving along the cylinder axis. Time is normalized by the viscous time τ_ν. Inset: Log-log plot of the asymptotic decay of the *vacf*

---

[5] It is also interesting to note that the interaction of the solvent and the interfaces is sensitive not only to the confinement itself but to the particular way in which the dynamic interaction takes place. Hence, whether one has stick or slip boundary conditions will modify the dynamic response of the suspension. In the case of one wall, the exponents characterizing the algebraic decay of the *vacf* will be different for a solid or fluid interface.

dependence is clearly non-exponential: The inset shows that the asymptotic decay is algebraic [50]. Moreover, contrary to what happened in an unbounded fluid, the *vacf* changes sign, which means that after some time the particle will reverse its direction of motion. This whole dependence is at odds with a simple Langevin picture, indicating that it cannot be rationalized simply in terms of vorticity diffusion. Hence, we must conclude that confinement induces a qualitatively different hydrodynamic coupling from the one ruling unbounded suspensions.

Physical Picture. If one looks at the evolution of the solvent density and velocity profiles, Fig. 4, one can see how vorticity is indeed damped due to the interaction with the walls. However, after this, one can still see how density perturbations decay slowly as a result of their interaction with the confining walls. The fluid flow this decay gives rise to is opposed to the initial direction of motion of the colloid. Due to the coupling of the particle velocity to this flow field, the colloid will end up moving in the direction opposed to its initial velocity. Hence, this indicates that it is the coupling of the colloid velocity to the compressible solvent flow which determines the colloid velocity relaxation.

It is possible to rationalize the features of the *vacf* observed in Fig. 3 by assuming that the interaction of the density perturbations with the confining walls gives rise to a diffusive density mode. Let us assume that the colloidal particle at position $\mathbf{r}_0$ is moving initially along the axis of the tube, which we consider to coincide with the x direction. The initial motion of the particle will induce a dipolar perturbation of the solvent density, which can be taken as localized. Therefore, if the relevant contribution of the density perturbation decays diffusively, its profile at later times will be of the form

$$\rho(\mathbf{r}, t) = \frac{\pi x e^{-(\mathbf{r}-\mathbf{r}_0)^2/(4D^*t)}}{(2\pi D^*t)^{d^*/2+1}},$$

where d^* is the number of directions that are not limited by the solid boundaries (and which are the directions along which density will be able to diffuse away, e.g. $d^* = 1$ for a cylindrical tube or two slits in two dimensions, $d^* = 2$ for two parallel plates in

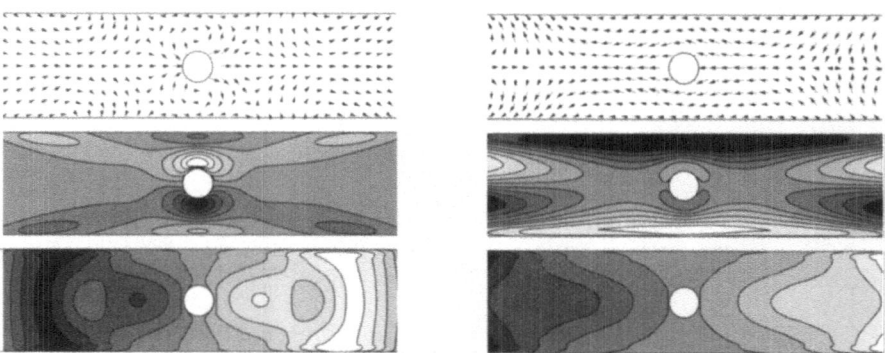

Fig. 4. Flow field, vorticity and density (top to bottom) for a fluid in a cylindrical tube in the presence of a colloidal particle moving initially along the cylinder axis to the right. Hydrodynamic fields are computed at the $z=0$ plane and shown at times $t/\tau_\nu=0.47$ (left) and $t/\tau_\nu = 0.78$ (right)

three dimensions). D^* refers to the diffusion coefficient that characterizes the assumed density diffusion (see (22) below for an explicit expression). The decaying diffusive density will give rise to a mass flux

$$\mathbf{j}_d = -D^* \nabla \rho \sim -1/(2\pi D^* t)^{d^*/2+1}$$

at the position where the colloid is. By simply assuming that the colloidal particle couples to its surrounding fluid flow (as it was the case in the unbounded suspension), one can understand why the colloid reverses its direction of motion (the relaxational flow induced by a diffusive density perturbation), why the time dependence is asymptotically algebraic (the diffusive character of the density decay) as well as its particular exponent (related to the dipolar nature of the relevant solvent perturbation and to the number of unconstrained directions provided by the geometrical constraints).

This intuitive picture can be derived from a more rigorous theoretical formulation, based on linearized hydrodynamics [50,51]. The basic ingredient in the hydrodynamic calculation is the fact that the presence of solid walls relaxes velocity gradients along the confined directions, such that for the tube geometry described above $\partial_x v_x$ becomes the dominant velocity gradient. However, at the same time, one cannot neglect the fact that the longitudinal velocity component v_x depends on the transverse spatial coordinates since otherwise stick boundary conditions cannot be fulfilled. With these two ingredients one can show that purely diffusive modes can be excited in a tube if $4c_s^2 k_x^2 < (\Gamma k_x^2 + \nu k_\perp^2)^2$, where $\Gamma = \nu + \xi$ is the sound wave damping coefficient, while k_x and $k_\perp$ refer to the wave vectors corresponding to excitations along and perpendicular to the tube axis respectively. Accordingly, in the hydrodynamic limit (for small enough values of k_x that keep the relation ensuring pure damping modes), the relevant hydrodynamic modes are

$$\omega_1 \sim \mathrm{i}\nu k_\perp^2 + \mathrm{i}\left(\Gamma - \frac{c_s^2}{\nu k_\perp^2}\right) k_x^2 \quad \text{and} \quad \omega_2 \sim \mathrm{i}\frac{c_s^2}{\nu k_\perp^2} k_x^2,$$

which show that for a confined geometry there exists a dynamical regime with two diffusive modes. While the first is related to the decay of shear fluid perturbations, as usual, the second one signals the appearance of sound diffusive modes. The previous equation for ω_2 also tells that the decay of this new diffusive mode is controlled by a diffusion coefficient that scales as

$$D^* \sim \frac{c_s^2 H^2}{\nu} \tag{21}$$

with H being the characteristic width of the tube. This shows that density diffusion is intrinsically linked to fluid compressibility. LB can also be used to perform a careful check of this diffusive mode. For example, one can consider a tube initially filled with fluid at rest in the absence of colloidal particles. If one introduces a localized density perturbation, analyzing the density profile as a function of time provides a good case to study sound diffusion. In Fig. 5 I show the time decay of the density profile along the direction of the axis of the tube. One can see how, after a transient characterized by sound propagation (signaled by the appearance of two peaks moving away from each other), once sound waves have died away there is still a remaining part of the density that decays more slowly. It is then possible to compute the width of this distribution as a

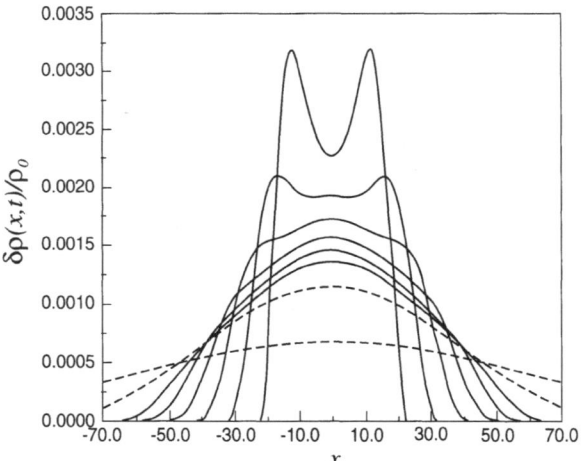

Fig. 5. Time evolution of an initial localized density perturbation in a 2D slit of width $H = 10$ lattice spacings for a fluid with $\nu = 1/2$. $\rho_0 = 24$ is the reference density, and the initial density perturbation is $\delta\rho = 0.1 \times \rho_0$. Density profiles are shown at times $t/\tau_\nu = 0.1$, 0.15, 0.2, 0.25, 0.3 and 0.35 with continuous lines. Times 0.5 and 1.5 are drawn with dashed lines

function of time. It increases linearly with time, which allows one to define a diffusion coefficient; such a diffusion coefficient is found to behave as

$$D^* = \frac{c_s^2(H^2 - 1/4)}{3\nu} + \frac{1}{2}, \tag{22}$$

hence exhibiting the expected dependence predicted in (21). The numerical factors $1/2$ and $1/4$ are lattice artifacts related to the intrinsic lattice diffusion and the precise location of the tube walls, respectively.

This analysis shows that in highly confined, elongated geometries, pressure relaxation gives rise to a new family of long-time tails, which have a physically different origin form the well-known long-time tails in unbounded fluids. Although I have concentrated on simple geometries, the behavior of the *vacf* can be shown to have the generic form

$$C(t) \sim -\frac{1}{t^{1+d^*/2}}.$$

They have the same time-dependence as the long-time tails predicted for a Lorentz gas [52] although they originate from different physical mechanisms.

The behavior of the *vacf* shows that, contrary to the expectations, the fact that momentum is not a conserved variable, does not mean that mode relaxation is necessarily exponential. In fact, only the decay of the angular velocity of a colloid centered in a tube will be exponential [51]. However, due to the coupling between translation and rotation, any rotating off-centered colloid will decay asymptotically with the same long-time tail as a translating colloid.

Suspension Dynamics. Under which conditions can such overdamped modes be observed? Sound diffusive modes are expected to be present for geometries that allow

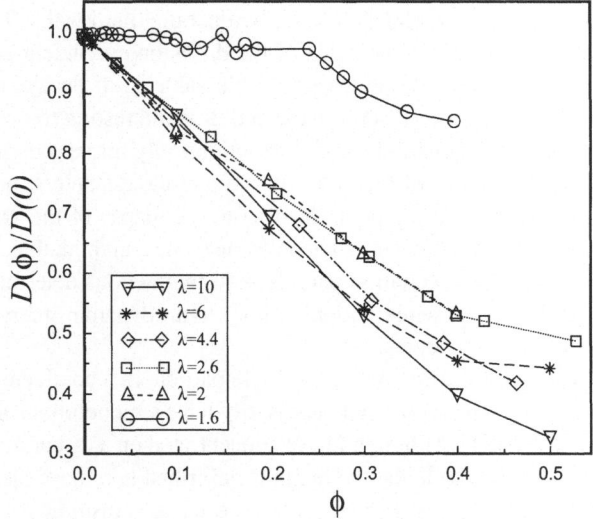

Fig. 6. Normalized diffusion coefficient as a function of the volume fraction for a confined suspension of hard spheres of radius $R = 2.5$ lattice units, as a function of the volume fraction for different widths as prescribed by the dimensionless parameter $\lambda \equiv 2R/H$

modes with a ratio satisfying

$$\frac{k_\perp^2}{k_x} \geq \frac{2c_s}{\nu}.$$

If one considers a colloidal suspension with particles of radius R in a tube of width H and length L, then the previous bound can be rewritten as

$$\left(\frac{2R}{H}\right)^2 \geq \frac{4}{\pi}\frac{c_s R}{\nu}\frac{R}{L},$$

which can be fulfilled more easily the more viscous the solvent is. For example, having an aspect ratio $R/L \sim 10^{-4}$ and a relatively narrow pore $(2R/H \sim 1/3)$, for water one needs colloids of size $R \leq 5 \cdot 10^{-5}$ cm to observe sound diffusion. For a more viscous solvent, such as olive oil, the bound shifts up to $R \leq 5 \cdot 10^{-3}$ cm. So, was it hard to find these long-time tails out using LB? In fact LB, as, in general, all coarse grained mesoscopic models for complex fluid flow have a large compressibility (the speed of sound is not too large). This unrealistic enhanced compressibility favors the appearance of diffusive sound modes, and makes it easy to analyze with LB.

How do these diffusive modes influence colloidal dynamics? The fact that the algebraic decay is related to the compressibility of the solvent implies that such a decay will not modify the diffusion coefficients predicted by a theory based on an incompressible solvent. It only warns you that it is necessary to resolve the dynamic decay carefully to be able to get quantitative predictions on transport coefficients in confined geometries using Green–Kubo expressions. By doing this, it has been possible to assess how confinement affects the diffusion coefficient of a colloidal suspension. Figure 6 displays the volume-fraction dependence of the normalized self-diffusion coefficient for a hard sphere colloidal suspension confined between two parallel hard walls. The normalization factor, $D(0)$, corresponds to the value of the diffusion coefficient at the center of the two walls at a vanishing volume fraction. This value itself decays as the gap between the two walls decreases, as can be expected. It is interesting to observe in Fig. 6 that diffusion is controlled by particle-wall interactions only for very high confinement. As soon as the width becomes of the order of two particle diameters, particle-particle hydrodynamic interactions are dominant leading to a collapse of the volume-fraction dependence of the scaled diffusion coefficient. Hence, one can think that, except at high confinement, to a good approximation particle-wall interactions determine the overall magnitude of the diffusivity, while colloid-colloid hydrodynamic interactions control the detailed volume-fraction dependence.

A careful analysis of the diffusivity of a single particle in a confined geometry also shows that the diffusion tensor does not necessarily have a monotonous dependence on the transverse direction [51]. Although I have concentrated on a discussion of the short-time dynamics of confined colloids, the features described here have also an influence on colloidal long-time diffusion. For example, already at vanishing volume fractions, the position dependence of the diffusion tensor gives rise to substantial non-Markovian effects in the diffusion process [53]. It is interesting to note that recent experiments have started to address colloidal diffusion in conditions analogous to the ones described here [54].

Finally, it is worth stressing the difference between the effect of confinement itself and the particular role played by boundary conditions. I have focused on the case of solid/fluid interfaces, where momentum satisfies *stick* boundary conditions. It would have also been possible to consider confined fluid layers (for example a tube of water surrounded by an immiscible fluid) with *slip* boundary conditions instead. In this situation there is no diffusive sound and one gets the usual algebraic decay for the *vacf* (and in particular an exponential decay in a tube-like geometry) [51]. Hence, rather than confinement itself, the features of the particle/solvent coupling are controlled by the particular mechanisms of momentum exchange at the interface. In this respect, the results described in this section are not only relevant for colloidal hydrodynamics, but have a wider scope. For example, solid membranes suspended in a solvent should exhibit analogous modes [55], and there exist theories describing the effects of overdamped sound in porous matrices and gels [56].

4   Non-ideal Fluids: A Binary Mixture

The description of colloidal suspensions in LB is based on the idea that there exists a gap between the typical size of the colloidal particles and that of the solvent molecules. As a result, the individual colloids interact with the collective modes of the solvent. In the absence of such a gap, other strategies are needed to incorporate the effective interactions in complex fluids. Again, the basic idea is based on introducing coarse-grained models that capture the basic interactions and collective dynamics and reproduce the dynamics of these systems at relevant length and time scales.

The traditional way to incorporate interactions in LGCA has been to consider different species of particles with effective interactions (accomplished by modifying the collision tables) [57]. This approach has also been used in LB [58]. However, in this section I will concentrate on the procedure developed by the Oxford group [59], which modifies the equilibrium distribution, rather than the collision matrix as a way to introduce the non-ideality of the fluid. In the next subsection I will describe the basic ideas underlying this approach which builds on the basic model discussed in Sect. 2. As a specific example of this approach, we analyze the kinetics of a binary mixture.

4.1 The Model

In order to simulate a non-ideal fluid, it seems enough to modify the expression for the pressure tensor $\mathbf{P}^{\mathrm{th}}$ appearing in the equilibrium distribution of f_i, as expressed in (7). The basic idea is then to select the pressure tensor in such a way that it is thermodynamically consistent . One way to achieve this is to derive the pressure from a given free-energy model, and control the thermodynamics the LB model will describe. Since phase coexistence implies that interfaces may be present, it is necessary that the free-energy model includes the energetic cost of sustaining them – hence, a free energy functional is needed [60]. Although in principle detailed, or microscopically motivated, functionals can be considered since LB will reproduce the collective hydrodynamic modes of the fluid, there is no point in introducing functionals with too detailed microscopic features, and usually a simple gradient expansion is considered.

In order to describe a generic pressure tensor, $\mathbf{P}^{\mathrm{th}}$, one needs to generalize the form of the equilibrium distribution given by (7). It is sufficient to consider an equilibrium distribution of form

$$f_i^{\mathrm{eq}} = a^\nu \left[A_\nu + \frac{\mathbf{j} \cdot \mathbf{c}_i}{c_s^2} + \frac{\rho \mathbf{v} \mathbf{v} : (\mathbf{c}_i \mathbf{c}_i - c_s^2 \mathbf{1})}{c_s^4} + \mathbf{G} : \mathbf{c}_i \mathbf{c}_i \right], \qquad (23)$$

where the expansion coefficients depend non-linearly on the density, temperature, and other thermodynamically relevant parameters. The additional tensor $\mathbf{G}$ accounts for the spatial gradients of the density fields. The presence of density gradients induces the breakdown of Galilean invariance in simple implementations of these models. However, more recent refinements (see e.g. [61]) indicate how to correct the equilibrium distributions to achieve Galilean invariance.

This approach has been applied to model different kinds of complex fluids, such a van der Waals model that exhibits a liquid/gas transition [59], a model fluid for binary mixtures or amphiphilic systems [62], and also to nematogenic fluids [38]. Except for the first example, the rest require the introduction of additional thermodynamic variables – local concentration or liquid crystalline order.

In order to describe how additional variables can be introduced, I will discuss the case of a binary mixture in detail. To describe its dynamics, in addition to the mean density ρ and the baricentric velocity $\mathbf{v}$ (derivable from the distribution function f_i as described in Sect. 2), it is also necessary to account for the local density difference between the two species, $\phi(\mathbf{r}, t)$. This is accomplished by introducing a second distribution function $g_i(\mathbf{r}, t)$, which is related to the new conserved variable, and which can be understood as the local density difference moving with velocity $\mathbf{c}_i$ at position $\mathbf{r}$ at time t. The connection between the new distribution and the relevant hydrodynamic variables parallels to that in a single fluid (cf. (5)), namely

$$\sum_k g_k^{\text{eq}}(\mathbf{r}_i, t) \quad = \sum_k g_k(\mathbf{r}_i, t) = \phi(\mathbf{r}_i, t)$$

$$\sum_k \mathbf{c}_k g_k(\mathbf{r}_i, t) \quad = \phi(\mathbf{r}_i, t)\mathbf{v}(\mathbf{r}_i, t)$$

$$\sum_k \mathbf{c}_k \mathbf{c}_k g_k(\mathbf{r}_i, t) = \phi(\mathbf{r}_i, t)\mathbf{v}(\mathbf{r}_i, t)\mathbf{v}(\mathbf{r}_i, t) + M\mu(\mathbf{r}_i, t)\mathbf{1}, \tag{24}$$

where μ is the chemical potential, which should be derived from the same free energy functional used to obtain the pressure tensor in order to ensure thermodynamic consistency. The parameter M is related to the diffusivity of the mixture. The equilibrium distribution function, g_i^{eq}, has the same form as (23) with a different set of constants determined by the previous constraints.

This second distribution function will evolve following a linear kinetic equation, analogous to the one obeyed by f_i, (4), i.e.,

$$g_j(\mathbf{r}_i + \mathbf{c}_j, t+1) = g_j(\mathbf{r}_i, t) + \sum_k L_{jk}(g_k(\mathbf{r}_i, t) - g_k^{\text{eq}}(\mathbf{r}_i, t)). \tag{25}$$

However, contrary to f_i, in order to recover the convection-diffusion equation that characterizes the evolution of ϕ in the hydrodynamic limit, it is enough to require that the local concentration is a conserved variable, as has been expressed in (24). Hence, the collision matrix of this second kinetic equation will not conserve the first moment of g_i, related to the concentration flux.

Following the steps of Sect. 2.4, it is possible to derive the hydrodynamic equations for a LB model with two distribution functions f_i and g_i. The first distribution function gives rise to the continuity and Navier–Stokes equations as was derived previously. For the distribution g_i, if we consider a BGK model with $L_{ij} = \gamma\delta_{ij}$, it is possible to show that the relevant hydrodynamic equation is a convection-diffusion equation

$$\frac{\partial\phi}{\partial t} + \nabla \cdot (\phi\mathbf{v}) = \left(\frac{1}{\gamma} - \frac{1}{2}\right)\left[\nabla^2(M\mu) - \nabla \cdot \left(\frac{\phi}{\rho}\nabla \cdot \mathbf{P}^{\text{th}}\right)\right],$$

where the relaxation coefficient γ gives essentially the mobility. The last term in the previous equation is small whenever compressibility does not play a relevant role. Nonetheless, appropriate minor modifications in the equilibrium distribution may be enough to remove such an spurious term [63].

As a result, this simple LB with two densities gives rise to the appropriate hydrodynamics for a binary mixture with viscosities that are the same for both species (the relaxation of f_i toward equilibrium does not depend on the local concentration). Again, it is possible to account for an asymmetry in the transport coefficients assuming Newtonian fluids.

The binary mixture model has been applied to a great variety of situations, including bidimensional spinodal decomposition [64], droplet dynamics under shear flow [65] and the flow in the presence of patterned surfaces [66]. In the next section I will focus on basic issues related to kinetics.

4.2 Spinodal Decomposition

Here, I will concentrate on the analysis of the kinetics of phase separation when a binary mixture is quenched to a temperature below its critical temperature, where it becomes immiscible. The absence of thermal fluctuations in LB will induce the subsequent spinodal decomposition of an initial homogeneous mixture. Even if this is a classic example in non-equilibrium statistical physics, there are still a number of questions that remain open. Using LB to address this problem in detail shows the potential of this method and what can be achieved in comparison with other numerical techniques.

Dynamical Scaling. If an initial mixed state of a binary mixture is quenched below its stability critical point, local density fluctuations will grow creating domains with the local value of the concentration corresponding to one of the values of the two coexisting phases. After this first stage, once local domains are separated by well defined interfaces, the process of phase separation will continue through domain coarsening. Depending on the relative amount of the two components, droplets of the minority phase are formed at late stages, or a bicontinuous structure develops for similar amounts of the two components. For simplicity, I will concentrate in the later case. The theoretical understanding of the late stage coarsening is based on the idea that there is one single length scale that characterizes the domain sizes as a function of time $L(T)$ [67]. Hence, if we use the characteristic length and times defined on the basis of the relevant physical parameters of the fluid, i.e.,

$$L_0 = \frac{\eta^2}{\rho\sigma}, \quad T_0 = \frac{\eta^3}{\rho\sigma^2},$$

where σ is the surface tension, then at late times, any dimensionless structural length $l(t) \equiv L(T/T_0)/L_0$ will be determined by a universal evolution equation.

This scaling assumption is compatible with the hydrodynamic equations, and they predict how the dimensionless length will evolve as a function of the dimensionless time. In fact, a dimensional analysis of the Navier–Stokes equations predicts the existence of two different kinetic regimes. To perform a dimensional analysis, one has to take into account that the pressure gradients induced by the interface curvature act as the driving force of the phase separation since the system is trying to minimize the free energy. In the scaling scenario, this term can be expressed in terms of the characteristic structural length, $\nabla \cdot \mathbf{P}^{th} \sim \sigma/L^2$ [20]. This scaling expression can be rationalized by taking into account that pressure differences across a curved interface are given by the Laplace pressure in equilibrium. This driving term, in the limiting regimes, will be balanced by either the viscous or the inertial mechanisms appearing in the Navier–Stokes equations. This gives rise to two characteristic scaling regimes. If viscous dissipation is dominant, balancing pressure gradients against viscous dissipation (scales as $\nu\nabla^2\mathbf{v} \sim \nu\dot{L}/L^2$) leads to a domain growth $L \sim T$ and it is referred to as the *viscous regime*. In the opposite limit where the relevant hydrodynamic contribution scales as $\mathbf{v}\cdot\nabla\mathbf{v} \sim \dot{L}^2/L$, domain growth will evolve as $L \sim T^{2/3}$. This is the *inertial regime*. The corresponding universal curve will then have two different regimes; a viscous regime where $L/L_0 = b_1 T/T_0$, and an inertial regime where $L/L_0 = b_{2/3}(T/T_0)^{2/3}$. The amplitudes b_1 and $b_{2/3}$ will also be universal (once a procedure to measure the length is prescribed). These two regimes will be separated by a crossover region, which is also universal.

The existence of scaling also means that any length scale one measures in the system will have the same time dependence. There are different ways to identify the relevant structural length [67]. One of the most common procedures consists of taking the first moment of the static structure factor,

$$L(t) = 2\pi \frac{\int \mathrm{d}k \, S(k,t)}{\int \mathrm{d}k \, kS(k,t)},$$

where $k = |\mathbf{k}|$ is the modulus of the wave vector. The spherically averaged static structure factor is defined as

$$S(k,t) = \langle \phi(\mathbf{k},t)\phi(-\mathbf{k},t)\rangle$$

with $\phi(\mathbf{k},t)$ being the Fourier transform of the concentration. This procedure is particularly popular since the structure factor is easily computed experimentally. Another practical procedure computes characteristic lengths from the eigenvalues of $\langle \nabla\phi(\mathbf{r})\nabla\phi(\mathbf{r})\rangle$. The sum of the three eigenvalues provides a measure of the inverse characteristic domain size.

Free Energy Model. Here, I will present results obtained with LB for a binary mixture which is described by the free energy functional

$$F[\phi,\rho] = \int \mathrm{d}\mathbf{r} \left\{ \frac{A}{2}\phi^2 + \frac{B}{4}\phi^4 + \frac{\kappa}{2}|\nabla\phi(\mathbf{r})|^2 + \frac{1}{3}\rho(\mathbf{r})[\ln(\rho(\mathbf{r})) - 1)] \right\} \quad (26)$$

corresponding to a ϕ^4-free energy model with a square gradient term. The thermodynamic pressure tensor and the chemical potential are derived from (26) in the usual way through functional differentiation,

$$\mu = A\phi + B\phi^3 - \kappa\nabla^2\phi$$

$$\mathbf{P} = \left\{ \rho + \frac{A}{2}\phi^2 + \frac{3B}{4}\phi^4 - \frac{\kappa}{2}|\nabla\phi(\mathbf{r})|^2 \right\} \mathbf{1} + \kappa\nabla\nabla\phi,$$

where $\mathbf{1}$ is the identity matrix. Inserting these expressions into the appropriate relations, one then determines the coefficients of the equilibrium distributions, f_i^{eq} and g_i^{eq}, that define the LB model.

In this free energy model there is no interaction associated to global density variations. Hence, there is no condensation and only a demixing transition will take place at low temperatures. This can be viewed as a minimalist free energy model that has a second order phase transition and sustaining interfaces of finite width. A, B and κ are parameters that define the thermodynamics of the model, and B needs to be positive to ensure thermodynamic stability. $A \geq 0$ corresponds to the high temperature phase where the two species are miscible, while for $A \leq 0$ there is a phase separation with two equilibrium phases with concentrations $\phi_{eq} = \pm\sqrt{-A/B}$. Finally, κ determines the interfacial width ξ when the two phases coexist at low temperatures, $\xi = \sqrt{-\kappa/(2A)}$, as well as the surface tension, $\sigma = \sqrt{-8\kappa A^3/(8B^2)}$. The gradient term in the free energy model can be viewed as a low wave vector expansion of an underlying, more microscopic functional [68].

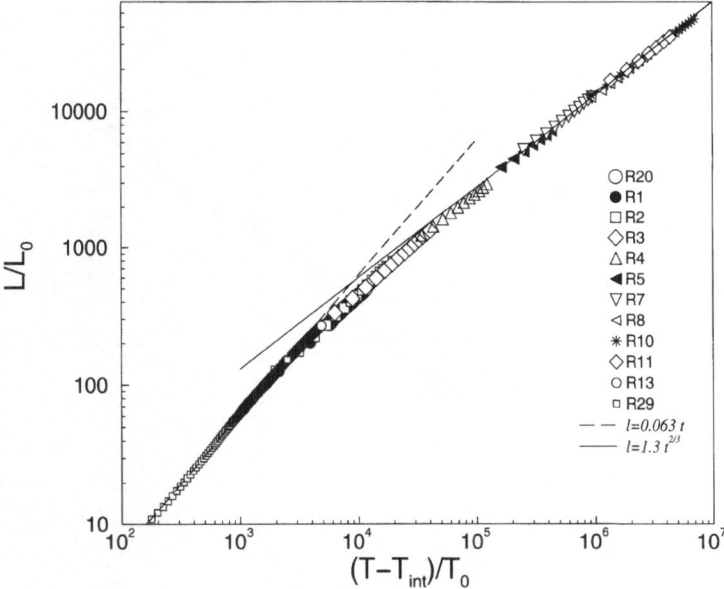

Fig. 7. Universal scaling plot of domain size vs time in dimensionless units. The time offset is related to the time needed in LB before well defined domains start to appear from the initial homogeneous condition (see [20,71]). Different symbols correspond to different physical parameters characterizing the LB mixture. See [71] for parameter details

Simulation Results. LB has proved to be an extremely valuable tool for establishing the existence of a universal curve. While there exists evidence of the viscous regime for symmetric quenches, a proper characterization of the inertial regime was missing. In fact, even if there were a number of simulation results indicating the plausibility of an inertial scaling regime, a proper plot of these data when expressed in dimensionless units showed that all of them (obtained with different simulation techniques) could be placed consistently at most in the crossover region between the viscous and inertial regimes [20].

The use of a parallel version of LB [69] has allowed sampling of 5 orders of magnitude in reduced lengths and 7 orders of magnitude in dimensionless times. In this way, it has been possible to enter the inertial regime and to establish its existence in an uncontroversial way [70]. In Fig. 7, I show the universal scaling curve as obtained from a number of LB runs. The different symbols correspond to different runs. In fact, it is not possible to sample long portions of this curve with a single simulation. Rather, one has to make use of the dynamic scaling idea and to change the parameters of the binary mixture (η and σ) to modify the characteristic parameters, L_0 and T_0, in order to reproduce different parts of the universal curve. By superimposing the different runs it is possible to reconstruct the universal curve. The different runs that appear in the legends of Figs. 7–8 refer precisely to different parameters of the fluid defining the LB model. See [71] for parameter details.

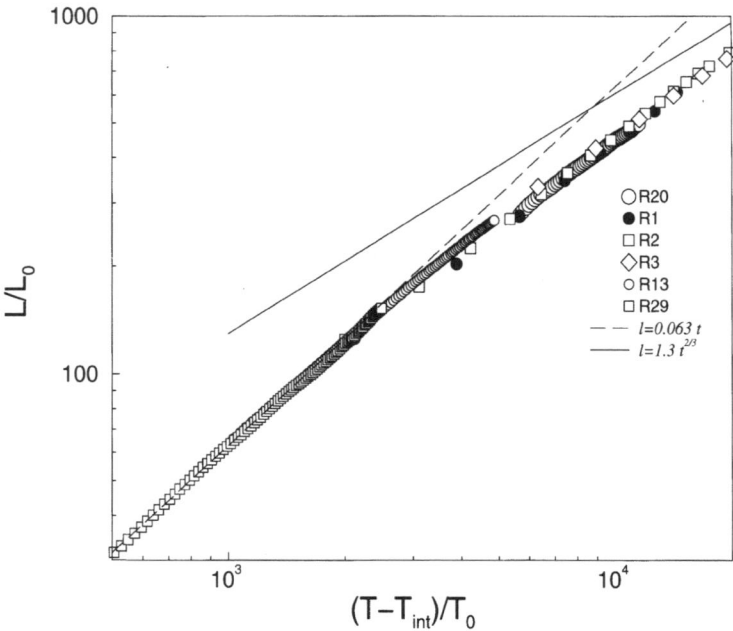

Fig. 8. The same as Fig. 7, but zoomed into the crossover region

The data obtained with LB have also been useful in addressing a number of different issues that were controversial in the literature. In particular, the crossover region has been analyzed in detail with LB, showing that there is scaling in this region as well, as depicted in Fig. 8. In fact, on the basis of DPD simulations it had been argued that the process of handle breaking (the mechanism by which domains coarsen in the symmetric, interconnected textures, see Fig. 9) could introduce a new length scale that would interfere with the simple scaling picture [72]. The results of Fig. 8 show that simulations are compatible with the simple scaling picture, and that the breaking process of handles does not introduce new relevant length scales.

There have also been speculations that in the inertial regime domain coarsening is controlled by capillary-wave dissipation [73] rather than by the transfer of energy into inertial motion [74]. This would lead to an exponent that would differ from 2/3. Again, Fig. 7 shows that there is no evidence of such a deviation from inertial scaling. Such a scaling is in fact compatible with the existence of capillary waves. In fact, from the LB simulation data, animated graphics have been produced to study graphically the evolution of domains during the coarsening process [75]. Such a study has been useful since it gives a clear insight of the textures that develop during spinodal decomposition, as opposed as analyzing only a moment of the structure factor, or the evolution of the structure factor itself, which is the routine procedure in spinodal decomposition. The analysis of interfacial motion shows how the different dynamic mechanisms enter and affect domain evolution. For example, it is clearly visible how interface relaxation is overdamped in the viscous regime, while in the inertial regime wavy interfaces develop. However, as I have mentioned before, the development of such interfacial waves is

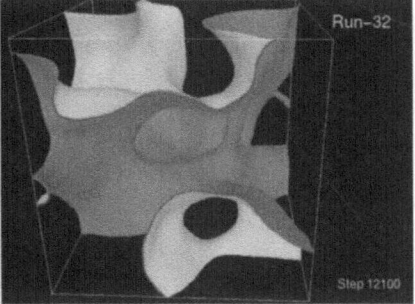

Fig. 9. Snapshot of the fluid/fluid interface between to immiscible fluids during spinodal decomposition. Left: Viscous regime; right: Inertial regime. See [75] for parameters

compatible with the inertial scaling described at the beginning of this section. Hence, while interfacial waves affect the morphology of the domains, they do not interfere with their growth which is controlled by the transfer of energy from the interface to domain bulk through convective flows.

Despite the understanding of the role of the inertial regime in the spinodal decomposition process, it is still unclear whether this is the final asymptotic regime in spinodal decomposition. It could happen that as Reynolds number (Re) increases (in the simulations the maximum Re is of order 10^3) capillary waves become sufficiently large so that they could break domains, hindering, or slowing down, their further growth. Up to the Reynolds numbers studied up to present day this process is not relevant as interfacial wave amplitudes seem to grow in pace with domain size. If this eventually takes place, it would suggest that turbulence becomes the dominant mechanism in the final stages of the demixing process as suggested by Grant and Elder [76]. Their claim is that according to simple dynamical scaling, Re is computed on the basis of domain size and domain velocity grows in time as $Re \sim t^{1/6}$. Such a diverging Re means that turbulence will control the (very) final stage of spinodal decomposition. Such a conclusion relies on the idea of a dynamic scaling where there is only one single characteristic length scale related to the domain size. In the previous paragraphs I have advocated for the plausibility of such a picture, and shown simulational evidence that rules out previously proposed departures from simple scaling. So, what is wrong with the idea of the final relevance of turbulence?

There still exists the possibility that a new length scale enters into the dynamic scaling because velocity gradients (which are related to the way in which momentum will be transferred and dissipated in the domains) may evolve on a length scale different from the domain size itself. In fact, it has been shown that such a new length scale will not interfere with the domain size growth [77]; it will only modify the scaling of those physical quantities that depend on velocity gradients. However, this has implications for the Reynolds number since one can now propose different definitions for Re. One important consequence is that Re is defined on the basis of velocity scales does not diverge with time [20]. This scenario would imply that the inertial regime is the final asymptotic regime in spinodal decomposition. The statistics gathered with LB simulations do not yet yield a completely consistent picture that sustains the idea of a separate scaling based on a single new length scale. While it seems clear that simple dynamic scaling on the basis of a unique length scale $L(T)$ is insufficient to account for the growth of averaged velocity gradients, numerical results are not conclusive on the existence of a single additional relevant length scale [20]. Hence, larger scale simulations are needed to settle these open issues concerning the complete scaling picture and its implications on the final asymptotic stages of spinodal decomposition.

5 Conclusions

The study of systems containing a variety of characteristic length scales require development of specific techniques that allow to sample the corresponding meaningful time scales. In this respect, LB is a competitive tool to study complex fluids at a coarse-grained scale.

The basis of the model is the simulation of solvent dynamics at the level of a simplified Boltzmann equation (in the sense that the collision operator is linearized around a reference equilibrium state), ensuring that the solvent dynamics reproduces the relevant collective dynamics. Moreover, deviations from hydrodynamic behavior – either due to the physical conditions or numerical artifacts – have a physically meaningful support (in the sense that one will enter into a generalized hydrodynamic regime, or into the kinetic regime described by the underlying Boltzmann equation) and can be quantified.

On top of this basic model, one can study complex fluids from different perspectives. As I have illustrated, one can carry out a detailed description of the dynamics of suspended objects if these are large enough so that they interact directly with the collective modes of the solvent they are suspended in. This is the case for colloidal suspensions, and also for other type of suspended objects, such as polymers. In the absence of a clear length scale separation, it is still possible to study non-ideal fluids by modeling their interactions at a coarse grained level on the basis of a free-energy functional.

For both approaches LB is a flexible technique that can be used to study different physical problems under a great variety of external conditions. In this respect, the flexibility in introducing generic boundary conditions is very important. From a numerical point of view, the locality of the model makes LB easily parallelizable, hence making it possible to reach parameters ranges that are currently out of scope with other techniques. I have discussed a few questions that have been addressed with LB to gain insight into different questions related to the dynamics and kinetics of complex fluids. These examples show the capabilities of the model and the kind of information that can be easily extracted from it.

Acknowledgements

It is a pleasure to acknowledge D. Frenkel, who introduced me in lattice Boltzmann and to C. P. Lowe and M. H. Hagen, from whom I learned how to use it. I am also grateful to M. E. Cates, J. C. Desplat, V. Kendon and A. J. Wagner, with whom I have learned how to combine the numerical abilities of LB with a sharper physical insight, and to the support provided by EPCC. I also want to thank A. J. C. Ladd, J. M. Yeomans and P. B. Warren for illuminating discussions.

References

1. R.G. Larson: *The Structure and Rheology of Complex Fluids* (Oxford University Press, New York 1999)
2. T. Witten: Rev. Mod. Phys. **71**, S367 (1999)
3. D. Frenkel. In: *Soft and Fragile Matter: Nonequilibrium Dynamics, Metastability and Flow*, ed. by M.E. Cates and M.R. Evans (SUSSP Publications and Institute of Physics Publishing, Bristol 2000)
4. A.P. Lyubartsev, A. Laaksonen: Phys. Rev. E **52**, 3730 (1995)
5. A.P. Lyubartsev, A. Laaksonen: On the Reduction of Molecular Degrees of Freedom in Computer Simulations, Lect. Notes Phys. **640**, 215 (2004)
6. A.P. Lybartsev, M. Karttunen, I. Vattulainen, A. Laaksonen: Soft Materials **1**, 121 (2003)
7. E.G. Flekkøy, P.V. Coveney: Phys. Rev. Lett. **83**, 1775 (1999)
8. P.J. Hoogerbrugge, J.M.V.A. Koelman: Europhys. Lett. **19**, 155 (1992)
9. U. Frisch, B. Hasslacher, Y. Pomeau: Phys. Rev. Lett. **56**, 1505 (1986)
10. R. Benzi, S. Succi, M. Vergassola: Phys. Rep. **222**, 145 (1992)
11. A. Malevanets, R. Kapral: Europhys. Lett. **45**, 552 (1998)
12. A. Malevanets, R. Kapral: Mesoscopic Multi-particle Collision Model for Fluid Flow and Molecular Dynamics, Lect. Notes Phys. **640**, 112 (2004)
13. P. Español, M. Serrano, H. C. Öttinger: Phys. Rev. Lett. **83**, 4542 (1999)
14. S. Wolfram: J. Stat. Phys. **45**, 471 (1986)
15. F. Higuera, S. Succi, R. Benzi: Europhys. Lett. **9**, 345 (1989)
16. Y. H. Qian, D. d'Humieres, P. Lallemand: Europhys. Lett. **17**, 479 (1992)
17. G. McNamara, B. Alder: Physica A **194**, 218 (1993)
18. L.-S. Luo: Phys. Rev. E **62**, 4982 (2000)
19. R. Verberg, A. J. C. Ladd: Phys. Rev. E **65**, 016701 (2001)
20. V. M. Kendon, M. E. Cates, I. Pagonabarraga, J.-C. Desplat, P. Bladon: J. Fluid. Mech. **440**, 147 (2001)
21. A. J. Wagner: PhD thesis, Oxford University, Oxford (1996)
22. O. Behrend, R. Harris, P. B. Warren: Phys. Rev. E **50**, 4586 (1994)
23. A. J. Wagner: cond-mat/0105067
24. A. J. C. Ladd: J. Fluid Mech. **271**, 285 (1994); J. Fluid Mech. **271**, 311 (1994)
25. A. J. Wagner, I. Pagonabarraga: J. Stat. Phys. **107**, 521 (2002)
26. J. A. Kaandorp, C. P. Lowe: Phys. Rev. Lett. **77**, 2328 (1996)
27. C. P. Lowe, D. Frenkel: Phys. Rev. Lett. **77**, 4552 (1996)
28. M. H. Hagen, D. Frenkel, C. P. Lowe: Physica A **272**, 376 (1996)
29. H. Chen, C. Teixeira, K. Molvig: Int. J. Mod. Phys. C **9**, 1281 (1998)
30. M. W. Heemels, M. H. J. Hagen, C. P. Lowe: J. Comput. Phys. **164**, 48 (2000)
31. C. P. Lowe, D. Frenkel, A. J. Masters: J. Chem. Phys. **76**, 1582 (1995)
32. P. Ahlrichs, B. Dünweg: J. Chem. Phys. **111**, 8225 (1999)
33. R. Verberg, A. J. C. Ladd: J. Stat. Phys. **104**, 1191 (2001)
34. R. Zwanzig, M. Bixon: Phys. Rev. A **2**, 2005 (1970)
35. J. F. Brady, G. Bossis: Annu. Rev. Fluid. Mech. **20**, 111 (1988)
36. A. J. C. Ladd: Phys. Rev. Lett. **88**, 048301 (2002)
37. D. W. Qi: J. Fluid. Mech. **385**, 41 (1999)
38. C. Denniston, E. Orlandini, J. M. Yeomans: Phys. Rev. E **63**, 056702 (2001); G. Tóth, C. Denniston, J. M. Yeomans: Phys. Rev. Lett. **88**, 105504 (2002)
39. B. H. Weigl, P. Yager: Science **283**, 346 (1999)
40. A. D. Stroock, A. K. W. Dertinger, A. Ajdari, I. Mezic, H. A. Stone, G. M. Whitesides: Science **295**, 647 (2002); A. D. Stroock, M. Weck, D. T. Chiu, W. T. S. Huck, P. J. A. Kenis, R. F. Ismagilov, G. M. Whitesides: Phys. Rev. Lett. **84**, 3314 (2000)

41. A. Terray, J. Oakey, D. W. M. Marr: Science **296**, 1841 (2002)
42. B. J. Alder, T. E. Wainwright: Phys. Rev. A **1**, 18 (1970)
43. τ_ξ can be made much larger than τ_ν by increasing the colloid density, ρ_p. In this limit the colloid relaxation is uncoupled to the solvent dynamics. However, it corresponds to an unphysical limit, since buoyant suspensions require $\rho_p \sim \rho$. Requiring $\tau_\xi \sim \tau_\nu$ couples the fluid and colloid velocoity evolution, and it is the physical mechanism leading to the instability of the explicit update of the colloid particles when executing the bounce-back
44. C. P. Lowe, D. Frenkel: Phys. Rev. E **54**, 2704 (1996)
45. J. X. Zhu, D. J. Durian, D. A. Weitz, D. J. Pine: Phys. Rev. Lett. **68**, 2559 (1992); A. J. C. Ladd, H. Gang, J. X. Zhu, D. A. Weitz: Phys. Rev. E **52**, 6550 (1995)
46. A. F. Bakker, C. P. Lowe: J. Chem. Phys. **116**, 5817 (2002)
47. I. Pagonabarraga: preprint
48. L. Bocquet, J.-L. Barrat: Europhys. Lett. **31**, 455 (1995)
49. I. Pagonabarraga, M. H. J. Hagen, C. P. Lowe, D. Frenkel: Phys. Rev. E **59**, 4458 (1999)
50. M. H. J. Hagen, I. Pagonabarraga, C. P. Lowe, D. Frenkel: Phys. Rev. Lett. **78**, 3785 (1997)
51. I. Pagonabarraga, M. H. J. Hagen, C. P. Lowe, D. Frenkel: Phys. Rev. E **58**, 7288 (1998) .
52. M. H. Ernst, A. Weyland: Phys. Lett. **34A**, 39 (1971)
53. M. H. J. Hagen: Diffusion of Confined Colloidal Particles. PhD thesis, University of Utrecht (1997)
54. B. Lin, B. Cui, J.-H. Lee, J. Yu: Europhys. Lett. **57**, 724 (2002); B. Lin, J. Yu, S. A. Rice: Phys. Rev. E **62**, 3909 (2000)
55. W. Cai, T. C. Lubensky: Phys. Rev. lett. **73**, 1186 (1994)
56. M. A. Biot: J. Acoust. Soc. Am. **28**, 168 (1956); D. L. Johnson: J. Chem. Phys. **77**, 831 (1972)
57. D. H. Rothman, S. Zaleski: *Lattice-Gas Cellular Automata: Simple Models of Complex Hydrodynamics* (Cambridge University Press, Cambridge 1997)
58. E. Flekkøy: Phys. Rev. E **47**, 4247 (1993); X. Shan, H. Chen: Phys. Rev. E **49**, 2941 (1994)
59. M. R. Swift, W. R. Osborn, J. M. Yeomans: Phys. Rev. Lett. **75**, 830 (1995); M. R. Swift, E. Orlandini, W. R. Osborn, J. M. Yeomans: Phys. Rev. E **54**, 5041 (1996)
60. The idea of deriving effective interactions from a free-energy functional has also been explored for other coarse grained models; for the particular case of DPD see e.g. I. Pagonabarraga, D. Frenkel: J. Chem. Phys. **115**, 5015 (2001)
61. D. J. Holdych, D. Rovas, J. G. Georgiadis, R. O. Buckius: Int. J. Mod. Phys. C **9**, 1393 (1998)
62. O. Theissen, G. Gompper, D. M. Kroll: Europhys. Lett. **42**, 419 (1998)
63. A. J. Wagner: private communication
64. A. J. Wagner, J. M. Yeomans: Phys. Rev. Lett. **80**, 1429 (1998); G. Gonnella, E. Orlandini, J. M. Yeomans: Phys. Rev. Lett. **78**, 1695 (1997); W. R. Osborn, E. Orlandini, M. R. Swift, J. M. Yeomans, J. R. Banavar: Phys. Rev. Lett. **75**, 4031 (1995)
65. A. J. Wagner, J. M. Yeomans: Int. J. Mod. Phys C **7**, 773 (1998)
66. O. Kuksenok, J. M. Yeomans, A. C. Balazs: Phys. Rev. E **65**, 031502 (2002)
67. A. J. Bray: Adv. Phys. **43**, 357 (1994)
68. R. Evans: Density functionals in the theory of nonuniform fluids. In: *Fundamentals of Inhomogeneous Fluids*, ed. by D. Henderson (Marcel Dekker Inc., New York 1992) p. 85
69. J.-C. Desplat, I. Pagonabarraga , P. Bladon: Comput. Phys. Commun. **134**, 273 (2001)
70. V. M. Kendon, J.-C. Desplat, P. Bladon, M. E. Cates: Phys. Rev. Lett. **83**, 576 (1999)
71. I. Pagonabarraga, A. J. Wagner, M. E. Cates: J. Stat. Phys. **107**, 39 (2002)
72. S. I. Jury, P. Bladon, S. Krishna, M. E. Cates: Phys. Rev. E **59**, R2535 (1999)
73. F. J. Solis, M. Olvera de la Cruz: Phys. Rev. Lett. **84**, 3350 (2000)
74. Note that in the inertial regime energy stored at the interface is transported by convective flow to the bulk of the domains, where it is eventually dissipated through viscous heating

75. I. Pagonabarraga, J.-C. Desplat, A. J. Wagner, M. E. Cates: New J. of Phys. **3**, 91 (2001); Movies showing how the interfaces evolve, and how handles break in the viscous and inertial regime are available at the journal web page
76. M. Grant, K. R. Elder: Phys. Rev. Lett. **82**, 14 (1999)
77. V. M. Kendon: Phys. Rev. E **61**, R6071 (2000)

Reverse Non-equilibrium Molecular Dynamics

Florian Müller-Plathe[1,2] and Patrice Bordat[1]

[1] Max-Planck-Institut für Polymerforschung, Ackermannweg 10, 55128 Mainz, Germany
[2] International University Bremen, P.O. Box 750561, 28725 Bremen, Germany

Abstract. We review non-equilibrium methods for calculating transport coefficients with emphasis on the reverse non-equilibrium molecular dynamics (RNEMD) method. It has fundamental and technical advantages over previous equilibrium and non-equilibrium techniques. For example, it applies the perturbation in a microcanonical way (no thermostat needed) and its raw data are well defined and robust gradients, rather than, fluxes which are often difficult to define and to calculate with sufficient accuracy. The method has so far been applied to the calculation of viscosities, thermal conductivities and Soret coefficients. Finally, the status and future potential of the RNEMD method are discussed.

1 Introduction

The reverse non-equilibrium molecular dynamics (RNEMD) method for calculating transport coefficients is based on the reversal of the experimental cause-and-effect picture. In the calculation of viscosities, for example, the effect, the momentum flux or stress, is imposed, whereas the cause, the velocity gradient or shear rate, is obtained from simulation. The RNEMD method of this work differs from other Norton-ensemble methods by the way, in which the steady-state fluxes are maintained. It involves simple exchanges of particle momenta, which are easy to implement and analyse. Moreover, it can be made to conserve the total energy as well as the total linear momentum, such that no external thermostatting is needed. The resulting raw data are robust and rapidly converging. The method has been tested on the calculation of the shear viscosity, the thermal conductivity and the Soret coefficient (thermal diffusion). Recently, it has been extended to molecular fluids. The algorithm should work not only together with Newtonian mechanics but with all equations of motion that locally conserve energy and linear momentum, for example dissipative particle dynamics (DPD) [1,2] or the Malevanets-Kapral method [3] .

2 Introduction: Transport Coefficients

Linear response is often found experimentally in transport through condensed media. Phenomenological equations relate a flux j (e.g. matter, energy, momentum) to a driving force or field E, which usually is a gradient of some quantity (e.g. concentration, temperature, flow velocity). In steady state, the two are proportional and the proportionality constant is the corresponding transport coefficient κ,

$$j = -\kappa E.$$

F. Müller-Plathe and P. Bordat, Reverse Non-equilibrium Molecular Dynamics, Lect. Notes Phys. **640**, 310–326 (2004)
http://www.springerlink.com/

The flux is defined as the amount of the quantity transported per time through an area perpendicular to the flux direction. We note that, in an anisotropic medium, the directions of j and E need not be colinear, in which case their vectorial nature has to be taken into consideration and κ is a tensor. Without loss of generality, we confine the discussion in this paper to isotropic media and we use the scalar form (2). The best known example of a linear response equation is probably Ohm's law $I = U/R$, where the charge current I is the flux, the voltage U the driving force and the electrical conductivity $S = 1/R$ the transport coefficient (there is no minus sign because of the sign conventions in electricity). The voltage is the gradient of the electric potential.

Equation (2) may also be generalised to connect multiple fields and fluxes. This is most conveniently done in the framework of Onsager's linear-response theory whose general form is

$$j_\alpha = -\sum_\beta L_{\alpha\beta} X_\beta, \qquad \alpha, \beta = 1, 2, \ldots . \tag{1}$$

The j_α of (1) are the same fluxes as the j in the phenomenological equations (2). The X_β, however, are thermodynamic forces that need to be chosen in such a way that the products $j_\alpha X_\beta$ have the dimensions of an entropy production rate. In general, they do differ from the phenomenological fields. The thermodynamic force of the Onsager theory that drives the diffusion of species 1, for example, is the gradient in the chemical potential of 1, rather than the concentration gradient in the phenomenological Fick's law. Correspondingly, the elements of the Onsager matrix $\mathbf{L}$ are related to but not identical with experimentally measured transport coefficients. Onsager theory includes the coupling between different types of transport (e.g. diffusion driven by a temperature gradient: the Ludwig–Soret effect, cf. Sect. 6). While it is universal, clean and symmetric ($L_{\beta\alpha} = L_{\alpha\beta}$), it has the disadvantage that the quantities of interest (Onsager coefficients and thermodynamic forces) are often complicated to determine.

Transport coefficients can be calculated by equilibrium molecular dynamics (MD) simulations using the appropriate Green–Kubo

$$\kappa \propto \int_0^\infty \langle j(t) \cdot j(0) \rangle \, \mathrm{d}t \tag{2}$$

or Einstein relations [4–6]

$$\kappa \propto \lim_{t \to \infty} \frac{\mathrm{d}}{\mathrm{d}t} \left\langle [F(t) - F(0)]^2 \right\rangle,$$

where F is related to the corresponding flux by

$$j = \frac{\mathrm{d}F}{\mathrm{d}t}.$$

The alternative route are non-equilibrium molecular dynamics (NEMD) simulations [6,7]. These usually usually similar in spirit to real experiments: The cause is an appropriate field E or gradient which is imposed on the system. This is not necessarily the same as in experiment, but it can be shown to generate the same response [7]. Then, the ensemble average of the effect, the resulting flux $\langle j \rangle$, is measured and the ratio of flux and field gives the transport coefficient κ.

Both equilibrium and non-equilibrium methods have their problems when it comes to practical calculations. The fluxes j and their time integrals F are sometimes not easily and unambiguously defined on a microscopic scale (for example, the heat flux). They have to be calculated as time or ensemble averages $\langle j \rangle$ and convergence is often slow (for example, the interdiffusion current). The convergence of time correlation functions $\langle j(t) \cdot j(0) \rangle$ can be even slower. The generalised mean-square displacements

$$\left\langle [F(t) - F(0)]^2 \right\rangle$$

may take a long time to reach the domain where they are linear in time. Some non-equilibrium methods have the ugly property of introducing external walls into the simulation, so that large systems are necessary to avoid surface effects. Almost all NEMD methods constantly feed energy into the system. This energy needs to be removed by a thermostat. Many thermostats, in turn, introduce their own problems, such as non-conservation of momentum (problems for viscosities), unclear statistical-mechanical ensembles, or an unphysically fast local removal of energy (corresponding to a too high effective thermal conductivity). All these features may compromise the calculated transport coefficients. For these difficulties, it is not surprising that no single method is in practice used for all transport coefficients.

The first ingredient of the reverse non-equilibrium molecular dynamics (RNEMD) method described here is the reversal of cause and effect: The flux is imposed and the corresponding field is measured. Hence, the flux is now the cause and the gradient the effect. This idea is not new [7–10]: Such methods, known as Norton-ensemble methods have their advantages in cases where the flux is difficult to define microscopically in an unambiguous way or is slowly converging. In addition, there can be technical reasons for reversing cause and effect. Still, Norton-ensemble methods have never been as popular as direct NEMD (Thevenet ensemble) methods.

The second ingredient of our RNEMD algorithm is a swap of certain particle velocities [11–15]. Methods similar in spirit have also been used by Hafskjold et al. [16–18]. Our methods have several attractive features:

- They conserve, under certain conditions, total energy and total linear momentum and, hence, need no thermostatting.
- They do not introduce artificial walls into the system and are compatible with the usual periodic boundary conditions.
- The resulting raw data are well-defined, rapidly converging, and easily analysed gradients, rather than, ambiguous fluxes.
- They are easy to implement.

3   Illustration of the RNEMD Method: Calculating Shear Viscosity

The shear viscosity connects a shear field (the perturbation) with a flux of transverse linear momentum [19]. The shear field is a gradient of one component of the average fluid velocity, say the x direction, with respect to a perpendicular direction, say the

z direction, $\partial \bar{v}_x / \partial z$. It is also called the shear rate. The transverse momentum flux $j_z(p_x)$ is colinear in an isotropic medium: It is the x component of the momentum p_x transported in z direction through a perpendicular surface of area A per time t, see Fig. 1. It can also be regarded as an off-diagonal (xz) component of the stress tensor. The proportionality coefficient is the shear viscosity η,

$$j_z(p_x) = -\eta \frac{\partial \bar{v}_x}{\partial z}. \tag{3}$$

In RNEMD, the momentum flux is imposed on the system in an unphysical way. The periodic simulation box is subdivided into slabs along the z coordinate (Fig. 2). The atoms inside the slab at $z = 0$ and its period images are propelled in $+x$ direction, those inside the slab at $z = L_z/2$ (with L_z the box length in z direction) and its periodic images in $-x$ direction (Fig. 3). This is accomplished by finding the atom moving most *against* the desired slab movement: In the slab moving in $+x$ direction ($z = 0$), the atom with the largest momentum component in $-x$ direction (= the atom with the smallest p_x) is found. Likewise, in the slab moving in $-x$ direction ($z = L_z/2$), the atom with the largest momentum component in $+x$ direction (= the atom with the largest p_x) is found. Then the p_x of the two atoms are interchanged. If both atoms have the same mass $m_1 = m_2 = m$, the unphysical momentum swap conserves both linear momentum and kinetic energy of the system as a whole. The sum of the momenta of the two atoms participating in the exchange is $m_1 \mathbf{v}_1 + m_2 \mathbf{v}_2$ before the swap and $m_1 \mathbf{v}_2 + m_2 \mathbf{v}_1$ thereafter. The momenta of the other atoms do not change in the swap and cancel in the momentum balance. The difference of total momentum before and after $(m_1 - m_2)(\mathbf{v}_1 - \mathbf{v}_2)$ is zero for the non-trivial case $\mathbf{v}_1 \neq \mathbf{v}_2$, only if $m_1 = m_2$. The same is true for the difference of

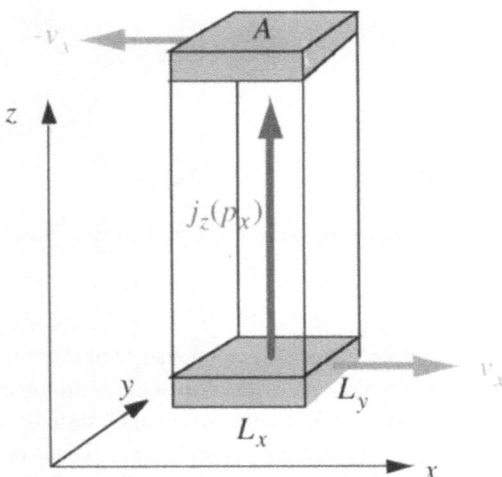

Fig. 1. Geometry of the experimental non-equilibrium situation. A gradient in v_x is set up in z direction by shearing the liquid. As a result, x momentum flows in z direction, giving rise to a momentum flux $j_z(p_x)$ through the xy plane of area A

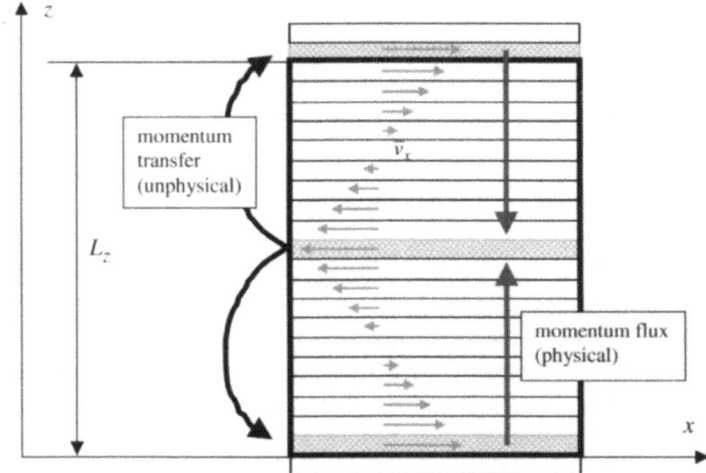

Fig. 2. Schematic view of the periodic simulation box. The reverse non-equilibrium molecular dynamics algorithm artificially transfers transverse momentum from the central slab ($z = L_z/2$) and its periodic images to the first slab ($z = 0$) and its periodic images. The momentum flows back through the fluid by friction, thereby causing a gradient in the transverse velocity

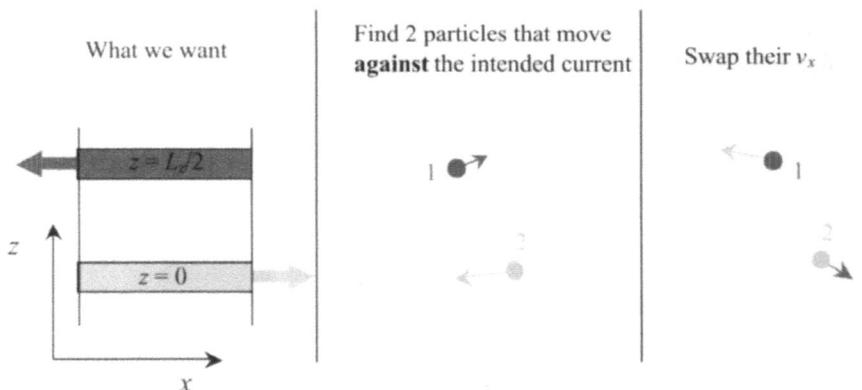

Fig. 3. The velocity exchange algorithm for transferring linear momentum in viscosity calculations

the corresponding kinetic energies $(m_1 - m_2)(v_1^2 - v_2^2)$. Since atom positions are not affected, the potential energy and, hence, the total energy of the system is conserved.

Three parenthetical remarks before continuation of the development of the algorithm: (i) The perturbation applied by the velocity swap is small. Usually, the motion of only two particles out of several hundred or more is interfered with once in 50 – 1000 MD time steps. (ii) We have found it sufficient to apply the perturbation in an impulse fashion, as described above. It has been pointed out that the impulse form of the perturbation leads to a time dependent structure formation in the fluid [20]. This does not seem to affect the numerical value of the calculated viscosity, though: If the momentum is transferred at the same rate, but by transferring a small amount in every time step (this can be accomplished by selecting other atoms for the swap), the calculated viscosities are the same to within their error bars [21]. (iii) Energy and momentum conservation are only fulfilled up to the limits dictated by the order of the MD integration algorithm. The Verlet algorithm, for example, assumes that the force acting on an atom is constant over the duration of the time step, leading to a continuous trajectory. Continuity is obviously violated for the two atoms whose velocities are abruptly swapped. The resulting drift of the total energy is minute but can be observed in very long RNEMD simulations. It can be countered easily with a thermostat so mild that it does not affect the value of the viscosity.

The amount of momentum $\Delta p_x = m(v_{x,1} - v_{x,2})$ transferred in one swap from the $z = L_z/2$ slab to the $z = 0$ slab is, thus, precisely known. If momentum swaps are repeated periodically, the total momentum transferred in a simulation P_x is the sum of the Δp_x. The system responds to the non-equilibrium situation by letting momentum flow in the opposite direction via a physical mechanism, friction. In steady state, the rate of momentum transferred unphysically by momentum swaps is equal to that of momentum flowing back through the fluid by friction. Hence, the momentum flux $\langle j_z(p_x) \rangle$ is given by

$$\langle j_z(p_x) \rangle = \frac{P_x}{2tA}, \tag{4}$$

where t is the length of the simulation and $A = L_x L_y$ (cf. Fig. 1). The factor of two arises from the periodicity of the system [11]. The angular brackets denote that $\langle j_z(p_x) \rangle$ is obtained as an average, as the amount of momentum transferred in every swap varies. Moreover, normally the swaps are not executed at every time step. The value of the average is, however, precisely known.

The physical momentum current gives rise to a velocity profile in the fluid (cf. Fig. 2). The flow velocity $\bar{v}_x$ in x direction in every slab is calculated as the average of the $v_{x,i}$ of all atoms i in that slab. If the momentum flux is not too large the velocity profile is approximately linear and and its slope (including error) can be obtained by a linear regression. In practice, more stable profiles have been obtained when (i) the two exchange slabs where excluded and (ii) the time steps in which the exchanges take place were omitted from the average. Viscosity is then given by (3). Its error can be estimated using the rules for error propagation from the error in the velocity gradient, the value for $\langle j_z(p_x) \rangle$ at steady state being known. If the velocity profile is not linear this means that the efficiency of the momentum transfer (η) is not uniform across the system. This indicates that the imposed momentum flux is too large, that the mechanism of momentum transfer is no longer uniform, and that the linear-response regime has been left.

Velocity profiles for a single-component Lennard–Jones fluid near the triple point are shown in Fig. 4 as an example. The interval between velocity exchanges W is varied between 3 and 1200 time steps Δt. The applied momentum flux (proportional to $1/W$) thus varies by more than two orders of magnitude. Except for the highest momentum fluxes ($15W \leq \Delta t$), the velocity gradient is uniform throughout the system. At low momentum flux (high W), the velocity gradient is difficult to measure due to noise, which leads to scattered results for the viscosity (Fig. 5). At high momentum flux, the decrease of the apparent viscosity is a physical nonlinearity known as shear thinning.

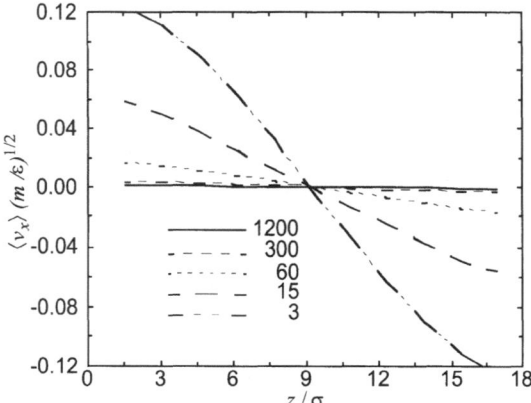

Fig. 4. Velocity profiles in the simulation cell (only one half is shown) for different intervals of momentum interchange (number of time steps W between momentum exchanges): 2592 Lennard–Jones atoms near the triple point [12], reduced units

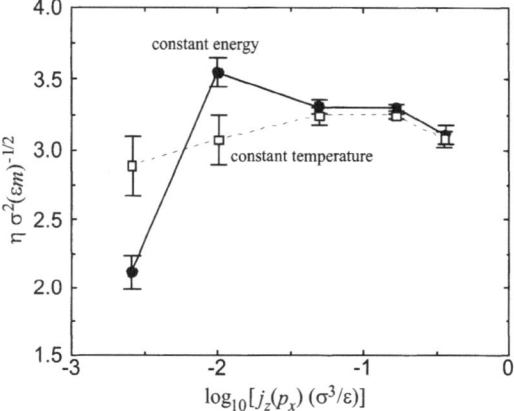

Fig. 5. Viscosity of the Lennard–Jones fluid near the triple point as a function of the applied perturbation (imposed momentum flux), reduced units. At small perturbations, statistical errors enter. At the highest perturbation, there is evidence of shear thinning. The plateau value agrees with available literature data [12]

In between, there is a range of momentum flux rates which are large enough to have a useful signal-to-noise ratio, and are yet small enough to leave the simulation in the linear-response regime. This is visible as a plateau in the shear viscosity. The plateau value agrees well with literature data obtained by other methods for the Lennard–Jones system at the same conditions [22].

At this point, one might ask how can there be shear flow apparently without viscous heating: On one hand, friction heats up any system and on the other hand the algorithm conserves energy. The answer to this dilemma is that there is viscous heating but the excess heat is drained by the momentum exchange mechanism itself. It acts like an internal thermostat and removes the heat generated by friction. The algorithm, which in

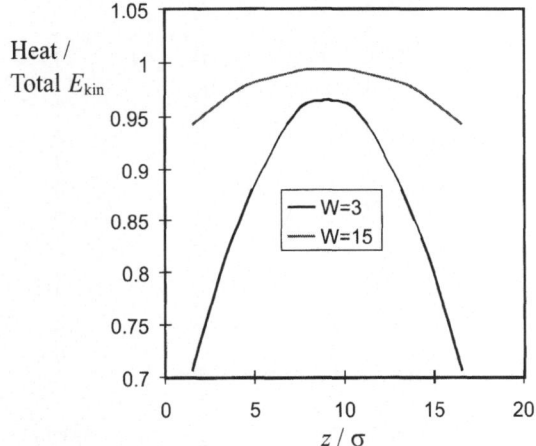

Fig. 6. Temperature profile in the triple-point Lennard–Jones system at high shear. Shown is the heat fraction of the total kinetic energy [12]

fact is a Maxwell demon, selectively picks particles with the largest velocity component $\mathbf{u}_i = \mathbf{v}_i - \langle \mathbf{v} \rangle$ against the flow direction of the slab $\langle \mathbf{v} \rangle$. The *peculiar velocities* $\mathbf{u}_i$, rather than the *absolute* velocities $\mathbf{v}_i$, define the slab temperature. After the exchange, the peculiar velocity of the particle is in the direction of the local flow and, although the absolute velocity is on average unchanged, the peculiar velocity and, thus, the temperature has decreased. The Maxwell demon has transformed undirected motion (heat) into directed motion (flow). Therefore, the two slabs in which the momentum exchange takes place are the heat sinks where the heat generated by friction is disposed of. As a consequence, a temperature profile across the box is expected. This is indeed found in Fig. 6, where we show the fraction of kinetic energy calculated from the peculiar velocities,

$$\frac{1}{2} \sum_{\substack{\text{atoms } i \\ \text{in slab}}} m_i u_i^2 \, \Big/ \, \frac{1}{2} \sum_{\substack{\text{atoms } i \\ \text{in slab}}} m_i v_i^2 .$$

For the strongest perturbations ($W = 3, 15$) the temperature profiles are parabolic. This is intuitively understood, since a linear flow velocity profile causes a quadratic kinetic energy profile associated with the flow. For smaller perturbations (not shown) the expected parabola are obscured by statistical noise [12]. This means that, in practical calculations, the perturbation can be made small enough for the collateral temperature profile not to compromise the calculated viscosities.

4 Modification of the RNEMD Method: Thermal Conductivity

The thermal conductivity λ can be calculated by RNEMD in a manner very similar to the shear viscosity [11]. Experimentally, the cause for energy flow is a temperature gradient

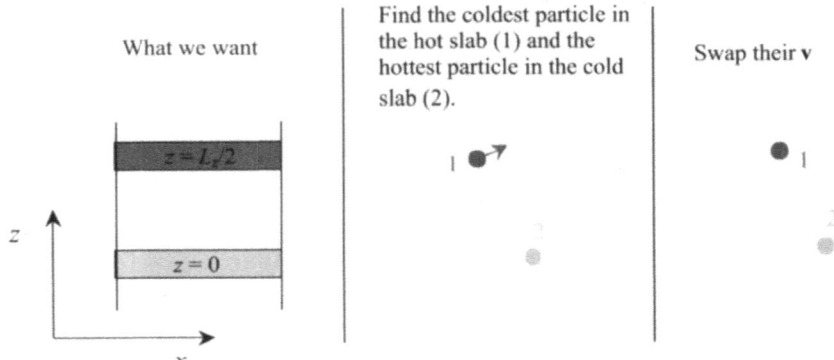

Fig. 7. The velocity exchange algorithm for transferring kinetic energy in thermal conductivity and thermal diffusion calculations

$\partial T/\partial z$ (taking the z direction again as the direction of field and flux); the effect is a flux of heat in the same direction $j_z(Q)$,

$$J_z(Q) = -\lambda \frac{\partial T}{\partial z}. \tag{5}$$

Cause and effect are again reversed (Fig. 7): An energy flow is imposed, and the resulting temperature gradient is calculated. The simulation box is subdivided into slabs along the z coordinate. The $z = 0$ slab is defined as the cold slab, the $z = L_z/2$ as the hot slab. Energy is transferred unphysically from the cold to the hot slab: The hottest atom in the cold slab [largest kinetic energy $m(v_x^2 + v_y^2 + v_z^2)/2$] and the coldest atom in the hot slab (smallest kinetic energy) are identified. The velocities (all Cartesian components!) of these two atoms are interchanged if the hottest atom of the cold slab has a higher kinetic energy than the coldest atom in the hot slab. We have always found this to be the case, since the Maxwell-Boltzmann distribution of atomic velocities is much wider than the difference of the mean velocities in the two slabs. Note that, as for the shear viscosity, conservation of total energy and total linear momentum is satisfied, provided both atoms have the same mass.

The unphysically transferred energy, calculated analogously to (4), flows through the system in the opposite direction by a physical mechanism, thermal conduction. When the steady state has been reached both fluxes are equal in magnitude. The unphysical flux is known exactly, since the energy transferred in each velocity swap is known. In steady state, a temperature profile T_{slab} is established through the system (Fig. 8). It is calculated from the kinetic energy of the atoms in every slab,

$$\frac{3Nk_B T_{\text{slab}}}{2} = \frac{1}{2} \left\langle \sum_{\substack{\text{atoms } i \\ \text{in slab}}}^{N} m_i v_i^2 \right\rangle,$$

and it is a rapidly converging property. The average of its gradient is obtained by linear regression together with an error estimate. Finally, λ is calculated from (5). Note that

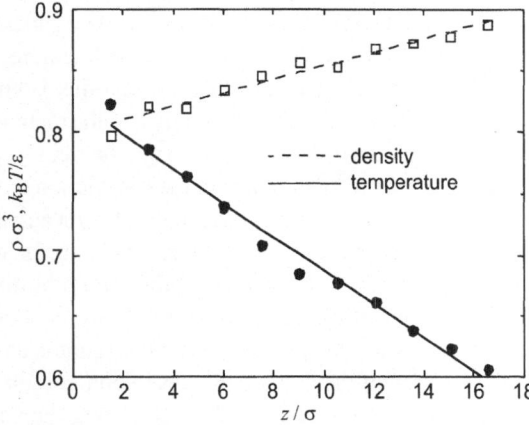

Fig. 8. Simulation with a steady energy flux: Resulting temperature and density profiles, reduced units. A Lennard–Jones system with 2592 particles is simulated, velocity are exchanged every 30 time steps [11]

the temperature profile leads to a concomitant density profile. For the relatively strong perturbation used in the example of Fig. 8 (Lennard–Jones fluid near triple point) the density profile is clearly visible. The perturbation has to be made small enough so that the density variation does not affect numerically the calculated thermal conductivity. As the RNEMD method for the thermal conductivity is similar to that for the viscosity, analogous considerations apply in the practical implementation and analysis.

5   Digression: Features of the RNEMD Method

For the shear viscosity, as for the thermal conductivity, there exist alternative methods, both of equilibrium and non-equilibrium type. So why would one want to use RNEMD instead? In addition to its simplicity there are other advantages of RNEMD. Different features make RNEMD attractive in the two cases.

RNEMD conserves the total energy and the total linear momentum of the system as a whole as long as one swaps velocities only among atoms (or molecules) of equal mass. This means that the algorithm does not destroy the microcanonicity of Newton's equation of motion. Hence, in contrast to other NEMD protocols, no thermostatting is required. This feature may be important for the calculation of the *shear viscosity*. Some thermostats introduce artificial dissipation of momentum, so the calculated viscosities may carry an intrinsic error.

Thermostatting problems do not arise in NEMD calculations of the *thermal conductivity*. The presence of two temperature baths leads, in steady state, to a constant mean temperature of the system as a whole, so all these methods are self-thermostatting. However, in all equilibrium and conventional non-equilibrium methods for the thermal conductivity, the heat flux $j(Q)$ has to be calculated explicitly. Whilst a heat flux is well understood macroscopically, there is considerable ambiguity about how to define it on a molecular scale: How is the energy of two interacting atom to be localised? Does one have to consider a flux of energy or enthalpy (the temperature gradient necessarily leads

to a concomitant density gradient, cf. Fig. 8)? How to calculate a local microscopic enthalpy? These questions have been intensely debated in the literature, see e.g. [7]. Our
RNEMD method offers the distinct advantage that the heat flux is known: The energy
transported in every velocity swap is calculated exactly and the method does not care in
what form and by what mechanism it flows back through the fluid.

A fundamental limitation of the method lies in the assumption that, in steady state,
the unphysical transfer is balanced by a physical flow. This means that only flows of
conserved quantities can be set up in this way. In this article, the quantities are energy
and linear momentum. In principle, one could also think of establishing flows of mass
(diffusion) or charge (ionic conductivity) by RNEMD. This would, however, be impractical, as the necessary swap of particle identities (force field parameters, charges) would
destroy the energy conservation of the method. Fluxes of non-conserved quantities like
the director of a nematic liquid crystal (useful for calculating the so-called rotational
viscosity of nematics) cannot be imposed at all.

6   The RNEMD and Higher-Order Transport Coefficients: The Ludwig–Soret Effect as an Example

Mass transport may also be driven by a temperature gradient. The phenomenon is known
as thermal diffusion or the Ludwig–Soret effect [23,24]. The phenomenological law is
traditionally written in terms of weight fractions $w_1 = M_1/(M_1 + M_2)$, for a binary
fluid and assuming field and fluxes in z direction.

$$ j_1 = D_{12}\varrho \left[\left(\frac{\partial w_1}{\partial z} \right) + S_T w_1 \left(1 - w_1 \right) \left(\frac{\partial T}{\partial z} \right) \right] \tag{6} $$

Here, j_1 is the flux of species 1, D_{12} the mutual diffusion coefficient, ϱ the average
mass density, and $S_T = D_T/D_{12}$ the Soret coefficient (in $1/K$) which is the ratio of the
thermal diffusion coefficient D_T and the mutual diffusion coefficient D_{12}. The Soret coefficient thus describes the relative importance of concentration-driven and temperature-
driven diffusion. If a temperature gradient is maintained across the system, for example
by the RNEMD method of Sect. 5, the fluid responds by forming a concentration gradient. When steady state is reached, energy is constantly flowing through the system, but
there is no mass flux, i.e. $j_1 = 0$. Therefore, (6) can be simplified and S_T is obtained
from the temperature and weight-fraction gradients in the system,

$$ S_T = \frac{1}{w_1(1 - w_1)} \left(\frac{\partial w_1}{\partial z} \right) \left(\frac{\partial T}{\partial z} \right)^{-1} . \tag{7} $$

For the special case of an equimolar mixture, (7) reduces to

$$ S_T = -2 \left(\frac{m_1}{m_2} + 1 \right) \left(\frac{\partial x_1}{\partial z} \right) \left(\frac{\partial T}{\partial z} \right)^{-1} , \tag{8} $$

with x_1 denoting the mole fraction of species 1. Equation (8) shows what is needed
for the calculation of the Soret coefficient. The RNEMD method for the thermal conductivity (Sect. 5) needs to be slightly modified. Velocity swaps now only take place

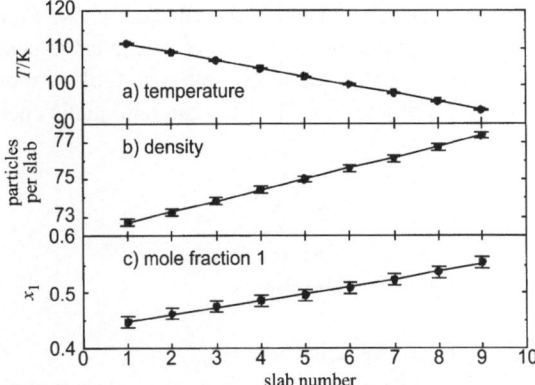

Fig. 9. Coupled heat and mass transport. By exchanging velocities, a constant energy flux is maintained through a binary equimolar isotopic Lennard–Jones fluid ($T^* = 0.81$, $\varrho^* = 0.85$, $m_1/m_2 = 6$). This leads to linear profiles of (**a**) temperature, (**b**) density, and (**c**) composition. From these gradients, the Soret coefficient can be calculated [14,15]

between atoms of like masses. In addition to the temperature profile, a profile of the mole fraction is calculated. This is done by simply counting the atoms of both species in each slab. The mole-fraction gradient is obtained from the mole-fraction profile by a linear regression. The typical appearance of a temperature profile, a density profile and a concentration profile while maintaining a steady-state heat flux through the system is shown in Fig. 9. From the temperature and concentration profiles gradients are obtained by linear regression and used in (8) to calculate the Soret coefficient.

7   Molecular Fluids and the RNEMD Method

The RNEMD method as outlined in Sects. 3 and 4 is was originally developed for atomic fluids. It can be used for molecular fluids, too, if the molecules are fully flexible and no constraints are present. In this case, the appropriate velocity exchanges are applied to individual atoms of the molecules. Energy and momentum conservation demand, as before, that the two atoms have the same mass. The analysis of the velocity or temperature profiles is also done on an atom-by-atom basis.

For molecules with holonomic constraints, the method has to be amended if energy and momentum conservation are to be kept [21,25]. The description that follows is our implementation into the atomistic MD program YASP [26]. The velocity exchange is now done between like molecules, rather than between atoms. Otherwise, the atomic velocities would not comply with the constraints after exchange. The selection of the two molecules, whose velocities are to be swapped, is based on their centre-of-mass momenta,

$$\mathbf{P} = \underset{\substack{\text{atoms } i \text{ in}\\ \text{molecule}}}{\sum} m_i \mathbf{v}_i = \mathbf{V} \underset{\substack{\text{atoms } i \text{ in}\\ \text{molecule}}}{\sum} m_i \qquad (9)$$

To establish a shear flow (η), two molecules are selected whose P_x are most directed against the desired flow directions of their respective exchange slabs. To establish a temperature gradient (λ, S_T), the translationally hottest and coldest molecules in the cold and hot slabs, respectively, are selected. The translational kinetic energy of a molecule is defined as

$$\frac{1}{2}V^2 \sum_{\substack{\text{atoms } i \text{ in} \\ \text{molecule}}} m_i \, ,$$

with $\mathbf{V}$ being its centre-of-mass velocity from (9).

For both temperature gradient and shear flow, the entire centre-of-mass velocity vectors $\mathbf{V}$ of the two selected molecules are exchanged. Thus, the velocity $\mathbf{v}'_i$ of atom i of molecule 1 after the exchange is given in terms of velocities before the exchange,

$$\mathbf{v}'_i = \mathbf{v}_i - \mathbf{V}_1 + \mathbf{V}_2.$$

As only centre-of-mass velocities are exchanged, the relative velocities of atoms within each molecule remain unchanged. Hence, the velocities after the exchange are compatible with the application of a constraint algorithm. It is also not a problem that in the case of shear flow the entire velocity vectors and not just the x components are swapped. The fact that the Maxwell demon selects on the x components is sufficient for establishing shear flow. Finally, the exchange of molecular centre-of-mass velocities instead of atomic velocities can also be done for flexible molecules, even if it is not necessary.

The presence of constraints complicates the calculation of the temperature profile, too. The temperature in a molecular dynamics simulation with constraints is given by the equipartition theorem

$$\left(\frac{3N}{2} - C\right) k_B T = \frac{1}{2} \left\langle \sum_{\text{atoms } i} m_i v_i^2 \right\rangle ,$$

where k_B is Boltzmann's constant and C is the number of constraints in the system. If a local (slab) temperature is to be determined, the angular brackets denote time averaging only over the atoms within the slab, and C refers to the number of constraints in the slab. This number can be estimated analytically only if the constraints can be assumed to be uniformly distributed throughout the slab. This is the case only if (i) the density is uniform (small perturbation) and (ii) the composition is uniform (only one molecular species, no Soret effect). In all other cases, C must be evaluated in the simulation. In YASP, bond constraints are solved for by a modified SHAKE procedure [27]. The constraints are stored as a list of pairs of atoms whose distance is to be kept constant. The program goes through this list and determines in which slab(s) the two atoms defining a constraint reside. For every atom, the constraint counter in its slab is incremented by 1/2. If an atom is part of n constraints, its contribution to the constraints in the slab is $n/2$. In this way, constraints spanning two slabs are split between the slabs. As a consequence, the scheme also allows for molecules extending over more than one slab.

8   A Bibliography of Applications of the RNEMD Method

8.1   Shear Flow, Viscosity

The RNEMD algorithm was developed using the Lennard–Jones fluid near the triple point as a known example [12]. The agreement between the viscosity from RNEMD and values from other methods was satisfactory. Recently, RNEMD has been used to study the breakdown of the Stokes–Einstein relation in a non-additive binary Lennard–Jones liquid, which is a known glass former [28]. The method has been changed to allow for molecular fluids with and without constraints [21]. Agreement between experimental and calculated results is generally within the experimental error bars (Fig. 10). A second real-world example concerns aqueous solutions of saccharose (Fig. 11), where again a good agreement with experiment was achieved [29]. The RNEMD algorithm has also been used to study structure formation upon shearing a simple model of block copolymers [30,31]. These authors also combined the RNEMD method with the equations of motion and the thermostat of dissipative particle dynamics.

8.2   Thermal Conductivity

As for shear flow, the RNEMD method for the thermal conductivity was developed with the triple-point Lennard–Jones fluid as a test case [11]. The molecular version has been used to study the thermal conductivity of liquid n-butane and two water models and good agreement with experiment was obtained [25]. The method has subsequently been used to study the temperature dependence of the thermal conductivity in the high explosive octahydro-1,3,5,7-tetranitro-1,3,5,7-tetrazocine (HMX) [32]. A related method has been used to determine the thermal conductivity of carbon nanotubes [33].

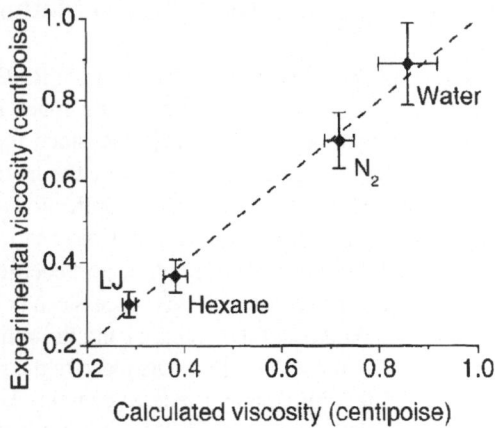

Fig. 10. Performance of the molecular version of the RNEMD method as implemented in YASP. The experimental and simulated viscosities of molecular fluids with and without constraints are compared [24]

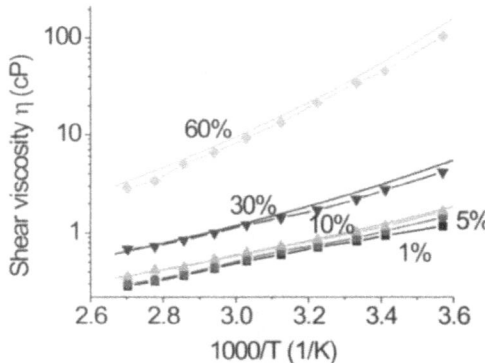

Fig. 11. Temperature and concentration (wt.% saccharose) dependence of the viscosity of aqueous solutions of saccharose (sucrose). Symbols denote simulation results, solid lines experiments, where available [28]

8.3   Thermal Diffusion, Soret Coefficient

The RNEMD method was used to study the molecular origin of the Ludwig–Soret effect and to calculate Soret coefficients. This was done using equimolar binary Lennard–Jones systems where the interaction strength, the diameter and the mass of the two species was independently varied [13–15,34]. The results were later experimentally confirmed [35]. The calculated Soret coefficients are very sensitive to thermodynamic conditions, but also details in the interaction potentials [36]. An apparent disagreement between results of different groups [14,15,37] could be resolved [36].

9   Status and Future Potential of the RNEMD Method

The fundamental advantages and disadvantages inherent in the RNEMD method have been discussed in Sect. 5. In terms of practical applications, much has changed since the initial development of the method in 1997 [11], and since the last synopsis in 1999 [14,15]. In particularly the molecular formulation of the method has increased its scope. This is evident in the bibliography (Sect. 8). Applications to both shear viscosity and thermal conductivity calculations for real fluids have shown an encouraging level of agreement between calculated and measured transport coefficients. More simulations are to be expected here. It is not yet clear if this success can be repeated for Soret coefficients, as they depend on subtleties of the potential and the simulation protocol.

At the same time the RNEMD method has found its place in the study of the fundamentals of transport in fluids. The bibliography (Sect. 8) contains examples both from thermal transport and shear, which use model fluids. Most useful in this context are the robustness and the ease of implementation of the method as well as the fact that it can be combined with any particle-based simulation method that locally conserves linear momentum or energy, respectively. This includes both the dissipative-particle method and the Malevanets–Kapral method (cf. Groot [1] and Español [2], and R. Kapral [3] in this volume, respectively). Applications of RNEMD on the mesoscale are therefore foreseeable.

Acknowledgements

Fruitful discussions with Dirk Reith are gratefully acknowledged, as are the lively remarks by Giovanni Ciccotti during the conference.

References

1. R.D. Groot: Applications of Dissipative Particle Dynamics, Lect. Notes Phys. **640**, 1 (2004)
2. P. Español: Statistical Mechanics of Coarse-Graining, Lect. Notes Phys. **640**, 65 (2004)
3. A. Malevanets and R. Kapral: Mesoscopic Multi-particle Collision Model for Fluid Flow and Molecular Dynamics, Lect. Notes Phys. **640**, 112 (2004)
4. M.P. Allen, D.J. Tildesley: *Computer Simulation of Liquids* (Clarendon, Oxford 1987)
5. D.A. McQuarrie: *Statistical Mechanics* (Harper Collins, NY 1976)
6. R. Kubo, M. Toda, N. Hashitsume: *Statistical Physics II* (Springer, Berlin 1985)
7. D.J. Evans, G.P. Morriss: *Statistical Mechanics of Nonequilibrium Liquids* (Academic-Press, London 1990)
8. D. Brown, J.H.R. Clarke: Phys. Rev. A **34**, 2093 (1986)
9. D.J. Evans, J.F. Ely: Mol. Phys. **59**, 1043 (1986)
10. L.J. Hood, D.J. Evans, G.P. Morriss: Mol. Phys. **62**, 419 (1987)
11. F. Müller-Plathe: J. Chem. Phys. **106**, 6082 (1997)
12. F. Müller-Plathe: Phys. Rev. E **59**, 4894 (1999)
13. D. Reith: Diploma thesis, University of Mainz (1998)
14. D. Reith, F. Müller-Plathe: J. Chem. Phys. **112**, 2436 (2000)
15. F. Müller-Plathe, D. Reith: Comput. Theor. Polym. Sci. **9**, 203 (1999)
16. B. Hafskjold, T. Ikeshoji, S. Ratkje: Mol. Phys. **80**, 1389 (1993)
17. B. Hafskjold, T. Ikeshoji: Mol. Phys. **81**, 251 (1994)
18. B. Hafskjold, S. Ratkje: J. Stat. Phys. **78**, 463 (1995)
19. P.W. Atkins: *Physical Chemistry*, 5th edn. (Oxford University Press, Oxford 1994)
20. C.P. Calderon, W.T. Ashurst: Phys. Rev. E **66**, 013201 (2002)
21. P. Bordat, F. Müller-Plathe: J. Chem. Phys. **116**, 3362 (2002)
22. B.J. Palmer: Phys. Rev. E **49**, 359 (1986)
23. C. Ludwig: Sitzungsberichte der Akademie der Wissenschaften Wien, Math.-Naturw. Kl. **20** (1856)
24. C. Soret: Arch. Genève **3**, 48 (1879)
25. D. Bedrov, G.D. Smith: J. Chem. Phys. **113**, 8080 (2000)
26. F. Müller-Plathe: Comput. Phys. Commun. **78**, 77 (1993)
27. F. Müller-Plathe, D. Brown: Comput. Phys. Commun. **64**, 7 (1991)
28. P. Bordat, F. Müller-Plathe: Europhys. Lett. (Submitted)
29. P. Bordat, R. Faller, R.R. de Melo Moreno, S. Canuto, F. Müller-Plathe: (in preparation)
30. T. Soddemann: PhD thesis, University of Mainz (2000)
31. T. Soddemann, B. Dünweg, K. Kremer: Eur. Phys. J. **6**, 409 (2001)
32. D. Bedrov, G.D. Smith, T.D. Sewell: Chem. Phys. Lett. **324**, 64 (2000)
33. M.A. Osman, D. Srivastava: Nanotechnology **12**, 21 (2001)

34. F. Müller-Plathe: In: *Thermal Nonequilibrium Phenomena in Fluid Mixtures*, ed. by W. Köhler, S. Wiegang, Lecture Notes in Physics **584**, 184 (Springer, Berlin 2002)
35. C. Debuschewitz, W. Köhler: Phys. Rev. Lett. **87**, 055901 (2001)
36. P. Bordat, D. Reith, F. Müller-Plathe: J. Chem. Phys. **115**, 8978 (2001)
37. B. Hafskjold. In: *Thermal Nonequilibrium Phenomena in Fluid Mixtures*, ed. by W. Köhler, S. Wiegang, Lecture Notes in Physics **584**, 3 (Springer, Berlin 2002)

Coarse-Graining in Polymer Simulations

Séverine Girard[1] and Florian Müller-Plathe[1,2]

[1] Max-Planck-Institut für Polymerforschung, Ackermannweg 10, 55128 Mainz, Germany
[2] International University Bremen, P.O. Box 750561, 28725 Bremen, Germany

Abstract. Polymers exhibit physical properties in a broad range of length and time scales. Molecular simulation techniques, individually, are restricted to much narrower ranges. Therefore, one needs to simulate polymers with models of several different scales, in order to have a complete picture of their properties. Such multiscale simulations need bridges between the different levels of simulation. Coarse-graining consists of deriving a simpler model from a more detailed one. Fine-graining is the reverse process. Here, we present methods to interlink the quantum chemical, atomistic, and mesoscopic levels.

1 Introduction

During the last decades, many powerful computational methods have been developed to simulate molecules and polymers. They can be classified by their basic degrees of freedom: electrons (quantum chemistry), atoms (force field), monomers or groups of monomers (mesoscopic models), entire polymer chains (soft fluids), or volume elements (finite elements). All these methods, and many others, have been applied side by side to polymers. Indeed, the properties of polymers are connected with various time and length scales: Ångströms and sub-picoseconds for the vibrations of atomic bonds to millimetres and seconds for crack propagation in polymer composites. Each scale is associated with at least one computational method (Fig. 1). Unfortunately it is impractical to build a model suitable for all scales. Models with all the sub-atomic details are computationally very demanding and these details are irrelevant for many applications. For example, studying chemical reactions requires the presence of the electronic wave functions; therefore a fragment of a few dozen of atoms may be simulated for a few picoseconds with a quantum chemical method. For studying polymer permeation, e.g., the electrons are irrelevant and a bigger system with longer simulation time is necessary: one may simulate a system of a few thousand atoms during a few nanoseconds with atomistic methods.

However, many polymer properties cannot be viewed on one length scale alone. One instructive example is the temperature dependence of the viscosity of bisphenol-A polycarbonate melts [1]: Decreasing the processing temperature from 500 to 470 K increases the viscosity by a factor of 10. This increase could also be brought about by a suitable modification of the monomer with a resulting increase in the activation energy equivalent to a temperature drop of 30 K. This would be a *local* change of the chemical nature of the polymer. On the other hand, the viscosity increase can also be achieved by changing the *global* structure of the polymer leaving its chemical nature intact. In the present case, doubling the length of the linear polymer increases the viscosity by a factor of 10, since the viscosity is proportional to MW [2,3] – MW is the molecular

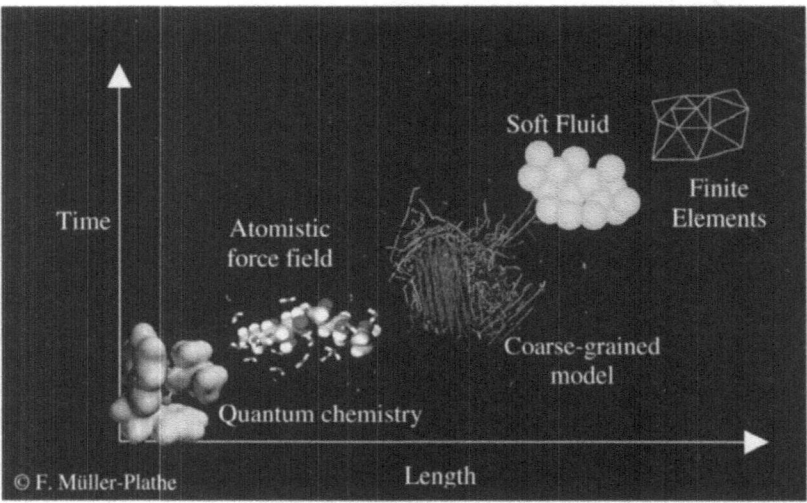

Fig. 1. Length and time scales, and associated computational methods in polymer simulations

weight of a polymer chain. This example shows that both detailed material-specific information of the chemical composition (i.e. atomistic model) and generic chain length and connectivity information (i.e. mesoscopic model) are necessary for a complete view of the melt viscosity. There are many more polymer properties, for which a combined approach is necessary. A few examples of how architectural features on different length scales influence polymer properties are collected in Fig. 2.

Bridges between different simulation methods are necessary to study the properties of polymers in a broad range of time and length scales. At the same time, multiscale simulation is a particularly challenging task and the paths from one scale to another are neither unique nor trivial. Coarse graining is the zoom-out method, which passes from a detailed model to a simpler one. It requires defining a new type of elementary particle, and the interactions between these new particles. In this paper, we will focus on coarse graining of atomistic models into mesoscopic ones, where the elementary particle is a group of atoms, typically a monomer. By simplifying the model in a rigorous way, it is possible to study large-scale polymer properties from first principles in a indirect way. This is the first purpose of multiscale simulations.

A good coarse-graining method allows the reverse process, also denoted as fine-graining or reverse mapping. To give an example, it is obvious how to go from an atomistic description of a polymer to a quantum-chemical description: keep the atom positions and re-insert basis functions and electrons. This is necessary in order to calculate for example non-linear optical properties of the polymer after structural relaxation. It is much more complicated to fine-grain a mesoscopic model into an atomistic one, as the back-mapping is not unique. Remapping indicates a second use of multiscale simulation. The coarse-grained model is used as a computational detour to faster generate samples for fine-grained analysis.

Chemical detail

Polyisobutylene:
amorphous rubber
(T_g=200 K, T_m=320 K)

Poly(vinylidene chloride):
semicrystalline, brittle
(T_g=255 K, T_m=470 K)

Tacticity (Stereochemistry)

Cellulose:
crystalline, insoluble

Starch (amylose):
soluble in water

Sequence

-B-B-S-B-S-S-S-B-B-S-S-

-B-B-B-B-B-B-S-S-S-S-S-S-

Random-poly(butadiene-co-styrene):
synthetic rubber

Block-poly(butadiene-co-styrene):
high-impact polystyrene (HIPS)

Topology

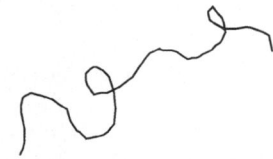

Low-density polyethylene:
branched, shopping bags

High-density polyethylene:
linear, bullet-proof vests

Fig. 2. Structural differences on different scales can influence the macroscopic properties

2   From Quantum Chemistry to Atomistic Simulation

To illustrate the concepts and numerical procedures used in automatic coarse-graining, we take a short detour and describe the generation of atomistic models from quantum-chemical and experimental information. The procedures are more easily introduced in this way, and also in our work, the automatic parameterisation of atomistic force fields has preceded the systematic coarse graining.

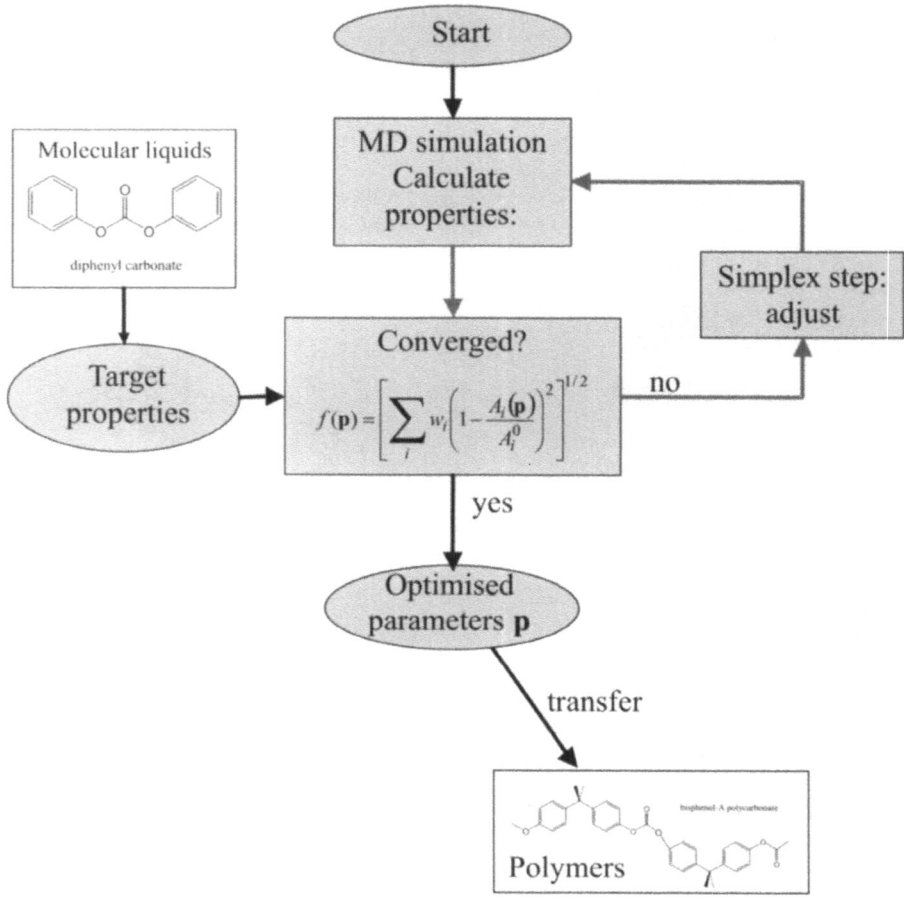

Fig. 3. Procedure for optimising force fields with the simplex algorithm. For details, see Sect. 2.3

Ab initio quantum chemistry deals with systems of the order of a few dozen atoms. It is inappropriate for macromolecules but, however, it can be used for small molecules whose chemical structure is similar to the repeat unit of a polymer, for example diphenyl carbonate, whose structure is similar to bisphenol-A-polycarbonate (Fig. 3). It provides important information, which is used to prepare an atomistic model, about the basic structure of these molecules: bond lengths, bond angles, torsions, and associated force constant, as well as partial charges. At the atomistic level, we need to define an empirical potential. For clarity we will discuss this potential in two parts: the bonded potential and the non-bonded potential.

2.1 The Bonded-Potential

The bonded potential contains usually three or four terms for bond stretching, angle bending, torsions and when necessary improper dihedral angles. Atom pairs forming a chemical bonds can be kept at a fixed distance (r_0) by a rigid constraint using the SHAKE algorithm [4] or, alternatively, close to it by a harmonic potential

$$V_{\text{bond}} = \frac{k_{\text{r}}}{2}(r - r_0)^2, \tag{1}$$

where r is the distance between the bonded atoms, r_0 its equilibrium value, and k_{r} is the force constant. To maintain bond angles close to their equilibrium values, one of the following forms of potential can be used:

$$V_{\text{angle}}(\theta) = \frac{k_{\theta}}{2}(\theta - \theta_0)^2 \tag{2}$$

$$= \frac{k_{\theta}}{2}(\cos\theta - \cos\theta_0)^2 \tag{3}$$

$$= \frac{k_{\theta}}{2}(1 - \cos(\theta - \theta_0))^2, \tag{4}$$

where θ is the bond angle, θ_0 its equilibrium value, and k_{θ} the force constant. Equation (4) is for the special case $\theta_0 = 180°$.

The rotations around bonds are restricted with potentials of the shape

$$V_{\text{torsion}}(\tau) = \frac{k_{\tau}}{2}(1 - \cos p(\tau - \tau_0)) \tag{5}$$

$$= \sum_{m=0}^{6} C_m \cos^m(\tau), \tag{6}$$

where τ is the torsional angle, τ_0 its equilibrium value, k_{τ} the force constant, and C_m the coefficients of a polynomial of order 6. When no rotation is allowed around the bond, a dihedral angle can be maintained by a harmonic potential

$$V_{\text{dihedral}}(\tau) = \frac{k_{\tau}}{2}(\tau - \tau_0)^2. \tag{7}$$

The bonded potential is the sum of all these contributions, and is a parametric function of the bond distances, bond angles, torsions and dihedral angles, the parameters being the force constants and the equilibrium values of these variables.

2.2 The Non-bonded Potential

The non-bonded potential contains two contributions, the electrostatic interactions between partial charges and the van der Waals interactions.

The electrostatics are typically represented by a Coulombic potential:

$$V_{\text{coulomb}}(r_{ij}) = \frac{1}{4\pi\varepsilon_0}\frac{q_i q_j}{r_{ij}}, \tag{8}$$

where q_i is the partial charge of atom i, r_{ij} is the distance between atoms i and j, and ε_0 is the vacuum permittivity.

The van der Waals interactions can adopt various functional forms. The most often used are Lennard–Jones, Weeks–Chandler–Andersen and Buckingham:

$$V_{\mathrm{LJ}}(r_{ij}) \qquad = 4\varepsilon_{ij}\left[\left(\frac{\sigma_{ij}}{r_{ij}}\right)^{12} - \left(\frac{\sigma_{ij}}{r_{ij}}\right)^{6}\right] \tag{9}$$

$$V_{\mathrm{WCA}}(r_{ij}) \qquad = \begin{cases} 4\varepsilon_{ij}\left[\left(\frac{\sigma_{ij}}{r_{ij}}\right)^{12} - \left(\frac{\sigma_{ij}}{r_{ij}}\right)^{6}\right] + \varepsilon & \text{if } r_{ij} \le 2^{\frac{1}{6}}\sigma_{ij} \\ 0 & \text{otherwise} \end{cases} \tag{10}$$

$$V_{\mathrm{Buckingham}}(r_{ij}) = A\,\exp\left(-\frac{r_{ij}}{B}\right) - \frac{C}{r_{ij}^{6}} \tag{11}$$

The parameters associated with the non-bonded potential are ε and σ for Lennard–Jones or Weeks–Chandler–Andersen formulations, and A, B and C for Buckingham formulation.

2.3   Force Field Optimisation

Quantum chemistry provides values for many of the parameters described above. However, adjustments are needed in order to reproduce satisfactorily the physical properties of the molecule under study. This task can be done automatically by a computer in a more efficient and systematic way than the usual manual trial and error method.

Let us consider the physical properties calculated in a simulation as multidimensional functions of the force field parameters. We define a penalty function that compares the calculated values of selected properties with their target values from experiment

$$f(\mathbf{p}) = \left[\sum_{i} w_{i}\left(1 - \frac{A_{i}(\mathbf{p})}{A_{i,\text{target}}}\right)^{2}\right]^{1/2}, \tag{12}$$

where $\mathbf{p}$ are the force field parameters, A_{i} the observable physical properties, $A_{i,\text{target}}$ their experimental values and w_{i} the statistical weight of A_{i}. Each evaluation of $f(\mathbf{p})$ requires a whole MD simulation, and we use the simplex algorithm to minimise this function by adjusting the force field parameters [2].

In order to start the simplex in a parameter space of dimension N, we need $N + 1$ starting points. So we perform $N + 1$ preliminary molecular dynamics simulations with slightly different starting parameter sets (derived from the quantum chemistry results for instance). Then we calculate physical properties of interest and the penalty function for each parameter combination. If for one of this sets the value of the penalty function is below a certain user-defined threshold (typically 0.01), the corresponding force field is supposed to be satisfactory. Otherwise, the optimisation begins, and a simplex move provides a new parameter set, which is used in a new molecular dynamics simulation. This process is repeated until the penalty function converges (Fig. 3).

In practice, this method converges slowly in a high-dimensional space, so only one or two properties are taken into account in the penalty function, and only two or three parameters are simultaneously adjusted. It is often more efficient to make several successive simplex minimisations involving different sub-sets rather than to try to adjust

Table 1. Physical properties of liquid DMSO at 298 K and 0.1013 MPa

Property	1995 model	This work	Experiment
Density (kg/m^3)	1099	1095 (2)	1095
Heat of vaporization (kJ/mol)	52.87	52.42 (0.05)	52.88
Diffusion coefficient (10^{-5} cm^2/s)	1.1	0.88 (0.02)	0.8
Rotational correlation time (ps)	3.9	3.50 (0.01)	5.2
Thermal expansion coefficient (10^{-3} K^{-1})	0.91 (0.11)	0.87 (0.09)	0.928
Isothermal compressibility (10^{-7} Torr^{-1})	0.90 (0.10)	0.67 (0.07)	0.70
Specific heat (J/(mol K))	108 (7.7)	101 (6)	118.28
Excess Helmholtz free energy (kJ/mol)	-29.7 (0.8)	-29.8 (0.2)	-29.7
Static dielectric constant	30	49 (4)	46
Shear viscosity (cP)	$1.26-1.29$ (0.24)	1.74 (0.14)	1.991

all parameters at the same time. Note also that the bonded parameters and the partial charges can often be obtained directly from quantum chemistry. Therefore, often only the van der Waals parameters have to be optimised here.

In order to illustrate the performance of the simplex method, we will make a small detour to molecular liquids. In 1995 a force field for dimethyl sulfoxide ((CH_3)$_2$SO) (DMSO) was published [3]. Unfortunately the density predicted by this model was too high. Therefore, we have applied the simplex method in order to re-optimise this force field [5]. This is a rigid united-atom model where the van der Waals interactions were of Lennard–Jones form (9). The parameters, if we exclude bonded ones, are ε_O, ε_S, ε_{CH3}, σ_O, σ_S, σ_{CH3}, q_O, q_S, and q_{CH3}. The penalty function was:

$$f(\mathbf{p}) = \left(\frac{|\varrho(\mathbf{p}) - \varrho_{\text{target}}|}{\varrho_{\text{target}}}\right) + \left(\frac{|\Delta H(\mathbf{p}) - \Delta H_{\text{target}}|}{\Delta H_{\text{target}}}\right), \qquad (13)$$

where ϱ is the density and ΔH the heat of vaporisation. The level of convergence was fixed at 0.02. The optimisation was carried out adjusting σ_{CH3}, because it influences the volume of the molecule and, hence, the global density, and q_O with a compensation on $q_S = -(q_O + 2q_{CH3})$, as the molecular dipole moment governs the attraction and the heat of vaporisation.

Approximately forty steps were needed for the simplex to converge. In the 1995 model, σ_{CH3} was 0.366 nm and q_O was -0.459 e, and they became 0.374 nm and -0.437 e, respectively, after optimisation. Physical properties calculated with both parameter sets are listed in Table 1. We clearly see improvements in the density, diffusion coefficient, isothermal compressibility, static dielectric constant and shear viscosity. The others properties calculated are not much affected.

Table 2. Parameter sets and physical properties of different models for liquid THF. ε (kJ/mol), σ (nm), ϱ (kg/m^3), ΔH (kJ/mol), D (10^{-5} cm^2/s), and K_T (GPa^{-1})

	1	1'	2	2'	3	Exp
ε_C	0.290	0.290	0.494	0.200	0.190	
ε_O	0.628	0.628	0.712	0.500	0.360	
ε_H	0.170	0.170	0.0001	0.150	0.150	
σ_C	0.306	0.295	0.385	0.343	0.385	
σ_O	0.216	0.300	0.300	0.300	0.350	
σ_H	0.250	0.240	0.271	0.236	0.190	
ϱ	881±6	884±5	894±3	888±6	887±5	884 [7], 889 [8], 873 [9]
ΔH	32.65±0.37	31.27±0.38	31.75±0.27	32.27±0.13	31.78±0.33	31.80 [7], 31.99 [8], 31.86 [10]
D	2.18±0.04	2.19±0.08	1.75±0.12	1.92±0.06	2.01±0.13	2.45 [11], 3.40 [12]
K_T	1.38±0.02	1.22±0.02	1.05±0.03	1.04±0.04	0.84±0.04	0.95 [7]

Another illustration of the possibilities and limitations of the simplex method is the optimisation of parameter sets for liquid tetrahydrofurane [6] (THF). The van der Waals parameters of this molecule have been optimised against the density (ϱ) and the heat of vaporisation (ΔH). Self-diffusion coefficient (D) and isothermal compressibility (K_T) have been also calculated for each model, but not included in the penalty function. The parameter sets and the corresponding physical properties are listed in Table 2.

Model 1 reproduced ϱ, ΔH and D well, but K_T was too high and the centre-of-mass radial distribution function (RDF) for this model showed a double peak structure (not shown), which seems unphysical. This was due to the fact that σ_O was too small. Therefore this parameter was fixed to 0.300 nm, and the other sigmas were reoptimised. This led to model 1' which still has too high K_T, but whose RDF is correct.

Model 2 originated from another simplex optimisation, started in a completely different region of the parameter space. This time the simplex converged to a local minimum where ε_H was nearly zero. As this corresponds to a united atom model, we fixed ε_H to 0.150 kJ/mol and reoptimised the other parameters. This led to model 2', which reproduces well all physical properties calculated except the diffusion coefficient.

After having optimised several parameter sets for THF with simplex, we acquired some knowledge of how the different parameters influence the different physical properties calculated. It was then easier and faster to optimise the last parameter set by hand. This is model 3, which reproduced all physical properties calculated satisfactorily.

The THF example illustrates well the pros and cons of automatic parameterisation schemes. They allow the simulator to quickly try out new force field concepts and have each of them optimised such that they can be compared on an equal footing. On the other hand, the simulator should always evaluate critically the results of optimisation. Simplex can converge to minima where the parameters are not physical. Then, the human hand has to interfere and put the simplex in another region of the parameter space.

3   Coarse-Graining from Atomistic to Mesoscopic Models

A mesoscopic model of a molecule is made up of super-atoms, which represent groups of atoms, monomers or parts thereof, or even several monomers. Interactions are only defined between the super-atoms. They are effective interactions as they incorporate in a mean-field way the effects of the atomistic degrees of freedom neglected. Thus, a reduction in the number of degrees of freedom is achieved. Typically it is of the order of 10:1, i.e. about 10 real atoms are collected into one super-atom. The reduction in the number of pair interactions to calculate is thus two orders of magnitude in a dense polymer melt. (It can be much more for a polymer solution, since, in the coarse-graining process, the explicit solvent is being disposed of as well.) In addition, the resulting coarse-grained potentials are softer than atomistic potentials such that larger time steps can be used without the MD integrator becoming unstable. A typical increase in the time step is between 10 and 100. In total, the efficiency increases by at least 3 to 4 orders of magnitude.

Interactions between super-atoms are broken down in almost the same way as inter-actions between atoms. In the cases studied so far, we could get rid of the electrostatic potential, as either our super-atoms were uncharged, or the dielectric screening was strong. If they had been charged the Coulombic form of the potential (8) would have had to be kept. In many cases we can also ignore the torsional potential, inasmuch as the influence of the torsions is often already covered by the van der Waals potential. In many applications, we are thus left with having to define and parameterise the bond stretching, angle bending and van der Waals interactions (Fig. 4).

The mesoscopic force field is constructed following more or less the same principles as in the atomistic case with one important difference. When optimising the atomistic force field we were trying to reproduce experimental properties of the fluid. In principle,

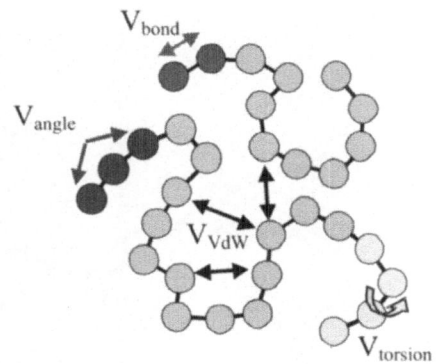

Fig. 4. Internal interactions in a polymer chain [13]

this could be done also at the mesoscopic level, but it would not help us to bridge the different simulation scales. As we want to be able to reintroduce atomistic details afterwards, we build a mesoscopic model by reference to an atomistic model. To this end, we amend the simplex method with a penalty function, which not only includes physical properties of the system but also structural properties of the atomistic model, such as radial distribution functions.

Before starting the coarse-graining procedure we need reference data from a fully equilibrated atomistic simulation of the same molecules. In the case of polymers, short oligomers are simulated at the atomistic level, because long chains do not relax in a reasonable amount of computer time. Then the same oligomeric system is simulated at the coarse-grained level in order to develop the coarse-grained force field. Finally, longer chains are studied at the mesoscopic level to calculate the large-scale properties. The oligomers used in the matching of models have to be long enough to allow the mesoscopic force field to predict the correct structure of longer chains, yet small enough to be relaxed and sufficiently sampled at the atomistic level.

The coarse-graining procedure contains three steps: Choice of the super-atoms and the functional form of the potential, and the optimisation of the potential parameters.

3.1 The Super-Atoms

As we want to limit the number of interactions in order to gain computer time, we try to build a mesoscopic model with as few super-atoms as possible. In the case of polymers, this often means trying first to map an entire repeat unit to one super-atom. Depending on the structure and the flexibility of the molecule, as well as on the purpose of the coarse-grained model, more super-atoms may be needed. The number of rotatable torsions in the backbone can provide a guideline as to how many super-atoms are required.

For instance, in bisphenyl-A polycarbonate ($-[C(CH_3)_2-C_6H_4O-(CO)-OC_6H_4]-)_n$ (Fig. 5), the phenylene groups are rigid, therefore the distance between the carbonate group and the isopropylidene group is almost constant. That means that the phenylenes are acting like bonds between these groups. This is why a 2-site model, with one site on the carbonate, and the other one on the isopropylidene group, has been used to map this polymer [12].

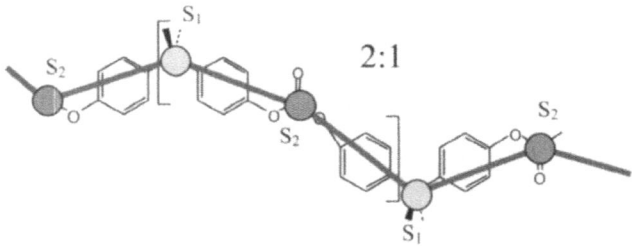

Fig. 5. Coarse graining of bisphenol-A polycarbonate. 2:1 scheme (2 super-atoms per chemical repeat unit) [12]

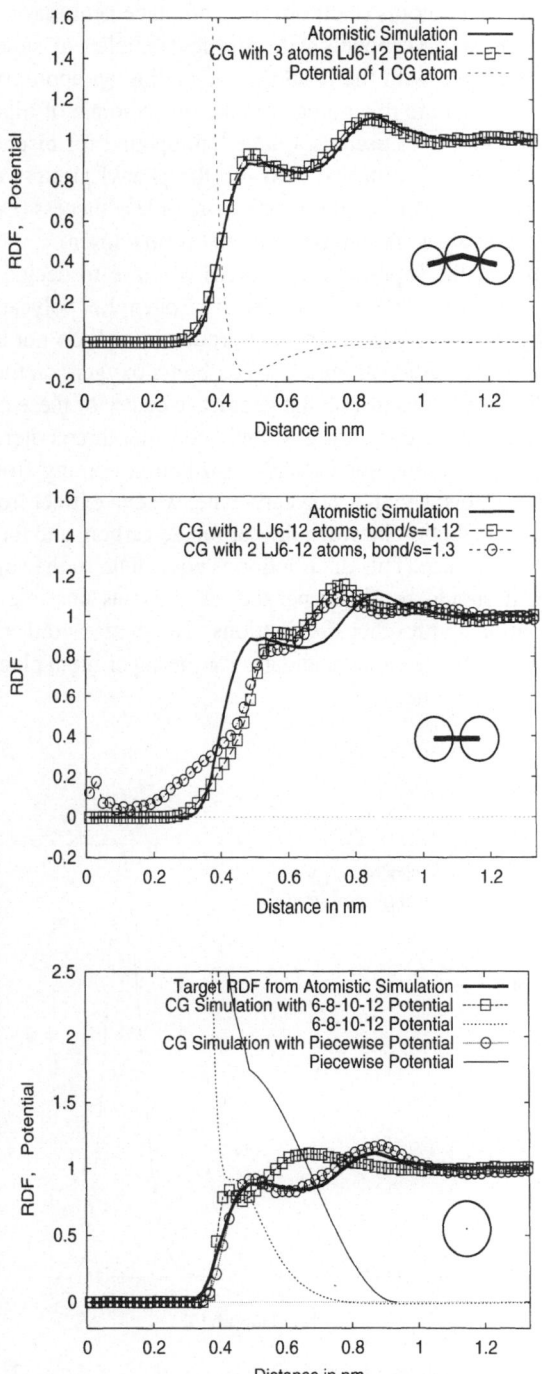

Fig. 6. Coarse-graining diphenyl carbonate: (**a**) 3-site model, (**b**) 2-site model, (**c**) two 1-site models. Shown is the centre-of-mass radial distribution function and, where useful, the corresponding coarse-grained potential (from [14])

The choice of the super-atoms (their number and placement) has consequences for the potential used, and several models with different number of super-atoms can be built for the same molecule with the condition of finding an appropriate potential in each case. In order to illustrate this point, we take the example of diphenyl carbonate $((C_6H_5O)_2CO)$ [14] , which we used as a model compound for bisphenol-A polycarbonate. It was coarse grained with one, two (on the phenyl groups) or three (on the phenyl and carbonate groups) super-atoms with more or less success (Fig. 6) which will be described in Sect. 3.2 together with the potential form chosen.

Once the number of the super-atoms is fixed, we have to decide where to place them. If we have agreed, for instance in the case of bisphenyl-A polycarbonate, that one super-atom should be placed on the carbonate group, we still do not know whether it is better to centre it on the carbon atom, on the carbonyl oxygen, on the centre of mass of the carbon and the three oxygens, on the geometric centre of these atoms, or, in fact somewhere else. To help us in this choice, we will consider three criteria.

The first is the super-atoms' bond length distribution coming from the atomistic trajectory. In the case of bisphenyl-A polycarbonate, we can extract from the atomistic trajectory a distance distribution between the carbonate carbon, and the isopropylidene backbone carbon, for instance. This distribution is equivalent to the super-atoms' bond length distribution. It should be single-peaked, as sharp as possible, and preferably free of cross-correlations with other distributions. The coarse-grainer should prepare several distributions for the different candidates for the super-atom placement, and give a preference to the sharpest one.

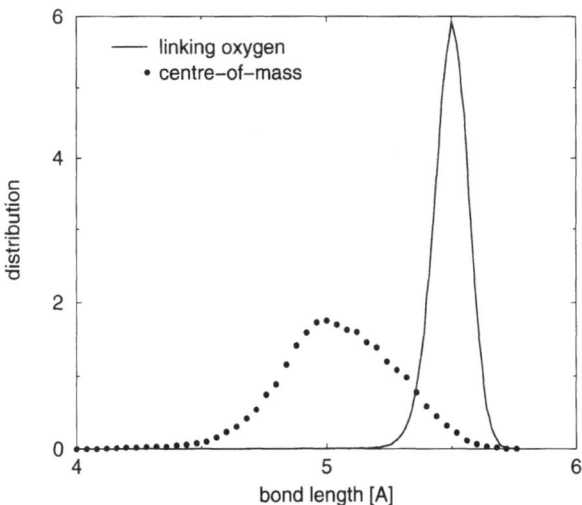

Fig. 7. Bond distribution of cellulose for different choices of super-atoms. The distribution for the super-atom centred on the linking oxygen is much sharper than the distribution for the super-atom centred on the center-of-mass of the glucose ring. Therefore, we have chosen to centre the super-atom on the linking oxygen

In the case of cellulose, the two candidates for the centre of the super-atom are the centre of mass of the ring and the oxygen linking two rings. The bond distribution of the latter is much sharper and therefore the super-atom was centred on the oxygen (Fig. 7).

A counter-example is sodium polyacrylate $(-CH_2-CH(COO^-)-)_n$ in aqueous solution. In Fig. 8a the sharpest distribution corresponds to a mapping with the super-atom on the backbone carbon carrying the carboxylate group, rather than on the carbon of the carboxylate group or on the centre of mass of the repeat unit. However, in order to allow for the influence of the side group, this polymer has been mapped [15] with a super-atom located on the centre of mass of the repeat unit. Its bond length peak is sufficiently sharp and its angular distribution includes some of the behaviour of the side group.

The second criterion concerns the excluded volume. Using spherical beads it is impossible to give a perfect representation of the excluded volume, but it should be kept as realistic as possible. As a consequence, the super-atom should be placed at the centre of an approximately spherical moiety. A bad choice of the super-atom position can alter much the excluded volume envelope of the chain. In the case of polycarbonate, a bad example is to place the super-atom on the carbonyl oxygen. At the mesoscopic level we would have to exclude a large spherical volume to make sure to reproduce van der Waals excluded volume of the atoms, which are concentrated only on one side of the sphere. A better choice is the center of mass of the atoms of the carbonate, or the backbone carbon.

In the case of polyisoprene $(-CH_2-CH=CH(CH_3)-CH_2-)_n$ [13], owing to the double bond, the carbons of the repeat unit are coplanar. Mapping this system with a spherical super-atom on the centre of mass of all atoms will again misrepresent the excluded-volume envelope of the molecule. Instead, It has been found better to place the super-atom at the centre of the single bond linking two repeating units (Fig. 9).

The third criterion is the ease of back-mapping. It is much less crucial for the choice than the other two criteria, but from two possible candidates, which already fulfill the other requirements, the more convenient one may be selected. We favour a super-atom centred on an atom, rather than on the centre of mass of a group of atoms, because the re-insertion of the missing atoms can be facilitated if we already know the position of two atoms. Note that it is often not possible to find super-atom positions that fully satisfy all requirements simultaneously. In such a case a compromise must be sought.

3.2   Definition and Optimisation of the Potential

Sometimes we can use the same functional forms for the coarse-grained and the atomistic case. Harmonic potentials work well for bonds but rarely for angles, because coarse-grained angle distributions mostly show multiple peaks. If the distribution contains several well-defined peaks, one may use a more complicated analytical form. Often, however, a numerical tabulated potential (see below) is the better alternative. Note, that all coarse-grained force fields are applicable only at conditions (in particular temperature, concentrations) of the atomistic reference simulation. A new force field has to be optimised for each new temperature investigated.

For example, in the case of polyacrylate with the super-atom centred on the backbone carbon carrying the carboxylate group, the angle distribution (Fig. 8b) shows two overlapping peaks. Here a numerical potential is advantageous. The Boltzmann inversion of

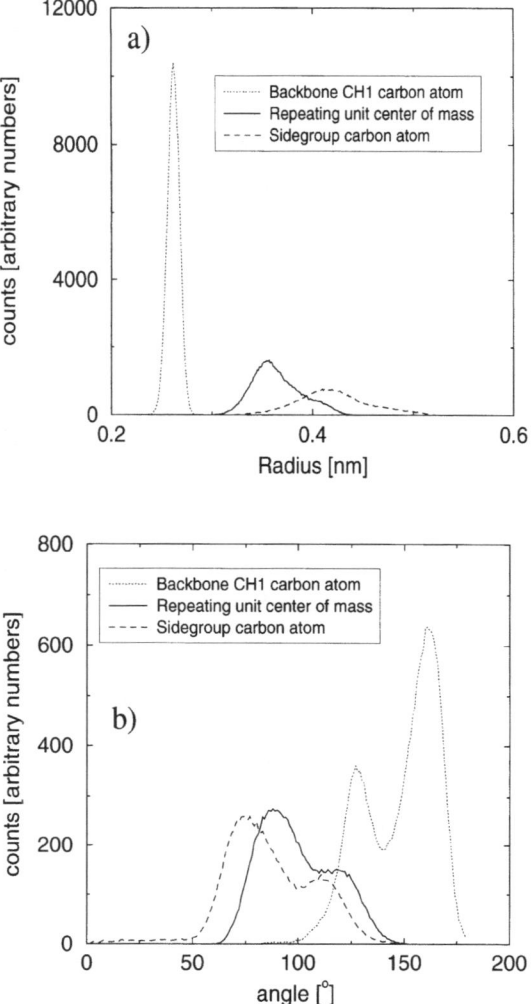

Fig. 8. Coarse-graining of sodium polyacrylate. One super-atom is defined per atomistic monomer. Its placement affects the distribution of structural features such as (**a**) the nearest neighbour ("bond") distance and (**b**) the angle formed by three subsequent super-atoms ("bond angles"): On the backbone carbon carrying the carboxylate group (black), on the carbon of the carboxylate group (blue), on the centre of mass of the chemical repeat unit (red)

the angle distribution gives the potential of mean force

$$V_{\text{angle}}(\theta) = k_{\text{B}} T \ln\left(P_{\text{ato}}(\theta)\right), \tag{14}$$

where k_{B} is the Boltzmann constant, T is the temperature, θ is the super-atom bond angle, and $P_{\text{ato}}(\theta)$ is its distribution obtained from the atomistic trajectory. Strictly speaking, a potential of mean force cannot be used as a potential energy function, because it is a free energy. It can be a good approximation, though, if the entropy can be neglected

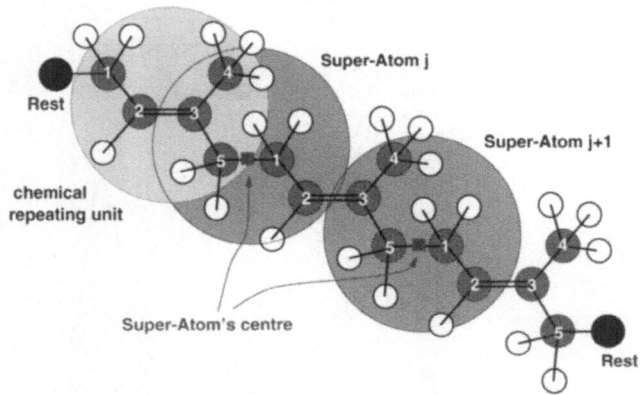

Fig. 9. Coarse-graining of polyisoprene. The super-atoms are centred in the middle of the bond linking two chemical repeat units (rather than at the centre of a chemical repeat unit), in order represent more realistically the excluded volume [13]

and if the distribution used to produce it is not influenced by other interactions within the system. In order to check the applicability of the approximation, we simulate the mesoscopic system with the potential of mean force, then we compute the bond angle distribution and compare it with the distribution obtained with the atomistic data. If they match, the potential is satisfactory. If not, we modify the potential iteratively as

$$V_{\text{angle}}^{n+1}(\theta) = V_{\text{angle}}^{n}(\theta) + k_{\text{B}} T \ln \left(\frac{P^{n}(\theta)}{P_{\text{ato}}(\theta)} \right), \tag{15}$$

where V_{angle}^{n} is the potential of the nth iteration, $P^{n}(\theta)$ the angle distribution produced with V_{angle}^{n}. This method, known as the iterative Boltzmann inversion method [13,16], converges rapidly. For example, only five iterations (Fig. 10) were needed to optimise the angle potential of poly(vinylalcohol)-[CH(OH)-CH$_2$]$_n$-, modelled with one super-atom per repeat unit placed on the backbone carbon carrying the hydroxyl group [13].

In the case of polyacrylate, it has been found necessary to include also a potential for super-atom torsions in order to lower the attractive part of the van der Waals potential (which was responsible of the collapse of long chains) [15]. The torsional potential was built the same way than the angular potential, using the potential of mean force created with the atomistic distribution as a starting guess, and optimising it with the iterative Boltzmann inversion method. This potential improved the reproduction of the first peak of the radial distribution function (Fig. 11a) obtained with a piecewise van der Waals potential.

For van der Waals interactions the same analytical forms as for the atomistic potential can sometimes be used. More complicated forms are often needed in order to reproduce the complicated structure of the super-atom radial distribution functions. In any case, an optimisation with the simplex algorithm is necessary, and this time the penalty function (cf. 12) includes a reference to the radial distribution function produced with the atomistic

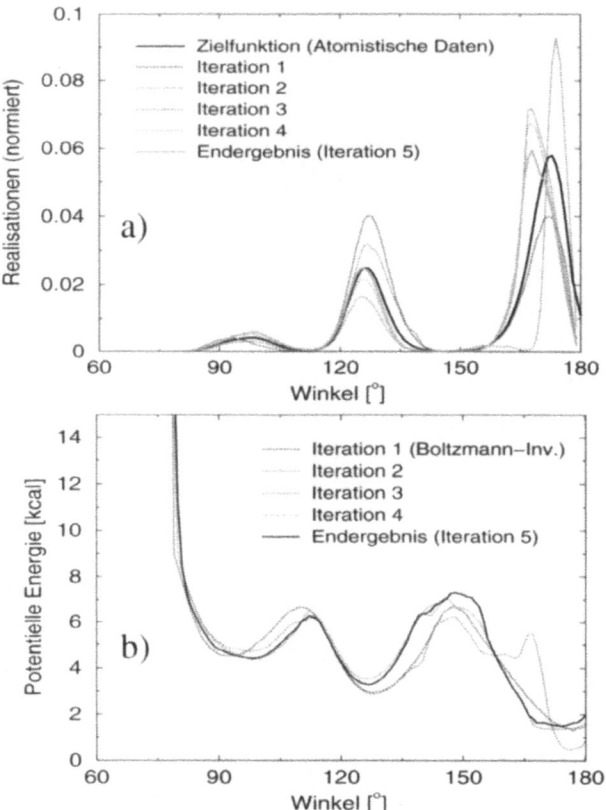

Fig. 10. Bond angle optimisation of poly(vinyl alcohol) using the iterative Boltzmann inversion method (from [13]). (**a**) The bond angle distributions of different iterations converge to the atomistic target distribution. The corresponding potentials are shown in (**b**)

trajectory:

$$f(\mathbf{p}) = \int w(r)[g(r,\mathbf{p}) - g_{\mathrm{ato}}(r)]^2 \, dr + \left[\sum_i w_i \left(1 - \frac{A_i(\mathbf{p})}{A_{i,\mathrm{ato}}} \right)^2 \right]^{1/2}, \qquad (16)$$

where $\mathbf{p}$ represents the mesoscopic force-field parameters, r the distance between two super-atoms, $g(r, \mathbf{p})$ is the mesoscopic radial distribution function produced with the parameter set $\mathbf{p}$, $g_{\mathrm{ato}}(r)$ the target radial distribution function produced with the atomistic trajectory, $w(r)$ is the weight function associated with g, A_i the observable physical properties, $A_{i,\mathrm{ato}}$ their experimental values and w_i the weight associated with A_i.

For the optimisation, the total radial distribution functions are rarely used. Partial radial distributions functions (where the contributions of the first or second bonded neighbours are disregarded, cf. Fig. 12), or intermolecular radial distribution functions

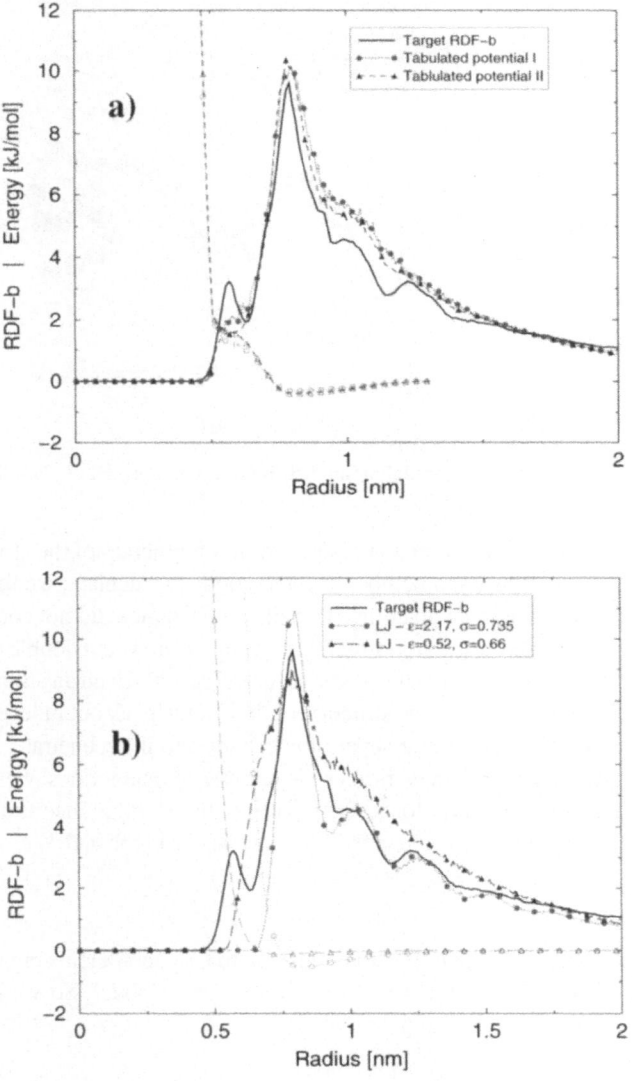

Fig. 11. Optimisation of the non-bonded potential of sodium polyacrylate against the atomistic radial distribution function (with two nearest neighbours excluded, "target RDF-b"). Open symbols denote potentials, filled symbols the corresponding RDF. (**a**) Trial with a piecewise analytical potential, with (I) or without (II) torsional potential (**b**) two attempts with Lennard–Jones potential

are used instead, as peaks from first and second-neighbour pairs are usually accounted for in the bonded, rather than the nonbonded, part of the potential.

We will take again the example of diphenyl carbonate for illustration. We first simulated the 3-site model (described in Sect. 2.2) with a Lennard–Jones potential (9). We used the simplex method to optimise the parameters ε and σ. It converged in a few

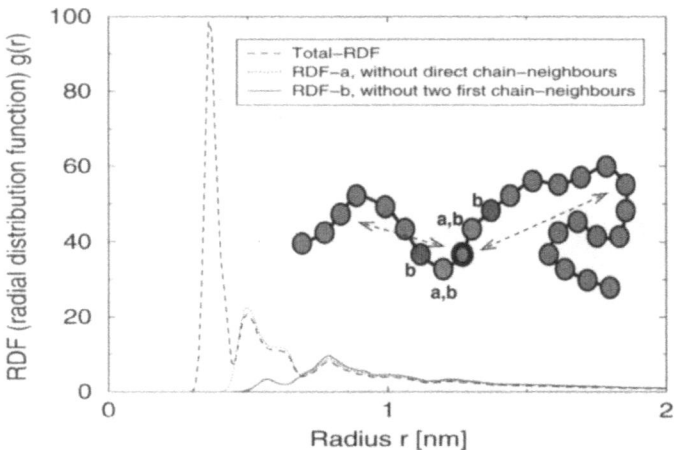

Fig. 12. Definition of the partial radial distribution functions: RDF-a and RDF-b

steps, reproducing the centre of mass radial distribution function of the atomistic model (Fig. 6a). In order to reduce the number of interactions to calculate, we then simulated the 2-site model with the same potential. This time the simplex did not completely converge, and the best radial distribution function obtained showed a double peak but was quantitatively wrong compared to the atomistic one (Fig. 6b). Even an attempt to change the bond length did not improve the structure of the double peak but induced unphysical behaviour at short distance, where super-atoms started to interpenetrate. To get rid of this interpenetration, and to reduce further the number of interactions, we tried a single site model. The usual Lennard–Jones form was unable to reproduce the double peak with a single site model, so we expanded it with terms in $1/r^{10}$ and $1/r^8$

$$V(r) = \frac{C_{12}}{r^{12}} + \frac{C_{10}}{r^{10}} + \frac{C_8}{r^8} + \frac{C_6}{r^6}. \tag{17}$$

But there again the simplex optimisation of the C_6 to C_{12} coefficients did not converge, and the peak structure was still worse than in the 2-site model. So we changed to a complicated piecewise analytical potential

$$V(r) = \begin{cases} \varepsilon_1 \left[\left(\frac{\sigma_1}{r}\right)^8 - \left(\frac{\sigma_1}{r}\right)^6 \right], & r < \sigma_1 \\[2mm] \varepsilon_2 \left[\sin \frac{(\sigma_1 - r)\pi}{(\sigma_2 - \sigma_1)2} \right], & \sigma_1 \leq r < \sigma_2 \\[2mm] \varepsilon_3 \left[\cos \frac{(r - \sigma_2)\pi}{(\sigma_3 - \sigma_2)} - 1 \right] - \varepsilon_2, & \sigma_2 \leq r < \sigma_3 \\[2mm] \varepsilon_4 \left[-\cos \frac{(r - \sigma_3)\pi}{(\sigma_4 - \sigma_3)} + 1 \right] - \varepsilon_2 - 2\varepsilon_3, & \sigma_3 \leq r < \sigma_4 \equiv r_{\text{cutoff}} \end{cases} \tag{18}$$

which has 4 different functional forms in 4 regions, and optimised all the ε and σ, and finally the centre-of mass RDF was reproduced very well (Fig. 6c).

Figure 6c highlights well both the advantages and the dangers of coarse-grained models in connection with automated parameterisation schemes. It shows that it is indeed possible to reproduce the intermolecular structure of a moderately complicated molecular liquid with only one interaction site per molecule. However, in the example of DPC this is achieved by spherically averaging the "reason" for the structure. The real reason for the double peak in the centre-of-mass RDF is the anisometry of the DPC molecule. It leads to two distinct closest packings of DPC in the liquid [17]. In the single-site (not the 3-site) model of DPC the anisometry is being replaced by a spherical potential that leads to two populations of spherically averaged distances. All angular information is lost. The kinkiness of the potential needed for this is very visible in Fig. 6c. When choosing a coarse-graining strategy, one therefore has to keep in mind which structural features have to be kept at the coarse-grained stage in order to incorporate the essential physics of the system. This holds for polymeric systems as well. Avoiding conceptual mistakes is the responsibility of the coarse-grainer. The automatic parameterisation procedures will, for each choice of coarse-graining strategy, deliver the optimum potential, however unphysical.

It seems obvious that when we need such complicated forms as the one of the (18), a tabulated numerical potential would be easier to handle even if it lacks parameters with a physical meaning. The first molecular example of this kind of approach goes back to Soper [18,19], who constructed a numerical atomistic force field for liquid water reproducing OO, OH and HH radial distribution functions from neutron scattering. The transfer of this method to mesoscopic force fields uses the iterative Boltzmann inversion scheme, described above for bond angles, to optimise the numerical van der Waals potential:

$$V_{\text{VdW}}^{n+1}(r) = V_{\text{VdW}}^{n}(r) + k_{\text{B}}T \ln \left(\frac{g^{n}(r)}{g_{\text{ato}}(r)} \right), \qquad (19)$$

where V_{VdW}^{n} is the van der Waals potential at the nth iteration, $g^{n}(r)$ the radial distribution function obtained with this potential, and $g_{\text{ato}}(r)$ the target radial distribution function from the atomistic trajectory. We can start the iterative process for example with the Boltzmann inverted radial distribution from the atomistic trajectory as the first trial potential $V_{\text{VdW}}^{0}(r)$, or with an analytical potential.

In order to illustrate the optimisation of the van der Waals potential, and the necessity of using numerical potentials, we take the example of polyacrylate [15], with the super-atom centred on the centre of mass of the repeat unit. The first attempt is a Lennard–Jones potential (9). A simplex optimisation of the parameters led to $\sigma = 0.735$, and $\varepsilon = 2.17$. The radial distribution function produced with this potential reproduced well the last peaks of the atomistic target RDF but completely missed the first one (Fig. 11b). A second simplex optimisation was started in another region of the parameter space, and the final parameters were $\sigma = 0.660$, and $\varepsilon = 0.52$. The first peak was again not properly reproduced, and the other ones were worse than with the first set of parameters. Then a piecewise analytical potential (18) was tried. This gives a much better representation of the first peak and an acceptable representation of the other ones. A torsional potential was then added, which improved the first peak a little without affecting the other ones (Fig. 11a). Finally, a numerical potential (Fig. 13) was developed

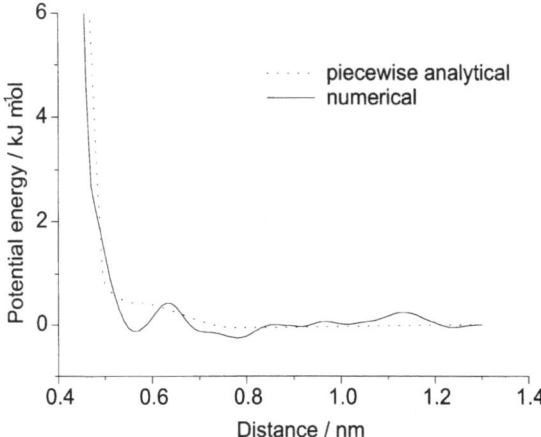

Fig. 13. Piecewise analytical potential (*dotted*) and numerical potential from iterative Boltzmann inversion (*solid*) for the nonbonded interactions (beyond first and second neighbours) of poly(acrylic acid) in solution. The piecewise analytical potential reproduces the corresponding RDF as shown in Fig. 11a. The numerical potential reproduces it to within line thickness

(19), and the corresponding radial distribution function reproduced the atomistic target to within line thickness (not shown in Fig. 11 for clarity).

An alternative method for optimising numerical van der Waals potentials was developed by Lyubartsev and Laaksonen (cf. their article in this volume). The update mechanism has a non-local component. The potential at distance r_i is updated not only in response to the deviation of the actual RDF from the target RDF at the same distance as in the Soper method, but at all distances r_j. One would expect, therefore, a faster convergence for Lyubartsev–Laaksonen, since in the Soper method any deviation takes several iterations to filter through to the point of correction. However, in practice the number of iterations necessary seems to be similar for both methods. Both methods share the same fundamental advantages and disadvantages with respect to analytical potentials. The numerical potentials achieve a perfect match of the corresponding RDFs or other structural distribution functions. However, they lack adjustable parameters with physical interpretation. (Whether parameters of the kind used in (18) have a meaningful interpretation is different question altogether.) In addition, there is the problem of non-uniqueness. In principle, there is a one-to-one correspondence between a pair potential and the RDF generated by it, at least for simple fluids. Our experiments, however, have shown that visibly different potentials, for examples from different iterations of (19) give rise to RDFs which are identical to within line-thickness [13]. That is to say, these RDFs have residual deviations of less than 0.1%, much smaller than the accuracy of the target RDF itself. In particular, this applies to constant-volume simulations, as the volume constraint helps to keep RDF peaks at their proper positions. We have experimented with test cases where the Soper algorithm had to converge to an *a priori* known potential given the corresponding RDF as a target. While all solutions were good enough for practical simulations for both Lennard–Jones and Weeks–Chandler–Andersen fluids, differences to the real potentials were still visible. It turned out that potentials are particularly susceptible to artefacts at long range: Non-zero values of the potential beyond the cutoff introduced by long-range features of the RDF, which are really due to the short-range part of the potential, are very resilient and take many iterations to disappear, even if the target RDF is already well reproduced. We would, therefore, advocate to start the iterative Boltzmann inversion with a potential as short-ranged as possible and only extend its range when long-range features remain [13].

4   Reverse Mapping

As stated above, an important purpose of coarse-grained models is the generation of well-equilibrated atomistic amorphous polymer structures. The relaxation of long chains at the atomistic level is computationally unfeasible, even using parallel computers. The detour via the mesoscopic level speeds up the process, even if we have to develop a new force field for the mesoscopic model and a back-mapping procedure. The advantage of reverse mapping is that different kinds of analyses can be done at different levels, as illustrated in Fig. 14.

The bisphenol-A polycarbonate reverse mapping is the example most thoroughly examined to date. The back-mapping procedure of Tschöp et al. consists of comparing a whole mesoscopic chain with a similar atomistic chain, and iteratively adjusting the torsional angles of the atomistic chain until the contours of both chains coincide. The criterion for the coincidence was the maximum distance between the centre of the atom groups of the atomistic chain, and the corresponding centres of the super-atoms. This method is transferable to any kind of polymer. The polycarbonate example shows

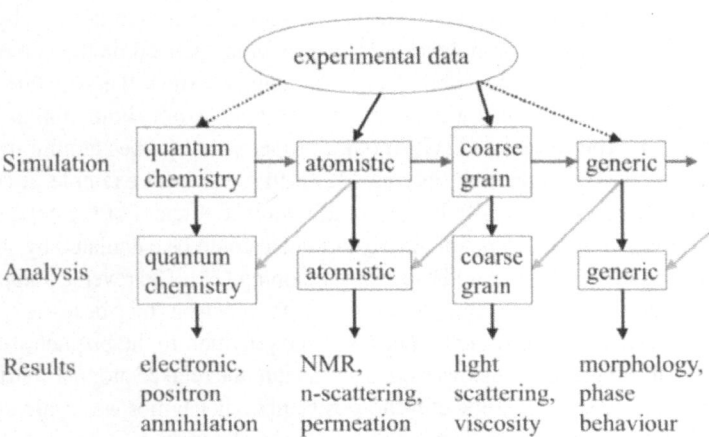

Fig. 14. Multiscale simulations and analysis. Different levels of simulation are associated with different levels of analysis. Coarse-graining procedures build a force field for a coarser level from the results of a finer level. Fine-graining procedures reintroduce lost details. In the case of polymers, coarse-graining is used in order to speed up the relaxation of the chains, and successive fine-graining are used in order to analyse the different properties

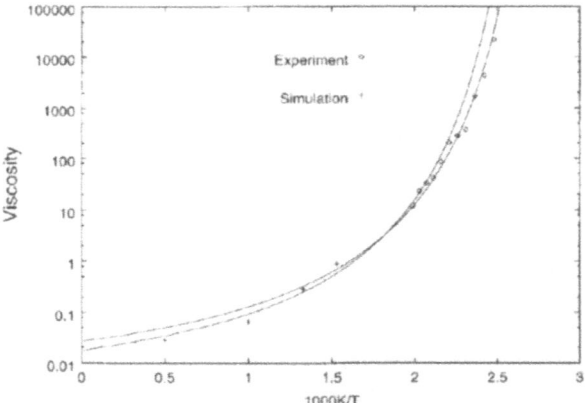

Fig. 15. Comparison of the experimental viscosity (*diamonds*) of bisphenol-A polycarbonate and the results of the simulation (*crosses*). A Vogel–Fulcher fit to the simulation data reproduces experiment and vice versa [12]. (Fig. 20 from [12])

that coarse-graining/fine-graining is a reliable method to generate atomistic structures, at least as probed by neutrons, positronium and small penetrants. This approach has been followed in the multiscale treatment of various polycarbonates [12,20]. Not only "coarse-grained" properties, which depend on the behaviour of entire chains, could be reproduced correctly: For example, the Vogel–Fulcher temperature of short chain bisphenol-A polycarbonate was calculated as 361 K (exper. 387 K) and the temperature-dependence of the viscosity matched that of experiment (Fig. 15). After threading an atomistic model through the coarse-grained chains followed by local relaxation by 200 ps of molecular dynamics, also "atomistic" properties, which depend on the behaviour of individual atoms, could be calculated. As an example, we show the structure factors of differently deuterated bisphenol-A polycarbonate melts from simulation as well as from neutron scattering (Fig. 16) [21]. Another property that probes atomic motion is the diffusion of penetrants through polymers [22]. In the present example, it could be shown that also the technologically important diffusion coefficient of the condensation by-product phenol through bisphenol-A polycarbonate could be calculated by atomistic MD with starting structures prepared by reverse mapping [23]. The reverse mapping can even be carried one step further: After the atoms also the electrons may be re-inserted into the structure to calculate electronic properties of the polymer. In the bisphenol-A poly-carbonate example, electron densities were inserted into the relaxed atomistic structures, and from them the average life time of ortho-positronium in cavities was evaluated [24]. This first-principles value (2.4 ns) agreed well with the experiment (2.1 ns) [25].

Kotelyanskii et al. [26] proposed an other fine-graining procedure in order to intro-duce atoms into their lattice model of polystyrene. They used a coordinate template of the building block and placed it on top of the lattice site (equivalent to a super-atom) they wanted to fine-grain. Then they rotated it taking into account the orientation with respect to its two nearest neighbours along the chain. For polystyrene they used two

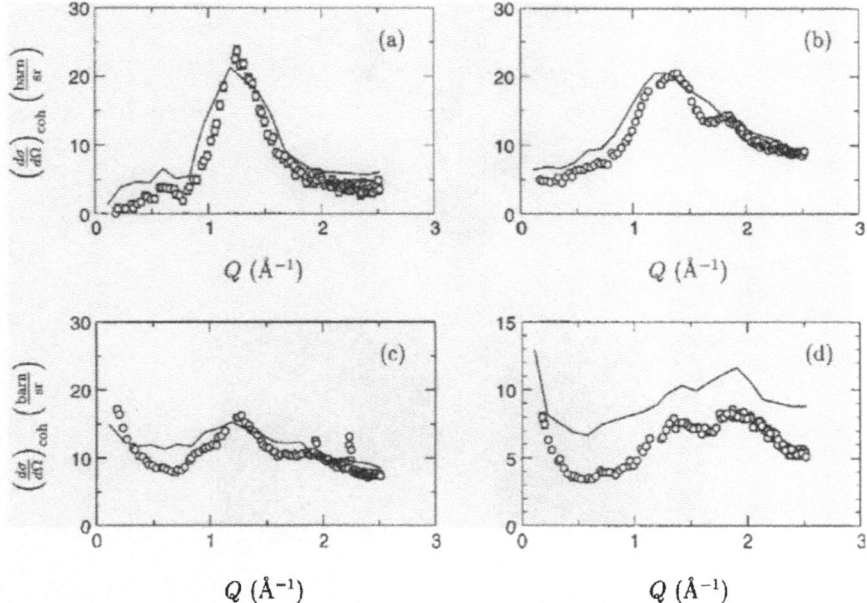

Fig. 16. Structure factor of bisphenol-A polycarbonate from simulation (*solid line*) and neutron scattering (*open symbols*), (**a**) fully protonated, (**b**) methyl group deuterated, (**c**) phenyl rings deuterated, (**d**) fully deuterated [21]

templates, one for the chiral carbon and the phenylene ring, and the other one for the methylene group. The structure obtained needed to be energy minimised, as the bond length and angles were usually far from their equilibrium values. This minimisation was done using the full potential needed for further MD runs, initially with a small cutoff of 4 Å. This method appears to be easily tranferable to other systems as well.

5 Relation to Other Coarse-Graining Procedures

So far we have presented our own coarse-graining procedure, which links atomistic and mesoscopic scales. Other ones have been developed and are described in the literature [27]. Among them are a few models where particle co-ordinates were restricted to lattice positions. They are sometimes easier to implement and analyse than continuum models, even if they are not significantly faster.

5.1 Lattice Models

One of the earliest mappings is the bond-fluctuation model (Fig. 17a) [28]. The underlying lattice is cubic primitive. An excluded-volume super-atom occupies 8 lattice sites. Bond distances and angles are allowed to vary between different discrete possibilities. The possibilities are restricted such that different polymer chains cannot pass through each other. While the non-bonded interactions are usually limited to excluded volume

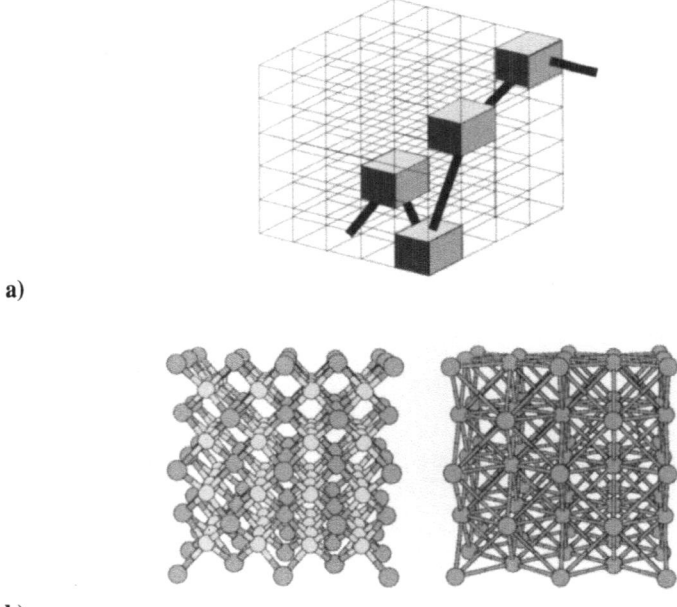

a)

b)

Fig. 17. Lattice models used in coarse-graining. (**a**) Bond-fluctuation model [28]. (**b**) Diamond lattice and the face-centered-cubic ("second nearest neighbour diamond") lattice derived from it by omitting every second lattice position

interactions, the bonded terms can be parameterised to reflect the chemistry of the polymer under study [29]. The bond-fluctuation model has been tried as a coarse-grained model for polyethylene and bisphenol-A polycarbonate as early as in 1991 [30,31] with considerable improvements later. Coarse-graining using the bond-fluctuation model has been carefully reviewed several times [27,32], so we omit it here.

Mattice et al. have developed a mapping from a rotational isomeric state (RIS) description [33] of a polymer onto a lattice model suitable for polymer melts, solutions and surfaces. The underlying lattice is face-centred cubic (fcc). A polyolefin chain with tetrahedral bond angles and only trans ($180°$) and gauche ($\pm 60°$) conformers can be folded onto a diamond lattice (which is a fcc lattice with all tetrahedral interstitials occupied, Fig. 17b). As every second backbone carbon can be omitted without loss of important detail, only the fcc lattice position are needed [34].

The RIS model, which is "atomistic", defines the intramolecular interactions. It involves rotational conformers of consecutive C-C bonds in the polyolefin backbone. The RIS weights are derived from experimental static chain properties and/or from calculated (for example by quantum chemistry) conformational equilibria of oligomers [33]. The first step of the coarse-graining procedure is a mapping from the atomistic RIS model to a slightly coarser one involving rotational conformers around a hypothetical bond between second-nearest neighbour backbone carbons. The resulting RIS transfer matrix is more complicated than the atomistic one (9×9 instead of 3×3), but it can be obtained analytically by enumerating the underlying atomistic rotational states [35].

The intermolecular potential of the original model for polyethylene goes beyond the excluded volume approximation, as it introduces finite repulsive interactions between non-bonded super-atoms occupying neighbouring lattice sites and attractive interactions for super-atoms occupying second-nearest neighbour lattice sites. The strengths of these interactions are balanced to give a second virial coefficient of zero, which is a necessary condition for θ conditions (θ solvent or melt). The non-bonded interactions are fine-tuned until the radius of gyration of the chain *with* non-bonded interactions equals that of the unperturbed chain *without* non-bonded interactions [35,36]. This concept has later been further refined in order to link the interactions of the lattice model to atomistic infor-mation [37]. As a super-atom on the lattice corresponds to 2 methylene units, the known effective Lennard–Jones ε and σ of C_2H_4 are used as a basis (9). Lattice neighbours are grouped into 5 shells according to the distance, i.e. also symmetrically inequiva-lent neighbours belong to the same shell, if they are at similar distances. According to their average distance, all the members of one shell receive the same interaction energy, which is mapped from the atomistic Lennard–Jones potential. The interaction energies for the 5-shell polyethylene model derived from $\varepsilon/k_B = 185$ K and $\sigma = 0.44$ nm are $+12.3654$, $+0.1660$, -0.5443, -0.1219, -0.0316 kJ/mol, i.e. the first shell is strongly and the second mildly repulsive, the outer ones show a decreasing attraction, like in the parent Lennard–Jones potential.

The non-bonded potentials used with the second nearest neighbour diamond lattice are spherically symmetric apart from the discretisation by the lattice. As in the derivation from the Lennard–Jones potential a detour via second virial coefficients is used, they are temperature dependent. For a polyethylene melt at a density of around 1 g/cm^3, only about one quarter of the lattice positions are occupied. Hence, local Monte Carlo moves of crankshaft type have a high probability of finding an empty lattice site. By modi-fication of the intramolecular potential to allow for asymmetric torsions and a longer-ranged non-bonded interaction (up to 9 shells), the model accommodated polypropylene too [38,39]. It was able to discriminate correctly between melts of isotactic and syndio-tactic polypropylene, both in the their static (characteristic ratio) and dynamic behaviour (decorrelation of end-to-end vector). Even the known phase separation of isotactic and syndiotactic, but not of isotactic and atactic, polypropylene could be qualitatively repro-duced [40]. The introduction of multiple-bead crankshaft-type Monte Carlo moves led to a significant speed-up for non-symmetric potentials [41]. Meanwhile, the method has been extended to different topologies (diffusion of linear vs. cyclic polyethylene [42]), geometries (polyethylene films, rather than, bulk [43]) and processes (structure forma-tion upon cooling [44]).

Another coarse-grained model using a fcc lattice has been proposed by Haire et al. [45] for polyethylene. It differs from the approach of Mattice and coworkers in sev-eral aspects. First, the mapping is coarser: There are four, rather than two, CH_2 groups per lattice point (super-atom). This leads to a lower concentration of vacant lattice sites ($10-30\%$, rather than 75%). Second, the non-bonded interactions are limited to a hard core without attractions. An additional repulsion of $E/k_B = 500$ K between neighbour-ing lattice sites is introduced *ad hoc* for technical reasons. Third, the intramolecular interaction is limited to an angle-bending interaction which is adjusted to reproduce the known characteristic ratio of $C_n = \langle R_{ee}^2 \rangle / Nl^2$ of C_{102} chains in the melt, with N the

(a)

(b)

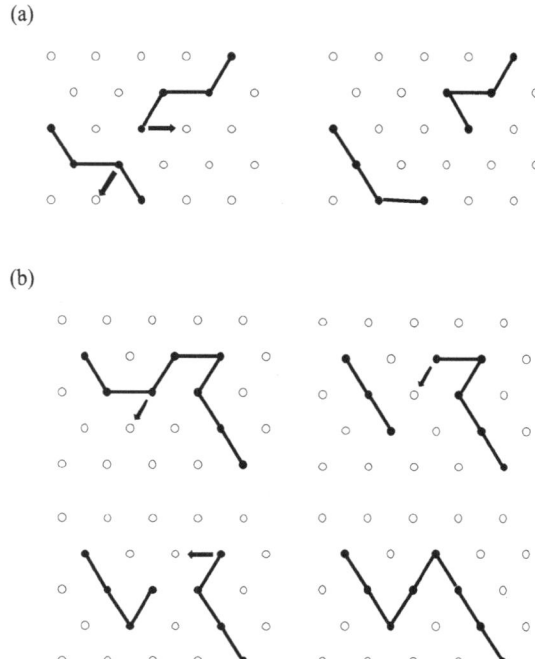

Fig. 18. Monte Carlo moves for the fcc lattice model [45]. (**a**) Single-bead moves, (**b**) Multiple-bead moves (here: 3 beads as an example)

number of monomers and l the bond distance between monomers. This quantity is not really atomistic, as it is derived from the end-to-end distance R_{ee}, which is a property of an entire chain. In the present case, the value happens to originate from atomistic literature data [46], however. Fourth, in contrast to the Mattice lattice method, the Monte Carlo scheme allows multiple-bead moves in addition to the standard end rotations and crankshaft moves (Fig. 18a). An example of a 3-bead move is shown in Fig. 18b: In the trial move the chain is allowed to break initially. The move is only accepted if, after the internal reptation of a segment of a few monomers, the chain can be reconnected again. If no limits are set to the length of the reptating segment, 2-bead moves are the most probable. As expected, a dependence of the dynamics on the maximum size of the moves is found: The mean-square displacement (for example of a central monomer, Fig. 19) is slowest if only single-bead moves are allowed and highest for unlimited moves. It is interesting, however, that only the speed of the dynamics is affected, not its other characteristics. The same scaling regimes are found in all cases and the mean-square displacements of Fig. 19 can be superposed by a shift in $\log t$. This is probably due to the multiple-bead move still being local when compared to the entanglement length. A consequence is that, after empirical determination of the shift, the simulations can be carried out with the multiple-bead moves, which is about one order of magnitude faster.

It is most interesting that the diffusion of the lattice polyethylene shows the same characteristics as those obtained from an atomistic molecular dynamics calculation. A melt of united-atom $C_{101}H_{204}$ chains was simulated by molecular dynamics. Although

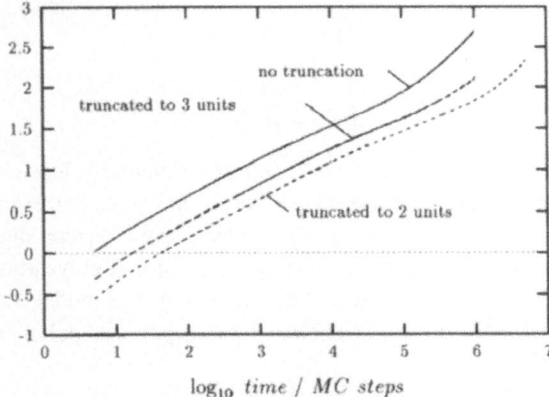

Fig. 19. Mean-square displacement of the central monomers of polyethylene chains in the melt modelled with the fcc model [45]

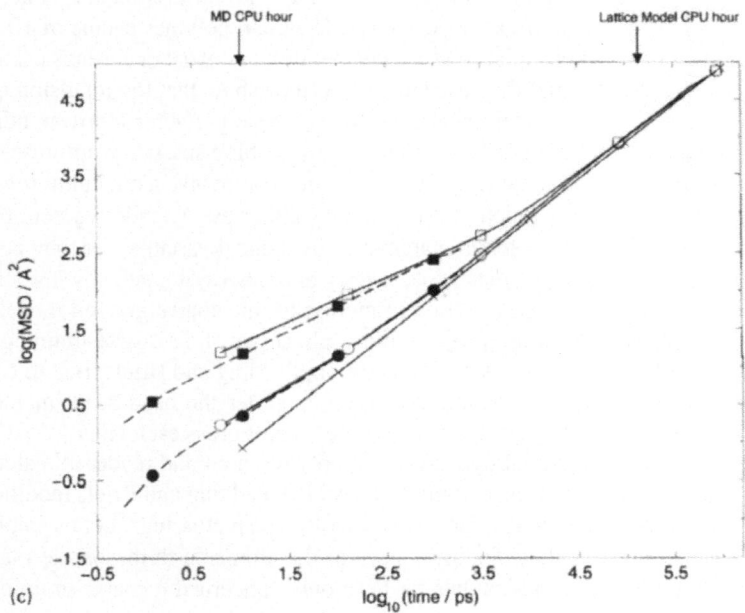

Fig. 20. Mean-square displacement of the central monomers (*squares*) and the centres of mass (*circles*) of polyethylene chains in a melt [45]. Closed symbols denote results from atomistic molecular dynamics simulations, open symbols results from lattice simulations after shifting in log t to bring both sets of curves into coincidence. The straight line is derived from the experimental diffusion coefficient

the simulation did not reach the regime of normal diffusion, the anomalous (subdiffusive) regimes of the mean-square displacement of the atomistic MD simulation and the lattice model could be superposed (Fig. 20).

Then the mean-square displacement of the linear (Einstein) regime of the lattice model was used to obtain a diffusion coefficient.

$$D = \frac{1}{6} \lim_{t \to \infty} \frac{\partial}{\partial t} \left\langle (\mathbf{r}(t) - \mathbf{r}(0))^2 \right\rangle \qquad (20)$$

The diffusion coefficient agreed well with experimental values as well as with other, much longer, atomistic simulations [47]. This is one of the few examples where through the route of multiscaling a calculated *dynamic* property reproduced experiment (the other one being the diffusion coefficient of bisphenol-A polycarbonate in the melt and the shear viscosity derived from it [12], see below). The model was subsequently used to simulate the interdiffusion of chains across a polymer-polymer interface (polymer welding) [48].

5.2   Coarse-Graining Further: From Super-Atoms to Blobs

Akkermans and Briels have developed an automatic procedure similar in spirit to our simplex method to coarse-grain a mesoscopic model into an even coarser one. They have coarse-grained a melt of excluded-volume Gaussian polymer chains of 10 beads into either one or two 'blobs' placed at the centres of mass of 10 or 5 beads each [49]. As an ansatz for the potential they use Gaussians (they show that the repulsion of two Gaussian chains is approximately Gaussian, too.), whose prefactor (corresponding to the interaction energy) and width (corresponding to the blob size) they optimise. Their optimisation strategy is different from ours, as it does not involve a direct minimisation of a structure-based penalty function. Instead, they either use extended-system molecular dynamics making their potential parameters dynamical variables, or they perform Metropolis Monte Carlo in parameter space. They have also systematically investigated ways to derive a Langevin-type equation of motion for the coarse-grained model from microscopic molecular dynamics [50] and the implications of the coarse-graining strategy for the calculated pressure [49,51]. In addition, Padding and Briels tried to combat the known effect that the coarser the model is, the softer the bead-bead interactions become, and the more likely it is that two polymer chains cross each other [52]. Chain-crossing would violate essential dynamics of a polymer melt and render all calculated transport properties invalid. Hence, it must be avoided. Padding and Briels modified the equations of motion by a term that detects and avoids chain crossings. Let us emphasise at this point that most of their findings apply to the atomistic-to-mesoscale mapping as well, although the applications thus far have only concerned a coarse-graining at a higher level in the hierarchy.

6   Conclusions

Polymers, in the melt, solvent-swollen or in solution, are exhibiting various properties on very different length and time scales. In order to build a complete picture of these properties with theoretical models, multiscale simulations are necessary. The path between the different scales is not trivial, and not unique. Efforts have been made to develop procedures, which systematically map upwards or downwards from a given model. The automatisation of such processes is of the utmost practical importance, as a simulator does not want to waste his/her time changing parameters by hand, when such a menial task can be left to a computer.

In the future, the development of coarse-graining procedure will look towards the reproduction of dynamic and transport properties with mesoscale models. We still need to make sure that the mesoscopic models can correctly predict such properties as melt viscosity.

Acknowledgements

We thank Cameron Abrams, Oliver Biermann, David Brown, Roland Faller, Oliver Hahn, Kurt Kremer, Hendrik Meyer, Mathias Pütz, Dirk Reith, Heiko Schmitz and Luigi delle Site for their contributions to the coarse-graining efforts in our group. Part of this work has been supported by the "Kompetenzzentrum Werkstoffmodellierung" of the German Ministry of Education and Research (BMBF), the Marie-Curie Training Site project HPMT-CT-2000-00015, the International Max-Planck Research School in Polymer Materials Science, and the EURODOC program.

References

1. K. Kremer, F. Müller-Plathe: Mater. Res. Bull. **26**, 169 (2001)
2. R. Faller, H. Schmitz, O. Biermann, F. Müller-Plathe: J. Comput. Chem. **20**, 1009 (1999)
3. H. Liu, F. Müller-Plathe, W.F. van Gunsteren: J. Am. Chem. Soc. **117**, 4363 (1995)
4. J.P. Ryckaert, G. Ciccotti, H.J.C. Berendsen: J. Comput. Phys. **23**, 327 (1977); see also Ciccotti and Kalibaeva: Molecular Dynamics of Complex Systems: Non-Hamiltonian, Constrained, Quantum-Classical, Lect. Notes Phys. **640**, 146 (2004)
5. P. Bordat, J. Sacristan, D. Reith, S. Girard, A. Glättli, F. Müller-Plathe: J. Comput. Phys. (Submitted, 2002)
6. S. Girard, F. Müller-Plathe: J. Comput. Phys. (Submitted, 2002)
7. Y. Marcus: *The Properties of Solvents* (Chichester, Wiley 1998)
8. *CRC Handbook of Chemistry and Physics*, 76th edn., ed. by D.T. Lide (CRC press 1995)
9. *Techniques of Chemistry. Organic Solvents*, 4th edn. (New-York, John Wiley & Sons 1999)
10. F. Franks, M.A.J. Quickenden, D.S. Reid, B. Watson: Trans. Faraday Soc. **66**, 582 (1970)
11. D.J. Gisser, B.S. Johnson, M.D. Ediger, E.D. Meerwaal: Macromolecules **26**, 512 (1993)
12. W. Tschöp, K. Kremer, J. Batoulis, T. Bürger, O. Hahn: Acta Polymer. **49**, 61 (1998)
13. D. Reith: Neue Methoden zur Computersimulation von Polymersystemen auf verschiedenen Längenskalen und ihre Anwendungen. PhD thesis, University of Mainz (2001)
14. H. Meyer, O. Biermann, R. Faller, D. Reith, F. Müller-Plathe: J. Chem. Phys. **113**, 6265 (2000)
15. D. Reith, H. Meyer, F. Müller-Plathe: Macromolecules **34**, 2235 (2001)
16. D. Reith, R. Faller, M. Pütz, F. Müller-Plathe: J. Chem. Phys. (in preparation 2002)
17. H. Meyer, O. Hahn, F. Müller-Plathe: J. Chem. Phys. **103**, 10591 (1999)
18. A.K. Soper: Chem. Phys. **202**, 295 (1996)
19. A.K. Soper: J. Phys.: Condens. Matter **9**, 2717 (1997)
20. W. Tschöp, K. Kremer, J. Batoulis, T. Bürger, O. Hahn: Acta Polymer. **49**, 75 (1998)
21. J. Eilhard, A. Zirkel, W. Tschöp, O. Hahn, K. Kremer, O. Schärpf, D. Richter, U. Buchenau: J. Chem. Phys. **110**, 1819 (1999)
22. F. Müller-Plathe: Acta Polymer. **45**, 259 (1994)
23. O. Hahn, D. A. Mooney, F. Müller-Plathe, K. Kremer: J. Chem. Phys. **111**, 6061 (1999)

24. H. Schmitz, F. Müller-Plathe: J. Chem. Phys. **112**, 1040 (2000)
25. H. Schmitz: Computersimulation von Positronium-Annihilation in Polymeren. PhD thesis, University of Mainz (1999)
26. M. Kotelyanskii, N.J. Wagner, M.E. Paulaitis: Macromolecules **29**, 8497 (1996)
27. J. Baschnagel, K. Binder, P. Doruker, A.A. Gusev, O. Hahn, K. Kremer, W.L. Mattice, F. Müller-Plathe, M. Murat, W. Paul, S. Santos, U.W. Suter, V. Tries: Adv. Polym. Sci. **152**, 41 (2000)
28. I. Carmesin, K. Kremer: Macromolecules **21**, 2819 (1988)
29. W. Paul, N. Pistoor: Macromolecules **27**, 1249 (1994)
30. W. Paul, K. Binder, K. Kremer, D.W. Heermann: Macromolecules **24**, 6332 (1991)
31. J. Baschnagel, K. Binder, W. Paul, M. Laso, U.W. Suter, I. Batoulis, W. Jilge, T. Bürger: J. Chem. Phys. **95**, 6014 (1991)
32. K. Binder, W. Paul, S. Santos, U.W. Suter: *Coarse-Graining Techniques*. ed. by M.J. Kotelyanskii, D.N. Theodorou (Marcel Dekker, NY in press)
33. W.L. Mattice, U.W. Suter: *Conformational Theory of Large Molecules. The Rotational Isomeric State Model in Macromolecular Systems* (Wiley, NY 1994)
34. R.F. Rapold, W.L. Mattice: J. Chem. Soc. Faraday Trans. **91**, 2435 (1995)
35. R.F. Rapold, W.L. Mattice: Macromolecules **29**, 2457 (1996)
36. P. Doruker, W.L. Mattice: Macromolecules **30**, 5520 (1997)
37. J. Cho, W.L. Mattice: Macromolecules **30**, 637 (1997)
38. T. Haliloglu, W.L. Mattice: Macromolecules **108**, 6898 (1998)
39. T. Haliloglu, J. Cho, W.L. Mattice: Macromol. Theory Simul. **7**, 613 (1998)
40. T. Haliloglu, W.L. Mattice: J. Chem. Phys. **111**, 4327 (1999)
41. T.C. Clancy, W.L. Mattice: J. Chem. Phys. **112**, 10049 (2000)
42. R. Ozisik, E.D. v. Meerwall, W.L. Mattice: Polymer **43**, 629 (2002)
43. P. Doruker: Polymer **43**, 425 (2002)
44. G. Xu, W.L. Mattice: Comput. Theor. Polym. Sci. **11**, 405 (2001)
45. K.R. Haire, T.J. Carver, A.H. Windle: Comput. Theor. Polym. Sci. **11**, 17 (2001)
46. W. Paul, G.D. Smith, D.Y. Yoon: Macromolecules **30**, 7772 (1997)
47. V.A. Hermandaris, V.G. Mavrantzas, D.N. Theodorou: Macromolecules **31**, 7934 (1998)
48. K.R. Haire, A.H. Windle: Comput. Theor. Polym. Sci. **11**, 227 (2001)
49. R.L.C. Akkermans, W.J. Briels: J. Chem. Phys. **114**, 1020 (2001)
50. R.L.C. Akkermans, W.J. Briels: J. Chem. Phys. **113**, 6409 (2000)
51. W.J. Briels, R.L.C. Akkermans: Mol. Sim. **28**, 145 (2002)
52. J.T. Padding, W.J. Briels: J. Chem. Phys. **115**, 2846 (2001)

Phase-Field Modeling of Dynamical Interface Phenomena in Fluids

Tapio Ala-Nissila[1], Sami Majaniemi[1], and Ken Elder[2]

[1] Laboratory of Physics, Helsinki University of Technology, P. O. Box 1100,
    02015 HUT, Finland
[2] Department of Physics, Oakland University, Rochester, MI 48309-4401, U.S.A.

Abstract. We discuss recent developments and applications of phase-field models to describe interface phenomena far from equilibrium. The basic idea is to coarse-grain over microscopic degrees of freedom which yields an effective description in terms of continuous macro-variables describing different physical phases or states of the system. In principle, and in some cases even in practice, the coefficients in this effective description can be related to the microscopic physical variables. Typically, phase-field models are constructed to describe (non-critical) bulk phases. However, interfaces between phases and non-equilibrium conditions can be easily implemented through appropriate boundary and initial conditions. The main advantages of the phase field models are (i) such models are relatively easy to construct using simple symmetry arguments and conservation laws; (ii) systematic development of the continuum limit which makes mesoscopic and macroscopic time and length scales accessible; (iii) natural emergence of interfaces and non-equilibrium conditions; (iv) relative ease of numerical implementation; (v) possibility for analytic work through projection techniques. In this review, we focus on some recent applications of phase-field models to interface dynamics and kinetic roughening of wetting fronts in random media.

1   Introduction to Coarse-Graining

The theory of equilibrium critical phenomena is well developed, offering a variety of powerful analytic and numerical tools to study phase transitions and related phenomena [1]. The central idea in the modern theory of critical phenomena is that of *coarse-graining*: some of the microscopic degrees of freedom in a given system are integrated out, leaving an effective system (characterized by an effective free energy or Hamiltonian) with fewer degrees of freedom embodied by the coarse-grained *order parameter* (block magnetization in the Ising spin model, density difference for simple liquids etc.). This coarse-graining leads naturally to the concept of universality of phase transitions, where only a few relevant physical quantities such as symmetry and spatial dimension play a decisive role.

   Let us try to demonstrate the central ideas behind coarse-graining by considering as a simple example the Ising model which describes the effect of temperature on ordering of spins of atoms in a crystal lattice. According to the basic principles of statistical mechanics, all static thermodynamic quantities such as the specific heat can be calculated from the partition function Z (or the free energy F)

$$Z \equiv \sum_{\{s_i\}} e^{-\beta H_{\mathrm{m}}} \equiv e^{-\beta F(T)}, \tag{1}$$

T. Ala-Nissila, S. Majaniemi, and K. Elder, Phase-Field Modeling of Dynamical Interface Phenomena in Fluids, Lect. Notes Phys. **640**, 357–388 (2004)
http://www.springerlink.com/

where $\beta \equiv 1/k_B T$. The microscopic Ising Hamiltonian is defined as

$$H_m \equiv \sum_{i,j} w_{i,j}\, s_i s_j. \tag{2}$$

The sum in (1) runs over all spin configurations $\{s_i\}$ with $s_i = \pm 1$. Coupling of the spins is usually assumed to be short-ranged giving rise to a constant interaction energy $w_{i,j} = W$ if two spins s_i and s_j occupy nearest-neighbor sites, and zero otherwise.

Equation (1) represents the microscopic starting point for coarse-graining. Actually it is an effective model, too, since quantum mechanical effects (such as electronic wave functions, coordinates of the nuclei, lattice phonons etc.) have been coarse-grained out. To move up to the next level in the chain of effective theories valid on larger scales, we divide the lattice into N_b cells each of which contains N_c spins: $N_b N_c = N$, where N is the total number of spins in the system. When the number of spins in a cell increases, their average value i.e. the block spin or magnetization $M(\mathbf{x}_n)$ in a cell around $\mathbf{x}_n$,

$$M(\mathbf{x}_n) \equiv \frac{1}{N_c} \sum_{i=1}^{N_c} s_i,$$

approaches a continuous function taking values in the interval $[-1, 1]$. By summing over only those microscopic spin configurations which give rise to a *given* block spin (field) variable $\phi(\mathbf{x}_n)$, we obtain

$$Z' = \sum_{\{s_i\}} \prod_{n=1}^{N_b} \delta[\phi(\mathbf{x}_n) - M(\mathbf{x}_n)]\, e^{-\beta H_m}, \tag{3}$$

where Z' denotes the restricted partition function which is different from Z. When calculating Z we sum over *all* microscopic spin configurations, whereas in the case of Z' we only include those configurations (through the action of the delta function δ) which are consistent with the given value of $\phi(\mathbf{x}_n)$. In other words, at the level of the coarse-grained partition function Z' the microscopic spin degrees of freedom have vanished and we are left with the coarse-grained block spin field ϕ (magnetization) that can take a different value in each cell.

In analogy to (1) we can express the restricted partition function in terms of a coarse-grained free energy F_c which is the fundamental ingredient of the phase-field methodology as will be explained in Sect. 2:

$$Z' \propto e^{-\beta F_c([\phi], T)}. \tag{4}$$

The square brackets indicate that $F_c[\phi]$ is a *functional* of the continuous block spin field $\phi(x)$. For Ising type of models, this functional can be shown to be of the form (when the order parameter is small i.e. close to the critical point)

$$F_c[\phi] \approx \int_\Omega d\mathbf{x} \left(\frac{K}{2}|\nabla\phi(\mathbf{x})|^2 + \frac{r}{2}\phi^2(\mathbf{x}) + \frac{u}{4}\phi^4(\mathbf{x}) \right), \tag{5}$$

where Ω is the volume of the system. The effective free energy $F_c[\phi]$, or more properly a coarse-grained effective Hamiltonian, is known as the Ginzburg–Landau–Wilson (LGW) or ϕ^4 free energy. This form can also be easily derived from the mean-field approximation for the equation of state of the Ising model [1]. The coarse-grained parameters of the model κ, r, and u are temperature dependent and their role is discussed in more detail in the next section. We note that (4) gives the probability weight for each individual magnetization profile $\phi(\mathbf{x})$ in the same way as $\exp(-\beta \sum_{i,j} w_{i,j} s_i s_j)$ gives the weight of each microscopic spin configuration $\{s_i\}$. To recover the true thermodynamic free energy $F(T)$ defined in (1) we have to sum over all fields $\phi(x)$:

$$\int \mathcal{D}[\phi(\mathbf{x})] \, \mathrm{e}^{-\beta F_c[\phi]} = \mathrm{e}^{-\beta F(T)}.$$

The functional integration measure with proper normalization is denoted by $\mathcal{D}[\phi(\mathbf{x})]$.

To summarize, we can perform the summation over all microscopic spin degrees of freedom in one go, arriving directly at the thermodynamic free energy F, or we can perform the configurational sum in two (or more) stages:

$$Z = \sum_{\{s_i\}} \mathrm{e}^{-\beta H_m} = \sum_{\{\phi(\mathbf{x}_m)\}} \left(\sum_{\{s_i\}'} \mathrm{e}^{-\beta H_m} \right) = \int \mathcal{D}[\phi(\mathbf{x})] \, \mathrm{e}^{-\beta F_c[\phi]} = \mathrm{e}^{-\beta F(T)}, \quad (6)$$

where we have defined the summation symbols as

$$\sum_{\{s_i\}'} \equiv \sum_{\{s_i\}} \prod_{n=1}^{N_b} \delta(\phi(\mathbf{x}_n) - M(\mathbf{x}_n)) \quad ; \quad \sum_{\{\phi(\mathbf{x}_n)\}} \equiv \int \mathcal{D}[\phi(\mathbf{x})].$$

For further details on how one performs the summation constrained by the delta function in (3), we refer the reader to [1] and [2] which also discuss more rigorously the conditions for the general coarse-graining operators restricting the summation over microscopic degrees of freedom. Continuing the partial summation chain in (6) by introducing more and more coarse-grained block spins it is possible to derive a full hierarchy of effective theories which can be related to each other through a process called renormalization [3]. Leaving the final summation over the block spins undone in (6) (i.e. not performing the integral $\int \mathcal{D}[\phi(\mathbf{x})]$) gives us the possibility to model systems having a spatially non-uniform field $\phi(\mathbf{x})$. For example, with the aid of the free energy functional $F_c[\phi]$ given in (5) we can model a system which has different magnetization in different spatial regions separated by domain walls. This naturally leads to the appearance of interfaces as explained in the next section.

2 Phase-Field Modeling

There are two methods which can be used for constructing the coarse-grained free energy $F_c[\phi]$ which is the main input of the phase-field formalism. In Sect. 1 we demonstrated by direct but somewhat formal coarse-graining how one can derive F_c. This is done by removing microscopic degrees of freedom (individual spins) by integrating (summing) them out such that only larger scale macroscopic degrees of freedom remain (the block spins). Through this method it is possible to obtain the parameters of the coarse-grained free energy (K, r, u in (5)) in terms of the parameters (T) and coupling constants

($w_{i,j}$) of the more microscopic theory (see (1)). While this procedure provides a direct link between the microscopic and macroscopic descriptions, it can only be carried out for relatively simple systems [4]. In practice it is usually much easier to determine the general structure of the free energy F_c directly from the known phase diagram of the system. The price to be paid is that the model parameters cannot be rigorously related to the microscopic quantities but have to be fitted. Below we shall demonstrate how the construction of the free energy functional F_c can be done based on the information contained in the (known) phase diagram of the system.

2.1 Construction of Free Energy

It should be noted that the form of F_c given in (5) is only valid close to a critical point. More generally, F_c can be written as

$$F_c[\phi] \approx \int d\mathbf{x} \left(\frac{K}{2} |\nabla \phi(\mathbf{x})|^2 + V(\phi(\mathbf{x})) \right), \tag{7}$$

where the potential V is a function (not a functional) of the field ϕ. In principle other types of derivatives of ϕ and even non-local terms could appear but (7) suffices for the following discussion. The gradient term in the free energy accounts for the free energy cost of spatial inhomogeneities and leads to a surface tension between two coexisting phases. The term originates from the interaction energy $w_{i,j}$ between neighboring spins of the Ising Hamiltonian (cf. (2)). Equilibrium information is contained in the form of the potential V.

Each absolute minimum of the potential V corresponds to an equilibrium phase of the system. For example, in a ferromagnetic material (e.g. the Ising Model) V will have one well above some critical temperature T_c and two wells below T_c. Above T_c the single minimum of V will be located at $\phi = 0$ since this is a paramagnetic state with no net magnetization. This is depicted in Fig. 1. For low temperatures ($T < T_c$), in the absence of external magnetic fields, the system starts ordering locally such that there can be ferromagnetic domains with positive or negative magnetization. Consequently, V will acquire two symmetric potential wells as shown in Fig. 1. In this example the wells must be symmetric since the two phases (i.e., one with spins up and the other with spins down) are identical in all aspects, except direction of the magnetization. This is an example of how the symmetry of the system dictates the general form of the (Landau) free energy F_c.

Let us next construct the simplest possible form of the potential V which fulfills the requirements mentioned above. Close to the critical temperature, V can be expanded in a Taylor series as in (5), κ and u are taken to be monotonic, positive functions of the temperature and r should behave as

$$r(T) = r_0(T - T_c),$$

where r_0 is a positive temperature independent coefficient. These choices ensure that the potential

$$V(\phi) \approx \frac{r(T)}{2} \phi^2 + \frac{u(T)}{4} \phi^4$$

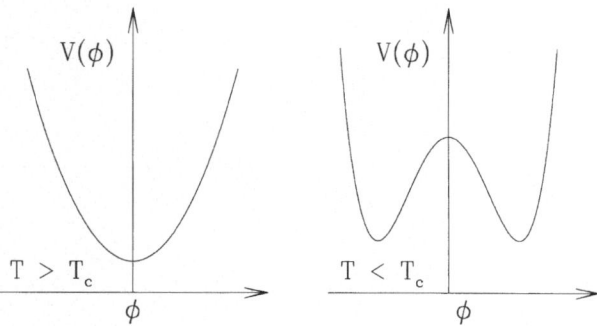

Fig. 1. Bulk free energy above and below the critical point T_c

has one single well at $\phi = 0$ above T_c and two symmetric wells below T_c. In the latter case the two minima can be determined from $\partial V/\partial\phi = 0$ and are at $\phi = \pm\sqrt{|r|/u} \sim |T - T_c|^{1/2}$, which gives the mean-field critical exponent $\beta = 1/2$.

2.2 The Phase-Field

In the previous section we found the minima of the free energy by simply solving for the (spatially constant) values of ϕ which satisfy

$$\partial F_c/\partial\phi = \Omega\partial V/\partial\phi = 0. \tag{8}$$

As we have seen above, simple analysis reveals the existence of an order-disorder transition at the critical point T_c and in fact gives the well-known *mean field results* for the static critical exponents. The field ϕ can be called an *order parameter* field since it characterizes the degree of order in the (spin) system. When $T > T_c$ there is no order (no net magnetization) and the order parameter ϕ takes value zero because the spins are pointing in random directions. When $T < T_c$, the order parameter takes a non-zero value in each of the two ferromagnetic phases. We can also call ϕ a *phase-field* since it assigns a different value or label to each of the phases (zero or a finite value).

Things become considerably more interesting when we consider a spatially inhomogeneus field $\phi(\mathbf{x})$. This allows us to differentiate between several coexisting phases. In those spatial regions where magnetization $\phi(\mathbf{x}) = +1$ (when properly normalized) all the spins are pointing up (phase A), and where $\phi(\mathbf{x}) = -1$ the spins are pointing down (phase B). Since the disordered phase cannot coexist with the ordered ones, the $\mathbf{x}$ dependence of the phase field for $T > T_c$ is quite boring: $\phi(\mathbf{x}) = 0$ everywhere. Taking into account spatial inhomogeneities means that we have to include the gradient term in the expression of the free energy. Also, to get the minimum of the free energy we need to use functional derivative. Thus, (8) will be replaced by

$$\frac{\delta F_c[\phi]}{\delta\phi(\mathbf{x})} = 0,$$

where the *functional derivative* (variation) is defined by

$$\frac{\delta F_c[\phi(\mathbf{x}')]}{\delta\phi(\mathbf{x})} \equiv \lim_{\varepsilon\to 0} \frac{F_c[\phi(\mathbf{x}') + \varepsilon\delta(\mathbf{x}' - \mathbf{x})] - F_c[\phi(\mathbf{x}')]}{\varepsilon}. \tag{9}$$

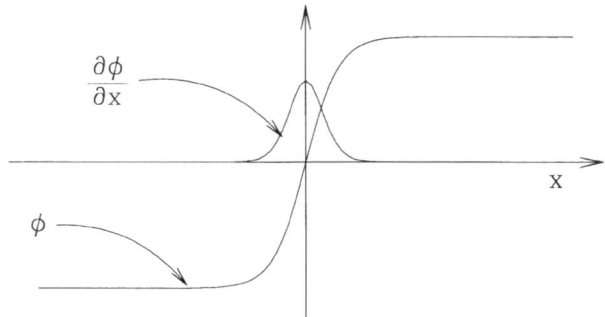

Fig. 2. One-dimensional domain wall and its derivative for the ϕ^4 model

Equation (9) works almost like an ordinary derivative, e.g. $\delta \int dx' \, [\phi(x')]^n / \delta\phi(x) = n[\phi(x)]^{n-1}$, and $\delta\phi(x')/\delta\phi(x) = \delta(x - x')$. The only nontrivial result that is needed here is that

$$\frac{\delta}{\delta\phi} \int dx' \, (\nabla'\phi)^2 = -2\nabla^2\phi,$$

which can be seen by partial integration. In a one-dimensional (1D) case the configuration which minimizes the free energy has to satisfy the nonlinear differential equation

$$\frac{\delta F_c[\phi]}{\delta\phi(x)} = -K\partial_x^2\phi(x) + r\phi(x) + u\phi^3(x) = 0, \tag{10}$$

which can be solved exactly with special boundary conditions. If we require that $\phi(x)$ approaches the *same* constant for $x \to +\infty$ and for $x \to -\infty$ we just recover the mean field results above. However, somewhat more interesting spatial information can be obtained by changing the boundary conditions for (10). For example, choosing that the two degenerate ordered ground states must be different at the boundaries, i.e. $\phi(x \to +\infty) = +(|r|/u)^{1/2}$ and $\phi(x \to -\infty) = -(|r|/u)^{1/2}$ ($T < T_c$) the static solution of (10) gives the nontrivial phase field or order parameter profile

$$\phi(x) = \sqrt{\frac{|r|}{u}} \tanh\left(\frac{x}{2\xi}\right), \tag{11}$$

where $\xi = \sqrt{K/2r}$ is the correlation length and gives a measure of the "width" of the interface (domain wall) which is now forced to appear in the system, see Fig. 2. This is the famous domain wall soliton solution. There are more complicated solutions to (10) which can be systematically classified into a hierarchy of excitations for the LGW type of Hamiltonians [5]. Such equations typically possess a rich solitonic structure corresponding to different excitations from the ground state.

Generalizing the concept of the phase field further we can make it time dependent by assuming that the system is initially in a non-equilibrium state at time t_0, in other words that $\phi(x, t_0)$ does not satisfy (10). Given that the system is then driven towards its equilibrium state by (free) energy dissipation, in the simplest case of the so-called **Model**

A of critical dynamics (see Sect. 5), relaxational dynamics is introduced by postulating

$$\frac{\partial \phi(\mathbf{x}, t)}{\partial t} = -\Gamma \frac{\delta}{\delta \phi(\mathbf{x}, t)} F_{\text{c}}[\phi] + \eta = \Gamma \left(K \nabla^2 \phi - \frac{\partial V}{\partial \phi} \right) + \eta, \tag{12}$$

where $\eta = \eta(\mathbf{x}, t)$ is a Gaussian stochastic noise that describes the random (thermal) effects of the environment and ensures the relaxation to correct stationary equilibrium distribution and has correlations

$$\langle \eta(\mathbf{x}, t) \rangle = 0$$

$$\langle \eta(\mathbf{x}, t) \eta \rangle \eta(\mathbf{x}', t') \rangle = \Gamma k_{\text{B}} T \delta(\mathbf{x} - \mathbf{x}') \delta(t - t'). \tag{13}$$

The notation $\langle \cdot \rangle$ means an average over the realizations of the noise field η. The coefficient Γ determines the rate of relaxation towards the equilibrium state. Equation (12) describes the evolution of a non-conserved order parameter. As will be described in Sect. 5 many other models have been developed to incorporate such features as conserved fields, multiple fields and hydrodynamics.

The specific form of the evolution equation (cf. (12)) is easily seen to guarantee the relaxation to a minimum energy ground state. One can think of it as analogous to the overdamped 1D Newton's equation for a point particle in a potential V:

$$\partial_t x = -\Gamma \frac{\partial}{\partial x} V(x).$$

If the particle starts at some nonequilibrium position x' it will start sliding into the minimum of the potential V. For purely relaxational dynamics, we can write down a Markovian time evolution for the probability distribution function $P[\phi, t]$ (the Smoluchowski equation) instead of the Langevin equation for the field variable ϕ. It should be mentioned that Langevin equations of the type of (12) can be in principle derived from microscopics using (Mori) projection operator techniques [6]. In their phenomenological disguise, they are also called time-dependent Ginzburg–Landau (TDGL) equations. Further examples of such equations will be discussed in the forthcoming sections.

In the examples discussed here we do not focus on the bulk behaviour of the phases but rather concentrate on the phenomena taking place at the *boundaries* between the different phases. With the aid of the so-called projection operator methods [7,8] we can integrate out the bulk degrees of freedom and derive an effective model of lower dimensionality which describes the kinetics of the phase boundaries. In other words we can trade the phase field ϕ which lives on a d dimensional space, to a collective coordinate denoted by h or R in Sect. 3 which lives in a $d - 1$ dimensional space and describes how the position of the boundary or interface changes with time. The projected equations of motion are typically nonlinear Langevin type of equations similar to (12). In the next section we derive the equation of motion for the interface position by starting from the bulk evolution equation for the phase field $\phi(\mathbf{x}, t)$ (see (12)). For simplicity the dynamics of the radius, R, of a single droplet of one phase immersed in another is first calculated to determine whether the droplet will shrink or grow. Following this calculation (see Sect. 3.2) we extend our derivation by taking into account a randomly curved front and thermal fluctuations. Thermal fluctuations are crucial since they often lead to interfaces that roughen in time. In fact, it will be shown that the nonlinear Langevin equation derived for the interface position reduces to well known models of interface roughening, namely the Edwards–Wilkinson and Kardar–Parisi–Zhang (KPZ) equations.

3   Sharp Interface Limit

When a single phase state $(T > T_c)$ is rapidly quenched to a two phase state $(T < T_c)$ small regions of each phase emerge. The subsequent dynamics are controlled by the motion of domain walls or interfaces that separate the regions. In Model A, below T_c, the double well structure of the potential allows the coexistence of two phases ($\phi = \pm 1$ in our example of the Ising ferromagnet) and the gradient energy term assigns an energy (or surface tension) to the domain wall which have a finite thickness as illustrated in Fig. 2. Since the domain walls cost energy the dynamics described by (12) seek to reduce the total amount of domains and in turn increase their average size. Thus in the phase field approach, interfaces arise naturally and no boundary conditions are needed to control their motion.

An alternative approach to studying the dynamics is to consider *sharp interface* models. In such an approach boundary conditions are proposed for the interfaces and then matched with equations proposed for the bulk phases which are treated as single phase equilibrium states. Many of these models are well established and have provided a great deal of insight in many phenomena including coarsening in order-disorder transitions (Allen–Cahn equation, Kardar–Parisi–Zhang equation), spinodal decomposition, and dendritic growth (Gibbs–Thompson condition). The advantage of such an approach is that interface equations are in one less dimension that the original equations. Unfortunately in many cases the sharp interfaces are very difficult to simulate (i.e., numerically stiff).

The sharp interface models can be recovered from the phase field models in the limit that the interface thickness (ξ) is much smaller than all other length scales in the problem. Other length scales include the radius of curvature of a front and diffusion lengths in conserved systems. Equations of motion for the interfaces can be obtained from the full phase field equations using projection operator methods [7,8]. In such an approach the interface equations are obtained by expanding around a flat planar stationary solution ϕ^s (such as the one described by (11)). Then the interface dynamics are obtained by first multiplying (12) by $\partial \phi^s (u - h)/\partial u$, where h is the position of the interface (height measured with respect to some reference plane) and u is a coordinate normal to the interface position, and then integrating over u to obtain an equation for h. This technique "projects" out the interface dynamics since $\partial \phi^s (u - h)/\partial u$ is a function that is sharply peaked around the interface position as shown in Fig. 2. For concreteness this method will be illustrated below for the simple model discussed in the previous section. A more detailed analysis of coupled non-conserved and conserved fields is given in [8].

3.1   Droplets

Before rigorously deriving the equation of motion of some arbitrary interface, it is instructive to first consider a single droplet of one phase immersed in another phase. In the following derivation a perturbation expansion around a *stationary* planar interface is presented. If the two states (phases) have different energy then the lower energy state will invade the higher energy state and the interface cannot be stationary, thus the potential V (see (12)) must be split into symmetric (V_s) and asymmetric (V_a) parts, i.e. $V = V_s + V_a$. To proceed it will be assumed that the order parameter can be written as

$$\phi(x, y, z, t) \approx \phi^{1d}(r - R(t)),$$

where the normal coordinate u to the interface is the radial coordinate: $u \equiv r$. The position of the interface h is the radius $R(t)$ of the droplet. Finally, ϕ^{1d} is the solution of a stationary planar front (e.g., for the ϕ^4 energy described by (5) ϕ^{1d} would be given by (11)). In spherical coordinates (12) gives

$$\frac{\partial \phi}{\partial t} = -\frac{dR}{dt}\frac{\partial \phi^{1d}}{\partial r} = \Gamma\left(-\frac{\partial V_s}{\partial \phi^{1d}} + K\frac{\partial^2 \phi^{1d}}{\partial r^2} + K\frac{(d-1)}{r}\frac{\partial \phi^{1d}}{\partial r} - \frac{\partial V_a}{\partial \phi^{1d}}\right).$$

The first two terms on the right hand side cancel because ϕ^{1d} is defined to be the profile which satisfies the following stationarity condition involving the symmetric part of the potential:

$$-\frac{\partial V_s}{\partial \phi^{1d}} + K\frac{\partial^2 \phi^{1d}}{\partial r^2} = 0.$$

Thus, we are left with

$$-\frac{dR}{dt}\frac{\partial \phi^{1d}}{\partial r} = \Gamma\left(K\frac{(d-1)}{r}\frac{\partial \phi^{1d}}{\partial r} - \frac{\partial V_a}{\partial \phi^{1d}}\right).$$

To obtain an equation for the interface position $R(t)$, the above equation is multiplied by $\partial \phi^{1d}/\partial r$ (since this function is sharply peaked near the interface), and integrated over r, the normal coordinate to the interface:

$$\frac{dR}{dt}\int_0^\infty dr \left(\frac{\partial \phi^{1d}}{\partial r}\right)^2 = \Gamma(1-d)\int dr \left[\frac{K}{r}\left(\frac{\partial \phi^{1d}}{\partial r}\right)^2\right] + \Gamma \Delta V, \qquad (14)$$

where $\Delta V \equiv V(\infty) - V(0)$ is the difference in energy of the phase inside the droplet and outside the droplet. The first term on the right hand side of (14) can be evaluated by expanding r around R, since $\partial \phi^{1d}/\partial r$ is sharply peaked around $r = R$, i.e. $1/r = 1/(R + \delta r) = (1/R)(1 - \delta r/R + \cdots)$. Thus to lowest order (14) becomes

$$\frac{\partial R}{\partial t} = \Gamma K\left(\frac{\Delta V}{\sigma} - \frac{(d-1)}{R}\right),$$

where $\sigma \equiv K\int dr\,(\partial \phi^{1d}/\partial \phi)^2$ is the surface tension. (Note that in some cases K may depend on ϕ and the above formula would have to modified). The above result can be written as

$$\frac{\partial R}{\partial t} = -\frac{A}{R} + B. \qquad (15)$$

For simplicity consider the solution of this equation in the limits $A = 0$ and $B = 0$:

$$A = 0: \quad R = R_0 + Bt;$$
$$B = 0: \quad R^2 = R_0^2 - 2At,$$

where R_0 is a constant of integration. Thus, a curved surface will flatten at a rate $R^2 \sim t$ and an interface between phases of unequal energies will advance at a constant velocity. As will become more apparent in the next section the $1/R$ term gives the Allen–Cahn equation [9] and the second term in (15) is responsible for the non-linear term in the Kardar–Parisi–Zhang equation [10].

3.2 Dynamics of Gently Curved Fronts

The above calculations will now be generalized to an arbitrarily curved interface and conducted in a more rigorous fashion. We will also consider the effects of thermal noise in this section. To begin with, it is convenient to introduce a curvilinear coordinate system (u, s) such that u is perpendicular to the interface and s is tangential (and is the arc length). Details of the curvilinear coordinate system are given in Appendix. The phase field profile ϕ can now be expanded around the flat stationary noiseless planar solution ϕ^{1d}, i.e,

$$\phi(\mathbf{x}, t) = \phi^{1d}(u) + \delta\phi(u, s).$$

It is now convenient to introduce a small parameter ε such that when $\varepsilon \to 0$, $\phi(u, s, t) \to \phi^{1d}(u)$. The parameter ε is therefore proportional to everything that takes ϕ away from a planar stationary interface. This includes the curvature κ and the difference in energy between the two phases due to the asymmetric part of the potential V_a. For simplicity V_a will be written as $V_a = \varepsilon \bar{V}_a$, which highlights the fact that the expansion is around a stationary interface (since the interface always moves when one phase has a lower energy than the other). In principle two small parameters could be introduced, one for each small quantity, but this can always be done in the end. Since $\delta\phi = 0$ when $\varepsilon = 0$ it will be expanded as follows:

$$\delta\phi = \varepsilon\delta\phi^{(1)} + \varepsilon^2\delta\phi^{(2)} + \cdots .$$

Near the interface derivatives with respect to u are finite, but of the order ε with respect to s. In addition, derivatives with respect to t are zero for $\varepsilon = 0$ because the velocity of a straight interface must be zero (in the absence of noise and potential asymmetry). Therefore, it is convenient to introduce the new dimensionless coordinates $(\bar{u}, \bar{s}, \bar{t}) = (u/\xi, s\varepsilon/\xi, t\varepsilon/\tau_{GL})$, where ξ is the correlation length and $\tau_{GL} = 1/(\Gamma r)$ is the Ginzburg–Landau time (parameter r defined in Sect. 2.2). Equation (12) then becomes (see Appendix for expansion of ∇^2 in these coordinates)

$$
\frac{1}{\tau_{GL}}\left[\varepsilon\frac{\partial\phi^{1d}}{\partial\bar{u}}\frac{\partial\bar{u}}{\partial\bar{t}} + \cdots\right] = \Gamma\left[-\left(\frac{\partial V_s}{\partial\phi}\bigg|_{\phi^{1d}} + \varepsilon\frac{\partial^2 V_s}{\partial\phi^2}\bigg|_{\phi^{1d}}\delta\phi^{(1)} + \cdots\right)\right.
$$
$$
- \left(\varepsilon\frac{\partial\bar{V}_a}{\partial\phi}\bigg|_{\phi^{1d}} + \cdots\right) + \frac{K}{\xi^2}\left(\frac{\partial^2\phi^{1d}}{\partial\bar{u}^2} + \varepsilon\bar{\kappa}\frac{\partial\phi^{1d}}{\partial\bar{u}}\right.
$$
$$
\left.\left. + \varepsilon\frac{\partial^2\delta\phi^{(1)}}{\partial\bar{u}^2} + \mathcal{O}(\varepsilon^2) + \cdots\right)\right] + \eta, \tag{16}
$$

where $\bar{\kappa} = \kappa\xi/\varepsilon$. Equating powers of ε gives to order ε^0,

$$
-\frac{\partial V_s}{\partial\phi}\bigg|_{\phi^{1d}} + \frac{K}{\xi^2}\frac{\partial\phi^{1d}}{\partial\bar{u}^2} = 0,
$$

which is satisfied by the definition of ϕ^{1d}. To order ε^1 (16) becomes

$$
\frac{1}{\tau_{GL}}\frac{\partial\phi^{1d}}{\partial\bar{u}}\frac{\partial\bar{u}}{\partial\bar{t}} = \bar{\kappa}\Gamma\frac{K}{\xi^2}\frac{\partial\phi^{1d}}{\partial\bar{u}} - \Gamma\frac{\partial\bar{V}_a}{\partial\phi^{1d}} + \Gamma\left(-\frac{\partial^2 V_s}{\partial\phi^2}\bigg|_{\phi^{1d}} + \frac{K}{\xi^2}\frac{\partial^2}{\partial\bar{u}^2}\right)\delta\phi^{(1)} + \eta.
$$

The next step is to project out the interface equations by multiplying both sides of the previous equation with $\partial \phi^{1d}/\partial u$, just as in (14), and integrating over the normal coordinate u: integrating over $\partial \phi^{1d}/\partial \bar{u}$ gives

$$\frac{1}{\tau_{GL}} \frac{\partial \bar{u}}{\partial t} \int_{-\infty}^{\infty} d\bar{u} \left(\frac{\partial \phi^{1d}}{\partial \bar{u}} \right)^2 = \bar{\kappa} \Gamma \frac{K}{\xi^2} \int_{-\infty}^{\infty} d\bar{u} \left(\frac{\partial \phi^{1d}}{\partial \bar{u}} \right)^2 - \Gamma \Delta \bar{V}_a$$

$$+ \Gamma \int_{-\infty}^{\infty} d\bar{u} \frac{\partial \phi^{1d}}{\partial \bar{u}} \left(-\frac{\partial^2 V_s}{\partial \phi^2} \bigg|_{\phi^{1d}} + \frac{K}{\xi^2} \frac{\partial^2}{\partial \bar{u}^2} \right) \delta \phi^{(1)}$$

$$+ \int_{-\infty}^{\infty} d\bar{u} \frac{\partial \phi^{1d}}{\partial \bar{u}} \eta.$$

The third term can be shown to be identically zero by integrating by parts twice.
Reorganizing the remaining terms and changing back to dimensional units gives

$$\frac{\partial u}{\partial t} = \Gamma K \left(\kappa - \frac{\Delta V_a}{\sigma} \right) + \nu. \tag{17}$$

In the new coordinate system the correlation function of the noise becomes

$$\langle \nu(s,t)\nu(s',t') \rangle = \frac{1}{\sigma^2} \int_{-\infty}^{+\infty} du \int_{-\infty}^{+\infty} du' \frac{\partial \phi^{1d}}{\partial u} \frac{\partial \phi^{1d}}{\partial u'} \langle \eta(u,s,t)\eta(u',s',t') \rangle.$$

To lowest order in ε, $\langle \eta(u,s,t)\eta(u',s',t') \rangle = 2\Gamma k_B T \delta(u-u')\delta(s-s')\delta(t-t')$ (see Appendix) so that

$$\langle \nu(s,t)\nu(s',t') \rangle = \frac{2k_B T \Gamma K}{\sigma} \delta(s-s')\delta(t-t').$$

The quantity $\partial u/\partial t$ in (17) is equal to $-v_n$, where v_n is the normal velocity of the interface, since u decreases if the interface moves in the positive u direction. Thus (17) can be written as

$$v_n = -D\kappa + \lambda + \nu, \tag{18}$$

where $D \equiv \Gamma K$ and $\lambda = \Gamma K \Delta V_a/\sigma$. Equation (18) contains important physics: for $\lambda = \nu = 0$ it becomes the Allen–Cahn equation. For $\lambda, \nu \neq 0$, (18) is the well-known Kardar–Parisi–Zhang (KPZ) equation the relevance of which will be discussed in Sect. 3.3. To show how we can get the KPZ equation from (18) it is convenient to move back to Cartesian coordinates, such that the interface position is described by $h(x,t)$ and assume that both $\partial h/\partial x$ and h are small, see Fig. 3. In this 1D coordinate system the curvature becomes

$$\kappa = -\frac{\partial^2 h/\partial x^2}{(\sqrt{1+(\partial h/\partial x)^2})^3},$$

and the normal velocity becomes

$$v_n = -\frac{\partial u}{\partial t} = \frac{1}{\sqrt{1+(\partial h/\partial x)^2}} \frac{\partial h}{\partial t}.$$

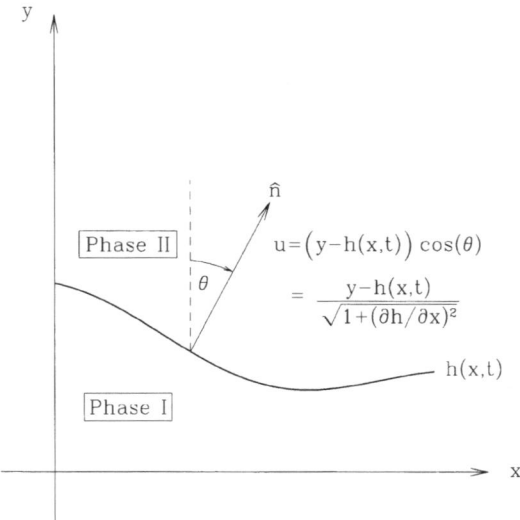

Fig. 3. Connection between Cartesian and curvilinear coordinates

Collecting terms in the small $\partial h/\partial x$ limit gives

$$\frac{\partial h'}{\partial t} = D\frac{\partial^2 h'}{\partial x^2} + \frac{\lambda}{2}\left(\frac{\partial h'}{\partial x}\right)^2 + \nu, \qquad (19)$$

where $h' = h - \lambda t$. Equation (19) is the KPZ equation [10] and has been used to characterize the kinetic roughening of driven interfaces. We note that in the absence of external driving fields the nonlinear gradient term would be missing from (19) without the asymmetry in the full potential V.

3.3 Kinetic Roughening

To show that the KPZ equation (19) leads to a "rough" interface (i.e. one with a diverging width to be defined below) it is sufficient to study the linear equation with $\lambda = 0$. In this instance h' satisfies a simple stochastic diffusion equation:

$$\frac{\partial h(\mathbf{x}, t)}{\partial t} = D\nabla^2 h(\mathbf{x}, t) + \nu,$$

which is known as the Edwards–Wilkinson (EW) equation (with white noise) [11]. We have simplified the notation by replacing h' with h. Fourier transforming in 1D space, this equation admits the solution

$$h(q, t) = e^{-Dq^2 t} h(q, 0) + e^{-Dq^2 t}\int_0^t dt' e^{Dq^2 t'}\nu(q, t'). \qquad (20)$$

The first term is a function of $qt^{1/2}$, giving a surface relaxation exponent of $1/z = 1/2$ and describing the relaxation of a curved interface to a flat interface (in the absence of noise). This implies that if a planar interface is perturbed on a length scale L, it will take a time of $t \sim L^z = L^2$ to relax. Not surprisingly, this exponent is the same as the domain growth exponent $R(t) \sim t^{1/2}$ for Model A.

If the noise term is non-zero then the interface will never be flat, due to the second term in (20). This term describes the roughening of an interface due to thermal fluctuations. This can be seen by calculating the time dependence of the width of the interface $w(t)$ which is defined to be

$$w(t) = \sqrt{\langle \int \mathrm{d}x (h(x,t)^2) \rangle} = \sqrt{\int \mathrm{d}q \langle |h(q,t)|^2 \rangle},$$

where the last equality arises for $\langle h \rangle = 0$. The solution for the width is then (assuming for simplicity $h(x,0) = 0$), $w^2 = (Dt)^{1/2}$ which means that the width diverges in time as $w(t) \sim t^{1/4} = t^\beta$. The exponent $\beta = 1/4$ is called the dynamic roughening (growth) exponent.

From a more general point of view, kinetic roughening of driven fronts is a fundamental problem in non-equilibrium statistical physics [12]. It has important applications in other problems in addition to surface growth, namely in such cases as dynamics of flux lines in random media, slow combustion of paper, and wetting of random medium (imbibition) which will be one of the topics dealt with in this paper. Comprehensive reviews of kinetic roughening can be found in [12,13].

As discussed above the kinetic roughening of a driven interface is usually characterized by the surface width which has been found in general to satisfy the following power law relationships:

$$w(t) \sim t^\beta; \quad w(L) \sim L^\chi,$$

where L is the (linear) size of the system, and the quantities β and χ define the *growth* and *roughness* exponents, respectively. This is called Family–Vicsek (FV) scaling [14]. In many cases of interest, kinetic roughening can be described by a (local) stochastic partial differential equations (PDE's) of the type

$$\frac{\partial h(\mathbf{x}, t)}{\partial t} = f[\nabla h(\mathbf{x}, t)] + v_e + \eta, \tag{21}$$

where f is some (local) function of the local, coarse-grained interface slopes ∇h, and v_e an external driving force (velocity). The stochastic noise term η may be white noise (cf. (13)) or it may depend explicitly on h itself. It is important to note that since there is no assumption about equilibrium here, in general the fluctuation-dissipation relation need not and will not be obeyed. However, for a finite system size L equations of the form given in (21) typically admit steady state solutions.

The EW equation discussed above [11] and the KPZ equation [10] are famous examples of roughening models that satisfy the general form of (21). In arbitrary dimension the EW equation gives $\beta = (2 - d)/4$ and $\chi = (2 - d)/2$ (for $d \leq 2$). For the KPZ equation $\beta = 1/3$ and $\chi = 1/2$ in 1D, while in higher dimensions there are no exact analytic results for the exponents ($\beta \approx 0.240$ in two dimensions (2D)).

One of the ingredients fuelling the immense amount of work done to study the KPZ equation is that it should be *the* generic equation of motion for many experimental cases of kinetic roughening. However, up to date experimental verifications of the KPZ scenario are scarce. Perhaps the clearest case is that of the kinetic roughening of slow combustion fronts in paper [15].

Underlying the simple FV scaling form there is an assumption that all the temporal and spatial properties of the interfaces can be described by the exponents β and χ, respectively. However, this does not hold in all cases even when the interface dynamics can be described by a local PDE of the type of (21). In particular, it is possible that both the temporal and spatial development of roughness is different on different time and length scales which implies the existence of many different exponents corresponding to different moments of the underlying probability distribution (or different correlation functions). A more detailed classification of the various cases of kinetic roughening has been presented recently by Ramasco et al. [16]. However, even more complicated cases of scaling are possible for cases which involve *nonlocality*, i.e. persistent temporal or spatial correlations in the front dynamics [17–20]. There are also important cases of kinetic roughening for which there may exist no explicit equation of motion for the interface variables [21,22].

3.4 Universality

Perhaps the most interesting feature of the sharp interface equation (18), derived for Model A in Sect. 3.2 is that the basic functional form is independent of the details of the free energy functional. The details only enter as constants such as D and λ. In this sense the results are *universal* since they do not depend on any microscopic details that would influence the form of the free energy. Just as in critical phenomena this universality is intimately related to massless fields, or Goldstone modes. To see this it is worth noting that the "interface equation" (19) can be written in the limit $\lambda = 0$ as

$$\frac{\partial h}{\partial t} = -D\frac{\delta \mathcal{H}}{\delta h} + \nu, \tag{22}$$

where $\mathcal{H}$ can be thought of as the interface free energy and is given by

$$\mathcal{H} = \int d\mathbf{x}\frac{1}{2}|\nabla h(\mathbf{x},t)|^2.$$

This free energy (to linear order in h) is identical to the free energy for a field at a critical point. Since there is no term proportional to h^2 the interface can be said to be *massless* or to contain a Goldstone mode. Physically this means that it costs very little energy to perturb the interface on long length scales. Finally, we note that the concept of "universality" for (local) growth equations of the type of (21) can be put on a somewhat more rigorous footing through the application of the so-called dynamical renormalization group method [12]. However, this approach is difficult to apply to cases where memory effects (and quenched noise) are present (not to mention cases where there is no equation of motion of the height variables in the first place), and thus the question of "universality classes" of kinetic roughening remains open.

The previous calculations have shown that the sharp interface equation associated with Model A is independent of the details of the phase field free energy functional. Although not shown here this turns out to be true for a wide variety of phase field models [8]. This makes the construction of phase field models relatively easy and is one of the advantages of this technique, over sharp interface models. Other advantages are discussed in the next section.

4   Advantages of the Phase-Field Approach

Among the central problems in modern physics is the question of multi-scale modeling, in other words how to reach mesoscopic and even macroscopic length and time scales starting from a microscopic approach. The phase-field formalism provides one possible bridge to connect microscopics to macroscopics. It is in principle – and in some cases even in practice – possible to systematically derive the coarse-grained phase-field variable(s) from a microscopic approach by systematic coarse-graining procedures, as we have outlined in Sect. 1. The Ising model example discussed in the first section is a case in point. On the other hand, it's also possible to identify the coefficients of the phenomenological TDGL approach with real physical parameters (such as surface tension, viscosity, and mobility) which in turn may be derived from microscopics. In this review, we don't intend to carry out such systematic programs, however, but simply list here the advantages of the phase-field approach, in particular as concerns interface dynamics.

- *Easy to construct.* Phase field models can be constructed using simple symmetry arguments and conservation laws.
- *Non-equilibrium conditions can be imposed by boundary and initial conditions.* By simply tuning the initial and/or boundary conditions (BC's) appropriately, the system can be prepared in a non-equilibrium state and its dynamics studied.
- *Interfaces emerge naturally.* By choosing appropriate BC's the system can be forced to have interfaces separating different bulk states. This means that usually one does not need any explicit BC's for the interfaces. This is a great advantage if the BC's are complicated and time-dependent.
- *Dynamics is "exact".* Away from bulk criticality, the LGW Hamiltonian and thus the whole phase-field approach should give essentially exact results for the bulk behavior. When looking at interface dynamics it turns out that usually all the relevant fluctuations are included if the sharp interface limit is valid.
- *Relatively easy to solve numerically.* Usually, as the dynamics of the TDGL type of equations is dissipative, it's relatively easy to solve the bulk equations of motion numerically.
- *Systematic analytic work possible.* Through the projection operator techniques, it's possible to derive equations of motion for the $d - 1$ (and lower) dimensional interfaces, starting from the bulk equation of motion for the phase-field in d dimensions. As mentioned above, such equations will usually contain all relevant fluctuations and give "exact" results away from bulk critical points.

Based on this list is should be clear that the phase-field approach has a vast amount of applications in different problems in physics. In particular, when one is concerned with systems having complicated boundary conditions stemming from e.g. randomness, time-dependence, or non-equilibrium conditions, it's usually worthwhile to consider applying phase-fields to the problem. Perhaps the most difficult problem is determining the physically relevant degrees of freedom, or slow variables in the system. In the Ising model example, the coarse-grained magnetization variable was a natural choice since it is also the order parameter. However, the phase-field variables need *not* be order parameters with respect to some symmetry-breaking transition in the system, but they can have a much more general physical interpretation. Some examples of phase field models are given in the next few sections.

5   Alphabet Soup of Models

In this section, we will review some well-known type of TDGL equations which have been extensively studied in the past. A natural application of such equations concerns critical dynamics close to a phase transition [23]. Another classical subject has been the kinetics of growth of droplets or domains for a system starting from a high-temperature disordered phase and quenched to a low-temperature ordered phase. A whole alphabet soup of models has been cooked up to describe various physical situations. They can be classified according to *conservation laws* for the order parameter and its *coupling to other relevant variables* [23,24].

- **Model A**. Perhaps the simplest case is that of Model A where there is a single non-conserved field (order parameter), and no conservation laws. This model is discussed in detail in Sects. 2 and 3 and summarized here. Model A describes order-disorder and magnetic phase transitions e.g. for the Ising model with spin-flip dynamics. The corresponding TDGL equation is given by

$$\frac{\partial \phi}{\partial t} = -\Gamma \frac{\delta F_{\rm c}[\phi]}{\delta \phi} + \eta(\mathbf{x}, t),$$

where $F_{\rm c}[\phi]$ is given by (5), Γ is a kinetic coefficient and η white (Gaussian) noise ("temperature") as given in (13). Here the average $\langle \cdot \rangle$ is again taken over the Gaussian noise distribution. The fact that Γ appears in the noise correlations is due to the second *fluctuation-dissipation* relation, i.e. that in equilibrium the system obeys the Gibbsian distribution [1]. This guarantees relaxation to the correct equilibrium state. Model A gives the universal Allen–Cahn (Kawasaki–Yalabik–Gunton) domain growth law for order-disorder dynamics, due to curvature-driven interface motion which leads to $t^{1/2}$ growth law for the average linear domain size, as shown in Sect. 3.1.
- **Model B**. If a binary alloy (or an Ising model with spin-exchange dynamics) is quenched below the coexistence line, spinodal decomposition occurs. The relevant description involves a locally conserved order parameter field (density or magnetization) and conserved noise (fluctuations) $\eta_{\rm c}$. The equation of motion becomes

$$\frac{\partial \phi}{\partial t} = D\nabla^2 \mu + \eta_{\rm c}(\mathbf{x}, t), \tag{23}$$

where the "chemical potential" $\mu(\mathbf{x}, t) = \delta F_{\rm c}[\phi]/\delta \phi$ and D is a "diffusion coefficient". The noise has zero average, but now

$$\langle \eta(\mathbf{x}, t)\eta(\mathbf{x}', t') \rangle = -2D k_{\rm B}T\nabla^2 \delta(\mathbf{x} - \mathbf{x}')\delta(t - t'). \tag{24}$$

Model B is easy to derive from arguments similar to those that lead to the ordinary diffusion equation. Namely, there is a conserved current $\mathbf{j} = -D\nabla\mu$. Local conservation law $\partial_t \phi = -\nabla \cdot \mathbf{j}$ then gives Model B immediately. Model B gives the universal Lifshitz–Slyozov $t^{1/3}$ growth law for linear domain size in the case of spinodal decomposition.
- **Model C**. If the system has coupling between a conserved field such as concentration $C(\mathbf{x}, t)$ and a non-conserved (order parameter) field $\phi(\mathbf{x}, t)$, such as sub-lattice

concentration, the relevant description is given by Model C. It describes e.g. order-disorder transitions in the presence of impurities. The equation of motion for the non-conserved field is as in Model A, namely

$$\frac{\partial \phi}{\partial t} = -\Gamma \frac{\delta F_1[C, \phi]}{\delta \phi} + \eta(\mathbf{x}, t),$$

with noise correlations for η given by (13). The equation of motion for the conserved field follows the case of Model B:

$$\frac{\partial C}{\partial t} = D\nabla^2 \frac{\delta F_1[C, \phi]}{\delta C} + \eta_c(\mathbf{x}, t),$$

where noise correlations for $\eta_c(\mathbf{x}, t)$ are in turn given in (24). The important fact is now that the free energy (Hamiltonian) has changed to include coupling between the two fields:

$$F_1[C, \phi] = F_\phi[\phi] + F_C[C, \phi], \tag{25}$$

where the new coupling part is given by

$$F_C = \int d\mathbf{x} \left[r_1 \phi(\mathbf{x})^2 C(\mathbf{x}) + \frac{1}{2} r_2 C^2(\mathbf{x}) + r_3 C(\mathbf{x}) \right].$$

Model C is applicable to a variety of physical cases. Modifications of it have been used to study eutectics [25] (systems with coexisting liquid phase, and two solid phases), and dendritic growth [26].

- **Model H.** A rather nontrivial application of the TDGL formalism is the case of dynamics of a pure liquid. The complication arises from the various hydrodynamic modes which must be included in the Hamiltonian. The simplest of such models is the case of Model H which is written down in terms of a phase-field variable $\phi(\mathbf{x})$ and conserved current $\mathbf{j}$. In Model H, sound waves are ignored, and only transversal modes are included into the fluctuations (the convective term $(\mathbf{j} \cdot \nabla)\mathbf{j}$ is also left out):

$$\frac{\partial \phi}{\partial t} = D_1 \nabla^2 \frac{\delta F_2[\mathbf{j}, \phi]}{\delta \phi} - g_0 \nabla \phi \cdot \frac{\delta F_2[\mathbf{j}, \phi]}{\delta \mathbf{j}} + \eta_h(\mathbf{x}, t),$$

and

$$\frac{\partial \mathbf{j}}{\partial t} = \tau \cdot [D_2 \nabla^2 \frac{\delta F_2[\mathbf{j}, \phi]}{\delta \mathbf{j}} + g_0 (\nabla \phi) \frac{\delta F_2[\mathbf{j}, \phi]}{\delta \phi}] + \zeta(\mathbf{x}, t), \tag{26}$$

where the free energy has changed to

$$F_2[\mathbf{j}, \phi(\mathbf{x})] = F_c[\phi(\mathbf{x})] + \int d\mathbf{x} \frac{1}{2} |\mathbf{j}(\mathbf{x})|^2. \tag{27}$$

The coefficient D_2 is essentially the kinematic viscosity, and the vector quantity τ is a projection operator which selects the transverse part of the vector on the right hand side of (26). The conserved noise terms satisfy

$$\langle \eta_h(\mathbf{x}, t) \eta_h(\mathbf{x}', t') \rangle = -2D_1 k_B T \nabla^2 \delta(\mathbf{x} - \mathbf{x}') \delta(t - t'),$$

and

$$\langle \zeta_\alpha(\mathbf{x},t)\zeta_\beta(\mathbf{x}',t')\rangle = -2D_2 k_\mathrm{B} T \nabla^2 \delta(\mathbf{x}-\mathbf{x}')\delta(t-t')\delta_{\alpha\beta},$$

where α, β denote the spatial indices of the noise vector ζ. At the end of this review, we will come back and discuss the range of validity of Model H in some more detail.

6   Application: Kinetic Roughening in Imbibition

To illustrate the usefulness of using a phase field approach, the phenomena of imbibition will be examined in detail in what follows. In particular, we will describe how a phase field model of imbibition can be constructed and analyzed using projection operator methods and direct numerical simulation. Emphasis will be placed on the physical interpretation of the results and comparison with experiments.

6.1   Background

The wetting of a disordered, porous medium is a challenging theoretical problem with important applications in such fields as oil recovery and paper industry. There are many experimental and theoretical attempts to describe such processes. What makes this problem theoretically difficult to handle is the disorder in the medium which makes it practically impossible to e.g. solve the Navier–Stokes equations of hydrodynamics with physically correct boundary conditions. Thus, this problem is a good candidate for phase-field modelling where the BC's can be handled much more easily.

A particularly interesting case of wetting is that of spontaneous wetting of paper (or a 2D disordered matrix) by a liquid through *capillary forces*. We call this case imbibition here. It possesses all the relevant physical features of the more general wetting problems, but is more amenable to experiments since the 1D interfaces of wetting fronts are easy to visualize and analyze. Paper is an interesting example of a disordered (but not always uniformly random [27]) medium, formed by a thin network of overlapping cellulose fibers. Careful experiments on the kinetic roughening of slow combustion fronts have in fact shown that on scales large compared to the fiber size (a few mm) paper sheets can be considered to be uniformly random 2D networks leading to Gaussian correlations in the effective noise [15].

Many imbibition experiments have been in fact carried out. The simplest one is to analyze the roughness of a vertically rising wetting front in paper which is eventually pinned by gravity and evaporation. Using ordinary paper and a water-ink solution, it was concluded by Buldyrev et al. [28] that the roughness of the pinned fronts can be described by the so-called Directed Percolation Depinning model, which gives the result $\chi = 0.633$ for the roughness exponent. This prediction agrees well with the experimental data presented in [28]. However, a more careful inspection of the spatial scaling regime presented in [28] reveals that scaling is observed at length scales smaller than the typical fiber size which is less than 3 mm. This raises a serious question about the meaning of the measured value of χ. In another similar experiment [29], ink front in paper was allowed to stop by evaporation, and qualitative connection to the Directed Percolation Depinning model with a gradient term in the disorder strength was made.

On the other hand, it has been known for decades that even without gravity or evaporation, the velocity of a liquid front rising in a capillary tube slows down according to the Washburn law [30]

$$v \propto H^{-1}, \tag{28}$$

where H is the (average) height of the liquid-gas or liquid-solid interface. This fact was used in another experiment where the paper sheet was wet horizontally and pulled down such that the average position of the wetting front remained stationary [31]. It was found that $v \propto H^{-1.6}$ (as opposed to (28)). From the temporal development of the second-order height-height autocorrelation function the authors also concluded that $\beta \approx 0.56$.

6.2 Phase-Field Model of Imbibition

The main problem with the previous approaches to imbibition dynamics described above is that the front dynamics has been interpreted within the standard local PDE approach (cf. (21)). However, such models neglect one of the most important physical facts in imbibition, namely the *local conservation of liquid* (forgetting about evaporation for the time being). In order for the liquid front to proceed further, additional liquid must be transported from the reservoir which leads to the slowing down as embodied by the Washburn law for a single capillary tube. Thus, any serious attempt to model imbibition must start from the basic conservation law.

From the point of view of phase-field modelling then, there are two main ingredients that must be included. The first one is a coarse-grained phase-field variable $\phi(\mathbf{x}, t)$ which differentiates between the "wet" and "dry" phases of paper. The second is the actual driving force for spontaneous wetting, namely that arising from the spatially inhomogenous distribution of capillary forces within the paper sheet. The simplest way to include these two effects is to construct a free energy functional for a two-state model, coupled to a spatially random, quenched field $\alpha(\mathbf{x})$ [19]:

$$F_i[\phi(\mathbf{x}, t)] = \int d\mathbf{x}[\frac{1}{2}(\nabla\phi)^2 - \frac{1}{2}\phi^2 + \frac{1}{4}\phi^4 - \alpha(\mathbf{x})\phi].$$

Here, the first part of the free energy is just the (dimensionless) ϕ^4 model with a double-well structure, and thus the two values of the field $\phi = \pm 1$ correspond to the wet and dry phases, respectively. The field α can be taken to be a random, Gaussian distributed local "capillary pressure" that determines the stochastic contribution to the local chemical potential through the standard formula

$$\mu = -\delta F_i / \delta\phi.$$

This field will have a non-zero average value $\langle\alpha\rangle = \bar{\alpha}$ which tilts the double-well potential, with Gaussian correlations $\langle\alpha(\mathbf{x})\alpha(\mathbf{x}')\rangle - \bar{\alpha}^2 \propto \delta(\mathbf{x} - \mathbf{x}')$. It drives the wet phase to become the eventual equilibrium state of the system. The actual equation of motion that corresponds to the wetting process follows then naturally from the local continuity equation

$$\partial_t\phi + \nabla \cdot \mathbf{j} = 0,$$

where the flux is given by

$$\mathbf{j}(\mathbf{x}, t) = -\nabla\mu.$$

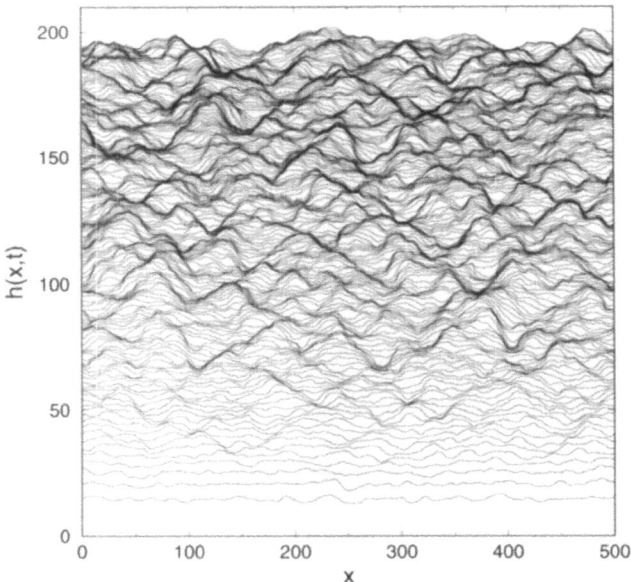

Fig. 4. Series of equal-time imbibition fronts obtained by integrating (29) numerically for the 2D geometry

This gives

$$\partial_t \phi = -\nabla^2 \left[\nabla^2 \phi + \phi - \phi^3 + \alpha(\mathbf{x}) \right]. \tag{29}$$

Depending on the boundary and initial conditions, different wetting scenarios can be studied with the aid of (29). The simplest case which we will consider first is that of the wetting of a 2D paper sheet which is dipped to an infinite reservoir of liquid located at $y \leq 0$. Thus, the wetting front starts at time $t_0 = 0$ at the bottom ($y = 0$) and advances along the "paper" on the xy plane. The position of the interface $h(x,t)$ is determined by the value where the phase-field ϕ changes sign as we move from the wet phase ($\phi = +1$) to the dry phase ($\phi = -1$). Thus, we can determine the interface position from the condition: $\phi(\mathbf{x},t) = 0$. The relevant boundary conditions are those corresponding to the reservoir, with $\mu(y = 0) = \alpha_0$ (where we can set $\alpha_0 = 0$), and the "dry" condition at the top which means that $\phi(y \rightarrow \infty) = -1$.

In Fig. 4 we show a series of interface configurations taken at equal time intervals, from numerically integrating (29) for the imbibition dynamics. There are two distinct features in the interfaces. First, kinetic roughening is clearly visible as the fronts become rough when they proceed. Second, the interface velocity is clearly slowing down as a function of time (height) which is what we expect to happen due to the local conservation law. Before presenting details of the numerical work, however, we will discuss an analytic approach to (29).

Analytic Results. A major advantage of the phase-field approach to interface dynamics is that it's possible to derive, starting from the equation of motion for the d dimensional bulk field $\phi(\mathbf{x}, t)$ an effective equation of motion for the $d-1$ dimensional *interface* that separates the two phases. This is essentially done by assuming a sharp, well-defined interface and *projecting* the bulk degrees of freedom at the interface by integrating out the spatial bulk variable perpendicular to the interface. The technical details of such calculations were given for Model A in Sect. 3. In this instance the calculations are complicated by the conservation law which implies that the bulk is coupled to the interface motion. For application of the projection operator method to conserved systems in general we refer the reader to [7,8] and to imbibition specifically to [19,20,32]. Unlike Model A, the equation of motion of interface for imbibition is not local due to the conservation law. Assuming that the interfaces are single-valued functions, and linearizing the equation of motion in h leads to the following equation for the spatial Fourier components $h_k(t)$ of $h(x, t)$ [19,33]:

$$\dot{h}_k \left(1 - e^{-2|k|H}\right) + |k|\dot{H}\, h_k \left(1 + e^{-2|k|H}\right) = |k|\left(\eta_k - \sigma k^2 h_k\right),\qquad(30)$$

where $H = h_0$ is the average interface position. It is interesting to note that while this linearized equation of motion is local in the Fourier space, it's non-local in real space due to the appearance of the $|k|$ terms. Even in this simplified form, this equation is difficult to analyze in detail. However, if we first examine the average behavior of the meniscus which does not contain any fluctuating Fourier components, we immediately get the result that

$$H(t) = (\bar{\alpha}t)^{1/2},$$

which is exactly the Washburn equation. This behavior should be contrasted with the local PDE's of the KPZ type, which for driven interfaces lead to a *linear* dependence of the average height on time (i.e. constant average interface velocity). This demonstrates the importance of the local conservation law for imbibition.

There is another interesting result which can be obtained from (30). For small k, to lowest order the long wavelength Fourier modes are dampened by the term $|k|\dot{H}h_k$. This can be interpreted to arise from the kinetics of the moving front itself, due to the conservation law according to which liquid must be transported from the reservoir for the interface to advance. On the other hand, for small wavelengths the fluctuations are dampened by the term $\sigma|k|^3 h_k$, which is due to the effect of surface tension which wants to smooth out roughness. When these terms are of equal magnitude, there is a *crossover scale*

$$\xi_\times \simeq (\sigma H/\bar{\alpha})^{1/2} \sim t^{1/4},$$

due to the fact that $H \propto t^{1/2}$. We will demonstrate below that $\xi_\times$ will in fact control and determine the development of roughness for imbibition fronts.

Numerical Results. We now present a summary of the numerical results obtained by integrating (29) for the 2D imbibition case. First, we have verified numerically that indeed $H(t) \propto t^{1/2}$. Second, we find that the surface width grows as

$$w(t) \sim t^\beta, \quad \beta = 0.32 \pm 0.02.$$

This value is seemingly close to the 1D KPZ result $\beta = 1/3$, but in fact it has a completely different origin. Namely, the role of the crossover length $\xi_\times$ is revealed in the steady-state structure factor $S(k, H) = \langle |\bar{h}_k(t)|^2 \rangle$ which shows that correlations extend only up to $\xi_\times$. This means that it controls the development of roughness, and thus the following scaling relation should be valid:

$$w(t) \sim \xi_\times^\chi \sim t^{\chi/4} \sim t^\beta, \qquad (31)$$

where χ is the (usual) global roughness exponent. From the structure factor, we obtain $\chi = 1.25 \pm 0.1$ which gives $\beta \approx 0.31$, in good agreement with the numerical estimate from the width. Furthermore, we have examined the scaling of the spatial correlation function

$$G_2(r, H) \equiv \langle \overline{[h(x+r,t) - h(x,t)]^2} \rangle^{1/2} \propto v^{-\chi/2} g(rv^{1/2}),$$

where the scaling function $g(u) = const.$ for $u \gg 1$, and $g(u) \sim u^{\chi_{loc}}$ for $u \ll 1$. Overbar denotes spatial averaging. The quantity χ_{loc} defines the *local* roughness exponent, and we find numerically that $\chi_{loc} \approx 1$. Thus, the imbibition fronts are *superrough* and show anomalous scaling.

To further analyze the temporal correlations of the fronts, we have considered the general q^{th} order time-dependent correlation functions

$$C_q(t, H) = \langle \overline{|h(x, t+s) - h(x, s)|^q} \rangle^{1/q} \sim t^{\beta_q}, \qquad (32)$$

where the different moments $q = 1, 2, 3, \ldots$ can now have different growth exponents β_q. A series of the functions are shown in Fig. 5. Numerically, we find that the values of β_q diminish with increasing q as

$$\beta_2 = 0.85 \pm 0.04;$$
$$\beta_4 = 0.76 \pm 0.04;$$
$$\beta_6 = 0.69 \pm 0.04.$$

This means that the fronts display *multiscaling* as well. It is also important to note that unlike for the local PDE's, $\beta_2 \neq \beta$ because β is determined through $\xi_\times$ in (31).

Experimentally it was measured for paper wetting that $\beta = 0.56$ using the correlation function $C_2(t)$ [31]. However, as discussed in [34], there are various complications with experiments involving paper sheets and a water-based wetting liquid which may in part explain the different results. There is also an attempt to explain the experiments by a pipe network model of fluid flow [35]; however, the results from the model are critically discussed in [32].

In the most recent experiments [36], imbibition was studied between two rough glass plates instead of a paper sheet. Washburn's law was clearly verified, and a crossover length was found which separated two regimes with different spatial roughness, with $\chi \approx 0.8$ and 0.6 for short and long scales, respectively. In [36] this was interpreted as a crossover from nonlocal effects at short distances to the local KPZ picture (with quenched noise) at larger scales. However, the value of $\chi = 0.8$ is not inconsistent with the phase-field prediction that $\chi_{loc} \approx 1$. This quantity is best measured from the structure function. Unfortunately, in [36] neither $S(k, t)$ nor the temporal development of the roughness were analyzed. We also note that one related problem of two-phase fluid flow in Hele–Shaw cells has been recently studied both experimentally and using a hydrodynamic model in [37–39].

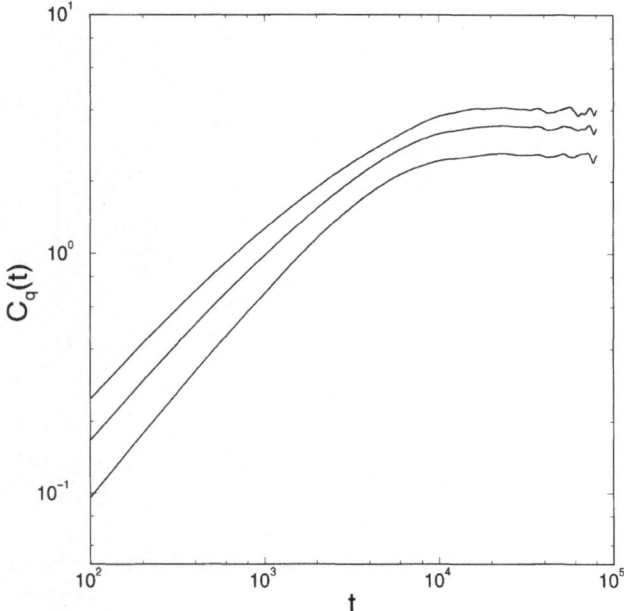

Fig. 5. Log-log plots of the time-dependent correlation function (32). Curves from bottom to top are for $q = 2, 4$, and 6, respectively, and show different slopes indicating temporal multiscaling

6.3 Evaporation and Gravity

Under realistic conditions of paper wetting, evaporation and gravity may play an important role depending on the experimental setup. This is the case e.g. in [28,29], where the rising ink front in paper was eventually pinned by these effects. Thus, the role of these two effects poses an interesting question regarding the roughness properties of the imbibition fronts. It is relatively simple to include both evaporation and gravity in the phase-field model of imbibition. Evaporation introduces a term in the equation of motion that violates the local conservation law (a sink term), while gravity introduces a convective term in the y direction (assuming the previous 2D geometry). The new equation of motion can be written as:

$$\frac{\partial \phi(\mathbf{x}, t)}{\partial t} - g \frac{\partial \phi(\mathbf{x}, t)}{\partial y} = \nabla^2 \mu - \frac{1}{2}\varepsilon[1 + \phi(\mathbf{x}, t)], \tag{33}$$

where μ is the same chemical potential as before. In this form, the strength of evaporation and gravity are described by the two dimensionless coupling constants ε and g, respectively.

The equation of motion (33) can be again analyzed assuming a sharp interface, and using projection operator techniques [7]. On the mean-field level, the equation of motion for the average interface height becomes

$$\frac{dH(t)}{dt} = \frac{\bar{\alpha}}{2H(t)} - g - \frac{1}{4}\varepsilon H(t). \tag{34}$$

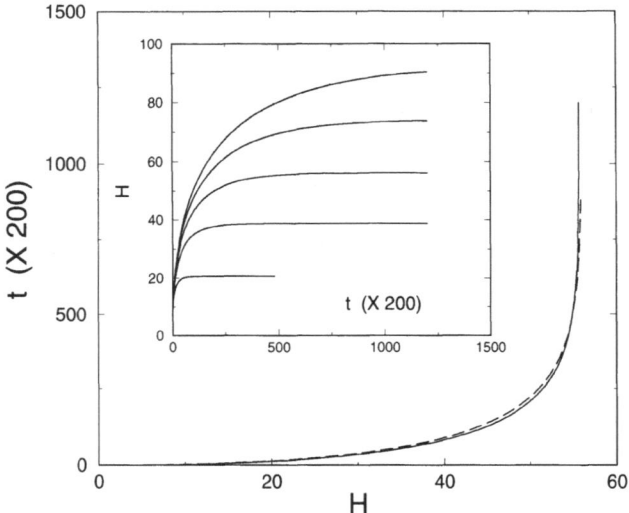

Fig. 6. Time development of $H(t)$ as given by the mean-field result of (35) for $\varepsilon = 0$ (*solid line*), compared to a numerical solution of (33) (*dashed line*). Inset shows $H(t)$ for values of g corresponding to different pinning heights (g increases from top to bottom)

This equation immediately gives the result that in the presence of gravity and evaporation there is a finite *pinning height* $H_p(\varepsilon, g)$ where the front stops propagating. Physically, this is of course what is expected. In the two limiting cases of either no evaporation or zero gravity, the pinning heights are given by

$$H_p(\varepsilon = 0, g) = \frac{\bar{\alpha}}{2g}; \quad H_p(\varepsilon, g = 0) = \sqrt{\frac{2\bar{\alpha}}{\varepsilon}},$$

respectively. What this means in practice is that for early times neither evaporation nor gravity play a role, i.e. $t \ll t_c \equiv \min(\bar{\alpha}g^{-2}, \varepsilon^{-1})$ the interface still follows the original Washburn law with $H(t) \propto t^{1/2}$. This is followed by a crossover regime around $t \approx t_c$ towards an eventual exponential approach in time towards the pinning limit for $t \gg t_c$. For zero gravity, this approach is given by

$$\left(\frac{H(t)}{H_p}\right)^2 = 1 - e^{-\varepsilon t/2},$$

and for no evaporation by the transcendental equation

$$\frac{H(t)}{H_p} + \ln\left(\frac{H_p - H(t)}{H_p}\right) = -\frac{2}{\bar{\alpha}}g^2 t, \tag{35}$$

for $H(t_0 = 0) = 0$. In Fig. 6 we show the time dependence of $H(t)$ as given by (35) and a full numerical solution of (33) with $\varepsilon = 0$, demonstrating the excellent agreement between the mean-field solution and the full equation of motion, respectively.

As shown by the analysis above, for early times the system still follows Washburn's law which means that the crossover length $\xi_\times(t)$ controls the extent of the fluctuations.

However, when the system crosses over to the evaporation or gravity dominated time regime, there is a new length scale that comes into play. For the gravity dominated case ($\varepsilon = 0$), this length scale is given by

$$\xi_g(t) = \frac{1}{2}\left(\frac{\sigma H(t)}{\bar{\alpha}}\right)^{1/2},$$

where $H(t)$ is the solution of (35). This means that $\xi_g(t)$ does not follow simple power-law behavior in time any more. At pinning $H = H_p$ this length scale is given by

$$\xi_g(H_p) = \left(\frac{\sigma}{2g}\right)^{1/2} \equiv \xi_\times(H_p).$$

This leads to scaling of the interface width at pinning with g, and we find numerically that the global roughness exponent is given by $\chi \approx 1.25$ which is the same value as for the pure imbibition case. This is in contrast with [28] where pinning was interpreted in terms of the local Directed Percolation Depinning picture, with $\chi = 0.633$.

In the same way, we can analyze the evaporation dominated case ($g = 0$). In this case, however, it's difficult to find a single correlation length describing the extent of fluctuations for $t > t_c$. Only at pinning can a clear-cut correlation length ξ_e be defined by

$$\sigma\xi_e^{-3} = 2\varepsilon(1 - e^{-2H_p/\xi_e}).$$

Again, by numerically computing the structure factor at pinning we find that $\chi \approx 1.25$. The case $\xi_e \ll H_p$ where $\xi_e \propto (\sigma/\varepsilon)^{1/3}$ corresponds to "weak" evaporation, while the opposite limit of $\xi_e \gg H_p$ where $\xi_e \propto [\sigma^2/(\bar{\alpha}\varepsilon)]^{1/4}$ corresponds to "strong" evaporation. Thus, we can relate the saturated width of the interface at pinning to the pinning height by $w_s \sim H_p^\gamma$, where $\gamma = 2\chi/3 \approx 0.83$ for weak and $\gamma = \chi/2 \approx 0.63$ for strong evaporation, respectively. This can be contrasted to [29], where again by mapping the evaporation case to the local Directed Percolation Depinning picture with a gradient term in the pinning density, the authors concluded that $\gamma = 0.523$. This is inconsistent with the phase-field model predictions. Most importantly, the analysis here shows that there seems to be no regime where the local KPZ type of description is valid, even with evaporation and gravity included.

6.4 Capillaries

The phase-field formalism can be extended to various other important cases of liquid front dynamics. A relatively straightforward extension of the model presented in this section is the case of capillary rise in a tube, either within a 2D or even full 3D geometry, when the liquid can only rise through diffusion (hydrodynamics will be discussed in the next section). For such a case, we need to explicitly differentiate between the gas phase and the solid wall(s) where the liquid is going up. To this end, we consider a case where there is a single capillary tube in contact with a liquid reservoir at the bottom. The corresponding free energy of the imbibition model is modified to:

$$F_m[\phi] = \int d\mathbf{x}\left[\frac{\gamma}{2}(\nabla\phi)^2 + V_m(s, \phi, A)\right], \tag{36}$$

where the *multi-well potential* V_m now assumes different shapes depending on the spatial region we are considering:

$$V_m \equiv \frac{1}{2}(1 + s)\frac{1}{4}(\phi^2 - 1)^2 + \frac{1}{2}(1 - s)\frac{K}{2}[\phi - A(\mathbf{x})]^2. \tag{37}$$

Those spatial regions where the characteristic function $s(\mathbf{x}) = +1$ will be occupied by fluid phase (liquid or gas) as V_m has a double well structure in variable ϕ there. The remaining regions where $s(\mathbf{x}) = -1$ are occupied by solid and V has only a single well as a function of ϕ. Additionally, it is possible to include chemical impurities on the tube wall by taking $A(\mathbf{x})$ which corresponds to the solid phase potential well, to be e.g. a Gaussian random variable with mean value $\bar{A}$ and variance σ_A. Depending on the BC's and initial conditions, the model of (36) can be used to study many different geometries such as the capillary tube case, influence of a single defect to capillary rise, and droplet spreading.

7 Phase-Field Models and Hydrodynamics

The model presented in the previous section explicitly assumed that the fluid diffuses into a random media (such as paper) ignoring hydrodynamics effects. Such an approximation is appropriate in imbibition since fluid is unlikely to "flow" into a piece of paper. In other situations hydrodynamics effects must be included. Before discussing the complications associated with hydrodynamics it is interesting to note that the diffusive model has many similarities with the hydrodynamics of capillary flow [19]. First, the average velocity of the meniscus is inversely proportional to the average height of the meniscus in agreement with the Washburn law for capillary rise. In addition, the pressure jump at the interface in the fully hydrodynamic description is similar to the Gibbs–Thomson condition for the chemical potential at phase boundary [40] in the phase-field model of imbibition. Moreover, the pressure distribution in a capillary tube resembles the distribution of the chemical potential making these two variables analogous to each other. We also note that a Darcy type of hydrodynamic approach to fluid flow in porous medium [41] produces an interfacial evolution equation which is very similar to our purely diffusive phase-field model (cf. Sect. 6.2). However, there are important differences. For example, in the Washburn law for capillary rise the average velocity of the meniscus increases with increasing capillary radii, while in the diffusive model it decreases.

There are several coarse-grained approaches to *explicitly* include hydrodynamic degrees of freedom to models with dissipative dynamics, such as the Lattice Boltzmann [42], Dissipative Particle Dynamics [43], and Malevanets–Kapral methods [44]. For further details and references, we refer the reader to some of the other articles in this book. For the present case of the phase-field evolution equation, it can be coupled to the continuum Navier–Stokes equations to include hydrodynamic modes. It is possible to derive the corresponding phase-field model from microscopics using the projection operator methods developed e.g. in [7]. Below we will end up with the same model by using phenomenological arguments.

Let the velocity field of the fluid be denoted by $\mathbf{v}$. The force equation of the fluid element (momentum balance equation) in the phase-field formulation of fluid dynamics will contain an extra force not present in the conventional formulation of hydrodynamics [45]:

$$\phi\left(\partial_t\mathbf{v} + (\mathbf{v}\cdot\nabla)\mathbf{v}\right) = -\nabla p - \nabla\cdot\Pi' - \phi\nabla\mu + \zeta_v. \tag{38}$$

In this equation the phase-field ϕ is interpreted as the density field of the fluid, p denotes pressure, Π' is the viscous stress tensor of the fluid, and μ the chemical potential: $\mu = \delta F_{\rm m}/\delta\phi$, where $F_{\rm m}$ is given in (36). Finally, $\zeta_{\rm v}$ denotes the stochastic momentum exchange due to thermal fluctuations. The term $\phi\nabla\mu$ is the capillary stress term which takes into account interfacial phenomena close to phase boundaries. To make it more obvious, we define a new pressure field:

$$p' \equiv p + \phi\mu.$$

In terms of p' we can rewrite (38) in the following form:

$$\phi\left(\partial_t\mathbf{v} + (\mathbf{v}\cdot\nabla)\mathbf{v}\right) = -\nabla p' - \nabla\cdot\Pi' + \mu\nabla\phi + \zeta_{\rm v}. \tag{39}$$

The term $\mu\nabla\phi$ is non-negligible only at the immediate vicinity of the phase boundary since only there the phase-field ϕ changes rapidly, while in the bulk we have $\nabla\phi \approx 0$. Considering static equilibrium with no fluctuations ($\mathbf{v} = 0$, $\zeta_{\rm v} = 0$), we have

$$-\nabla p' + \mu\nabla\phi = 0,$$

which integrated across the phase boundary yields immediately the static boundary condition in the normal direction $\mathbf{n}$ for the stress tensor:

$$\Delta p' = \sigma\mathcal{K}. \tag{40}$$

Here σ is the surface tension and $\mathcal{K}$ stands for the local curvature of the boundary. The relation given in (40) is also known as the Laplace pressure drop across a curved surface. In the tangential direction $\mathbf{s}$ there is another force:

$$\int_{-\infty}^{\infty} {\rm d}u\,\mu\frac{\partial\phi}{\partial s} = \int_{-\infty}^{\infty} {\rm d}u\left(\frac{\delta\Delta F}{\delta\phi}\right)\frac{\partial\phi}{\partial s} \approx \frac{\partial\sigma}{\partial s}. \tag{41}$$

The difference in the free energy between a system having a single phase boundary and a pure system having no boundaries is denoted by ΔF. The double well form used in (5) actually guarantees that $\Delta F = F_{\rm c}$, which justifies the first equality in (41). The second relation in (41) follows from the fact that by definition $\sigma = \int {\rm d}u\,\Delta f$ (for a straight interface), where f is the free energy density per unit volume: $\Delta F \equiv \int {\rm d}V\,\Delta f$. Thus, we can see that there is an effective force term active at the phase boundary in the Navier–Stokes momentum equation, which originates from the capillary stress term $\mu\nabla\phi$:

$$\mu\nabla\phi \approx \left(\sigma\mathcal{K}\mathbf{n} + \frac{\partial\sigma}{\partial s}\mathbf{s}\right)\delta_{\Sigma},$$

where the delta function δ_{Σ} is effective only in the vicinity of the phase boundary Σ. These extra force terms contribute to the boundary condition for the stress tensor which in the conventional hydrodynamics has to be set as an auxiliary condition which the various fields should satisfy at the boundaries. The phase-field formalism has made it possible to replace the boundary condition in the sharp interface model (that is, conventional hydrodynamics for two fluids separated by an infinitely thin interface) with an extra force term in the Navier–Stokes equation which is entirely expressible in terms of the *bulk* fields μ and ϕ.

If the viscosities of the two fluids on different sides of the phase boundary are not the same, the simplest way of including this effect is to make viscosity η (not to be confused with noise) appearing in the expression of the viscous stress tensor Π' a function of the phase-field: $\eta = \eta(\phi)$ such that $\eta_1 = \eta(\phi_1)$ and $\eta_2 = \eta(\phi_2)$, where η_i and ϕ_i $(i = 1, 2)$ are the value of the viscosity and the bulk value of the density in fluid i. Similarly, one can impose the no-slip boundary condition at solid-fluid boundaries. Close to solid walls the viscosity should in principle become infinite (in reality very large, with a finite cut-off) in order to make the fluid elements stick to the solid walls. Another possibility would be to use an additional force term in the velocity (momentum) equation (39) which would cancel the kinetic energy of the fluid elements close to the solid walls.

In addition, the phase-field itself has to be coupled back to the velocity field $\mathbf{v}$. This is done by the continuity equation:

$$\partial_t \phi + \nabla \cdot (\mathbf{J}_c + \mathbf{J}_d) = 0,$$

where the total mass flux is a sum of convective current ($\mathbf{J}_c \equiv \phi \mathbf{v}$) and diffusion current ($\mathbf{J}_d \equiv -D \nabla \mu$). Plugging in the currents, the evolution equation becomes

$$\partial_t \phi + \nabla \cdot (\phi \mathbf{v}) = D \nabla^2 \mu, \tag{42}$$

where the term on the right can be interpreted as the exchange current across the liquid-gas boundary. Numerically, the term $D \nabla^2 \mu$ is responsible for maintaining the interfacial region sharp enough such that the fluids do not mix completely. In the spirit of the critical dynamics models (Sect. 5) we should add a stochastic term η_c (cf. (23)) to accompany the dissipative term $D \nabla^2 \mu$. However, if strict mass conservation is required, i.e., when no sources or sinks, stochastic or deterministic, are allowed, then (42) reduces to

$$\partial_t \phi + \nabla \cdot (\phi \mathbf{v}) = 0, \tag{43}$$

which is just the standard continuity equation. Apart from the specific form of the free energy (chemical potential), the model defined by (38) and (43) is the same as the model introduced in [46]. If we decouple the noise sources and require that the fluids are incompressible, that is $\nabla \cdot \mathbf{v} = 0$, we can write the momentum balance equation in the form

$$\varrho (\partial_t \mathbf{v} + (\mathbf{v} \cdot \nabla) \mathbf{v}) = -\nabla p - \eta \nabla^2 \mathbf{v} + \mu \nabla \phi, \tag{44}$$

where ϱ is the density field. The phase-field ϕ appearing on the right hand side serves as marker field (concentration field) which serves to produce the right boundary condition for the stress tensor. Equation (44) together with (42) constitute the model of [47] (cf. also [48]). Finally, we note that selecting just the transverse momentum fluctuations and interpreting the phase-field as the entropy density instead of mass density (which is taken to be constant), (44) together with (42) supplemented with thermal noise give Model H of critical dynamics presented in Sect. 5. More detailed analysis of the these differences will be presented elsewhere.

8   Summary and Conclusions

In this review, we have tried to give a flavor of how and why the phase-field methodology is a powerful way of modeling many interface phenomena. Before concluding it is important to note the limitations of this approach. As discussed in Sect. 1 the parameters of the phase-field model can in principle be derived from a "microscopic" Hamiltonian. However, in practice this task is very onerous and usually it is more convenient to phenomenologically introduce a free energy (or more correctly, an effective Hamiltonian) that is consistent with some mean field equilibrium properties of the system, such as the phase diagram. Thus in practice phase fields cannot often *predict* equilibrium properties, such as phase diagrams. In essence the properties of the phase diagrams are used as input to the free energy functional. The main purpose of phase field models is to predict dynamical and critical behavior. Even in this regime, quantitative predictions can only be made in the long-wavelength, late-time limit, since short time and length scales have been coarse grained out. On the other hand, the phase-field approach allows one to reach mesoscopic and even macroscopic length and time scales, unlike most more microscopic approaches.

Perhaps the most attractive feature of the phase-field approach is that the models can be constructed without a deep knowledge of the precise microscopic details of a physical system and still obtain the correct long wavelength behavior. For example, it was shown in Sect. 3, that the Allen–Cahn and Kardar–Parisi–Zhang equations can be derived without reference to a specific free energy. Naturally this considerably simplifies the construction of the free energy functional needed in the phase field approach. In addition, specific forms of free energy functionals can be chosen for computational efficiency.

Phase-field models provide a simple and efficient method for studying pattern formation in a wide variety of phenomena, including polymerization, charge density waves, the solidification of pure and binary alloys, spinodal decomposition, order-disorder transitions, block co-polymers, crystallization and epitaxial growth. In this article we have provided a general but brief introduction into the development and analysis of phase field models and given a detailed description of a specific application, that of imbibition. At the end we have also outlined various problems where further work is necessary. We hope that this article in part inspires work in these directions.

Acknowledgements

We wish to thank useful discussions with many of our colleagues, including Martin Grant and Martin Dubé. This work was supported by a National Science Foundation grant, NSF–DMR Grant 0076054 (KRE), and by the Academy of Finland through it's Center of Excellence program.

Appendix: Curvilinear Coordinates

The curvilinear coordinates (u, s) are illustrated in Fig. 7. The Cartesian coordinates $\mathbf{r} = (x, y)$ are connected to the curvilinear ones through the relationship

$$\mathbf{r} = \mathbf{R}(s) + u\hat{n}(s),$$

where $\hat{n}$ is the normal vector and $\mathbf{R}$ is the distance from the origin of the (x, y) coordinate system to a position along the interface. The metric for coordinate transformation can be obtained by noting that

$$\frac{\partial \mathbf{r}}{\partial u} = \hat{n},$$

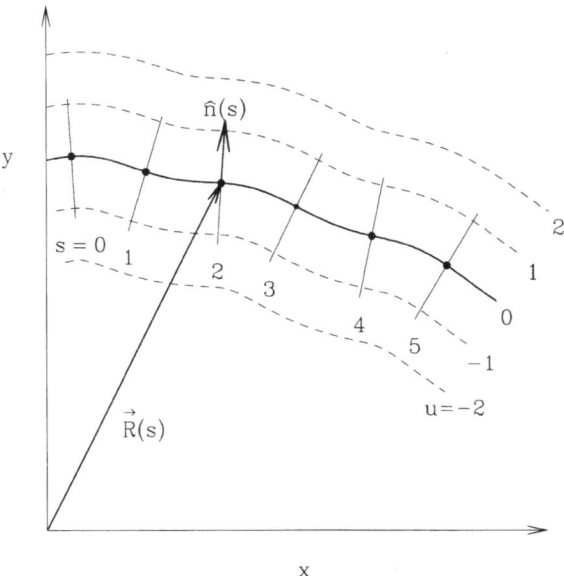

Fig. 7. Curvilinear coordinates

and

$$\frac{\partial \mathbf{r}}{\partial s} = \frac{\partial \mathbf{R}}{\partial s} + u\frac{\partial \hat{n}}{\partial s} = \hat{t} + u\left(\hat{x}\cos\left(\theta\right) - \hat{y}\sin\left(\theta\right)\right)\frac{\partial \theta}{\partial s} = \hat{t}(1 + u\kappa),$$

where $\hat{t}$ is tangent to the interface, θ is the angle between the y axis and the normal and κ is the curvature defined as $\partial\theta/\partial s$. The metric of the transformation is then

$$g_{ij} = \sum_k \frac{\partial x_k}{\partial u_i}\frac{\partial x_k}{\partial u_j},$$

where x_i's are the "old" coordinates (i.e. (x, y)) and u_i's are the new coordinates (i.e. (u, s)) so that

$$g_{11} = 1$$
$$g_{22} = \left(1 + u\kappa\right)^2$$
$$g_{12} = g_{21} = 0.$$

The Laplacian operator can now be obtained using the standard formula

$$\nabla^2 = \frac{1}{h_1 h_2}\left(\frac{\partial}{\partial u}\left(\frac{h_2}{h_1}\frac{\partial}{\partial u}\right) + \frac{\partial}{\partial s}\left(\frac{h_1}{h_2}\frac{\partial}{\partial s}\right)\right),$$

where $h_1 = \sqrt{g_{11}}$ and $h_2 = \sqrt{g_{22}}$. Thus the Laplacian in curvilinear coordinates becomes

$$\nabla^2 = \frac{\partial^2}{\partial u^2} + \frac{\kappa}{\left(1 + u\kappa\right)}\frac{\partial}{\partial u} + \frac{1}{\left(1 + u\kappa\right)^2}\frac{\partial^2}{\partial s^2} - \frac{u\kappa_s}{\left(1 + u\kappa\right)^3}\frac{\partial}{\partial s},$$

where $\kappa_s \equiv \partial\kappa/\partial s$.

It will be useful to consider the following dimensionless scaled parameters: $\bar{u} = u/\xi$, $\bar{s} = \varepsilon s/\xi$ and $\bar{\kappa} = \xi\kappa/\varepsilon$. The Laplacian becomes

$$\xi^2 \nabla^2 = \frac{\partial^2}{\partial \bar{u}^2} + \frac{\varepsilon\bar{\kappa}}{(1 + \varepsilon\bar{u}\bar{\kappa})} \frac{\partial}{\partial \bar{u}} + \frac{\varepsilon^2}{(1 + \varepsilon\bar{u}\bar{\kappa})^2} \frac{\partial^2}{\partial \bar{s}^2} - \frac{\varepsilon^3 \bar{u}\bar{\kappa}_{\bar{s}}}{(1 + \varepsilon\bar{u}\bar{\kappa})^3} \frac{\partial}{\partial \bar{s}}.$$

To lowest order in ε this reads

$$\xi^2 \nabla^2 = \left(\frac{\partial^2}{\partial \bar{u}^2} \right) + \varepsilon \left(\bar{\kappa} \frac{\partial}{\partial \bar{u}} \right) + \varepsilon^2 \left(\frac{\partial^2}{\partial \bar{s}^2} - \bar{\kappa}^2 \bar{u} \frac{\partial}{\partial \bar{u}} \right)$$

$$+ \varepsilon^2 \left(-2\bar{u}\bar{\kappa} \frac{\partial^2}{\partial \bar{s}^2} + \bar{\kappa}^3 \bar{u}^2 \frac{\partial}{\partial \bar{u}} - \bar{u}\bar{\kappa}_{\bar{s}} \frac{\partial}{\partial \bar{s}} \right) + \cdots .$$

Finally, it is also useful to determine how the delta functions transform:

$$\delta(\mathbf{r} - \mathbf{r}') = \delta(u - u')\delta(s - s')/J,$$

where

$$J = \frac{\partial x}{\partial u} \frac{\partial y}{\partial s} - \frac{\partial y}{\partial u} \frac{\partial x}{\partial s} = (1 + u\kappa);$$

$$\delta(\mathbf{r} - \mathbf{r}') = \delta(u - u')\delta(s - s')(1 - u\kappa + \cdots).$$

References

1. N. Goldenfeld: *Lectures on Phase Transitions and the Renormalization Group* (Addison-Wesley, Reading 1992).
2. M.E. Fisher: Rev. Mod. Phys. **46**, 597 (1974)
3. K.G. Wilson, J. Kogut: Phys. Rep. C **12**, 75 (1974)
4. K.R. Elder, O. Malis, K. Ludwig, B. Chakraborty, N. Goldenfeld: Europhys. Lett. **43**, 629 (1998)
5. J. Timonen, M. Stirland, D.J. Pilling, Y. Cheng, R.K. Bullough: Phys. Rev. Lett. **56**, 2233 (1986)
6. D. Forster: *Hydrodynamic Fluctuations, Broken Symmetry, and Correlation Functions* (W.A. Benjamin, Inc. 1975)
7. K. Kawasaki: Ann. Phys. **61**, 1 (1970)
8. K.R. Elder, M. Grant, N. Provatas, J.M. Kosterlitz: Phys. Rev. E **64**, 21604 (2001)
9. S.M. Allen, J.W. Cahn: Acta Metall. **27**, 1085 (1978)
10. M. Kardar, G. Parisi, Y.C. Zhang: Phys. Rev. Lett. **56**, 889 (1986)
11. S.F. Edwards, D.R. Wilkinson: Proc. R. Soc. London, Ser. A **381**, 17 (1982)
12. A.-L. Barabási, H.E. Stanley: *Fractal Concepts in Surface Growth* (Cambridge University Press, Cambridge 1995)
13. J. Krug: Adv. Phys. **46**, 139 (1997)
14. F. Family, T. Vicsek: J. Phys. A **18**, L75 (1985)
15. J. Maunuksela, M. Myllys, O.-P. Kähkönen, J. Timonen, N. Provatas, M.J. Alava, T. Ala-Nissila: Phys. Rev. Lett. **79**, 1515 (1997); M. Myllys, J. Maunuksela, M.J. Alava, T. Ala-Nissila, J. Timonen: Phys. Rev. Lett. **84**: 1946 (2000); M. Myllys, J. Maunuksela, M. Alava, T. Ala-Nissila, J. Merikoski, J. Timonen: Phys. Rev. E **64**, 036101 (2001)
16. J.J. Ramasco, J.M. López, M.A. Rodriguez: Phys. Rev. Lett. **84**, 2199 (2000)
17. T. Salditt, H. Spohn: Phys. Rev. E **47**, 3524 (1993)

18. J. Heinonen, I. Bukharev, T. Ala-Nissila, J.M. Kosterlitz: Phys. Rev. E **57**, 6851 (1998)
19. M. Dubé, M. Rost, K.R. Elder, M. Alava, S. Majaniemi, T. Ala-Nissila: Phys. Rev. Lett. **83**, 1628 (1999); Eur. Phys. J. B **15**, 701 (2000)
20. M. Dubé, S. Majaniemi, M. Rost, M.J. Alava, K.R. Elder, T. Ala-Nissila: Phys. Rev. E **64**, 051605 (2001)
21. M.-P. Kuittu, M. Haataja, N. Provatas, T. Ala-Nissila: Phys. Rev. E **58**, 1514 (1998); Phys. Rev. E **59**, 3774 (1999); M.-P. Kuittu, M. Haataja, T. Ala-Nissila: Phys. Rev. E **59**, 2677 (1999)
22. J. Asikainen, S. Majaniemi, M. Dubé, T. Ala-Nissila: Phys. Rev. E **65**, 052104 (2002)
23. P. Hohenberg, B. Halperin: Rev. Mod. Phys. **49**, 435 (1977)
24. P.M. Chaikin, T.C. Lubensky: *Principles of Condensed Matter Physics* (Cambridge 1995)
25. K.R. Elder, F. Drolet, J.M. Kosterlitz, M. Grant: Phys. Rev. Lett. **72**, 677 (1994)
26. N. Provatas, N. Goldenfeld, J. Dantzig: Phys. Rev. Lett. **80**, 3308 (1998)
27. N. Provatas, M.J. Alava, T. Ala-Nissila: Phys. Rev. E **54**, R36 (1996)
28. S. Buldyrev, A.-L. Barabási, F. Caserta, S. Havlin, H.E. Stanley, T. Vicsek: Phys. Rev. A **45**, R8313 (1992)
29. L.A.N. Amaral, A.-L. Barabási, S.V. Buldyrev, S. Havlin, H.E. Stanley: Phys. Rev. Lett. **72**, 641 (1994)
30. E.W. Washburn: Phys. Rev. **17**, 273 (1921)
31. V. Horvath, H.E. Stanley: Phys. Rev. E **52**, 5166 (1995)
32. M. Dubé, M. Rost, K.R. Elder, M. Alava, S. Majaniemi, T. Ala-Nissila: Phys. Rev. Lett. **86**, 6046 (2001)
33. A somewhat similar equation of motion has been derived (but not carefully analyzed) by using a completely different approach in Ref. [41]
34. M. Dubé, M. Rost, M. Alava: Eur. Phys. J. B **15**, 691 (2000)
35. C.-H. Lam, V.K. Horvath: Phys. Rev. Lett. **85**, 1238 (2000)
36. D. Geromichalos, F. Mugele, S. Herminghaus: Phys. Rev. Lett. **89**, 104503 (2002)
37. A. Hernández-Machado, J. Soriano, A.M. Lacasta, M.A. Rodríguez, L. Ramírez-Piscina, J. Ortín: Europhys. Lett. **55**, 194 (2001)
38. E. Alvarez-Lacalle, J. Casademunt, J. Ortín: Phys. Rev. E **64**, 016302 (2001)
39. J. Soriano, J.J. Ramasco, M.A. Rodríguez, A. Hernández-Machado, J. Ortín: Phys. Rev. Lett. **89**, 026102 (2002)
40. A.J. Bray: Adv. Phys. **43**, 357 (1994)
41. V. Ganesan, H. Brenner: Phys. Rev. Lett. **81**, 578 (1998)
42. S. Chen, G.D. Doolen: Annu. Rev. Fluid Mech. **30**, 329 (1998)
43. P.J. Hoogerbrugge, J.M.V.A. Koelman: Europhys. Lett. **19**, 155 (1992); P. Español, P. Warren: Europhys. Lett. **30**, 191 (1995)
44. A. Malevanets, R. Kapral: J. Chem. Phys. **110**, 8605 (1999); J. Chem. Phys. **112**, 7260 (2000)
45. L.D. Landau, E.M. Lifshitz: *Fluid Mechanics* (Pergamon Press, Oxford 1984)
46. S.P. Das, G.F. Mazenko: Phys. Rev. A **34**, 2265 (1986)
47. R. Chella, J. Viñals: Phys. Rev. E **53**, 3832 (1996)
48. H. Furukawa: Phys. Rev. A **31**, 1103 (1985)

Index